Neuropsychotherapy

How the Neurosciences Inform
Effective Psychotherapy

Counseling and Psychotherapy Investigating Practice
from Scientific, Historical, and Cultural Perspectives

Editor, Bruce E. Wampold, University of Wisconsin, Madison

This innovative new series is devoted to grasping the vast complexities of the practice of counseling and psychotherapy. As a set of healing practices delivered in a context shaped by health delivery systems and the attitudes and values of consumers, practitioners, and researchers, counseling and psychotherapy must be examined critically. By understanding the historical and cultural context of counseling and psychotherapy and by examining the extant research, these critical inquiries seek a deeper, richer understanding of what is a remarkably effective endeavor.

Published

Counseling and Therapy with Clients Who Abuse Alcohol or Other Drugs
Cynthia E. Glidden-Tracy

The Great Psychotherapy Debate
Bruce Wampold

The Psychology of Working: Implications for Career Development, Counseling, and Public Policy
David Blustein

Forthcoming

The Pharmacology and Treatment of Substance Abuse: Evidence and Outcomes Based Perspectives
Lee Cohen, Frank Collins, Alice Young, and Dennis McChargue

In Our Clients' Shoes: Theory and Techniques of Therapeutic Assessment
Stephen Finn

Making Treatment Count: Using Outcomes to Inform and Manage Therapy
Michael Lambert, Jeb Brown, Scott Miller, and Bruce Wampold

IDM Supervision: An Integrated Developmental Model for Supervising Counselors and Therapists, Third Edition
Cal Stoltenberg and Brian McNeill

The Great Psychotherapy Debate, Revised Edition
Bruce Wampold

Casebook for Multicultural Counseling
Miguel Gallardo and Brian McNeill

Culture and the Therapeutic Process: A Guide for Mental Health Professionals
Mark Leach and Jamie Aten

Neuropsychotherapy

How the Neurosciences Inform Effective Psychotherapy

Klaus Grawe

Psychology Press
Taylor & Francis Group

New York London

First published in 2007 by Psychology Press

Published 2021 by Routledge
2 Park Square, Milton Park, Abingdon, Oxon OX14 4RN
605 Third Avenue, New York, NY 10017

Routledge is an imprint of the Taylor & Francis Group, an informa business

Cover image: © Hogrefe Verlag GmbH & Co. KG
Cover design by Kathry Houghtaling-Lacey

Library of Congress Cataloging-in-Publication Data

Catalog record is available from the Library of Congress

ISBN 9780805861211 (hbk)
ISBN 9780805861228 (pbk)

CONTENTS

SERIES FOREWORD

This innovative new series is devoted to grasping the vast complexities of the practice of counseling and psychotherapy. As a set of healing practices delivered in a context shaped by health delivery systems and the attitudes and values of consumers, practitioners, and researchers, counseling and psychotherapy must be examined critically. By understanding the historical and cultural context of counseling and psychotherapy and by examining the extant research, these critical inquiries seek a deeper, richer understanding of what is a remarkably effective endeavor.

Klaus Grawe, before his untimely death in 2005, wrote one of the most popular German psychotherapy books. We are fortunate to have an English translation by Björn Meyer and Martin Grosse Holtforth of his work, which describes how the neurosciences inform the practice of psychotherapy. Psychotherapy traditionally has benefited from basic research in psychology and related disciplines. However, the field often loosely borrows concepts and fashions psychotherapy to fit with the zeitgeist without deep understanding, leaving us to wish for a more rigorous application of basic science. As the neurosciences are revolutionizing our understanding of so many human emotions, cognitions, and behaviors, both mundane and esoteric, it is vital that we get the psychotherapy/neuroscience link correct. In *Neuropsychotherapy: How the Neurosciences Inform Effective Psychotherapy*, Professor Grawe has rigorously integrated findings from neuroscience research to develop an approach to psychotherapy that is sound, informed, and most importantly effective. With this book, Professor Grawe has allowed the field to exist in the current scientific milieu and integrate science and practice at a proper level.

—*Bruce E. Wampold, PhD, ABPP, Series Editor*
University of Wisconsin, Madison

FOREWORD

Bernard D. Beitman, MD
Columbia, Missouri

Psychoanalysis dominated psychotherapy theory during the first half of the 20th century, until rebellions by cognitive-, behavioral-, and emotion-focused foes provided new theoretical perspectives. Family, systems, and multicultural models then further reshaped the theoretical landscape to include individuals in their social environments. We now recognize the obvious: Patients have inner lives composed of thoughts, feelings, and fantasies that are shaped by developmental experiences; these inner workings manifest in behaviors that are shaped not only by family influences but also by social and cultural contexts.

Then why does psychotherapy theory and practice not seem clear? Why do proponents of different schools still seem to fight with each other? Why do psychotherapy integrationists keep producing new forms of integration? Why are policy makers, students, and patients still struggling to define psychotherapy?

Klaus Grawe responds to these questions by shifting contexts from words to brain using neurobiological contributions to clarify, sharpen, and shape our concepts of psychotherapy. *Neuropsychotherapy* provides strong impetus for therapists to loosen their grip on the need for overarching theories and to pay attention to that part of the patient that we are trying to help them change: their brains. Focus on the brain provides the opportunity for neurobiological empathy, reduction in stigma, reduction in patient self-blame, and sharper conceptualizations of current theory and technique.

Neurobiological empathy emerges from therapists' realization that the brains of their patients with, for example, anxiety and depression contain hyperresponsive amygdalas. Therapists help patients to slow down their amygdalas. Stigma is reduced, the way the stigma of tuberculosis was reduced—by defining the biological processes that create the problem. Self-blame can be similarly reduced—once the neurobiology is understood and accepted, treatment focuses on correcting the brain disorder, not on blaming oneself for having caused it. Last, sharper conceptualizations of current techniques will emerge from imaging studies. For example, the ventromedial prefrontal cortex appears to be essential for inhibition to take place and hold. Such studies also indicate that extinction may be a theoretical ideal that is being replaced by inhibition. Exposure to create inhibition then has a brain target. Can we develop new ways to perhaps thicken the ventromedial prefrontal cortex?

Practitioners and students are likely to groan when facing the task they are about to undertake in reading this consequential book. However, if you persevere, you will change the way you think about patients. Patients will no longer be only mental selves, functioning interpersonally, but they also will become people with dysfunctional brain parts. You may find this concept dehumanizing, but it need not be. Just as people with cancer or heart disease or diabetes have dysfunctional organs, so do people with depression, posttraumatic stress disorder, and panic disorder. The depressed person who has difficulty experiencing pleasure, who plans poorly, and who cannot make decisions has dysfunctional anterior cingulate and prefrontal cortices. Perhaps the most robust finding in neuroimaging of patients with depression is that they have decreased functioning of the prefrontal cortex (PFC). The more severe the depressive symptoms, the less the functioning of the PFC.

Grawe, in his introductory chapter, helps us to make use of this concept in the consulting room. The depressed person with reduced function of areas involved in the planning, deciding, and action initiation will not be ready to work on changing ingrained interpersonal patterns until the PFC is functional again. Therefore, the behavioral activation suggested by cognitive therapists has a brain basis. Mastery and pleasure make PFCs function more effectively. Since self-awareness also seems to depend upon PFC activity, self-exploration can then be facilitated after increased frontal activation.

Grawe suggests that we learn about amygdalas, hippocampi, the anterior cingulate cortex and to think in the terms of the circuits of which each is a part. For example, amygdalas are hyperactive in depression and anxiety (for the most part), which leads Grawe to a rather lengthy discussion of how anxiety and depression are clinically related. Stress often reduces the functioning of the hippocampus, whose cells are highly vulnerable to cortisol, which is released through activation of a corticotropin-releasing hormone that a dysfunctional hippocampus cannot inhibit. The hippocampus then becomes increasingly more dysfunctional, making depression worse.

Grawe forces us to recognize that we must, as therapists, come to connect basic human psychological needs with specific, often overlapping brain circuits. He describes four basic needs, illustrating each of them around attitudes toward money:

> People who feel compelled to save money in order to maintain a maximum sphere of potential future options clearly have a strong need for control. Their sense that one is able to use this money "just in case" can be interpreted as the creation of a generalized sphere of behavioral flexibility. Other people's relationship with money is determined, instead, by their need to enjoy a maximum number of pleasurable experiences. They might spend their money, for example, for culinary delights, for journeys, or for other pleasurable events. Yet others use their money to satisfy their need for self-esteem enhancement, for example, by driving a prestigious car, wearing expensive clothes, or owning an impressive villa. Individuals who have to use money to satisfy their need for attachment are perhaps in a comparatively worse situation but such instances are, of course, far from rare. (p. 212)

Grawe's emphasis on attachment needs is critical to understanding human functioning. Because we are so vitally social and interpersonal, and because psychotherapy is
at its base interpersonal, we must start with attachment and its vagaries. Self-esteem
enhancement, to my mind is a form of attachment activity—we receive increases in
self-esteem from the looks in the eyes of those who are important to us. Grawe's
descriptions of the need for control also touch upon a central and overlooked aspect
of psychotherapy theorizing. The need for control may simply be described as a need
for expectations to match experiences. When they do not match, stress responses are
activated. When the stress response cannot lead to resolution of the discrepancy, psychiatric (and medical) problems are likely to ensue. The most problematic of these
mismatches involves interpersonal relationships. We come back again to attachment,
as well as the need for pleasure. Who can deny we are pleasure-seeking animals? First
we require the basic needs of food, clothing, and shelter, then the human rights of free
speech and freedom of assembly and freedom from persecution. Then good food,
comfortable clothing and shelter, entertainment, travel, and perhaps mild doses of
mind-altering substances like alcohol, coffee and tea. And yet I return to the greatest
of all pleasures—love. Love is still yet to be well defined, still pursued by most,
prayed for by many. Love can give us a sense of belonging—to someone, to some
group, or to some higher being. Attachment needs lead us to our spiritual selves.
Again, we have to return to the brain to understand how all these complexities exist
in human functioning; nonbrain theories are not sufficient.

With this emphasis upon attachment, both stated and implied, the next logical step in
the evoulation of neuropsychotherapy is to examine the interpersonal neurobiology of
the psychotherapeutic situation. Or said more baldly, the two brains in the room.
These brains are interacting—sometimes at high rates—with each other, influencing
each other. Keep in mind that amygdalas affect each other; prefrontal cortices affect
each other; various neurochemicals are squirting around in response to each brain;
gene expression is being modulated. The working alliance, so well known to be a
major variable in therapeutic outcomes, has its own neurobiology too, as I have
attempted to describe (Beitman, Viamontes, Soth, & Nittler, 2006). Unfortunately, our
colleague Klaus Grawe's untimely death denies us his informed perspectives on this
important topic; one of Grawe's legacies will be the continuation of theory and
research by others.

The psychotherapy of the near and distant future will be increasingly dependent upon
our ability to understand brain and mind interaction. We speak words, we emit body
language signals, perhaps even send out bioelectromagnetic energy to our patients.
Psychotherapy challenges us to comprehend the ethereal mind interacting with the
earthy brain.

PREFACE

I have followed the progress in the neurosciences with great interest for the past 10 years. To me, this area seems so exciting that I frequently noticed the emerging impulse to begin contributing myself. However, this does not mean that I have any intention of withdrawing from my original area of research, the domain of psychotherapy. On the contrary, my interest in the neurosciences is fueled by the questions that occupy me as a psychotherapy researcher. What is the basis of the changes that lead individuals to come to therapy? How can it be that a human brain would suddenly produce something qualitatively new, which we call a mental disorder? If mental disorders have neural foundations, would it not make sense that psychotherapy would be effective to the extent that it can influence these neural bases?

My research team and I have been investigating the mechanisms of action in psychotherapy for a long time. Is it not reasonable to assume that the mechanisms that we have identified based on analyses of the therapeutic processes would have a neural foundation in the functioning of the brain? Are there neuroscientific findings that can inform and illuminate our understanding of how psychotherapy achieves its effects? With these questions in mind, I have followed the progress in the neurosciences over the past few years.

Several years ago, when my interest in the neurosciences had just emerged, I attempted to integrate current psychotherapy research with basic psychological research and with the neurosciences, with the goal of evaluating whether these three areas in combination could provide the kind of solid empirical and theoretical foundation that would allow us to move beyond the arbitrary division of traditional therapy schools (Grawe,2004). This attempt resulted in the formulation of a theoretical framework that permitted the coherent integration of many results from all three research areas. This theoretical frame, which we termed consistency theory, has since then informed our research and clinical activity at the University of Bern. Today, 6 years later, many empirical studies have been reported to test the assumptions of consistency theory, and many clinical experiences have emerged as we attempted to translate the concepts of consistency theory into practice.

In the meantime, a great deal of progress has also been made in the neurosciences, and the rate of this progress sometimes seems astonishing. I waited eagerly for the past 6 years to find the time to thoroughly read and process all the material I have collected on these topics, and to relate this information to our own research and clinical efforts. A research semester in the summer of 2003 finally provided this opportunity. I withdrew from the world for half a year in order to engage fully with the neuroscientific research that has been published during the past 6 years. I sifted through this work with my questions about psychotherapy in mind at all times. This time, my excursions into neuroscientific research were even more fruitful than they had been 6 years ago. It is simply fascinating to learn about the recent studies and findings. Much of this work has immediate relevance for therapy, and I cannot imagine that psychotherapy will remain impervious to these findings in years to come.

However, a huge cleft remains at this point between the world of the neurosciences and that of psychotherapy. It is typically not easy for psychotherapists to pick up neuroscientific research articles and comprehend them with all their specific jargon. It is the exception that a psychotherapist would have such detailed knowledge of the brain that the findings could be integrated into a coherent context. Neuroscientists, on the other hand, may often have no interest in contemplating the therapeutic implications of their findings, and even if they do, they often have an antiquated conception of psychotherapy that has little in common with the current state of psychotherapy research and currently available therapy potentials.

Therefore, it was an appealing challenge for me to build a conceptual bridge between these two worlds. The first three chapters of this book attempt to provide an overview, in comprehensible terms, of the therapy-relevant findings that have been discovered in the neurosciences in recent years. In these chapters, I do not yet relate these findings to our own research on mental disorders and the psychotherapeutic process. Chapters 4 and 5, by contrast, are conceptual chapters. In chapter 4, I articulate a model of the emergence of mental disorders, and chapter 5 provides a model of the mechanisms underlying the effects of psychotherapy. Both chapters were informed not only by many neuroscientific findings but also by the results of our own research. This work has been guided in recent years, as I mentioned above, by the assumptions derived from consistency theory. These two chapters can be regarded as a substantially more detailed, revised, and new version of my consistency theory of psychotherapy.

The work on this book was quite exhausting at times, but it was equally enriching. Having acquainted myself thoroughly with the recent neuroscientific research, I have become convinced that psychotherapy can derive crucial innovative impulses from these findings, which will lead to an accelerated further development of psychotherapy. This book constitutes no more than the beginning of this process. The title of the book denotes an agenda rather than a finished product. Even in our own clinical work, we will need several more years before we have translated into action all of the suggestions contained in chapters 4 and 5. Neuropsychotherapy, in the sense that I propose it here, can also be practiced without explicitly using neuroscientific methods.

My thorough studies of the brain have, however, not brought me any closer to resolving the question of how one could elegantly and efficiently express that research findings apply equally well to both sexes. I am simply not obsessive or conscientious enough to use, each time, both the male and female grammatical form, even though this would often be more politically correct. I also find such an approach aesthetically unsatisfactory. Let me reassure the reader at this point, then, that the statements I make in this book about mental functioning and about psychotherapy apply equally to both genders, even though I often write only about the male case. This certainly allows me (as a man) to use a more consistent and less complicated form of expression. I admit, however, that this practice seems less and less appropriate in light of the increasing preponderance of women in the area of psychotherapy. My hope is that more competent persons might one day find a good solution to this problem. I also note that the English version of this book does tend to use the more politically correct—albeit more cumbersome—male as well as female form in most (but not all) cases.

PREFACE, ENGLISH TRANSLATION

When Klaus Grawe died unexpectedly at the age of 62, half of the English translation of his book *Neuropsychotherapie*, which was published in German in 2004, had been completed. Klaus Grawe had been very eager to make his insights accessible to his English-speaking colleagues to foster a fascinating dialogue between psychotherapy, the neurosciences, and other disciplines that have the potential to contribute to the continuous improvement of psychotherapy. *Neuropsychotherapy* bears testimony to Klaus's encyclopedic knowledge, his boundless inquisitiveness, his courage to cross conceptual boundaries, his firm grounding in the empirical sciences, and his determination and outstanding talent to connect heterogeneous pieces of knowledge in an integrative conception of psychotherapy. His coworkers at the University of Bern and the *Institute of Psychological Therapy* (now the *Klaus Grawe Institute*) admired, and benefited greatly from this unique coalescence of intellectual qualities, in addition to enjoying his outstanding qualities as a supervisor, mentor, and colleague. It was an honor and obligation to accept Mariann Grawe's request to help finish the English translation of *Neuropsychotherapie*. Björn Meyer, who had already translated Klaus Grawe's previous book *Psychological Therapy* (Grawe, 2004) did an admirable job in translating each chapter very accurately and efficiently. He not only found concise translations for complex concepts on the basis of a profound understanding of psychology and Klaus Grawe's thinking, but he also succeeded in letting Klaus Grawe's spirit shine through the English text. Mariann and I preserved the content of Klaus Grawe's writing completely, but in an effort to stay up to date, we added newer studies about neuroscientific correlates of psychotherapy in the second chapter. With Lawrence Erlbaum Associates (LEA) we found the perfect "home" for Klaus Grawe's book within the series *Counseling and Psychotherapy* (edited by Bruce Wampold). We are very grateful for the very efficient, smooth and enjoyable collaboration with Steve Rutter of LEA. In addition to the aforementioned, we would like to thank Karin Hofmann for her unceasing enthusiasm and admirable ability to multitask; Yvonne Egenolf, Daniel T. Gustafson, and Jerry W. Halsten for their very helpful comments on neuroscientific contents, especially in the first three chapters; and Louis G. Castonguay for mentoring the publication process.

Mariann and I hope that *Neuropsychotherapy* will inspire researchers and practitioners alike to proceed in the spirit of Klaus Grawe and integrate knowledge from the neurosciences, psychology, and other disciplines in order to advance psychotherapy research and practice.

—*Martin Grosse Holtforth*
Bern, Switzerland, Spring 2006

This publication in the English language reflects Klaus's vast and unprecedented efforts to bridge the gap between the neurosciences and the understanding of psychological disorders and their treatment.

Klaus considered Martin an immensely gifted and innovative young scientist with whom he enjoyed the stimulating exchange of ideas. He was convinced that Martin's original and independent thinking, together with his serious commitment to the improvement of the practice of psychotherapy, would enable him to one day become a significant figure in the field of psychotherapy research.

It was therefore clear that I would entrust Martin with the task of the completion of the English translation, and I am deeply grateful to him and the exceptional team that he has chosen, for their dedication in making this publication possible.

—*Mariann Grawe-Gerber*
Zürich, Switzerland, Spring 2006

Only a few weeks before Klaus Grawe died in the summer of 2005, he had expressed to me his enthusiasm about finishing the English translation of his *Neuropsychotherapy*. It has been a pleasure and honor to be involved in this project and to witness its posthumous publication. My hope is that the translation of this book preserves the spirit of Klaus Grawe's thinking while at the same time making his ideas accessible to a wider international audience. The breadth, depth, and integrative reach of Grawe's volume are impressive: Starting with an exposition of basic neuroscientific mechanisms, he goes on to review the neural foundations of major mental disorders, to articulate a comprehensive theory of psychopathology and psychotherapy—consistency theory—and to deduce from all this a set of principles and guidelines for psychotherapy practice. Even though *Neuropsychotherapy* is explicitly based on an understanding of the neural foundations of experience and behavior, it is not reductionistic, nor does it deny or downplay the importance of psychological processes. Indeed, consistency theory is firmly connected to other psychological theoretical approaches, from control theory to cognitive–behavioral models to basic need theories. Grawe's theory is also compatible with contemporary psychodynamic and cognitive models that emphasize the unconscious, implicit, or automatic nature of mental functioning. One of the main conclusions arising from *Neuropsychotherapy* is perhaps the idea that therapists must attend to and skillfully target the processes transpiring beyond patients' conscious awareness. Therapists must ensure that their patients repeatedly encounter experiences that shift them from avoidance modes toward approach modes of functioning and that satisfy their basic psychological needs. These experiences must occur sufficiently strongly and frequently so that the corresponding neural activation patterns become firmly established. *Neuropsychotherapy* is not about the simple administration of technical interventions, then. Becoming a skilled neuropsychotherapist requires the acquisition of a complex set of skills, but, as documented throughout this volume, such efforts are likely to bear fruit and result in a more effective and mutually enjoyable psychotherapy. *Neuropsychotherapy*, then, is an integrative approach that takes patients' experience seriously on all levels, from motivational constellations and need dispositions to symptom patterns and, ultimately, to the neural underpinnings of experience and behavior. Grawe's ambition in this volume is not to propose yet another school of therapy but to move the field and the profession beyond the divisive splintering into theoretical camps. As such, consistency theory and *Neuropsychotherapy* can be viewed as attempts to formulate a *grand theory* of psychopathology and psychotherapy. Klaus Grawe also emphasized, however, that he welcomed the further development and continual empirical updating of consistency theory. My hope is that the translation of this book will inspire researchers and clinicians to engage with Grawe's rich world of ideas and to follow his call to further refine his neuroscientifically informed integrative psychotherapy.

—Björn Meyer
London, Spring 2006

ACKNOWLEDGMENTS

Our research and clinical team at the University of Bern and at the Institute for Psychological Therapy in Zürich has contributed significantly to this book in two ways. Many of the findings that are discussed in chapters 4 and 5 have emerged from our research over the past several years. These studies have been supported over a long period by the Swiss National Science Foundation (Schweizerischer Nationalfonds für wissenschaftliche Forschung, Project Nos. 1114-52657.97 and 1113-67204.01). However, not everything that has been developed over the past years can be cited and referenced in the form that one would cite scientific studies. Much is based on ideas and clinical experiences that were significantly influenced by clinical collaborators whose work is not specifically cited in the text. Specifically, this includes Barbara Heiniger-Haldimann, Mariann Grawe-Gerber, Daniel Regli, Urs Jost, Franziska Zahrli-Veronesi, and Simon Itten. I would like to thank them, as well as my more narrowly defined research collaborators, for the exceedingly fruitful collaboration over the last years. The research work that is discussed in this book involved the following collaborators: Martin Grosse Holtforth, Daniel Regli, Emma Smith, Daniel Gassmann, Hansjörg Znoj, Anne Trösken, Christoph Stucki, Sonja Kohls, Alexander Fries, Özgür Tamcan, and Günter Wüsten. Roger Schmied supported my preparation of this book with great organizational talent, by ensuring that the neuroscientific literature that I reviewed for this book was made available in a timely and fantastically organized manner. Alexander Fries has supported my work by reviewing all the psychological literature that was broadly related to the concept of consistency.

I also owe considerable debt to my team for freeing me from my normal responsibilities for half a year, which enabled me to devote my time to the work on this book. This resulted in increased workload and responsibility for many colleagues. I would especially like to thank Professor Hansjörg Znoj, who assumed my leadership position during the time of my absence and fulfilled this responsibility most excellently. It should not at all be taken for granted that after half a year, one could simply return to work without finding a huge mountain of work accumulated during this time.

I would also like to thank the colleagues who took the time and effort to read the entire manuscript, or parts of it, and provide feedback. This includes Dietmar Schulte, Martin Grosse Holtforth, Barbara Heiniger-Haldimann, Daniel Gassmann, and

Hansjörg Znoj. I have received invaluable suggestions from them, which resulted not only in improvements in specific sections but also in a more successful overall presentation.

Marielle Sutter has conscientiously edited the entire manuscript, revised the reference section, and constructed the author and topical indexes. I thank her, as well, for many useful suggestions and corrections. Roger Schmied and Karin Hofmann have worked on the figures to turn them into printable objects. Their competent help in the many things that lie between the writing and the final version of this book is greatly appreciated. Hogrefe Publishers, and especially Kathrin Rothauge and Franziska Stolz, have also helped to ensure that the book could be published swiftly after its completion.

My deepest gratitude is extended to my wife and life partner Mariann, to whom I am connected by a deep bond of love. I could not have written a book such as this without her. She enabled me to concentrate fully on the book, not only by accepting the neglect of many of the activities that make our life together fulfilling, but also by attending to the many things that would otherwise occupy me, by managing our life together, by lovingly tending to our children and closing all the gaps that opened up because of my physical and mental absences during this time. The fact that this book could be produced in this way is, to a large extent, also a result of her effort. I dedicate the book to her in love and gratitude.

—*Klaus Grawe*
Zürich, December 2003

Neuropsychotherapy

How the Neurosciences Inform
Effective Psychotherapy

CHAPTER ONE

INTRODUCTION

1.1 THE INSIGHTS GAINED IN THE NEUROSCIENCES ARE RELEVANT FOR EACH OF US

The "decade of the brain" has come and gone, but a century of progress in the brain sciences is on the horizon. Over the past 15 years, groundbreaking research has begun to illuminate the neural foundations of experience and behavior. Several more decades will pass before we will fully comprehend the baffling complexity of the brain's basic processes, but even the knowledge that is already available suffices to significantly shake up our conventional understanding of ourselves as mind- and soul-bearing beings. The current neuroscientific paradigm shift is not linked to a single person, such as Copernicus or Darwin, but its implications are comparable to those earlier scientific revolutions.

The decade of the brain has massively accelerated the rate of progress in the neurosciences. The main reason for this acceleration has been the availability of new research tools, in particular, the newly available capacity to observe the brain in action, as it were. The recent development of neuroimaging and similar technologies has fueled an exhilarating dynamism in the neurosciences that at times appears truly mind-boggling. Indeed, nearly every month, renowned scientific journals such as *Nature* or *Science* publish neuroscientific research with exciting results that significantly extend our knowledge in this area. The only other scientific domain that can perhaps rival the dynamism and impact of the current neuroscientific knowledge revolution is genetics. Neuroscience overlaps, of course, with genetic research because neural plasticity is intrinsically linked with gene expression, as we shall see later (see chapter 2).

The current breakthroughs in the neurosciences did not emerge in a vacuum. They were preceded by a long series of important findings, recognized by 23 Nobel prizes awarded to neuroscientists in the 20th century. Indeed, these foundational insights reach back to the beginning of the past century. In 1906, the Spaniard Ramon y Cajal and the Italian Camillo Golgi—both neuroanatomists—were the first neuroscientists to be awarded the Nobel prize for medicine. In their day, a common assumption was still that the nervous system is a seamless network. Ramon y Cajal then developed the idea that the brain consists of single neural cells and that impulses are transmitted between neural contact points, which Sherrington later termed *synapses*. This key insight led many to herald Ramon y Cajal as the father of neuroscience. Until the 1950s, however, the view of the brain as a network of neurons and synapses remained a merely theoretical position. It was not until the development of the electron microscope that this position was confirmed as fact—a fact that we now often regard as self-evident.

Twenty-one Nobel prizes for other outstanding research followed these early insights, and each marked an important step toward 21st-century neuroscience. In the year 2000, Arvid Carlsson and Paul Greengard were given the Nobel prize for their work on the role of dopamine in neural signal transmission, as was Erick Kandel for his groundbreaking work on the neural foundations of learning. With Kandel, we have arrived in the 21st century. His research group at Columbia University in New York remains one of the most productive of our day. Throughout this book, we shall sample often from the exquisite neuroscientific delicacies that Kendal's team serves up seemingly effortlessly and at breathtaking speed.

What exactly are the neuroscientific revelations, then, that have the potential to permanently alter our human self-understanding—our conception of what it means to be human? I would argue that this isn't so much about singular findings but rather about the overall conclusion emerging from the recent neuroscientific research. In Kandel's words: "From these considerations it follows that all the brain processes—from the motoric regulation of movement to the most intimate lines of thought—are ultimately biological processes" (1996, p. 713). The brain scientist Joseph LeDoux has expressed this idea even more poignantly in the title of this last book: *Synaptic Self: How Our Brains Become Who We Are*. The last sentence in this book reads: "You are your synapses. They are who you are" (2002, p. 324).

LeDoux aims to show in his book *how* our synapses make us who we are. The question is no longer *whether* our personality emerges from neural structures and processes but rather, he is concerned with the specific mechanisms determining how we become ourselves. No topic, regardless of how thorny or controversial, is omitted. The question of free will becomes the question of how our brains accomplish the feat that we experience ourselves as the authors of our own feelings, thoughts, and actions. Indeed, conscious awareness has become somewhat of a favorite topic among leading neuroscientists. Their discussions do not concern themselves with the question of whether consciousness emerges from the brain, but rather, *how* specifically one should construe the emergence of consciousness from basic neural processes (Bachmann, 2000; Crick & Koch, 1990, 2003a, 2003b; Dehaene & Naccache, 2001; Edelman, 1989; Edelman & Tononi, 2000; Koch, 2003).

There is indeed a great deal for us to consider, then, if it is true that all that we think, know, believe, hope, suffer, decide, or do, is ultimately linked—down to the most minute detail—to the structures of neurons, synapses, and the processes among them.

If even concepts such as mind and soul are ultimately—in their existence as well as their particular characteristics—a product of neural networks (see chapter 2), then it is indeed time to reconsider our conception of what it means to be human. The process by which other sciences receive and absorb the recent neuroscientific revelations has only just begun. Considerably more time will pass, however, before the new conception of human self-understanding has penetrated societal awareness more generally.

Modern philosophers have quickly recognized the societal relevance and explosive potential of the newly emerging neuroscientific findings and claims. Suddenly, fact-oriented scientists are appearing on the scene with statements about domains traditionally claimed by philosophy—the question of free will, for instance, or that of human consciousness. In her 1986 book *Neurophilosophy*, Patricia Churchland was among the first to articulate how our very conception of the mind–body problem is altered by neuroscientific evidence. Churchland argued that traditional dualistic views had become untenable in light of the new evidence. Such earlier and now outdated views include, for example, the position advanced by brain scientist Paul Eccles (Popper & Eccles, 1977), who argued that the brain causally influences the mind and the mind *as an independent entity* also influences the brain. The new view, by contrast, holds that causality is unidirectional: From the brain to the mind, but not from the mind to the brain.

What are the implications for psychotherapy? Will it become a superfluous anachronism? Must psychotherapists surrender their turf to the psychopharmacologists? Not at all!

1.2 BRAIN, PSYCHOTHERAPHY, AND PSYCHOPHAMACOLOGY

If all mental processes are grounded in neural processes, then changes in mental processes should also be linked with detectable changes in neural processes. The evidence clearly indicates that mental processes can be effectively and permanently altered through psychotherapy. Inevitably, the effectiveness of psychotherapy—in cases when therapy *is* effective, that is—is mediated by its effects on the brain. When therapy doesn't alter the brain, it also *cannot* be effective. In LeDoux's words: "Psychotherapy is fundamentally a learning process for its patients, and as such is a way to rewire the brain. In this sense, psychotherapy ultimately uses biological mechanisms to treat mental illness" (LeDoux, 2002, p. 299).

This is quite an unusual line of thinking for most psychotherapists, and not just for them. Let's consider the position of Eric Kandel—who incidentally began his scientific career as a psychiatrist—in somewhat greater detail:

> A fascinating association in this context is that psychotherapy, as far as it leads to substantial behavior change, appears to achieve its effect through changes in gene expression at the neuronal level. An analogue line of thinking suggests, then, that neurotic disorders are linked with changes in neural structure and function, just as specific forms of

mental illness include structural (anatomical) brain changes. Consequently, any—successful—psychotherapeutic treatment of the neuroses and personality disorders would also trigger structural changes in the involved neurons. With the improved resolution of neuroimaging methods, we are now at the brink of the fascinating possibility to use these tools not merely for the diagnosis of mental illness but also to ensure the effectiveness of psychotherapies. (Kandel, 1996, p. 711)

To be sure, we are not there yet. But the direction toward which these developments can (and will likely) move is becoming increasingly clear. Indeed, it would seem a most attractive option to replace our current merely descriptive but nonexplanatory diagnostic systems (Diagnostic and Statistical Manual of Mental Disorders [DSM] and International Statistical Classification of Diseases and Related Health Problems [ICD])—which apart from their undeniable advantages are also plagued by a number of severe limitations (Beutler & Malik, 2002)—with a more functional taxonomy; one that classifies phenomena based on their pathogenesis. Truly immense progress would be achieved, indeed, if we were to determine with precision which aspects of the brain would have to be altered in order to attain specific desired changes in experience and behavior. The task of neuropsychotherapy research would then be to determine which kinds of events a patient must experience in order for the desired brain changes to follow. The task of the neuropsychotherapist, ultimately, would be to ensure that such theoretically needed experiences become concrete patient reality.

At this point, all of this remains utopian. To turn the utopia into reality, a lot more needs to be known; the research on the neural correlates of mental disorders has only just begun (see chapter 3). Nevertheless, it is abundantly clear that this knowledge will grow exponentially within the next decade because so many teams are now engaged in this quest. Psychopharmacologists are eagerly translating every knowledge gain into improved medication treatment. It is becoming clear, moreover, that targeted alterations of neurotransmitter balances in specific brain regions can lead to the same kinds of synaptic changes as those associated with learning experiences. The effects of psychopharmacological interventions would be massively improved, however, if they were not applied as indiscriminately—throughout the entire brain—as is currently the case (see Stahl, 1996).

One must remember as well, however, that the brain at all times remains dependent on experiences that are mediated by the senses. Even when the brain is under psychopharmacological influence, sensory experiences continue to have their effects upon the brain. When such experiences are negative, their consequences will be negative. Yet, a person's experiences and their subjective meaning critically depend on his or her motivation and actions. Most experiences, after all, are not simply passively experienced—not even those that create and maintain mental disorders. From the cradle to the grave, human beings are in a state of continuous motivation: constantly desiring to attain but also to avoid certain experiences. This principle holds true just as well for people with mental disorders. Regardless of how excellent their psychopharmacological treatment may be, they will also always benefit from instruction and support to bring about the needed increase in positive—and reduction in harmful—experiences. Once such concrete, positive life experiences are being realized, self-sustaining and healthier brain structures and processes can fall into place. Irrespective of the rate of pharmacological progress, then, we will continue to need a profession that has as its goal to take the time and develop the specialized expertise required to find out—in the individual patient case—which precise experiences are needed to attain positive change, a profession that aims to guide and support such a patient, to ensure that the needed experiences are turned into concrete reality. Thus, the neurosciences by no means render psychotherapy superfluous. On the contrary, the necessity of psychotherapy results from, and is clarified by, a neuroscientific perspective on mental disorders. I predict that even many of those who previously were skeptical toward therapy will thus come to view it as necessary and important.

Psychotherapy and pharmacological intervention are not just two alternative ways to achieve a largely equivalent outcome—one by a sensory and the other by a biochemical pathway. Their relationship can be characterized as neither alternative nor symmetrical. Without sensory experiences that satisfy their basic needs, human beings cannot live and attain happiness (see also chapter 4). The brain is designed for the purpose of experiencing need-satisfying events; human happiness is more than the right combination of neurotransmitters. An adequate neurotransmitter combination is undoubtedly important, and this biochemical prerequisite is fortunately also met for most people. When this precondition is not in place—regardless of what the causes might have been—it can be very helpful to pharmacologically reestablish this disturbed balance. In some cases, this might even be the only promising path. Once the balance is reestablished, the treated patient is then (again) on an equal level with those who are equipped by nature with the correct neurotransmitter mixture. But this equality is reestablished only in the biochemical aspect; nothing is decided about the future happiness or suffering of the person, which will continue to depend on the sensory events experienced by the person. And again, the principle holds true, even for the person treated with medication, that we are not just passive victims of our life experiences but also to a large extent actively determine them. In turn, the events that are experienced depend upon the person's motivational potentials, on the abilities, knowledge, and situation-specific reactions that have formed in the person over the course of his or her life history.

All of this information is stored on a neural level in distinct memory systems and can be altered through new experiences. But these specific experiences are critical and indeed indispensable. To be sure, neurotransmitters can influence the activation threshold of problematic thought, emotion, action, and reaction potentials, but such pharmacological alterations could not result in the emergence of new memory traces. To create new memory content, which can then change subsequent experience and behavior, the person needs to take in new sensory experiences that change old memory content. Such new experiences do not come about by chance, even when the right neurotransmitter balance is in place. From a neuroscientific perspective, psychopharmacological therapy that is not coordinated with a simultaneous, targeted alteration of the person's experiences cannot be justified. The widespread practice of prescribing psychoactive medication without assuming professional responsibility for the patient's concurrent experience is, from a neuroscientific view, equally irresponsible.

In many cases, psychotherapy alone (without any pharmacological intervention) can achieve neural changes that are linked with positive consequences for the person's experience and behavior. Neurotransmitter imbalances, then, can be corrected not only via pharmacotherapy but also via therapy alone (see chapter 3 for more detail on this point). Long before the advent of our artificially created psychoactive medications, naturally occurring life events were linked with fluctuations in serotonin and dopamine levels. Events that are specifically and professionally designed to bring about certain experiences can, of course, have similar effects. Thus, a neuroscientific perspective provides ample justification for psychotherapy alone, even when used without pharmacological cointervention. This holds true at least for those disorders and problems for which studies have provided solid evidence that psychotherapeutic interventions can effectively lead to improvement.

The use of pharmacotherapy alone—in the absence of the professional and competent structuring of the treated patient's life experience—is not justifiable from a neuroscientific perspective. Such treatment implicitly assumes that the patient will somehow manage to encounter the right kinds of events on his or her own, without professional guidance. Cases in which this happened to work well cannot be used, however, to scientifically substantiate this widespread practice. For instance, when depressed patients are treated only with medication—a current common practice—the short-term effects are often quite good but not really so impressive across all treated patients. Averaging across the different types of antidepressive medication, the effect size of pharmacotherapy appears to be half a unit above the effect size associated with placebo (Joffe, Sokolow, & Streiner, 1996). The long-term relapse rate, however, appears to be nearly 80% among patients who experienced immediate improvement following treatment (Elkin, 1994). With continued psychotherapeutic treatment, this long-term relapse rate appears to be noticeably reduced (Elkin, 1994; Rush & Thase, 1999). As soon as medication no longer augments or inhibits synaptic transmission, and therefore no longer protects from negative life experiences, the patient's life constellation that originally led to the depression can once again exert its noxious influence. This is even more likely because the patient on pharmacotherapy alone is unlikely to have learned what to do to be nondepressed without the protection offered by medication.

There are additional reasons for considering the patient's life situation when prescribing pharmacological agents. That is, the effect of medication can be completely different in different patients or even within the same patient at different times, depending on the person's particular life constellation. The evidence for this point comes from a study on crayfish, but it is nevertheless quite interesting and notable. Yeh, Fricke, and Edwards (1996) identified a specific crayfish neuron whose response to serotonin—the very neurotransmitter known to be critical in pharmacological therapy of depression—depended to a large degree on the animal's status within the social hierarchy. This neuron controls the tail-flip reflex, which is part of the crayfish's fight-or-flight reaction. Animals that are more dominant within the social hierarchy respond to serotonin with an increased action readiness of this neuron, whereas socially lower animals experienced the opposite effect: an inhibition of this neuron.

When two crayfish that are both low within the hierarchy are kept together, one of them soon takes on a dominant role. As soon as dominance is achieved, the previously inhibitory effect now reverts to its opposite: serotonin now activates instead of inhibits the neuron.

Is it plausible that our nervous system should have lost this sensitivity to contextual change over the evolutionary course of species differentiation? It seems unlikely. In light of the complexity of our nervous system, this context-responsiveness has probably increased rather than decreased over time. The more we find out about the interdependency of our nervous system with its social context, the more specific we will be able to be in our application of individualized pharmacological agents targeting unique patient contexts. The medication's positive effects in individual cases would likely improve immensely, compared to today's status quo. Considering that the decisions for a patient's medication dosage and duration are often still based on trial and error, or at best on uncontrolled clinical heuristics, it is perhaps not surprising that we currently do not exceed an average effect size of .50. But we don't have to be content or resign ourselves to this state of affairs. Pharmacotherapy, even more than psychotherapy, critically depends on the future progress we can expect from the neurosciences.

The neuroscientific view of how synaptic signal transduction can be effectively influenced suggests that psychopharmacologists and psychotherapists ought to work side by side much more than this is currently the case. Targeted pharmacological intervention can facilitate a patient's readiness to learn and can intensify the effects of specific learning experiences. This is evident, for instance, in a study by O'Carroll, Bryslade, Cahill, Shajahan, and Ebmeier (1999). They showed participants a series of emotionally engaging images. After one week, they tested how much participants had remembered. Participants who had been under the influence of a noradrenergic agonist when they first viewed the images were able to recall more details than the participants who had received a placebo. The participants in the placebo condition, in turn, remembered more details than those who had received a noradrenergic antagonist, which inhibits noradrenergic activity. Natural memory capacity (the participants had not been instructed to try to recall as much as possible, so this was not an explicit memory learning task), then, can be biochemically facilitated and inhibited. If patients are pharmacologically induced into such a state of heightened learning readiness, it becomes all the more important which types of events are subsequently encountered and encoded. This increases the responsibility incumbent upon the psychotherapist.

The psychotherapist's task becomes, thus, to clarify and facilitate the occurrence of specific learning experiences that are likely to exert a positive influence on the patient's problems and the neural structures underlying those problems. The efficacy of synapses coding for the learning experience has to be facilitated by the therapist in order to achieve the intended therapeutic effect. This process is quite similar to what happens between the synapses when new memories are formed. Neurotransmitters dock at exactly those locations on pre- and postsynaptic membranes that were involved in the previous action potential transmission, thereby elevating the subsequent signal transmission at the synapse (see chapter 2). What we know so far about the effects of pharmacotherapy and psychotherapy suggests that these interventions may potentiate each other's effectiveness. If such potentiation does indeed occur, and for whom, remains at this point an open research question. The few studies that have investigated this issue suggest an advantage for combined psychotherapy and pharmacotherapy, especially for patients with more severe symptoms (Thase, 1997). However, the available studies of combinatorial treatments cannot be regarded as adequate tests of the kinds of ideas I am outlining here. Previous studies in this area have been insufficiently specific in examining the reciprocal effects of therapy and pharmacotherapy. In these studies, one intervention is typically conducted in parallel to the other instead of being carefully coordinated so that one intervention could strategically facilitate the effects of the other. Such strategic treatment would require sound knowledge of both medication and psychotherapy, which is probably still rare in most contemporary treatment contexts.

1.3 NEUROSCIENCE AND PSYCHOTHERAPY

Neuroscientists and psychotherapists live in separate worlds that are quite removed from one another. It has only been a short time since they began to develop any degree of mutual interest. An area that could be a natural bridge, however, is their common interest in mental disorders. The interest of the neuroscientists in this area has been stimulated primarily by the discovery of the functional significance of neurotransmitters for mental disorders. This interest initially provided an obvious link to biological psychiatry rather than to psychotherapy, which is especially understandable when considering how psychiatrists quickly invested their hope into, and began to favor, pharmaceutical approaches to treating mental disorders. Psychotherapy was suddenly unfashionable, or at least it became something that could be left to the psychologists (rather than the psychiatrists). The overly simplified idea that mental disorders are largely hereditary further contributed to this changed climate.

Subsequently, however, neuroscientists discovered the enormous plasticity of the brain in response to environmental influences. It became increasingly clear that the genetic contribution to manifest mental disorders is far smaller than previously assumed, and that individual life experiences play a much larger role in determining gene expression. It also became clear that plasticity functions in two ways: facilitating disorders on the one hand but on the other hand also alleviating or compensating symptoms (see chapter 3). Since these discoveries, neuroscientific articles are increasingly ending with speculations on the possibilities to target neural structures and processes via psychotherapeutic means, or with speculations on how psychotherapy might be further developed from a neuroscientific perspective. Such ideas, however, tend to be found at the end of thick books that otherwise are silent on the topic; they are still unfulfilled visions of the future. One such vision—Kandel's—we have already briefly encountered earlier (p. 3). He further elaborated these ideas in two articles (Kandel, 1998, 1999).

Another example is Nancy Andreason's (2001) engaging book, *Brave New Brain: Conquering Mental Illness in the Era of the Genome*. After providing a very expert overview of recent neuroscientific research on the major mental disorders, she wrote:

> As shown in chapters 8 to 11, we know quite a bit about the brain systems affected by disorders such as depression, panic disorder, or posttraumatic stress disorder. These disorders, which frequently result from the brain's reaction to cumulative stress or unhealthy environments, are well suited for targeted cognitive interventions. Such interventions will increasingly be developed by competent scientists who are able to combine their knowledge about human behavior and its measurement with knowledge about brain systems and their malleability. Such approaches have already been applied successfully in interventions that teach dyslexic children to hear sounds and words more precisely, so that they can rewire their brain anew and write and read more effectively. Strategic attacks on the mood disorders could follow such models and target fundamental disorder aspects, such as reduced flexibility when confronted with a mishap, or the inability to regulate an intensive reaction to such situations appropriately. (p. 401)

When reading the last sentence, many cognitive behavior therapists might note that they are already doing some of the things that Andreasen envisions for the future. However, they would likely have to admit that their skills are not yet strategically combined with "knowledge about brain systems and their malleability." Therapists might also find in other parts of Andreasen's book that their field is somewhat misrepresented; for example, when Andreasen writes "Systematic desensitization is the most common behavioral intervention" (p. 379) or "historically, the development and application of psychotherapy has never been regulated systematically ..." (p. 401). In these formulations it is evident that psychotherapists and neuroscientists truly live in separate worlds. The fact that a psychotherapy oriented on scientific principles has emerged, beyond the psychoanalysis and behavior therapy of the 1960s, is easily lost on scientists working in other fields. This scientific psychotherapy easily remains invisible under the avalanche of nonsense being published on the topic of "psychotherapy" year after year, and it is apparently hard even for people such as Nancy Andreasen, editor of the *American Journal of Psychotherapy*, to detect the signal of this empirical psychotherapy. Fortunately, psychotherapy has continued to evolve far beyond what some neuroscientists sometimes suspect or insinuate. Empirically oriented psychotherapy has more to offer for the realization of the principles envisioned by Kandel, LeDoux, and Andreasen than these authors themselves suspect. Demonstrating this point shall be my task in this book.

Psychotherapists, vice versa, have also become increasingly interested in the neurosciences over the past 6 years. In my previous book (*Psychological Therapy*, published in 1998 in German and 2004 in English), I have attempted to develop an extensive psychological framework for therapy based on the then-extant knowledge in the neurosciences (and also based on extant psychotherapy research and basic psychological research). I had finished writing the book in 1997 and had integrated literature up to 1996. At that time, I had not been aware of any attempts on the part of psychotherapists to systematically review neuroscientific findings and derive practical implications for therapy. Since then, however, the decade of the brain has had clear effects on psychotherapists as well. There are by now a considerable number of publications that either generally review neuroscientific findings with an eye on potential therapeutic applications or that focus on specific research areas and deduce psychotherapeutic principles from them (in chronological order: Deneke, 1999; Liggan & Kay, 1999; Gabbard, 2000; Beutel, 2002; Bock & Braun, 2002; Förstl, 2002; Storch, 2002; Sulz, 2002; Westen & Gabbard, 2002 a, 2002b; Beutel, Stern, & Silberweig, 2003; Etkin, Cappas, Andres-Hyman, & Davidson, 2005; Pittenger, Polan, & Kandel, 2005; Grosjean, 2005). This movement resembles an emerging tidal wave. At the annual conference of the Society for Psychotherapy in Santa Barbara in 2002, psychotherapy researchers for the first time devoted a special forum to the question of what the implications of *affective neuroscience* (Panksepp, 1998) might be for therapy (Grawe, 2002; Hollon, 2002). The presidential address at this conference, given by my long-term colleague and friend, Franz Caspar, was also devoted to the topic of neuroscience and psychotherapy (Caspar, 2002). At the World Congress of Psychotherapy in Trondheim in 2002, a keynote lecture was held on the topic of "The revolution in the neurosciences: Implications for psychotherapy research and practice" (Gabbard, 2002). While writing this section, I am actually supposed to finish the introduction for a book on the topic of *The Neurobiology of Psychotherapy* (Schiepek, 2003). In short, over a very short period, neuroscience has emerged as a hot topic for psychotherapists.

I am all in favor of this development. But I also believe that the implications of neuroscientific research for psychotherapy are more far-reaching than is often acknowledged in publications and presentations on this topic. Sometimes one can already tell while reading or listening that everything will essentially continue as usual. No doubt, Freud's metapsychology, the concepts of the ego, id, and superego will surely have to be revised; transference interpretations that refer to early-life transactions between child and parents would likely be incompatible with a modern understanding of memory development, but—it is sometimes argued—neuroscience simply provides further evidence for Freud's essential accuracy with regard to the importance of early childhood experiences and the existence of unconscious processes. Many of those who traditionally distinguish among separate "schools of psychotherapy" will likely maintain their persuasion, regardless of neuroscientific discoveries. New findings that fit within the preexisting perspectives will be integrated; the rest will be gladly ignored. Some have even warned explicitly that our still limited knowledge prohibits a direct translation of neuroscientific evidence into scientifically based therapy strategies (Beutel, 2002, p. 9).

I would also argue that the advent of neuroscience does not necessitate a completely new start for psychotherapy or that all earlier findings will suddenly be outdated. One hundred years of practical experience have left us with a rich fund of knowledge, even if we limit ourselves to those findings that have passed the filter of empirical therapy research. According to the principles described earlier, therapeutic strategies that have proved to be effective for particular problems derive their impact from the specific changes they achieve in the brain. These effects could not be achieved without the presence of therapists who have the necessary training and knowledge. It would be impossible for a neuroscientist to achieve these effects, except if he or she were to complete the necessary training.

One does not have to be a neuroscientist in order to be able to achieve changes within the brain. Every good teacher, every good football coach, every expert who knows something about persuading others can achieve such effects. What is needed is knowledge about how to persuade or influence people and about the exercises that appear to be particularly effective in the relevant domain. There is little doubt that across various professions, good persuaders are characterized by a set of advantageous "human qualities," but such characteristics alone do not suffice to make a successful football coach. Beyond these general human qualities needed across professions, what is needed for success is occupation-specific expertise. Would a football coach perform better if he or she acquired detailed knowledge about brain functions? Hardly! Nothing we know suggests that now, after the decade of the brain, football coaches have suddenly become fascinated with neuroscience.

By contrast, psychotherapists have recently become much more interested in neuroscience. Why? Probably because they suspect that neuroscience might improve their domain-specific expertise. This would be a realistic appraisal of the current state of psychotherapy. In this respect, our situation is quite similar to that of the psychopharmacologists. We have a set of effective interventions, but we do not know precisely how they work because we don't know enough about that which they affect. What, after all, are the causes of mental disorders? If we knew this with precision, we would have diagnostic systems other than the DSM-IV and ICD 10. For good reasons, these classification systems did not include explicit statements about etiological factors. The experts could not reach consensus; questionable assumptions still outweigh the proven facts.

The gain associated with this exclusion of etiological concerns is that different judges can more easily agree on the presence of a diagnosis. This makes it easier to know with precision what the other person means when speaking about panic disorder, dysthymia, bulimia, etc. The current convention requires that a specific set of criteria must be fulfilled in order to make such diagnoses. But conventions are, indeed, *just* conventions, and not scientific facts. A person could memorize the entire DSM but still would not have *understood* the mental disorders. Yet it is a basic human need to strive for understanding (see also chapter 4). Psychotherapists seek to understand mental disorders and to understand how psychotherapy functions. This need to understand is so strong that it sometimes overrides common sense. Thus, many new therapists turn to therapy schools that claim to provide the dearly desired knowledge.

But even—or especially—experienced therapists have a renewed, intensive desire to understand (even more about) mental disorders and the function of psychotherapy. The prefabricated truisms of the therapy schools don't impress them anymore; they are beyond that stage. They have learned to work with the tools acquired from the therapy schools—the thought patterns and action repertoires—but they also know about the limitations of these tools. They tend to work either eclectically or they have a pronounced interest in the integration of those methods that they have experienced as valuable and effective. For such therapists, integration attempts that mesh with their experience or that provide entirely "fresh," not yet overused perspectives are particularly attractive to extend their therapeutic expertise. The recent neuroscientific insights constitute such fresh input. They clearly point to a road away from the well-trodden paths of the traditional psychotherapy schools. This also explains the suddenly growing interest in neuroscientific findings and insights whose relevance to psychotherapy is now becoming clear. It is only now, after the decade of the brain, that the neurosciences truly have something valuable to offer for the field of psychotherapy.

1.4   WHAT IS MEANT BY THE TERM *NEUROPSYCHOTHERAPY*?

Thus, we have arrived at the core topic of this book. It is about what neuroscience already has to offer for psychotherapy. In this context, it would not make sense to artificially separate neuroscience strictly from other basic psychological science.

> The boundary between neuroscience and cognitive science is fuzzy and ultimately arbitrary. It is not a natural boundary of subdisciplines but is indeed based on a lack of knowledge. With increasing knowledge, we see an increasingly common joining of the biological and psychological sciences. At exactly these contact points, our understanding of mental processes will be placed on an increasingly solid foundation. As modern cognitive psychology has demonstrated, the brain produces an internal representation of the perceived world. Neurobiology, in turn, has shown how these representations can be explained at the level of single neurons and their connections. The convergence of these disciplines has revealed entirely novel insights into the phenomena of perception, learning, and memory. (Kandel, 1996, p. 713)

I should add that this joining of disciplines applies not only to cognitive processes in a narrow sense, but that "cognitive neuroscience" has recently been joined by "affective neuroscience," thus leading to increasingly broad mergers. Indeed, the emotions have become an explicit emphasis within neuroscientific research (Damasio, 1999; LeDoux, 2004). The merging of psychology and neuroscience is expanding to include an increasing number of subfields; soon it will also encompass areas in which purely psychological research still dominates neuroscientific work today. This is the case, for example, in contemporary research on the motivational aspects of mental functioning.

In my book, *Psychological Therapy* (Grawe, 1998, 2004) I have already attempted to link the results of psychotherapy research with recent findings from psychological and neuroscientific research, and to make inferences for the practice of therapy. My emphasis in that book was on basic psychological research. The neuroscientific part had to be considerably smaller because neuroscientific research, at that point, simply did not offer much for psychotherapy. With the title of that book I attempted to show that psychotherapy today no longer needs to rely on the traditional therapy schools but actually has a fully sufficient conceptual basis in contemporary psychology.

With this book I am taking another step in the same direction. Since the 1998 book, the development in the neurosciences has been almost explosive. Many questions of interest for psychotherapists can now be answered from a neuroscientific perspective. The answers are anything but trivial. In some cases, they suggest conclusions that cast doubt on conventional therapeutic conceptions and practices. In this book, then, the relationship between neuroscientific and basic psychological research is the opposite of that in the previous book. I am relying here primarily on neuroscientific findings, but I include other areas of psychological research as far as they are relevant to the particular issues at hand. This shifting of emphasis is mirrored in the title of the book: *Neuropsychotherapy* emphasizes the neuroscientific foundations of psychotherapy, whereas *Psychological Therapy* emphasized genuinely psychological basic science. Nevertheless, my main concern has remained the same—the scientific justification of psychotherapy on the basis of contemporary basic research findings.

Neuropsychotherapy, therefore, is concerned with the same issues as *Psychological Therapy*. The books together can be viewed as complementary attempts to provide a theoretical foundation for a "generic psychotherapy" (Grawe, 1995, 1996; Grawe, Donati, & Bernauer, 1994).

This generic psychotherapy attempts to utilize the entire repertoire of validated, effective psychotherapeutic processes in order to achieve maximally positive therapy outcomes. Once the various therapeutic strategies and processes have been uncoupled from their original theoretical backgrounds, the question of therapeutic effectiveness can be asked afresh. In this book, I am attempting to explain the effectiveness of psychotherapy by relating therapeutic processes primarily to neuroscientific findings. In this context, I am making the explicit assumption that specific therapeutic strategies have been shown to differ in their relative effectiveness; that is, that psychotherapy research has empirically validated some strategies more than others. These strategies are then explained by relating them to lawful regularities on the level of neural functioning. Such a neuroscientific explanation of already existing therapeutic strategies does not result in the creation of a new form of psychotherapy but instead yields a new perspective on psychotherapy. The original links between intervention strategy and theoretical justification are disbanded and replaced by an integrative view that transcends the boundaries among traditional therapy forms. On the basis of such an independent, new perspective, it is possible to relate to each other, and combine with each other, strategies that previously seemed incompatible because of their mutually inconsistent theoretical justifications.

In my view, an appropriate foundation for such a generic psychotherapy would have to be maximally consistent with current psychotherapy research as well as with basic psychological and neuroscientific research. The problems faced in psychotherapy and the corresponding findings uncovered in psychotherapy research (which is my primary area of expertise) have served as a guiding heuristic as I sifted through volumes of current neuroscientific research. In this book, I highlight those findings that appeared to have particular relevance for psychotherapy. However, I have not limited myself here to merely summarizing findings but, as a step beyond that, I ask about the implications of the neuroscientific findings for the specific problems arising in psychotherapy. In chapter 5 I draw on neuroscientific research to propose a set of concrete guidelines for therapy planning and the therapeutic process. These implications can be termed *neuropsychotherapeutic* and, as such, they differ from the implications arising from other theoretical frameworks. On the one hand, then, neuropsychotherapy refers to a neuroscientific perspective on the problems in psychotherapy and, on the other hand, to the practical implications that emerge from this perspective.

Before elaborating further on the specifics of neuroscientific research, however, I would like to illustrate with a case example how a neuropsychotherapeutically informed and enriched psychotherapy might look in practice.

Each statement that is made in the following case description is based on the findings of one or, in some cases, several empirical studies. The specific references are of course important, but I omit them here in the introduction in the interest of flow and readability. In later chapters, however, I provide the needed detail to clarify the empirical basis for these clinical claims and strategies.

1.5   HOW MIGHT NEUROPSYCHOTHERAPY LOOK IN CONCRETE CLINICAL PRACTICE?

Let's imagine a therapy situation in which a therapist faces a depressive female patient, Ms. H. As in every therapy session, Ms. H. sits in her chair with a sad, fatigued facial expression, as if paralyzed, making no effort whatsoever to take initiative. Instead, she waits for the therapist's action. The therapist engages with her in a friendly and caring manner, asking her how she feels and if she is able to have the therapy session today. She responds with a bitter, muted tone of voice, stating that she obviously doesn't have a choice, that nothing would change anyway, and that there is no point to begin with. Similar exchanges have occurred in previous sessions; they are typical for this patient. They are also typical in interactions occurring outside of therapy.

Let us assume that the therapist's orientation is of the modern psychodynamic–interpersonal variant. He might relatively quickly have arrived at a conceptualization of this recurring interaction pattern. One of his goals would be to help this patient realize how her behavior repeatedly triggers disappointing experiences by making her interaction partners feel helpless and, ultimately, angry toward her. The therapist would attempt to clarify the fears and wishes that give rise to this self-damaging interactional pattern, and he would come up with strategies that might alleviate her fears. It probably wouldn't be easy for him to control his own feelings of helplessness and anger, which naturally arise in interactions with this patient, but he knows that this is important in order to avoid a renewed confirmation and strengthening of the patient's fears.

These are typical speculations that a well-trained modern psychotherapist might come up with when faced with such a clinical situation. We could easily imagine other therapists who, faced with the same scenario, would act less competently and sensitively, so that the patient would quickly terminate therapy. This, too, would not be far from current therapeutic reality.

A neuropsychotherapist would also consider all of these possibilities; the general approach would be quite familiar to him. However, other considerations would be added. He would think of the likelihood of Ms. H.'s enlarged and overactivated amygdala, which selectively and overly sensitively responds to negative situations. He would consider how the amygdala has particularly well-developed connections to the ventromedial regions of the right prefrontal cortex, whose activation is linked with emotional states. He would know about the wealth of firmly established projections between this area and the dorsolateral regions of the prefrontal cortex, which is critically involved in the activation of avoidance goals. In addition, he would realize how impoverished the corresponding areas in the left hemisphere are, due to their insufficient activation—areas that play an important role in positive emotions and the pursuit of approach goals (these points are elaborated in detail in chapters 3 and 4). The therapist would envision a wealth of elaborately developed synapses connecting the areas that represent avoidance goals with those responsible for the production and maintenance of negative emotions. These connected areas might be envisioned by the therapist as "brain swellings" because of their rich and elaborate development.

However, we must not envision these "brain swellings" as simply overactivated structures. Robert M. Post has explained the complexity of the situation as follows: "Areas of hypofunction or hyperfunction may involve either excitatory or inhibitory pathways, and may represent either pathological processes or compensatory adaptations triggered by pathology " (as cited in Mesulam, 2000, p. 408). Thus, restoring balance may involve either altering excitatory or inhibitory pathways, depending upon the mechanism of the pathology.

While the therapist is still thinking of the brain swellings and their consequences, he also begins to envision how a chronically elevated cortisol level is often associated with a noticeably damaged and shrunken hippocampus. A hippocampus in this state won't be of much use to Ms. H., even though it is critically important for the learning of new relationships, such as the relationship between her behavior and that of her interaction partners, or for the acquisition of new memory contents more generally. The therapist also considers how, for a large proportion of depressed patients, the anterior cingulate cortex may become difficult to activate. This region plays an important role in the active engagement with conflicts and ambiguous situations, and in the conscious experiencing of feelings. Both functions are therefore only partially available to the depressed patient.

The therapist realizes that it is not sensible to work with this patient directly on her problem behavior. First, he must rebuild the impoverished brain regions because their easy activation will be necessary to enable the patient to pursue positive goals in a self-initiated, self-governed manner; to enable her to experience joy and contentment; and to become open and accessible for the learning of relationships that she must understand in order to consciously regulate her interpersonal encounters in new patterns. The therapist has completely internalized the neuroscientific rule of thumb, "use it or lose it." He knows that the impoverished neurons and synapses must be activated in order for them to recover. Once this is accomplished, they will be more easily reactivated and ultimately can once again play an important role in the patient's thinking, feeling, and actions. She will then be reenabled to pursue positive goals and experience positive emotions.

Activating impoverished neurons is not easy, however, because they tend to resist such efforts. The synaptic connections are weak and must first be strengthened; therefore the connected neurons must be activated in order to facilitate the connecting pathway. This is the second rule of thumb, known as Hebb's principle, that the therapist has internalized: *Neurons that fire together wire together* (based on the Canadian psychologist Donald Hebb, who already in 1949 recognized and anticipated several principles of neuronal functioning that remain valid today). In the case of Ms. H., the neurons in the right dorsolateral and ventromedial prefrontal cortex have already fired together very frequently with those in the amygdala. This has led to a well-ingrained neuronal circuit (a *cell assembly*, in Donald Hebb's words). This circuit can be very easily activated, for example, by even the slightest sign of impatience in the voice of the therapist. The circuit will also recurrently reactivate in an automatic manner whenever Ms. H. is left on her own or inhibitory control is not maintained. Each activation is accompanied by the experience of depression and depressive behavior. With each activation, the projections among the connected areas are increasingly firmly ingrained.

The therapist thus realizes that he must block the activation of these hyperdeveloped connections and in turn activate the impoverished synapses in the left prefrontal cortex as often as possible. If he succeeds in this quest, he can expect that Ms. H. will once again become more active, that she will experience positive emotions more often, and that her previously dormant positive repertoire will once again move to the forefront—the repertoire that has always been present in the form of memory traces (neuronal circuits), even though excitability of these traces was previously weakened. From the perspective of the therapist, Ms. H. at this point is not able to behave more positively, given the current state of her brain. It is not resistance that prevents her from engaging more constructively in therapy. In her state, she cannot simply plan to view the world more positively or to self-initiate and engage with positive activities. The knowledge about Ms. H.'s neuronally mediated inability helps the therapist to not feel angry toward her, as has been the case with many others in her environment. The therapist also does not feel helpless and incapacitated by her because he knows how he can assist her.

The therapist knows that he must take the initiative and responsibility to become independent of her depressive interactional patterns. He must enable Ms. H. to experience events, as frequently as possible, that trigger positive emotions in her, or that—based on prior case conceptualization—are likely to be highly relevant for her motivational goals (even though these goals might be hard to recognize given her current depressive state). The therapist can trust in the reactivating "power" of these motivational goals because, over the course of Ms. H.'s lifetime, they have become even more deeply ingrained than the synaptic connections corresponding to Ms. H.'s current state. The therapist also knows in advance that his efforts will not be reinforced initially by changes in Ms. H.'s state. Because of this knowledge, however, he does not react with disappointment and impatience when his efforts at first appear to have no effect on her. He knows that the transcription process of gene expression, which is being stimulated by sufficiently frequent positive experiences, requires several weeks to manifest in a noticeably increased number of synapses. The key point to remember at this stage is simply that he must not be discouraged and stop the facilitation of frequent positive experiences. The creation of new and the restrengthening of already present synapses takes time.

It is more easily said than done to facilitate frequent positive experiences in Ms. H. while she is in her depressive state. In his training, the therapist had considerable difficulty translating into action what his mentors and supervisors called "resource activation," "motivational priming," or "complementary therapist behavior." By now, however, he views these as his most important tools, even more crucial than his problem-specific intervention strategies. (I do not elaborate here on this important technical aspect of therapeutic work. More detail is provided in chapter 5.)

Another aspect not to be neglected is the many negative thoughts and emotions that Ms. H. experiences frequently, especially at times when she is alone. As long as these negative patterns continue to take up a large proportion of total mental activity, there is simply not enough room for the positive activation patterns that the therapist wants to facilitate. The negative patterns must be reduced or blocked. On the neural level this means that the neural activation patterns must not be activated as frequently anymore or, should they become activated, that their activation must be disrupted or blocked as quickly as possible.

For this purpose, the therapist solicits the assistance of Ms. H.'s most important reference persons, her husband and her two adolescent children. After having conveyed to them his view of the situation, he explains that their wife/mother is feeling so poorly because three brain areas that chronically produce negative emotions have developed disproportionately. He shows them a picture of the brain that illustrates this process and responds to their questions. He emphasizes that this is not their wife's/mother's fault; that she cannot easily overcome this by herself and is not responsible for her state. Equally, he tells them there is much that can be done in order to help her, and that they can actively participate in this process. He notes that the brain cells are like muscles; if they are not used, they whither away, but if they are used continuously they become stronger. The neurons of their wife/mother in those three areas are like highly trained muscles, with the important difference that she cannot turn them on and off by sheer willpower. This switching must come from the outside, and they can help with it, by including her in as many positive activities as possible, by not letting her just sit there by herself, left with her ruminations, but instead by engaging her in ever new activities. They should feel free to interrupt Ms. H.'s ruminating and worrying at any time. The therapist discusses this with them in great detail and supports the family in translating these principles into action, by speaking with them every couple of days on the telephone and inquiring into how things are going and encouraging them to stick with the program.

A few weeks after the initiation of these resource-activating and problem-behavior-blocking interventions the first clear signs of improvement become evident in Ms. H. After 3 months, her depressive symptoms have largely disappeared. If one were to use functional neuroimaging tools at this point, the initially clear asymmetry in prefrontal cortex activation would have likely resolved. Even the size of her hippocampus may have normalized (see more detail on this point in chapter 3). The therapist now administers Beck's depression questionnaire and notes that Ms. H. scores within the range achieved by normal, nondepressed people.

Many therapists would find it quite obvious that therapy should be terminated at this point. After all, Ms. H. and her family regard therapy as a complete success; all are satisfied with the outcome. Ms. H.'s therapist, however, would not go with this option because he has arrived at a conceptualization of Ms. H.'s case that prevents him from being completely content at this stage. He remembers the scientific findings that suggest that Ms. H., over the course of the next 2 years, has a 60% to 80% chance of experiencing another depressive episode (Elkin, 1994) if he terminates therapy at this point.

Up until now, Ms. H.'s therapy was purely symptom-oriented. The point was to change the neural underpinnings of her depressive symptoms and to move her once again into the range of normal mental functioning. The therapist's intention to continue therapy is related to his understanding of how Ms. H.'s depression originated in the first place.

It had been Ms. H.'s first depressive episode. Everyone had agreed that she had not been depressed a year earlier, even though she had for some time already been nervous, anxious, and somehow more stressed than at other times, but certainly not depressed. Slowly over time everything seemed to become overwhelming for her; she lost all energy, until eventually she did not feel up to anything at all anymore.

Let's translate what we know about Ms. H.'s depression and its history onto the neural level. How did the changes in her brain come about over the course of the last year? New synapses grow and already existing synapses gain in strength when they are frequently activated, and they weaken and disappear when they are not activated for some time, which can be a consequence of an active blocking of the synapse. A year ago Ms. H. had still actively pursued goals and experienced positive emotions.

Her left prefrontal cortex had not been as hypoactive as it was at the beginning of therapy, and her right prefrontal cortex had not been as hyperactive at that time. The neurons in this area must have been recurrently activated since then, which facilitated their synaptic connections. Their activation was linked with negative emotions and with avoidance behavior. Thus, Ms. H. must over the course of the last year have experienced increasingly negative emotions and must have tended to avoid more and more situations. This probably led to an active blocking of left prefrontal cortex activation, such that the synaptic connections in that region got weaker. This, in turn, was associated with Ms. H.'s tendency to pursue fewer and fewer approach goals and her reduced ability to experience joy.

Negative emotions arise primarily when events occur that have negative implications for our goals. The more important the goal, the stronger the negative emotion. There must have been events in Ms. H.'s life over the last year—or even during the longer preceding period during which she appeared anxious and stressed—that constituted a threat or an obstacle to her important goals. There is no other way to explain the overdeveloped state of those brain regions associated with the experience of negative emotions at the beginning of therapy. These threat- or loss-related emotions must not have been consciously experienced by Ms. H. in the same way that she must not have consciously represented the threatened or lost goals. The activation of the corresponding brain areas must not have been linked with consciously represented experience. On the contrary, the conscious awareness that important goals are not being attained, or that something of great importance has been lost, produces in itself a pronounced cognitive dissonance, which, according to Festinger's well-validated dissonance theory, is usually avoided at all costs. Even stimuli and situations that trigger memories (i.e., that activate the same brain areas) are avoided. Because of this, there was no conscious engagement with the events and situations that for Ms. H. constituted a threat or a loss for her implicit goal system. This process led to the emergence of strong avoidance tendencies—so strong, indeed, that the dorsolateral part of the prefrontal cortex, which is implicated in the representation of avoidance goals, was hyperactive at the beginning of therapy because of its chronic activation. At the beginning of therapy, then, Ms. H. was in a state of generalized avoidance. She no longer engaged with her environment and no longer responded to challenges because she expected only unpleasant, overwhelming, and negative events to follow. From this perspective, her depressive state was a generalized protective reaction against the environment. This state is characterized by inactivity only on the level of outward appearances, only in terms of her actual exchanges with the environment. Despite this appearance of inactivity, the avoidance system is highly activated and along with it, the hypothalamic-pituitary-adrenal axis, which responds to stress with increased release of cortisol. Chronic stress is not at all healthy, and certainly not for the hippocampus, which is damaged by this high cortisol level to such a degree that its volume shrinks measurably.

Viewed from this perspective, Ms. H. has experienced over the last year a cascade of neural events, a chain reaction of positive feedback loops that resulted in easily excitable neuronal circuits. Even though the neurons participating in this cascade had returned to a deactivated state by the end of therapy, as soon as Ms. H.—after relinquishing the sick role and overcoming her acute depressive state, both of which provided some protection—is exposed once again to the same influences and experiences that triggered the cascade originally, we can expect that the synaptic modulations that led to the depressive state are already primed, ready to be reactivated. These synapses are easily activated again when the same stimuli, experiences, and life events are once again encountered. And why should they not? Nothing has changed in terms of her life situation, except that she is now no longer sick and will surely once again be confronted with her old set of demands and challenges. There is a high probability that the same events that led to the first depression will recur in a similar fashion. One difference is that this process should now transpire more quickly than the first time because of the neural priming. This is what happens quite regularly with depression; most depressions are indeed marked by recurring depressive episodes.

This is why Ms. H.'s therapist finds it important to continue with therapy despite her markedly improved state. He wants to build up resistance that prevents her from quickly sliding back into another depressive episode. These considerations of course don't only come up at the end of therapy; they have been contemplated by the therapist since the very first session. Which incongruences (discrepancies between perceptions and goals) have triggered the negative emotions that stood at the beginning of the cascade? Why was she unable to remove this incongruence? After all, she had many other incongruences over the course of her life and has been able to master them without becoming depressed. What was different this time? What was missing for her, which contingencies have prevented her independent mastery? In his search for the sources of these incongruences, the therapist must consider her objective life circumstances, her important interpersonal relationships, as well as her personal resources and deficits.

Could it have helped her if she had recognized and consciously confronted the emerging incongruences much earlier? What was it that prevented her from consciously turning toward these incongruences, making the necessary decisions, and following through with them? Could it be that an unclarified motivational constellation plays a role in this context? Are there motivational tendencies that conflict with one another and that prevent her from making a clear decision and using her powers to pursue positive goals? Did this motivational conflict occur beyond conscious awareness? Could it be that such an unclarified motivational conflict situation contributes to her unconsciously entering recurrent problematic constellations, for which there is then indeed no easy solution, and which is guaranteed to lead to negative emotions? Or does she have a lack of important personal resources and abilities, such as assertiveness, social skills, self-control, persistence, and coping strategies, that would help her to master emerging problems?

In his search for possible sources of these incongruences, the therapist does not rely solely on his personal impressions but uses specific psychological tests that are designed for this purpose. In addition, he completes the picture that he has gained from his conversations with the patient by consulting with important reference persons; in this case her husband and her children. All these sources of information result in a final composite image. At the very beginning the therapist gave Ms. H. an "incongruence questionnaire" (see chapter 4) that covers the overall range and intensity of mental incongruences. From this test he already knew at the beginning that Ms. H. is characterized by a high level of incongruence. This is, of course, what would be expected in the case of a depressed person. Even more important than this overall level were the clues that the therapist obtained from this questionnaire and from other sources; clues about the specific domains in which there were incongruences between perceptions and goals and about the sources responsible for these perceived incongruences. These sources are important leverage points for the therapist that he can use in his quest to attain change, and that go beyond acute symptom reduction and ensure long-term therapeutic success.

Even after a just a few therapy sessions, the therapist had come up with such a composite image and had made a treatment plan that went beyond the initially described resource-activating and depressive-behavior and experiencing-blocking interventions. The plan intended that the identified sources of incongruence would be tackled one at a time with appropriate interventions, but only once Ms. H. had attained a state that would allow her to engage constructively in therapy. The precise strategy with which such constructive collaboration with Ms. H. could be attained had already been contemplated by the therapist after the first session. The plan was instituted and already after a few sessions the therapist began to address specific sources of incongruence. He began with a rather small and circumscribed problem, which from past experience was known to yield a high probability of initial success. The point here was to facilitate a concrete experience that would convey to Ms. H. that positive change is possible at all. From this, he moved to another incongruence source, one for which Ms. H. would most likely be highly motivated and for which she would have the necessary resources, based on his prior case conceptualization. The rationale for sequencing the interventions in this motivational order—rather than in the order of problem severity—was related to the easy excitability of the involved neurons. The more one could expect that Ms. H. would be able to activate already existing motivational potentials, the more likely she would engage in a persistent, self-motivated fashion. After all, the motivation potentials being activated were her own, and the more these primed abilities were being utilized, the greater the probability that Ms. H. would experience herself as someone able and competent.

In this manner, some of the sources of incongruence had already been addressed by the time Ms. H. recovered from her acute depression, a time at which thoughts about therapy termination might have come up (see previous). Apart from the resource-activating interventions, the work on these sources of incongruence had probably also contributed to the improvement of her depressive state. However, not only the therapist but also Ms. H. realized at this point that their work together was not yet finished.

Thus, they continued to work on the remaining sources of incongruence, and it became possible to address the more thorny issues, which had previously been so threatening to Ms. H. that she had been unable to engage with them in a conscious, deliberate manner. At this point, however, Ms. H. had become more resilient. Moreover, Ms. H. by now relied on her therapist as a competent person whom she could easily trust with her problems, and she had repeatedly experienced first-hand that she was able to overcome problems successfully. With this foundation, she was now able and willing to expose herself even to experiences that would predictably trigger strong negative emotions. With this entirely altered context, however, this exposure now did not lead to avoidance, which in different circumstances could have initiated the depressogenic cascade. Instead, the exposure to unpleasant events triggered an active, albeit very painful and exhausting engagement with problems from which she had turned away a very long time ago.

In his work on the patient's sources of incongruence, the therapist was also guided repeatedly by neuroscientific principles. He was never content with a single problem solution, with a single instance of successful coping with a difficult situation, or with a single new insight that had been attained. He insisted that Ms. H. continue to expose herself to similar situations in order to cope successfully, to think her newly discovered adaptive thoughts repeatedly and use them in new contexts, even when she felt that she understood already or could do what was required. He realized that the better these new neural circuits were ingrained during therapy, the more easily they would be activated when they are needed later; for example, when Ms. H. encounters again her old demands or perhaps when she is confronted with new challenges. In those situations, the danger exists that the old pathways, whose activation triggered the last depression, will once again be activated. The question then becomes how easily the newly established pathways become activated. The more well-ingrained they are, the more easily they can be activated, and the more they will function as an effective barrier against the cascade leading into depression.

These considerations also contributed to the therapist's plan to not suddenly terminate therapy. Instead, after the most important therapy goals had been reached and Ms. H. began to view the future more confidently and felt that she no longer needed therapy, the therapist made quarter-yearly appointments for the next 2 years with her and made her promise that she would contact him immediately should she notice that she is once again slipping back into her old patterns. He wanted to make sure that the first signs of a relapse would be immediately noticed and addressed by continuing to work on the issues that had been targeted in treatment. The therapist wanted to prevent Ms. H. from becoming severely depressed again and from having to cope with the hassles associated with entering the medical care system. He knew that Ms. H. chances to remain free from future depressions would be greatly improved if she could manage to stay nondepressed for the next 2 years.

After Ms. H. continues to function well 2 years after therapy termination (and indeed even better than at the time of termination), the therapist knows that he has done good work. He has done much more than merely treat the depression. By working with her on the removal of the major sources of incongruence, he has not only removed problems but has set the foundation more generally for solid mental health. Removing incongruences means that a person's experiences, the perceptions he or she encodes, are aligned with his or her most important motivational goals. The motivational goals—those goals that truly influence a person's experiences and actions, regardless if consciously or unconsciously—that have emerged for a person over the course of a lifetime are his or her individual means to satisfy basic needs (more detail on this in chapter 4). In other words, reaching important motivational goals will be reflected in a low level of incongruence and will be associated with relative satisfaction of the person's basic needs.

Ms. H. completed another incongruence questionnaire—the same one she had completed prior to therapy—2 years after therapy had terminated. This time she scored very low, compared to her previous result. Low incongruence is typically associated with psychological well-being and good mental health, and this also holds true for Ms. H. A person whose basic needs are well satisfied has the lowest relative risk of developing a mental disorder. Because he knows about these processes, Ms. H.'s neuropsychotherapist generally strives for more than the mere alleviation of the manifest disorder. He realizes that an orderly brain—a smooth coordination or consistency among neural processes—is the best guarantee of good health.

1.6   STRUCTURE OF THE BOOK

Chapter 2 introduces "What Psychotherapists Should Know About the Brain." The chapter does not require prerequisite knowledge about the brain. It explicates the brain's fundamental functions, from the processes at single synapses and neurons to more complex neural circuits. Such an exposition of basic brain processes could have been structured quite differently, of course. Throughout the chapter, I have tried to keep in mind the concerns of the psychotherapist who wishes to know more about the brain. The chapter is intended to enable readers without prior knowledge so that they can understand subsequent chapters and comprehend the conclusions without undue difficulties.

Understanding the brain's functioning inevitably requires an in-depth appreciation of the processes related to the synaptic transmission between neurons. Changing a person's experience and behavior ultimately requires changes in the activation pattern of neural activity Learning, as well, is based on modulations in synaptic transmission within the neuronal network. It is particularly important for psychotherapists, therefore, to understand how synaptic connections can be changed effectively and enduringly. This question arises repeatedly throughout the book and is discussed in increasing depth. A first overview of these processes is provided at the very beginning of the second chapter because all subsequent information builds on this foundational knowledge. Understanding the process by which synaptic connections are altered during learning ultimately also requires an appreciation of the biochemical processes related to these changes. I was well aware while writing these sections that many psychotherapists might find such topics rather difficult, and I have therefore tried to describe these processes in a clear, straightforward manner. I would like to encourage the reader to not skip these sections, even if some terms might not be familiar or immediately understood. Once the basic principles of action potential transfer have been comprehended, all other neural principles and conclusions should be understood relatively easily.

The second chapter also has another goal: to convey to the reader the essential premise of this book, that all mental processes are based on neural processes. To demonstrate the general applicability of this premise, I illustrate this principle also with domains of experience outside the field of psychotherapy, such as the state of being moved by music or of being in love.

Even more relevant for psychotherapy, of course, is the state of anxiety. The neural circuits and brain regions involved in the experience of anxiety will be discussed in some detail. The example of anxiety also serves more generally to demonstrate the central importance of basic neuroscientific research in the context of psychotherapy. For example, the question will be addressed of whether anxiety reactions, once started, can be effectively stopped.

Consistent with the basic premise that all mental processes are based on neural ones, I also address in the second chapter the neural basis of goal-directed action, of consciousness and volitional decision making. I illustrate with some examples showing that specific mental activities covary closely with neural activity in specific, circumscribed brain regions.

The second chapter closes with a discussion of a central question for psychotherapists: To what extent can unfavorable neural structures, once they have developed, be corrected through psychological influence? I will illustrate with many impressive examples that one of the most remarkable characteristics of our neural system is its high degree of plasticity. Well into adulthood, the brain continues to respond remarkably sensitively to recurring stimulation of high intensity, and even in advanced age, the brain retains its ability to form new neural structures. Psychotherapy can utilize this high degree of neural plasticity. In order to do so, however, it is necessary to provide forms of stimulation that are sufficiently intense and frequent, so that new neural structures can be effectively ingrained. The findings from neural plasticity research suggest relatively clear conclusions regarding the necessary characteristics that would allow psychological influences to effectively facilitate the formation of enduring new neural structures.

Two aspects that are particularly relevant for psychotherapy are not covered in chapter 2: the neural foundations of mental disorders and the motivational aspects of mental functioning. These topics are each addressed in their own chapters.

Chapter 3 summarizes our current knowledge about the neural correlates of several mental disorders that are particularly relevant for psychotherapy. The most detail is devoted here to depression, simply because this disorder has generated the most empirical research. These sections also provide the scientific basis for the claims that I made in the case example of Ms. H. Compared to depression, other disorders continue to be relatively neglected. Four anxiety disorders—posttraumatic stress disorder, generalized anxiety disorder, panic disorder, and obsessive–compulsive disorder—have at least been examined in sufficient detail so that preliminary implications for psychotherapy can be discussed. The research basis for other disorders seemed still too narrow and controversial to draw specific conclusions at this point. Given the intensity of the research that is currently being conducted in this area, the statements in this chapter are necessarily preliminary in nature. It is safe to say that even in 10 years time it will be possible to base one's writings on the neural correlates of mental disorders on a much broader empirical foundation. Such detail will increasingly become an integral part of our disorder-specific knowledge, which every psychotherapist should know about. Today, however, we are still far from this point. And perhaps because of this still limited knowledge, it seemed particularly important to me to illustrate with a few selected mental disorders how such knowledge might look, in principle, and what kinds of implications will arise for the practice of psychotherapy.

Chapter 4 almost constitutes a book within a book, judging by its comprehensiveness. Its emphasis is on the motivational aspects of mental functioning. The chapter explicates how motivated mental functioning can give rise to the formation of mental disorders. In this chapter, as in others, I am assuming a neuroscientific perspective wherever possible. However, compared to chapters 2 and 3, the neuroscientific perspective here is relatively strongly complemented by psychological research findings.

In my book *Psychological Therapy* (Grawe, 1998/2004) I proceeded inductively; that is, I started with a large number of psychological and neuroscientific research findings and arrived at what was then a new perspective on mental functioning. At the center of this perspective was the construct I termed *consistency*, and therefore, I called this perspective *consistency theory*. It emerged as the result of an intensive engagement with all those phenomena related to the simultaneous nature of many parallel mental and neural processes.

The fourth chapter opens with this theory. I begin by summarizing the theory's most important premises in order to convey to the reader the overall context. Next, I discuss each of the premises in greater detail and report what can be said about these premises on the basis of current research. The research findings I summarize here are primarily new and had not yet been available at the time that *Psychological Therapy* was published. In engaging with this new research, I have learned much new and consequently had to revise some of my earlier ideas. However, I have also found a great number of new research findings that support some of the central tenets of consistency theory. Some of these new findings are indeed from empirical studies that were specifically inspired by the formulation of consistency theory. These studies, in particular, have been conducted in recent years by me and my colleagues with the explicit aim of testing certain assumptions of consistency theory. The fourth chapter therefore relies much more heavily on my own research work than the two preceding chapters. Chapter 4, then, reflects more directly my own scientific perspective. I try to make this explicit at the very beginning of the chapter by starting with a summary of the major assumptions of consistency theory.

The research findings that I summarize in the context of basic needs and consistency regulation, however, are also of considerable interest independent of this context. The totality of the research summarized in chapter 4 yields a picture of the emergence of mental disorders that has actually surprised me in its clarity. The research is showing relationships that were new to me and that also were not, as such, integrated in my original formulation of consistency theory. However, these new findings can be smoothly integrated into the theory. Chapter 4, then, can be viewed as a substantially improved new version of consistency theory.

Nevertheless, the theory remains partially incomplete in chapter 4. It contains only the part that speaks about the emergence and maintenance of mental disorders but is silent on the treatment of the disorders. This aspect, then, is addressed in the fifth chapter. Chapter 5 concerns that area of research with which I have long been actively associated and about which I know the most, compared to other areas. This area is known as psychotherapy research. For many years now, my colleagues and I have concerned ourselves with studies on the effectiveness of psychotherapy and the translation of new research findings into improved concrete therapeutic strategies. Apart from doing research, most of us are also actively working as therapists, which allows us to test these improved strategies in the clinic and, ultimately, to test the effectiveness of these strategies on a broader scale. In addition, at the Psychotherapy Outpatient Clinic ("Praxisstelle") in Bern and at the Institute for Psychological Therapy in Zurich, we provide training based on the current state of our knowledge for about 150 therapists at any given time, and the therapists are instructed to translate the findings into their own clinical practice. All therapies are accompanied by comprehensive assessments, prior to therapy, at therapy termination, at regular follow-up intervals, and at various time points during the process of therapy. Moreover, for over 20 years now it has been standard protocol to record the therapy sessions on video, so that they are available at a later point for process analyses.

For 7 years now, these therapies have been based on our research questions related to the consistency theoretical therapy conceptualization that I articulated in my book, *Psychological Therapy*. Seven years is indeed a long time for a highly motivated team of researchers and therapists. These 7 years have allowed us to test clinically and empirically many aspects of the therapy according to consistency theory, to formulate some of its aspects with greater precision, or to correct aspects that needed updating based on empirical findings or based on clinical experience. I have summarized the conclusions that have emerged from our research work over these years, as well as the conclusions arising from the totality of the preceding chapters, in a section in the fifth chapter entitled, "Conclusions for Psychotherapy."

The chapter is divided into three sections. The first section contains that part of consistency theory that was still missing in chapter 4. In this part I not only draw conclusions based on what was said in the preceding chapter, but I integrate results from our own therapy process research in recent years. This last part with the conclusions is strongly influenced by viewpoints that have crystallized in me over the past few years through our own clinical and research work.

The second part of the conclusions for psychotherapy, by contrast, has little to do with our own research work and my preexisting views. It is concerned with the neural mechanisms underlying therapeutic change as they can be deduced from the neuroscientific research covered in the preceding sections. I demonstrate that ultimately all therapeutic changes can be linked to the activation of existing and the facilitation of new neural activation patterns. Effective therapy is based on effective neural facilitation. Neuroscientific research, in turn, clearly suggests methods to make this facilitation particularly effective. This part shows, then, that neuroscientific research directly leads to practice-relevant conclusions about how psychotherapy ought to be conducted to maximize its effectiveness.

In the third part of the conclusions, I summarize all that can be deduced from the previously covered material for the practice of therapy. This concerns, on the one hand, guidelines for therapy planning and, on the other, guidelines for the conduct of therapy. Because these guidelines are all based on a substantial empirical foundation, I believe that therapies that are conducted along these guidelines will lead to a particularly good outcome. As long as this is not empirically tested, of course, it will remain an assumption. However, I believe that the foundation underlying these guidelines is already solid enough for these guidelines to be used in current therapeutic practice. Many of the guidelines are empirically as soundly substantiated as many of the traditional intervention recommendations that are offered by psychiatric or psychological associations for the treatment of specific disorders.

The guidelines in chapter 5 do not represent what happens in most current psychotherapies today. They cannot be categorized easily as belonging to any one particular therapy school. Because of this, they demonstrate how much room for improvement still remains in the field of psychotherapy, by showing how therapy can orient itself on current theoretical and empirical findings rather than on decade-old principles. This is all the more true because these guidelines are intended to be general in nature, not specific to any particular disorder. Whereas empirically based, disorder-specific treatment guidelines have become increasingly well established in recent years, the more general, disorder-transcending guidelines for treatment planning and conduct are primarily still oriented on the principles of the various therapy schools. In my view, the guidelines articulated in chapter 5 render such therapy school specific guidelines relatively obsolete. If complemented by disorder-specific knowledge, these guidelines should indeed suffice for most situations arising in therapy practice. Psychotherapists that actively orient their practice around these guidelines can probably expect to increase the effectiveness of their therapies substantially. Therapy training institutes could also increasingly aim to convey such scientifically justified guidelines and thereby contribute to a general updating, so that therapy practice will increasingly be based on current rather than decade-old knowledge.

Chapters 4 and 5 together provide a perspective on what a scientifically founded therapy practice could look like today. In chapter 6 I take a look into the future to speculate on how psychotherapy might be further developed, especially if it uses neurosci-

entific findings not only for general guidance but also in the concrete process of
diagnosis and therapy. This chapter is necessarily speculative. It shows some of the
developments that we can probably expect in the future, but it does not go into great
depth. I am primarily concerned in this book with what the neurosciences can contribute to psychotherapy today. It is indeed one of the most exciting questions to ask
which aspects of psychotherapy might be completely reinvented based on the continuing influence of the neurosciences, but answers to such questions would likely
resemble science fiction. At this point, the available facts appeared interesting enough
so that a desire to elaborate on the fictional aspects did not really arise.

CHAPTER TWO

WHAT PSYCHOTHERAPISTS SHOULD KNOW ABOUT THE BRAIN

2.1 THE BRAIN: THE EPITOME OF COMPLEXITY

Have you ever been astonished by how a monotonous sequence of magnetic potentials can create songs as beautiful as Schubert's *Symphony in C-major* or *Kind of Blue* by Miles Davis and John Coltrane? Listening to CDs has become such a mundane activity that most of us have stopped marveling at how such transformations occur. When we listen to music, it is as real to us as it could possibly be. Music cannot be reduced in its beauty to a mere sequence of sounds or magnetic states. It is music, not physics—qualitatively of a different kind, belonging to a different reality than the media on which it is stored and the devices that are required to elicit the sound. Yet, even though music is qualitatively totally different from a sequence of binary magnetic states, its existence and character are utterly reliant on it. Every change in the binary information sequence has an impact on that we perceive as music.

An analogous relationship exists between the neural structures and processes in our brains and our experiences and behaviors. When we feel, think, or do something, this is all based on specific neural activation patterns. For that which we experience or do, it only matters whether specific neurons are firing, how quickly they fire, and in combination with which other neurons they fire. This is a yes–no question and, as such, a form of digitized information. Is this really possible? Can it be that the multitude of all possible mental and psychological phenomena is reliant on such simple material processes? Is it not possible to have even the most secret thought, to exert free will and make choices, without this being elicited by a specific pattern of neural activation in our brains? Indeed, how is my brain supposed to know what I shall decide in the next moment? If I decide—at this moment—to continue working here on my computer, even though the sun is shining outside after days of rain (I am writing in perpetually rainy Hamburg), then am I making this decision because experience has resulted in the ingraining of synaptic pathways that connect particular neurons in my frontal lobe, whose activation is linked with the experience of making a decision? Could my willpower really be based entirely on the transmission of action potential at certain synapses?

It is sometimes hard to accept the findings discovered by the neuroscientists. The answers to some of the obvious questions that arise from these findings would have to be rather convincing if they are to be believed. But a token of conciliation for our offended human pride: The human brain is without a doubt the most complex entity that nature has created. In it, the accumulated wisdom of 500 million years of evolutionary history is encoded. When we contemplate the brain's structure, we cannot help but marvel in awe. Its complexity far exceeds our imagination. Even the sheer number of neurons is awe-inspiring. The number of core neurons is estimated at about 100 billion. This does not yet include the glia cells, which are another very predominant type of brain cell, about whose function we still do not know very much. There are probably another 10 to 15 times as many cells of this kind. It is rather unlikely that evolutionary pressures, with their harsh selectivity, would have allowed for the development of such a mass of these cells if they had not fulfilled an important function—especially when considering the space restrictions within the skull. Some of the functions of the glia cells are already known, but it is thought that more secrets about them are yet to be discovered.

Every neuron is connected to other neurons via up to 10,000 incoming as well as outgoing synapses. The total number of synapses is thought to be measured in trillions. According to Edelman (1992), a piece of brain tissue the size of the tip of match contains approximately 1 billion synapses. About 2 million to 3 million nerve fibers are thought to connect our perceptual cells with the brain. Each of those perceptual cells fires up to 300 impulses per second to the brain. How is it possible to maintain a functional order under such circumstances? If humans had really been created within a single day, then the design of the brain alone would have been a considerable challenge, even for a god. An enormous technical problem would be the question of how all this complexity with all its interconnected neurons should be housed within an enclosure measuring only 1,200 cubic centimeters, with a weight of less than 1.5 kilograms.

Fortunately, we do not have to worry about such details. If evolution had not already solved this most ingeniously, then we would not exist in the way we do exist today. Our continued existence is strong evidence for the brain's excellence in design. If we were to award a prize for the best evolutionary accomplishment, then the human brain would have to be the hottest candidate.

So far I have only highlighted the vast number of elements that constitute the brain. It is even more awe-inspiring, however, to consider the degree of qualitative differentiation within the brain. Every neuron contains in its nucleus the entire genome of that particular individual. This consists of approximately 35,000 individual genes. One is tempted to believe in a plan of divine creation when considering the process by which this genetic material is selectively activated and extracted over the course of ontogenesis such that each neuron ends up at the exact spot where it belongs in order to fulfill its function. The evolutionary alternative to such a divine plan is presented with great dramatic detail in chapters 14 to 21 of the most detailed of the neuroscientific textbooks, *Fundamental Neuroscience*, by Squire et al. (2003). However, in this neuroscientific drama, there is no single villain and the plot concludes with the happy ending that each neuron finds its position and understands its task—and this happy ending occurs despite (or rather because of) events as dramatic as programmed cell death and other neural fates, due to axonal pathfinding, local guidance, topographical maps, and neurotrophic factors that show up just at the right moment. Ultimately, each neuron has become a specialist within its domain. One neuron might respond to horizontal lines, another one to vertical ones, yet another one only to curved lines; this one responds to fast movement and that one to slow motions; this one to red, and that one to blue. Neurons that have landed in the visual cortex contribute to our perception of sight; those that end up in the auditory cortex, to that which we hear. Each neuron never fulfills its functions alone but works along with groups of neighboring neurons. The groups of neurons, in turn, interact with other groups that may be located in entirely different brain regions.

Recent neuroscience theories, such as that of Marcel Mesulam (2000) have attempted to explain how these groups of neurons form distributed large-scale networks. Each network is responsible for a different type of parallel information processing. Mesulam identified five such networks that are necessary for face/object recognition, linguistic processing, memory/emotional functioning, spatial attention, and comportment/executive functioning. Obviously, our focus will be the memory/emotional functioning and the comportment/executive functioning systems because these systems are most likely to function abnormally in various mental illnesses and require systematic alteration of activation patterns through psychotherapy. The memory/emotional functioning network can be further broken down into two regional epicenters, the amygdala and the hippocampal–entorhinal complex. Neuroscience research suggests that psychotherapy may help to regulate the amygdalocentric network and heighten the activity of the hippocampal–entorhinal complex necessary for learning new behavioral coping strategies. The comportment/executive functioning system has its two epicenters within: the orbitofrontal/medial frontal cortex and the dorsomedial prefrontal/posterior parietal cortex. Multiple functional neuroimaging studies have shown hypo-activation of frontal brain regions in a variety of psychiatric disorders such as depression, schizophrenia, and obsessive–compulsive disorder.

Within these distributed networks, each individual neuron must contribute its special function to a complex collaboration. Otherwise the corresponding mental function ceases to exist. When the neurons that specialize in color perception fail, the person is no longer able to see colors—he or she becomes color-blind. When a particular set of "gnostic neurons" fails, the person can no longer recognize faces. The dramatic effects that can result from such a malfunction of specific neurons have been described evocatively in Oliver Sacks's case history about *The Man Who Mistook His Wife for a Hat* (1987).

The collaboration of specialized neurons in neural circuits is made possible by the transmission of action potentials among the individual neurons. This potential is transmitted via the synapses, which connect two neurons with one another. Before we turn to the transactions among neurons on the level of the entire brain, it is critical to elaborate on the processes transpiring within the neurons themselves and at the synapses during the transmission of action potentials. These microlevel processes are the basis of all other processes within the brain. They determine what is and is not possible within the brain. If therapeutic changes achieve their effects via changes of the brain, then they are ultimately effective—on a microlevel—because they modulate synaptic efficacy. Let us take a closer look at the fundamental process underlying the transmission of action potentials to see whether relevant implications for psychotherapy can be derived.

2.2   WHAT EXACTLY HAPPENS DURING THE TRANSMISSION OF ACTION POTENTIALS BETWEEN NEURONS?

The functions of individual neurons depend on three factors: their localization within the brain, their connections with other neurons, and their individual characteristics. Even the internal organization and appearance of neurons can differ considerably. Figure 2–1 illustrates some of the more common types of neurons.

However, it is not only in their visible appearance (in the electron microscope) that neurons differ from one another. Neurons are designed to communicate with other neurons. They do this via electrical transmissions and chemical synapses. For the chemical transmission of signals, neurons use transmitters that they themselves produce. These are, narrowly defined, the neurotransmitters, neuromodulators, neuropeptides, and neurohormones—often all of these are collectively referred to as neurotransmitters. There are nine different kinds of just the classical neurotransmitters (glutamate, glycin, gamma amino butric acid [GABA], dopamine, norepinephrine, epinephrine, serotonin, histamine, and acetylcholine) and more than 50 types of neuroactive peptides. Until recently, it was assumed that each neuron produces only one specific neurotransmitter, but this assumption has by now been shown to be inaccurate. Most neurons produce multiple neurotransmitters, which is only reasonable, as this multiplies the neuron's options to influence other neurons. Of importance, the neurotansmitters work on very different time schedules. The effects of glutamate and GABA transpire within milliseconds, those of dopamine and serotonin within seconds or minutes, and those of the neuropeptides and neurohormones often within hours, days, or even weeks.

The production and release of neurotransmitters are, however, by no means the only contributions that a neuron makes to the transmission of action potentials between neurons. Neurotransmitters are transported in vesicles to axon terminals and then are released into the synaptic gap, which transmits the activation of the neurons to other, postsynaptically connected neurons. Then there are the receiving parts of the neurons—the neuro-receptors—via which the neuron takes in signals from other neurons. Most of these are located on the dendrites, but there are also some on the cell body and on the axon. Figure 2–2 graphically illustrates the connections among neurons, and Figure 2–3 shows a synapse with its basic elements.

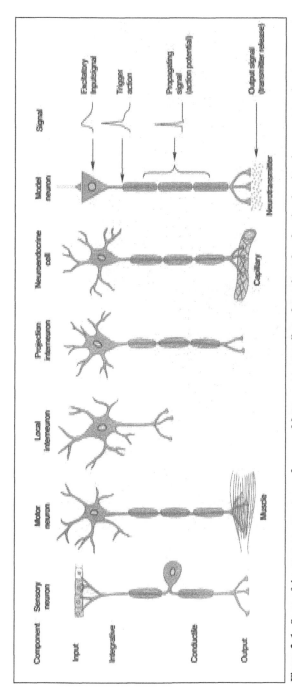

Figure 2–1. Some of the more common types of neurons. Most neurons, regardless of type, have four functional regions in common: an input component, a trigger or integrative component, a conductile component, and an output component. Thus, the functional organization of most neurons can be schematically represented by a model neuron. Each component produces a characteristic signal: the input, integrative, and conductile signals are all electrical, whereas the output signal consists of the release of a chemical transmitter in the synaptic cleft. Not all neurons share all of these features; for example, local interneurons often lace a conductile component.

Note. From *Neurowissenschaft* (p. 32), by E. R. Kandel, J. H. Schwartz, & T. M. Jessel, 1996, New York: Springer. Copyright. Reprinted with permission.

33

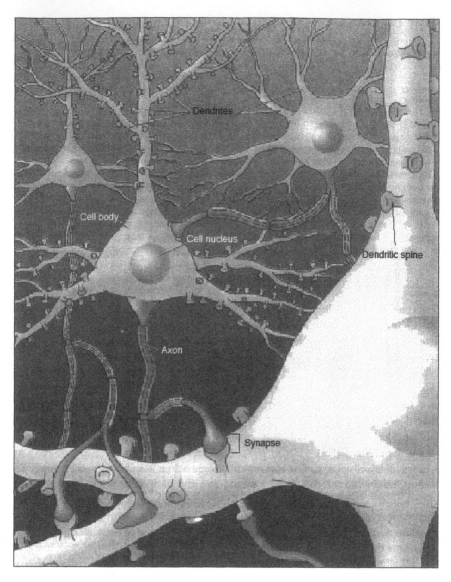

Figure 2–2. This is how a neuronal network can be envisioned. It is shown how neurons are connected with each other via dendrites for the incoming signals and axons for the outgoing signals. Note. From "Die Bedeutung neurowissenschaftlicher Forschungsansätze für die psychotherapeutische Praxis. Teil I: Theorie [The impact of neuro scientific research on psycho theraputic practice. Part 1. Theory]," by M. Storch, *Psychotherapie*, 7, p. 286. Adapted with permission.

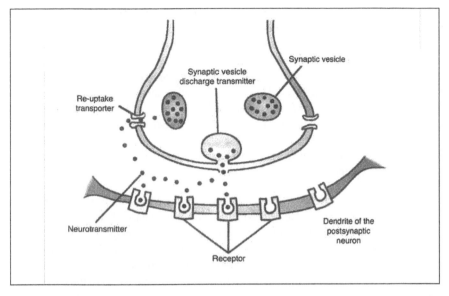

Figure 2–3. Simplified illustration of the structure and basic elements of a synapse.
Note. From *Brave New Brain: Conquering Mental Illness in the Era of the Genome* (p. 91), by N. Andreasen, 2001; New York: Oxford University Press. Copyright 2001 by Oxford University Press. Adapted with permission.

2.3   THE BIOCHEMICAL PROCESSES TRANSPIRING AT THE SYNAPSES AND WITHIN NEURONS

The receptors on the postsynaptic membrane have either an activating or an inhibiting function. They are specialized to bind particular kinds of neurotransmitters. Thus, there are dopamine receptors, serotonin receptors, glutamate receptors, and so forth. The neurotransmitters either open or close ion channels. The membrane of the neuron has a resting potential of –65 mV; that is, the number of negatively charged ions dominates inside the neuron. When ion channels are opened, positively charged ions can enter the neuron. Once their number has reached the point of a particular threshold, the membrane is depolarized and an electric action potential is created. The potential propagates along the axon and results in the release of neurotransmitters at the end of the axon.

There are a great many different ion channels. All of them can contribute to the alteration of the membrane potential. Some of them do so by opening up; others by closing. The conditions under which they do one or the other and the molecular mechanisms of these processes are still only partially known.

Every neuron also has a great many different kinds of receptors and ion channels. Whether an action potential is created depends on the balance between open versus closed ion channels; that is, how much activating versus inhibiting input the neuron simultaneously receives from the multitude of synaptic receptors. A neuron can there-fore be regarded as a kind of chemical calculator that converts a great number of ana-log signals into a digital signal—the action potential. At the output side of the neu-ron, the digital signal is once again converted to analog signals—the neurotransmit-ters. The neurotransmitters are released from vesicles, in which they are transported to the axon endings, in the form of quanta. The number of quanta depends on the speed with which action potentials are sequentially generated; that is, it depends on the firing rate of the neuron. The number of action potentials, in turn, determines the concentration of neurotransmitters in the synaptic gap. This, once again, is an analog signal that affects the postsynaptic receptors.

Neurotransmitter material that does not connect to the postsynaptic receptors imme-diately is taken up again by the presynaptic membrane and is then recycled. If this reuptake of the neurotransmitter into the presynaptic membrane is inhibited—for example in the treatment of depression via serotonin reuptake inhibitors—the result is that higher levels of the neurotransmitters (such as serotonin) are available in the gaps among the neurons. The freely available serotonin can connect to the postsynap-tic serotonin receptors as soon as space is available, which can then further modulate the synaptic transmission at this particular synapse (more about this is discussed later).

The number of receptors per neuron and their sensitivity for signal transmission depends, in turn, on a large number of chemical processes within the neuron—pro-cesses that are far more complicated than what I have described here in a greatly sim-plified manner.

There are two distinct kinds of neurotransmitter receptors. In the case of the first kind—so-called ionotropic receptors—the transmitter itself leads to the opening or closing of the ion channel. This process transpires within milliseconds, and these are primarily the (activating) Alpha-amino-3-hydroxy-5-methylisoxazole-4-proprionic acid (AMPA) receptors and the (inhibiting) GABA receptors. These receptor types, as well as the associated transmitters glutamate and GABA, are widespread through-out the entire brain. These receptors are primarily involved in processes of rapid sig-nal propagation. Along with the electrical synapses, which are also well suited for the quick transmission of action potentials (for simplicity, I do not elaborate on them in detail here), these receptors are the basis of any rapidly transpiring neural processes such as thoughts going through our minds, responding to sounds, formulating sen-tences, and so forth. These receptors could be called the "workhorses" of the brain. They do not result in enduring changes of synaptic properties but instead take advan-tage of the already existing ones.

There is also another important activating type of glutamate receptor, known as N-methyl D-aspartate (NMDA) receptor (see Figure 2–4). It responds more slowly than the other two types of ionotropic receptors because the ion channel, at the resting potential of the membrane, is blocked by magnesium and opens only when a depo-

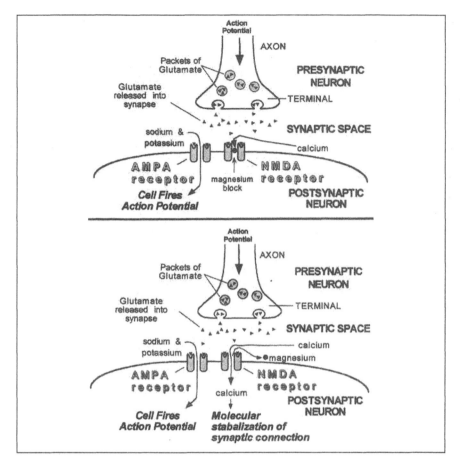

Figure 2–4. Illustration of the transmission of synaptic action potential via Alpha-amino-3-hydroxy-5-methylisoxazole-4-proprionic acid and N-methyl D-aspartate receptors. Explanation provided in the text. *Note*. From *The Emotional Brain* (p. 219), by J. E. LeDoux, 2004, London: Phoenix. Copyright 2004 by Phoenix. Adapted with permission.

larization of the postsynaptic membrane has occurred. The opening of the NMDA ion channel thus requires a prior activation of the postsynaptic neuron via another kind of activation. Only if this occurs can calcium flow into the cell (see Figure 2–5).

This calcium then activates a within-cell transmitter known as a *second messenger*, which in turn triggers a chemical cascade that ultimately leads to a selective strengthening of the transmission properties of the synapses that are involved in this specific activation. This process is known as long-term potentiation of the synapses. Long-term potentiation transpires over the course of seconds to minutes. This results not in the heightened excitability of the entire neuron but instead only affects the synapses that were involved in the prior activation. Therefore, connections between very

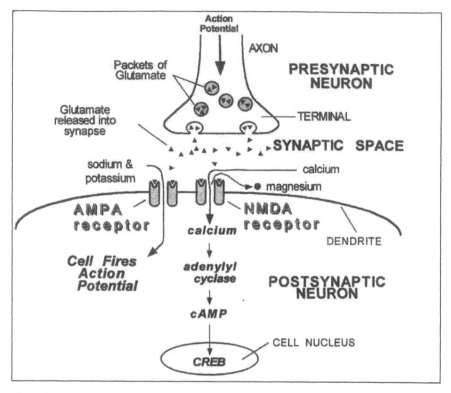

Figure 2–5. Second messenger cascade triggered by the influx of calcium via an N-methyl D-aspartate (NMDA) receptor. Explanations in the text.
Note. From *The Emotional Brain* (p. 223), by J. E. LeDoux, 2004, London: Phoenix. Copyright 2004 by Phoenix. Adapted with permission.

specific neurons are strengthened. This is what is meant when throughout this book I speak of the *facilitation* of neural connections.

The neurotransmitters that bind to the ionotropic receptors affect the cell from the outside. In the case of the second type of receptor, known as G-protein-associated receptors, the neurotransmitters affect the synapses in an indirect fashion. That is why these neurotransmitters are also called neuromodulators. The neuromodulators that dock onto the outer cell membrane activate a within-cell enzyme, and this enzyme in turn produces a secondary, within-cell transmitter type, also known as a second messenger. This messenger then triggers a cascade of follow-up reactions that influence the biochemical state of the neuron in such a manner that ion channels either open or close. These types of receptors therefore have an effect that modulates the membrane potential of the cell from within the cell itself. This effect is achieved via a temporary increase in the concentration of a second messenger in the postsynaptic neuron. Such modulation effects normally persist over the course of seconds or minutes. This mechanism, mediated by so called non-NMDA receptors, also results ultimately in long-term potentiation of the previously activated synapses (and only of those).

As a consequence of this process, for several minutes the neuron can be activated more easily via the ionotropic receptors. This means that the processes that are involved in the potentiated synapses are transpiring more easily for several seconds or minutes than would otherwise be the case. If during this period of heightened excitability the neuron continues to be stimulated intensely, then the activation or inhibition becomes progressively stronger because yet another process is beginning to engage.

This process, which has the greatest importance for psychotherapy, is a third kind of second messenger cascade, which involves even more biochemical steps than the previous ones. Interested readers will enjoy the detailed and nuanced discussion of these processes by Kandel et al. (1996). Briefly, a sufficiently long and intensive stimulation of ion-governed receptors leads to an increase in the second messenger cAMP (cyclical adenosinemonophosphate), which then activates protein kinase A (PKA). This, in turn, activates a protein called CREB (cAMP-response-element-binding protein). This protein is a so-called transcription factor—as it can activate the transcription of genes within the cell nucleus. The activated genes start synthesis of proteins begin to produce new proteins, which in turn can trigger three possible effects:

• Facilitate excitability of the previously activated synapses that had triggered this biochemical cascade.
• Enhance the production of neurotrophines. These are elements that lead to the formation of additional synapses around the previously activated synapses. The number of synapses therefore increases selectively, which strongly increases the future excitability of the neuron via this communication pathway. A small road, as it were, gradually grows into a multilane highway.
• Increase the production of so-called retrograde messengers. These are messengers that affect the presynaptic cell from the postsynaptic cell, thus ensuring that more neurotransmitters are being produced at the presynaptic cell, which are needed for the subsequently increased signal transmission.

In light of the fact that I have burdened my readers with quite a bit of biochemistry by now—even though they are probably primarily interested in psychotherapeutic questions—I would like to return now to a point that was raised earlier in the introduction. That is, the biochemical processes I described here typically do not come about as a result of pharmacological intervention but instead occur in the context of normal mental activity. Psychological processes and real-life experiences bring about increases or reductions in the synaptic transmission. That means that we, as therapists, ought to view these biochemical processes as opportunities that can be used in a targeted fashion. Therefore, it is important to have detailed knowledge of these processes. It may not suit everyone's taste, of course, to delve into the kinds of biochemical depths that would allow us to fully comprehend the specific molecular action involved in the biochemical cascades within the neurons. Indeed, it is not necessary to know what exactly is meant by "phosphorylization of an enzyme" in order to comprehend the implications arising from such descriptions for psychotherapy. These implications, however, are of utmost importance for psychotherapy because they underlie any long-term changes of the brain. That is why I aim to elucidate these processes in some detail.

I have attempted to describe the processes occurring at the cell level in such a manner that they can be understood without detailed previous knowledge. It was inevitable to draw a line at many points and not venture into even greater detail. Readers who wish to gain a more specific and contextualized understanding of these processes are strongly encouraged to study the textbook *Neurosciences* by Kandel et al. (1996). Indeed, I would argue that in addition to his Nobel Prize, Kandel deserves to receive a prize for this textbook. Those who desire even greater detail and wish to find answers to specific questions will find comprehensive information and many relevant references in the rather longer (more than 1,400 pages) textbook by Squire (2003).

What, then, is the relevance for psychotherapy of the different ways in which the transmission of action potentials can occur at the synapses?

2.4   IMPLICATIONS FOR PSYCHOTHERAPY

Let us imagine what occurs at the synapses of a patient during a therapeutic dialogue. The AMPA receptors are responsible for the fact that the patient can follow the statements and questions of the therapist while simultaneously engaging in the dialogue with his or her own concerns, ideas, and viewpoints. These receptors enable a lightning-fast activation transmission at the synapses that have previously been well facilitated. The more the dialogue moves along the same routes the patient is already used to, the more easily he or she can follow the therapist's suggestions and engage in the conversation. While patient and therapist are moving along these routes, however, it remains unlikely that long-term changes will be occurring at the synapses. The conversation might flow well and seem important and interesting to both parties, but long-term changes do not result from this process. Indeed, a lot of time can be wasted in therapy with such AMPA dialogues, as we might term them. This is consistent with findings by Schulte-Bahrenberg and Schulte (1993) and Schulte (2003), who reported that frequent changes in the therapist's intervention goals and methods were associated with poor therapy outcome.

Change requires that synaptic connections that are not yet well established become activated as often, intensively, and over a as long time period as possible. If one expects the therapeutic dialogue alone to facilitate change, then one ought to approach the problem intensively and persistently from all angles, so that all the neurons that are connected to the corresponding neural networks can be properly activated. This would then open the NMDA receptors and kick off a second messenger cascade. The emergence of such a cascade is even more likely if dopamine receptors are simultaneously activated. This tends to be the case to the degree that important goals are activated at the same time within the patient (more detail on this in chapter 4). Working on a problem should therefore always occur in the context of pursuing an important, currently activated approach goal held by the patient. As long as this requirement is not fulfilled—as long as it is primarily the therapist who pursues this goal while the patient is only partially engaged or even reluctant—one cannot expect the permanent facilitation of pathways relevant to the patient's goal. This implication, which is derived here from the processes related to the transmission of synaptic action potentials, is also consistent with recent findings from psychotherapy process research (Orlinsky, Grawe, & Parks, 1994; E. Smith & Grawe, 2001, 2005). If the therapy does not succeed in the activation of important approach goals held by the patient, and making these goals the engines that drive the therapeutic change process, then the therapist would be well advised to refrain from treating the problem.

It is also important to note that the emphasis of therapy should not be too much, or for too long, on the identification and activation of problems. Identifying and analyzing problems is only productive to the degree that this enables later change-oriented interventions. The point is, again, to facilitate changes in a positive direction, to establish the emergence of new neural activation patterns. Therefore, the emphasis must be predominantly on altering the problem, on facilitating new thoughts, behavior patterns, and emotions. This activation of new neural activation patterns must be repeated as often as possible, otherwise the neural connections will not be ingrained sufficiently strongly. If the therapy remains stuck at the point of identifying and analyzing problems, it is unlikely that new, more positive activation patterns will be formed. This is also consistent with the results of our own process analyses (Gassmann & Grawe, 2006; E. Smith & Grawe, in press).

If one considers the process of therapy from the perspective of synaptic transmission, an important question concerns the degree to which a therapeutic dialogue alone can be expected to affect the neural activation patterns in a way that ultimately leads to self-sustaining improvement. The neural activation patterns representing the problems of the patients take place in real life, not in the therapy session, and the newly emerging patterns must prove their value in real-life conditions. Given this, would it not be necessary to facilitate the new patterns in those real-life situations in which they later are expected to influence the patient's experience and behavior? Wouldn't it seem reasonable that a session in the therapist's office can be no more than preparatory in nature, to enable the later ingraining of new neural activation patterns in the reality of the patient's life outside of the therapy room? Shouldn't there be a close connection between the therapeutic interventions and the patient's concrete current situational context? How would it be possible, for instance, that a confrontation with one's own past—which would activate the contents of episodic memory—would lead to the facilitation of new neural patterns in situations where the patient experiences a panic attacks, such as in the case of an agoraphobic client in a crowded department store? If one considers such therapeutic actions from the perspective of synaptic activation transmission, considerable doubts begin to surface about the likelihood that such maneuvers can effectively treat these anxiety symptoms.

At this point within the book, we are still at the very beginning of our neuroscientific reflection on psychotherapy, and we can expect that many other points will arise later that might change the way we view psychotherapeutic dialogue. Nevertheless, it seems clear already that, from the perspective of changing synaptic transmission and connectivity, many interventions must be regarded with considerable skepticism.

One take-home message from this perspective is certainly that the permanent facilitation of new patterns of experience and behavior on the neural level requires a concentrated and long-lasting effort to establish and maintain these same experience and behavior patterns. Because these new ways of experiencing and behaving are not yet deeply ingrained, they are unlikely to occur spontaneously but must instead be specifically facilitated and actively supported by the therapist. This requires a targeted, directive stance on part of the therapist, and in order to direct effectively, the therapist must be able to envision clearly that which is to be facilitated. A therapy session with the attitude, "Let's see what the patient brings up today and take it from there," is really a wasted hour in terms of the likelihood that new neural activation patterns will be permanently ingrained.

2.5   IS IT REDUCTIONIST TO RELATE MENTAL PROCESSES TO THEIR NEURAL BASIS?

Based only on what I have described up to this point about the processes within neurons, it is already clear that each neuron can be viewed as a true miracle of nature. Each neuron is by itself a highly complex chemical power plant, equipped with a large reference library and huge storage facility (35,000 genes). Within each neuron, many different production processes are transpiring in perfect mutual coordination, and all of them respond flexibly and immediately to new challenges. The "neuron power plant" is also communicating with the environment via various channels of communication, and it accomplishes all of this in a self-organizing manner, without the supervisory control of an executive director.

Each neuron is a microcosm of its own. Yet the function of this power plant within its "societal context," in the universe of the neurons (i.e., the brain), is to communicate with other neurons. Its structure and working methods are all oriented in accordance with this, and this function can only be accomplished by the neuron in close collaboration with other neurons. The result of this requirement is that neurons are extremely interconnected in networks of trillions of synapses.

This degree of interconnectedness is the basis for the practically unlimited number of possible communication patterns among the neurons. These communication or neural activation patterns are the basis of our experience and behavior. Because their number is unlimited, the number of possible mental phenomena is also unlimited— of course within the limiting frame of the brain's basic organic structure. The neural activation patterns that develop within an individual brain are critically influenced by the sensory experiences that impact upon the individual over the course of his or her development (ontogenesis). The genes of the individual are the framework delineating the potential range within this process. The specific life experiences of the individual, in turn, are critically influenced by the culture, society, and social group within which the individual develops. The neural structures that develop within the individual therefore mirror the individual's cultural, societal, and familial context. It goes without saying that culture and society exist independently of any individual brain, as well as the mental products and the results of the actions of that individual, which have become part of a particular culture, society, or social group. Culture is the product of the collaboration of many individual brains across many generations, but the existence of culture no longer depends upon the existence of these individual brains. In turn, these individual brains were, in part, a product of this brain-independent culture.

Is this way of relating brain and culture inherently reductionistic? Is my participation in the mental and cultural world in any way compromised because my aesthetic and moral sentiments are based on neural activation patterns? Do I perceive Beethoven's music as less emotionally gripping because of this neural basis? Is my commitment to a worthy cause less valuable because it is determined by processes within my brain? Are my thoughts and decisions less my own because I now know that a few hundred milliseconds before I had made them, neural signals that announced these events could have already been detected?

Why should they be? My neurons still fulfill their functions just as reliably as they did previously, before I had any conscious awareness of these processes. The neurons still create these subjective experiences most reliably. I continue to feel as if I am the master directing my course of actions and my decisions. After all, it is still *my* brain. If I am a good friend to myself, then I can also continue to feel friendly toward my brain, which creates "my self." Woody Allen (in his movie, *The Sleeper*) aptly stated that "my brain is my second-favorite organ." The fact that I am reliant on my brain for all my subjective experiences and my behavior is no more a threat or insult to myself than the fact that I am reliant on my lungs for breathing and on my legs for walking.

I am explicitly saying this here because many people find that relating mental processes to neural processes is inherently reductionistic, and this often implies that reductionism is somehow intrinsically negative because it negates or reduces basic human dignity. None of this is new; it resembles the processes surrounding Darwin's theory over a hundred years ago. However, just as we did not regress to become apes because of the discovery of evolutionary principles, we also will not be reduced to brain tissue because of the principles uncovered by the neurosciences. Relating mental processes to neural processes in no way tarnishes or reduces the spheres of mental, cultural, or even spiritual experience. On the contrary: our mental and cultural worlds are significantly enriched by this new knowledge and awareness. Our view of the world and of ourselves as human beings within the world is significantly advanced by these insights, far beyond Copernicus and Darwin. This knowledge will help us facilitate human dignity within people's worlds, to increase how the world fits with people's inherent potentials. Psychotherapy is a good example for this process. By taking neuroscientific findings into consideration, psychotherapy can not only become more effective but can also facilitate people's awareness that therapy can (and should) be much more than the alleviation of disorders. Happy people—those who live in accordance with their needs and goals—do not develop mental disorders. This principle will become clearer from a neuroscientific perspective than from many other perspectives that are often assumed in clinical psychology. I have already attempted to show this in the introduction, and this implication of a neuroscientific point of view will become increasingly obvious throughout this book. Reduction occurs when people are reduced to their disorders, not when one recognizes that their mental disorders— just like any other mental process—are products of certain neural structures and processes.

2.6   NEURAL ACTIVITY TRANSPIRES IN PATTERNS OF ACTIVATION AND INHIBITION

A single neuron "knows" nothing. No meaning is stored or encoded within it. Meaning arises only from patterns of simultaneously activated neurons. The brain is, as it were, world champion in the creation of such patterns. How does the brain accomplish this?

If one considers all the different stimuli that impact from the environment and from within the body upon the brain, then one could assume that our brains must be chronically in a storm of stimulation overload. Disorganized activation storms within the brain, however, are a very rare exception. An epileptic seizure could perhaps be viewed as such an event, which shows how atypical they are. The processes within the brain normally transpire rather orderly. If one attempts to envision all the neurons and their interconnections simultaneously—and we are now dealing with magnitudes that, other than in the brain, are encountered perhaps only in outer space—then, at each moment, it is clear that only a very small proportion is simultaneously activated. Of the nonstimulated neurons, a subproportion is not activated because they are currently not being stimulated. But there is a second very important subproportion that is not currently activated because it is actively inhibited by other neurons. Inhibition is just as typical of brain activity as is activation. Inhibition is indeed necessary for the creation of order.

While I am writing this, my brain constantly receives a wealth of information from within my body and from the external environment, but my becoming conscious of this information is being actively inhibited. My buttock, leg, and back are consistently sending signals to my brain, after hours of writing, and these signals are clearly registered by my brain but are inhibited from becoming conscious. Only when my back begins to send stronger pain signals, do I begin to become aware of this, pay attention to it, and perhaps get up from my chair to take a few steps. This only works because on a lower level of information processing, my brain constantly monitors and processes the information originating from within my body. Because of this process it is possible to maintain a body position that enables me to write. As a result of this processing, I have changed my position several times over the course of several hours, without any of this entering the level of conscious awareness.

If I had recorded the various sounds near my desk over the past couple of hours, I would probably be very surprised upon listening at just how many distracting noises had impacted on me the entire time. The street noise that I now notice because I am devoting attention to it had not entered my conscious awareness earlier. Had I not heard this noise earlier? Indeed, I did hear it; it did reach my primary auditory cortex, but it never became the focus of my attention because this was being actively inhibited. If somebody on the street had called my name, I would have immediately noticed it, and I would have easily remembered it later. The brain—specifically, the thalamus—registers all of this information, but it does not relay all of it to other brain areas that are linked with conscious experience .

A particularly impressive example to demonstrate such inhibitory processes has been developed in the form of short film by the psychologist Daniel Simons. This film shows a baseball game, and the members of both teams are wearing black and white uniforms. The viewer sees how the ball passes back and forth between the players, and his or her task is to count how often the members of the white team pass the ball without it touching the ground. This requires a great deal of concentration because everything happens very quickly and involves a lot of motion. One has to pay close attention in order not to miss how the ball touches the ground. When the film is over, the participants in the experiment are asked how many times the ball made contact with the ground. It is impossible to be sure, but eventually one decides on a particular number. The experimenter then slightly nods and says that this number could be correct, but just to be sure, why not view the film once again together. This time the task is not to count but simply to watch the action.

And what happens now seems unbelievable: In the middle of the playing field, a person dressed up as a gorilla walks across the grass, stops, beats his chest like gorillas do, and then marches on. None of this had been noticed, despite the fact that this is all clearly visible and perfectly obvious. The typical reaction is that the viewer at first doesn't believe it and suspects that the experimenter has shown a different film the second time around. But it is true: The task to follow the ball very closely has so captured the attention on the first viewing that the gorilla—which was unrelated to the task—had been completely missed. Perceiving the gorilla had been actively inhibited by the brain. To be sure, the gorilla stimulated the ganglia cells of the retina, and these signals had been relayed to the visual cortex, but on their way to the areas that are linked with conscious awareness, the signals had somehow been blocked. The evaluation of incoming signals according to their meaning, and the inhibition of those signals that are currently evaluated as nonimportant, is a main function of the thalamus, which is connected to almost all other brain areas and is often described as a sort of relay station. Later we will encounter numerous other examples of such perceptions that, even though they are not linked with the quality of conscious awareness, can influence our behavior.

A SCIENTIFIC INTERLUDE

In the context of such perceptions that do not become conscious, we are faced with an interesting question: When dealing with such perceptions that do not become an object of our experience, is it reasonable to speak of a *mental process*? If not, what is it, then, that defines the attribute *mental*? Our experience is, by definition, excluded from the domain of the mental. Perhaps processes ought to be defined as mental if one can show that they have some specific consequences upon behavior, if not immediately, then at least delayed in the form of memory effects? (We would then be speaking about priming effects.)

Are our experience and behavior based upon mental processes or should experience and behavior themselves be regarded as two subproportions of mental phenomena? It is not easy to define what exactly is meant by the term *mental*. If one includes the neural level, the issue is clarified. Some neural processes lead to conscious experiencing, others have observable effects upon behavior, but a large proportion of neural processes are not expressed in either of these forms. Nevertheless, these processes exist. They can be documented with methods such as EEG (electroencephalography), neuroimaging technology, etc. It is much clearer to say that our experience and behavior are based on neural processes than to say that they are based on mental processes. That is because the existence of mental processes, without including the neural level, can only be documented by referring to experience and behavior. However, this would bring us dangerously close to circular reasoning. The existence of neural structures and processes can be defined without reference to experience and behavior. If we then relate neural processes to experience and behavior, we are not running the risk of circular reasoning. Therefore, it is logically advantageous, whenever possible, to relate neural rather than mental structures and processes to our experience and/or behavior. This is particularly crucial if our task is to make causal inferences. In part for this reason, I am convinced that the neuroscientific approach is a great gain for psychology. In terms of theory of science, this approach offers very obvious advantages.

This advantage in terms of statement-logicality that results from the inclusion of the neural level can be illustrated with the concept of the *reinforcer*. This term has been used widely as an explanatory construct throughout long phases of psychology's history and then later also in the context of behavior therapy. Skinner (1969) defined *reinforcers* operationally and functionally as stimuli that increase the frequency of specific behaviors. If the change in behavior is then, in turn, explained by referring to the reinforcer, a circular conclusion has occurred. However, if reinforcers are defined on a neural level—for example, as increased activity within the dopaminergic system (Berridge & Robinson, 1998; Rolls, 2000), then learning effects can be logically explained as the causal consequence of an increased dopaminergic activity without using circular logic. By including the neural level, psychology is enabled to offer more concrete statements than what had previously been possible. This also true in the context of defining and conceptualizing mental disorders and formulating statements about how to facilitate positive change in psychotherapy. These explanatory statements would probably be more convincing, especially for non-experts and for scientists from areas other than psychology, to the degree that they are not based solely on constructs that are inferred from behavior but instead on more concrete, tangible phenomena that can be appreciated with one's senses, such as PET (positron emission tomography) images of specific brain states. The fact that others would be more likely to appreciate such explanations has to do, in part, with this quality of sensory knowability, but from a theory of science perspective it is more relevant that yet another level, with its original, independent operational definitions, is included in our explanatory statements about experience and behavior.

The inhibitory processes about which I wrote earlier are sometimes also less well developed. This is the case, for example, in people with concentration difficulties or—in the case of serious underdevelopment, in people with attention deficit disorder. The state of mania is partially also characterized by the fact that the normal inhibitory processes are no longer functioning appropriately.

Active inhibitory processes are the prerequisite for orderly activity. If my conscious experience were constantly dominated by how my back feels at any this moment or by what kinds of noises are made by the cars outside, then I would not be able to produce much of this text. By actively inhibiting the relaying of this sensory information from within and outside my body, other neural processes can take center stage and allow my thoughts to be formulated and written down in an orderly fashion. But it is not just these inhibitory processes that form the basis of orderly, conscious action. The neural processes underlying that which I am now consciously thinking and writing are in turn accompanied by many other processes transpiring beyond consciousness, which also influence that which I write. The grammatical rules I use while writing these sentences are not conscious during the writing process, for instance, and my fingers manage to find the correct keys automatically, without conscious effort. The fact that I am sitting here and writing, rather than doing something else, is governed by activated motivational goals, which also have a specific neural representation within the brain. All these neural processes are simultaneously activated, but only a small proportion of them is consciously available to me during the writing process.

That which I experience clearly has a neural underpinning, yet this neural activity is only a small part of the total neural activity transpiring at any given moment. One of the most interesting questions psychology faces today concerns the factors that determine the contents of our consciousness, given that so many processes transpire on an unconscious level. Attentional control and even willpower certainly play a role in this process. But who or what determines the deployment of our attention and the direction of our will? We also cannot avoid confronting the question about the motivational determinants of neural functioning. What determines consciousness, attention, and free will? I will focus in more detail in the following section and in chapter 4 on these issues because of their centrality to our understanding of mental activity. These sections will discuss specifically how these systematizers of mental activity are neurally represented.

2.7   HOW DO NEURAL ACTIVATION PATTERNS ORIGINATE?

Before confronting such difficult questions, we should first revisit the question of how the brain manages at all to form orderly patterns that correspond to specific perceptual and experiential qualities. We have already learned about the critical role played in this context by inhibitory processes. They are often overlooked but they are the basic mechanisms that make us perceive and experience contrasts like a light figure on dark ground. But now let us turn toward the neural activation patterns that create the contents of our experience. These principles are the most easily illustrated with the processes underlying visual perceptions, which have also received the most research attention.

2.7.1   Neural Activation Patterns Form on the Basis of Cell Hierarchies

I had mentioned earlier that our perceptions are critically influenced by groups of highly specialized neurons that are simultaneously activated. A lot of our current knowledge about these processes goes back to the scientific breakthroughs made by the neurophysiologists Hubel and Wiesel (1959, 1962, 1968), who received the 1981 Nobel Prize for Medicine for their work. In one of Hubel and Wiesel's (1962) experiments, for example, they showed different visual stimulus patterns to cats whose eye muscles had been paralyzed so as to fixate their gaze on the patterns. With the help of microelectrodes, they analyzed the reaction patterns of more than 300 individual neurons from different areas within the visual cortex. Thus, they were able to determine how individual cells responded to various patterns of visual stimulation. Their observations showed that the single neurons had clearly distinguishable and differentially complex *receptive fields*, similar to the ganglia cells of the retina. Based on the relative ordering of center and periphery within these fields, they respond to various stimulus properties. One *response type* reacts, for example, to edges arranged in a certain orientation (angle) within the visual field, another one to edges in another angle, yet another type to movements in a certain direction, and so forth. Each neuron is a detector, as it were, for a very specific characteristic of each object that is presented.

Most cells have a simple receptive field and therefore were termed *simple cells* by Hubel and Wiesel. Others showed more complex response patterns (*complex cells*) and yet others even more complex response patterns (*hypercomplex cells*). The various cell types tended to concentrate in specific brain areas.

The cells work together in a sort of hierarchical organization: The characteristics of the simple cells result from the integration of the concentric ganglia cells within the retina, the characteristics of the complex cells stem from the integration of the simple cells, and the characteristics of the hypercomplex cells arise from integration of the complex cells. At the top of the hierarchy, finally, there are *gnostic* neurons, which respond to specific and highly complex objects such as hands or faces. Such cells were for a time jokingly referred to as *grandmother cells*. Figure 2–6 graphically depicts Hubel and Wiesel's hierarchical model of visual information processing.

Indeed, several independent research teams have documented the existence of face-specific neurons in the brains of monkeys (Perrett et al., 1984; Rolls, 1984; Young & Yamane, 1993). In monkeys, these neurons generally respond to faces, regardless of whether they are monkey faces or human faces. The responses are rather insensitive with regard to changes in the size, color, illumination, form, and location of the faces. They show a face invariance (Perrett, Mistlin, & Chitty, 1987). Some neurons also respond specifically to certain components of the face, such as the eyes, the mouth, or the facial expression. Because of the technical difficulties associated with measuring neural transmission, it has not yet been possible to document the existence of such single face neurons in humans, although face-specific cortex areas have been found (Grüsser, Naumann, & Seeck, 1990).

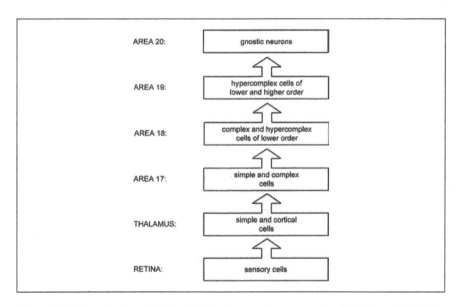

Figure 2–6. Hierarchical model of visual information processing according to Hubel and Wiesel. The complexity and specificity of the visual cells increases from the periphery (retina) to the highest visual centers in the cortex. While retinal ganglion cells with concentric receptive fields tend to respond to the simplest visual characteristics, gnostic neurons in the lower temporal lobe of the cortex respond to complex shapes (e.g., faces or hands). The response properties of higher-order cells are formed through the integration of the response properties of lower-order cells.
Note. From *Das Gehirn und Seine Wirklichkeit* [The Brain and Its Reality. Cognitive Neurobiology and Its Philosophical Consequences] (p. 141), by G. Roth, 1995, Frankfurt, Germany: Suhrkamp. Copyright 1995 by Suhrkamp. Adapted with permission.

Hubel and Wiesel's view of hierarchically organized neural groups as the basis of perception can, in large parts, still be viewed as valid today. However, it has been necessary to revise this model by the addition of another very important principle that illuminates that joining-together of neurons to neural groups. This is the binding-together of simultaneously activated neural groups into more complex neural activation patterns. The brain is characterized in large part by distributive-parallel information processing. Despite the existence of face neurons, the perception of faces is today viewed as being linked to *ensemble coding* and not to single detector cells (Young & Yamane, 1993). However, only a relatively small number of face neurons are required for the error-free identification of a face. The participation of several cells has obvious advantages. It enables the formation of invariances and enhanced recognition accuracy via the utilization of average-computations. In addition, it would be very risky for an organism to entrust such an important function to only a single cell.

2.7.2   Binding Together of Neural Activation Patterns via Synchronization

The basis enabling the binding together of neural groups that are distributed widely across the brain into a single activation pattern is probably linked to a *temporal integration mechanism*. The synchronization of their discharges allows for the process by which even neural groups that are remote from each other are linked together to form *cell assemblies*, in line with the Hebb's (1949) ideas.

Figure 2–7 shows these complementary mechanisms related to the formation of perceptual units. The upper part shows the hierarchical model by Hubel and Wiesel, whereas the lower part shows the binding together of cell assemblies via a mechanism of temporal coding.

How the synchronization process functions in detail continues to be a current topic of intensive study. The fact that the process occurs at all, however, can be regarded as clearly empirically substantiated (Engel, 1996; Singer, 1999). This evidence has also been documented by Wolfgang Miltner from the University of Jena (W. H. Miltner, Braun, Arnold, Witte, & Taub, 1999). In a classical conditioning experiment, a red floodlight that was turned on for 3 seconds was paired with an electrical shock administered to the middle finger of the right hand. These pairings were combined with pairings of green light—no shock. Since Pavlov's classic work, we know that after several such pairings, the red light alone will elicit the same reaction as the electrical shock. Conditioned stimulus (CS) and unconditioned stimulus (UCS) cluster together into a shared neural activation pattern. This is also what happened in this experiment. The electrical activity transpiring during the learning process was measured in different frequency bands; some electrodes were attached to the visual cortex (to measure the activity triggered by the color) and some were attached to the somatosensory cortex (to measure the activity triggered by the electrical shock)—both of these areas are clearly separated within the brain. The results showed an increasing coherence (synchronization) of brain activity in the gamma band between

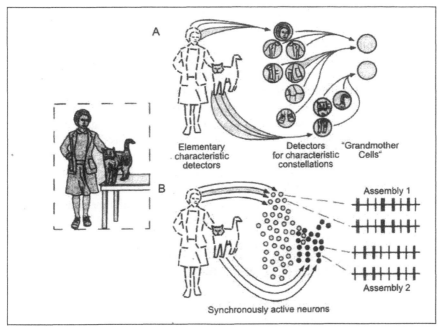

Figure 2–7. Two models for solving the problem of connection: (A) Integration through anatomic convergence. This model assumes that at the lower processing levels elementary object characteristics, such as the orientation of contours, are detected first. At the higher processing levels, cells with increasingly more specific response properties are then created through progressive convergence of the wiring. At the peak of the hierarchy, neurons (the so-called grandmother cells) are located, serving as specific detectors for entire objects, in this case for the woman and her cat. (B) Connecting of object characteristics through neural synchronization. The time-coding model assumes that objects in the visual cortex are represented through assemblies of synchronously firing neurons. In the example shown here, the woman and her cat would each be neurally represented through such an assembly (marked by open and filled-in symbols). These assemblies consist of neurons that detect elementary object characteristics. The association of the characteristics is illustrated by the temporal correlation between the neurons of an assembly (right). According to the time-coding hypothesis, all those neurons belonging to the same cell assembly each fire synchronously. Between the assemblies there is, however, no permanent temporal relationship.
Note. From "Prinzipien der Wahrnehmung: Das visuelle System [Principles of perception: The visual system]," by A. K. Engel, in G. Roth & W. Prinz (Eds.), *Kopf-Arbeit* (p. 202), 1996, Heidelberg, Germany: Spektrum Akademischer Verlag. Copyright 1996, Spektrum Akademischer Verlag. Adapted with permission.

37 and 43 Hz during the pairing of the red color and the electrical shock. This confirmed the accuracy of Hebb's assumption that "cells that fire together wire together," even when neurons are not directly connected via synapses. Of importance, the other version of the Hebbian rule states that when a presynaptic and a postsynaptic neuron are concurrently activated, the strength of the transmission between them should grow.

Psychotherapists tend to value simple rules very highly because they must apply their knowledge under highly complex conditions. Here is one such simple rule: That which is (re)activated recurrently (in the brain) will grow together. One activation pattern will trigger another one in the future. An example for the application of this rule: If I, as a therapist, permit a patient to recurrently experience predominantly negative emotions within the therapy sessions, then my person, the therapy office, and the next scheduled session will all come to function as a trigger of negative emotions for that patient, regardless of whether the patient or I had intended that. Negative emotions are accompanied by the activation of the avoidance system (see chapter 4 for further detail), which is the opposite of what I aim to achieve. The implication is that I must ensure that the patient will predominantly experience positive emotions during the sessions with me. How that can be achieved is a completely separate issue. The important thing is that I must achieve it if I want the patient to open up and engage with the therapy. All this has concrete implications for the strategies used to cultivate therapeutic relationships and for therapeutic actions more generally (specifically: for resource activation). More detail on these processes is provided in chapter 5.

2.8   THE NEURAL CONSTITUTION OF PERCEPTUAL UNITS

It is therefore primarily the recurrent simultaneous activation of the same neurons and neural groupings that determines the formation of neural activation patterns, neural groups, cell assemblies, neuronal assemblies, or whatever other name we should give them. This is the basis for the fact that we can easily and quickly identify the objects we encounter in daily life, such as tables, chairs, cars, people, letters, or numbers. They are all based on neural activation patterns that have been facilitated many thousands of times. These facilitations have occurred in widely varying conditions, such that the recognition of an object is based on that which remains constant in spite of these situational variations. We develop a concept or a schema, then, that is independent of each particular instance and the conditions in which it is encountered. We recognize a chair, even if it is flipped upside down. Thus, it is proverbially difficult to make someone mistake a cat for a dog.

As simple as the recognition of a chair appears to be, its perception is nevertheless based on rather complex processes. Figure 2–8 shows how many single characteristics have to be processed in parallel by specialized neural groups in order for the perceptual unit "chair" to emerge. According to Kandel, the emergence of even such simple perceptual phenomena is based on the participation of hundreds of thousands of neurons.

The neurobiologist Gerhard Roth writes in this context that "there is no neuron or neural group that could represent an object such as a chair, with all its detail *and* its various meanings. The perception of a concrete object requires the *simultaneous* activity of many cell groupings, which each encode only very limited aspects—such as detail aspects or category aspects—and these groupings are spread in various regions across the brain. Nowhere can we locate a single center into which all of these information aspects converge.

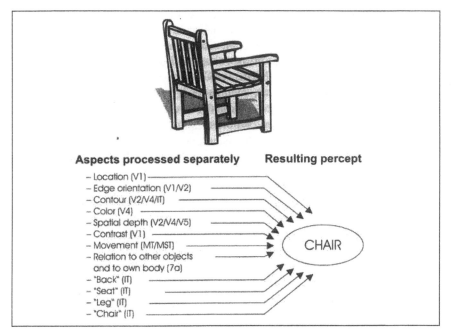

Aspects processed separately   Resulting percept

- Location (V1)
- Edge orientation (V1/V2)
- Contour (V2/V4/IT)
- Color (V4)
- Spatial depth (V2/V4/V5)
- Contrast (V1)
- Movement (MT/MST)
- Relation to other objects
  and to own body (7a)
- "Back" (IT)
- "Seat" (IT)
- "Leg" (IT)
- "Chair" (IT)

CHAIR

FIGURE 2–8. Integration of perception using the example of a chair. On the left, the different cortical areas are listed that are involved in the processing of the visual details, for example, in the identification of the object parts, respectively, and finally in the identification of the entire object as "chair."
The individual regions of the brain need not be detailed here in depth, because it is the principle that
is important here and not the specific example.
Note. From *Das Gehirn und Seine Wirklichkeit* [The Brain and Its Reality. Cognitive Neurobiology
and Its Philosophical Consequences] (p. 234), by G. Roth, 1995, Frankfurt, Germany: Suhrkamp.
Copyright 1995 by Suhrkamp. Adapted with permission.

This, however, radically clashes with our subjective experience. We do not perceive
the shape of a chair separate from its color, or the color separate from its shades of
light intensity and—when we move it—we don't perceive these aspects separately
from the chair's motion. Moreover, we certainly do not encode the meaning of "chair"
separately from the other aspects. Indeed, these components of the perceptual act
form a perceptual and consciousness unity.

But how is this subjectively perceived unity of perception, which we experience in
normal circumstances, created within the brain? This question is identical with the
question about the constitution of objects and scenes more generally. The rules
according to which this occurs, however, could not stem from the neurons or
synapses themselves because neurons and synapses do not possess any knowledge;
what they do is neutral in terms of the meaning of their activity.

The meaning-creating rules of perception emerge from the prior experience of the
cognitive system. This prior experience includes of course primarily the basic organi-

zation of the brain; how it evolved over the course of evolution, including the fact that the brain is that of a typical vertebrate, mammal, and primate. Further, it includes the fact that the sense organs typically become linked in similar fashion with specific brain areas among members of a single species, family, and so forth, and the fact that these brain areas then tend to become interconnected with each other in a similarly orderly fashion. All of this occurs, in part, independently of specific life experiences, even though it is not necessarily the case that specific genes predetermine each of these processes. Instead, there are certain genetically determined frameworks during the individual development of the brain, and these frameworks influence the structural order within the brain and thereby the system by which primary meanings are created in a self-organizing, *epigenetic* manner.

That which is perceived during adolescence and adulthood occurs within the frame of this basic constitution, which is partly genetically determined and partly acquired in early ontogenesis. This basic equipment determines to a substantial degree the perceptive and precognitive phase of perception; for instance, the way in which the visual system processes brightness, contrast, color, and movement, and the way in which depth perception develops. Most of these nonconscious and automatized processes cannot be modified by later individual experience (Roth, 1995, pp. 233–237).

The processes Roth described are the basis that enables us to establish contact with our environment, and the foundation that permits us to communicate about our surroundings as well as about subjective experiences. Because of these shared processes, we can assume that other people also see red when we perceive that color, and that rage feels similar in others as it does within ourselves. The extent to which such similarities really exist will always elude us because we have no criterion to judge the qualitative equivalence of subjective experiences. This problem of the "qualia" has plagued philosophers for many years (Nagel, 1974). On a practical level, however, our lives function quite smoothly under the implicit assumption that others refer to and know similar things when they speak of love, anger, impatience, happiness, and other such human experiences.

2.9   NEURAL CIRCUITS

Let us imagine that you are sitting in a train, reading a book. Your thoughts are wandering off; you are anticipating what it will be like when you arrive at home. You are looking forward to it. Outside, the scenery is quickly passing into the distance. You know this landscape well; you have once been here on a holiday and have fond memories of that time. In your compartment, three foreign tourists have a rather loud conversation about their travel experiences in your country, and they enjoy mocking some of the local customs. You listen for a while and begin to feel angry. You would like to step in and make a comment but then decide to hold back and remain silent. However, because of this you are now no longer able to concentrate on your book.

Such rich experiences correspond more closely to our daily reality than the perception of single objects, such as a chair. In reality, our brains are never occupied with the perception of only single objects. Even the contents of our current visual experience are much more complex. Within mere seconds, we see entire scenes with mountains, trees, houses, streets, cars, and people. Add to this our acoustic perceptions, smells, touch, whether we feel hungry or thirsty, whether we feel warm or cold, whether we are comfortable or feel pain. At the same time, our minds are flooded by memories, feelings arise, and action impulses emerge. At each moment we are also behaving in a particular way; our faces are conveying a distinct expression; our bodies are relaxed or tense; and we speak or remain silent. All of this continues to change, second by second.

All of these aspects of our experience and behavior are linked to complex *neural circuits*, which include specific brain areas and neural groupings. If we could monitor brain activity in such situations with functional magnetic resonance imaging (fMRI) technology, which optically differentiates more active brain regions from less active ones, we would notice profoundly different activity patterns changing by the second. In reality, unfortunately, we cannot examine such natural activities with imaging technology because this requires that the head be kept motionless in a tube-like cylinder, which is part of a larger apparatus that requires a great deal of space and makes considerable noise. For the time being, then, we can use such methods only to examine the kinds of human brain activities that can be elicited under these apparatus-specific conditions. Moreover, the temporal resolution of these methods is quite limited and does not yet suffice to study processes such as the ones I described previously.

What can be done, however, is to show study participants specific images while they are in the scanners, or to ask them to solve specific cognitive tests while the apparatus takes pictures showing which brain areas are particularly activated or deactivated during these tasks. Another option is to image the spontaneously occurring brain activity of various groups of people and to compare them with one another; such as groups of depressive or nondepressive people, or the same person at different points in time, such as schizophrenic patients when they are actively hallucinating and then later, when the hallucinations have stopped.

The many studies with imaging technology, which have by now been conducted with a great variety of research questions, show one thing with great clarity: There is no single area in the brain that corresponds to the specific location of any particular mental function. Every function, every perception, mental image, every feeling and act of the will is based on complex neural circuits that involve several brain areas. There is no such thing as "the location" of the will, of attention, of consciousness, fear, or of memory. However, there are specific brain regions whose capacity to function and activation are a necessary requirement for a specific mental function. For example, when the hippocampus is bilaterally functionally disabled, explicit memory contents can no longer be created. Under this circumstance, the person would no longer be able to remember that which is being experienced.

Most of our knowledge about the impairment of specific mental functions stems from observations and studies with brain damaged patients or with animals whose brains were lesioned in particular locations for the purpose of a study. Prior to the invention of imaging technology, such observation sometimes led to the assumption that the damaged area within the brain might really be the specific location of a particular function. The modern imaging technologies have corrected this mistaken view. They have shown that each respective mental function is linked not only to this one brain area but also to many other ones. For example, when memories are formed, it is not only the hippocampus that is actively involved but also other brain regions that are simultaneously activated. Brain areas that regularly work together are connected with each other by projection pathways. These are the foundation for the well-established neural circuits that bring specific order to the complexities of neural functioning.

In the following section, I will introduce several such neural circuits in somewhat greater detail. These sections are intended to fulfill several functions.

- They should convey an idea of the complexity of brain activity.
- They should illustrate more concretely what is meant by the concept of a neural circuit.
- They should, as a welcome side effect, illustrate the function of those brain areas that are most relevant for psychotherapy. I will elaborate the function of specific brain areas as these are relevant in the context of specific questions that arise. An alternative approach would be to proceed in the opposite direction, by showing which brain areas are responsible for which mental function. However, this approach tends to be rather tedious and could also lead to erroneous conclusions because no brain area can fulfill its function without close collaboration with other areas.
- They should show, by way of example, that each domain of human experience has its basis in the existence and functioning of specific neural circuits. For that reason, I have included two very specific examples that have nothing to do with clinical issues or with psychotherapy; namely, the feeling of being deeply moved by music and the experience of being in love with someone.
- The other neural circuits are far more complex. They are linked to entire mental functions, such as the formation of memory contents, the experience of fear and anxiety, intentional action, consciousness, and volitional acts. In contrast to the neural circuits mentioned before, these circuits are practically permanently activated, and they determine a large portion of our mental experience. One could write entire books about each of these areas. I focusing on selected aspects of the neural foundations of these functions, with the emphasis on the implications for psychotherapy.

Let us begin, then, with the neural circuit responsible for the formation of memory contents that can later be recalled.

2.9.1   The Formation of New Memory Contents

The important role of the hippocampus for memory became clear in a rather acciden-
tal fashion, in the context of a tragic mishap that occurred in the 1950s. A patient,
Henry M., who suffered from extreme epileptic seizures, underwent surgery during
which a large part of the temporal lobe that triggered the epileptic seizures was
removed unilaterally. The part that was removed also included the entire hippocampus. Prior to the operation, the physicians were not aware that the hippocampus on the
other side of the patient's brain had already atrophied almost entirely. As a result of
the surgery, the epileptic seizures improved, but there were unexpected side effects.

Post-surgery, Henry M. was unable to remember what he did and experienced. At first
the physicians thought that he had lost his entire memory. In time, however, it became
clear that the loss was of a very specific nature. The observations and specific tests
that were made with this patient ultimately led to a complete revision of our view of
how memory operates. Henry M. became one of the most intensely researched cases
in neurology and has become part of the history of the neurosciences. He contributed
much to our current knowledge about memory.

What was remarkable is that Henry M. had not lost his entire episodic memory. He
was still able to recall events that had taken place a long time ago—the longer the time
since the event, the better he was able to remember it. This was, in fact, the opposite
pattern from what is normally observed. People with healthy brains tend to have better recall for events that occurred not long ago. Henry M. also continued to have intact
knowledge about the world in general (semantic memory), his ability to speak
remained intact (semantic memory for word meaning and procedural memory for the
application of syntax and grammar), and his intelligence remained on a level that was
average for him. The major difference was that everything that happened after the
surgery continued to seem entirely new for him. He was able to remember things for
a few seconds, but he had lost the ability to remember anything for the longer term.
His view of himself was also permanently fixed in the past. He had lost the ability to
recognize himself in the mirror, but he could easily do so in old pictures.

Different researchers conducted various tests with Henry M. in the years following the
surgery, and their reports were subsequently published (N. J. Cohen & Corkin, 1981;
Corkin, 1968; Milner, 1965, 1967, 1972; Scoville & Milner, 1957). For example,
Brenda Milner asked Henry M. to copy a star of which he saw a mirror image. That
is not an easy task because one has to learn to move in the direction that is the opposite of the usual pattern. Henry M. was able to master this task as quickly as other people of his age, and he retained this newly acquired ability. He ultimately learned how
to draw in a mirror-image fashion, but each time he had no recollection of the fact that
he had ever done this before. The same pattern was observed with other tasks. He was
able to acquire complex motor skills and then retained these skills, but the tasks themselves appeared entirely new and unknown to him, as were the researchers themselves, whom he never recognized despite having met them repeatedly.

Later on, similar studies were conducted with other patients who had lesions in parts of the hippocampus. It became clear that among amnesic patients, the ability to acquire new skills remained intact not only for motor skills but also for cognitive activities such as reading mirror-image text or learning strategies for previously unknown games or solving puzzles (N. J. Cohen & Squire, 1980). Warrington and Weiskrantz (1973) were able to show that the process, which we now call *priming*, also remained intact among amnesic patients. The patients were presented with pictures or words in five different versions. At the one extreme, the pictures and words were presented in such an incomplete pattern that they could not be identified without being given additional information. At the other extreme, the images and words were normal and fully completed. Like other people, the amnesic patients were able to recognize the incomplete images and words more easily after having previously seen the complete versions. The ability to use the information well after the complete pictures and words had been presented also remained intact. In later studies, Weiskrantz and Warrington (1979) were also able to show that classical conditioning functions in the same way among amnesic patients as it does among normal people.

In sum, these and other studies led to the insight that we must distinguish among various different types of memory (Squire, Knowlton, & Musen, 1993; Squire & Zola, 1996, 1998). Squire introduced the distinction between *declarative memory* (because the memory contents can be articulated—or declared—verbally) and *procedural memory*. Today, the terms *explicit* and *implicit memory* are more commonly used. All those forms of memory that remained intact in Henry M. would today be considered part of implicit memory. The part that is lost through lesioning of the hippocampus belongs to that which today is referred to as *explicit memory*. Because of its special significance for psychotherapy, the important distinction between explicit and implicit memory will be further elaborated later.

Figure 2–9 provides an overview of the various brain areas that are involved in the formation of memory. This includes those that are not part of the circuit responsible for the formation of explicit memory. Each memory system—such as the one for the learning of movement patterns, for emotions, etc.—has its own neural circuit. The figure is intended to provide an initial overview of the localization of various brain regions, each of which we will revisit repeatedly later on.

It would be inaccurate to assume that the hippocampus is the sole location at which memory is stored; after all, memory content that has been stored a long time prior to a hippocampal lesion remains largely intact. Interestingly, such content remains relatively more intact to the degree that the encoding took place relatively longer before the damage to the hippocampus. The closer the events are to the time point of the lesion, the worse the person's memory for them. This suggests, then, that the hippocampus is not only critical for the acquisition of new memory contents but also is involved in the consolidation of explicit memory contents over longer periods (Squire, 1992; Squire & Kandel, 1999).

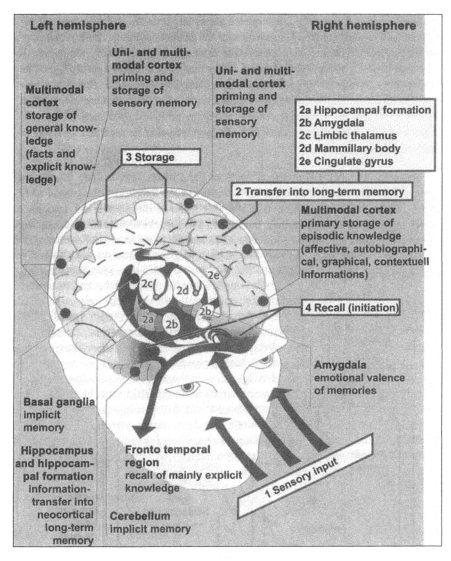

Figure 2–9. Overview of the brain regions involved in various memory systems (Adapted Figure from Markowitsch, 2002; p. 104).

Let us take a closer look at the neural circuit that is involved in the formation of new explicit memory contents (Figure 2–10): When we experience an event, our sense signals initially reach the visual, auditory, and somatosensory cortex. This is where sense-specific representations of the objects and events are being formed. These sense-specific representations form the first step within a complex processing chain. They are relayed to convergence zones in the rhinal cortex—a region that is in the immediate vicinity of the hippocampus, which is why it is sometimes called the *parahippocampal region.* The convergence zones integrate the various sense-specific signals into a multimodal representation of the current situation. These representations have a certain degree of independence from the original sensory modalities. It is these holistically perceived situations, rather than single sensory perceptions, that are ultimately consciously perceived and remembered by the person. This is why perceptions can also turn into concepts—abstract representations—that are independent of the specific sensory modalities. The hippocampus receives input from several convergence zones in the rhinal cortex and can therefore also be regarded as a sort of superconvergence zone.

The sense-specific cortical areas, as well as the various intermediate processing regions, also send signals at each step of the way to the thalamus—the central relay station within the brain—which then decides whether the signals are important enough to be processed any further. In order to make this decision, the thalamus has to be in continuous close connection with those regions in the prefrontal cortex, within which our current goals and enduring motivational tendencies are represented. If signals are deemed to be particularly significant or meaningful, then already at this

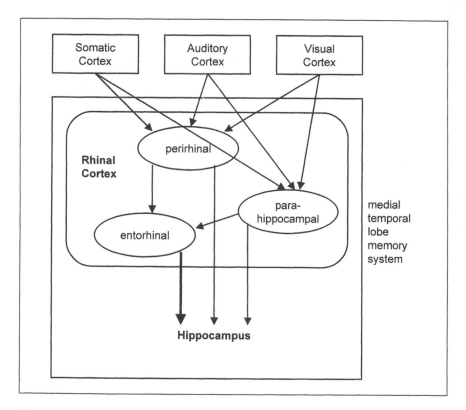

Figure 2–10. The rhinal cortex and the hippocampus as sequentially organized convergence zones, into which signals of the various sense organs are integrated. (Adapted from LeDoux, 2002; p. 105).

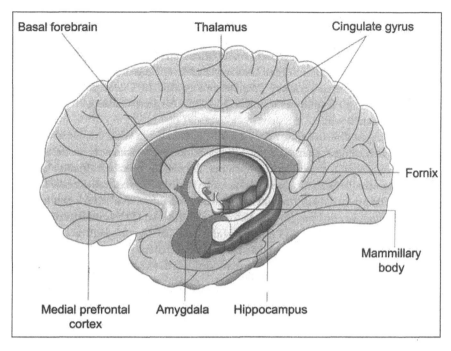

Figure 2–11. Localization of brain areas that are involved in the formation of recallable memory contents (Adapted from Deneke, 1999; p. 79).

stage regions will be activated that are associated with positive or negative feelings. These regions include, for example, the amygdala and the ventromedial regions of the prefrontal cortex. Figure 2–11 conveys an impression of the location within the brain of these regions, all of which are involved in the formation of memory.

The thalamus is continuously equally closely connected with the central working memory, which includes various regions in the prefrontal cortex. This working memory governs the assignment of attention to incoming signals—as long as no sudden, unexpected signals, which the thalamus judges to have utmost priority, demand immediate attention. The possibility that attention can at all times be "conquered from the outside" is essential for our survival because otherwise, while working memory is occupied with self-generated contents, one could end up in a situation of grave danger. The involvement of the brain areas that represent working memory in the next steps of the information processing chain is an essential requirement for the transformation of the perceptual input into explicit memory contents. Information that does not even reach this working memory buffer, whose signal processing is not linked with the quality of conscious awareness, is not being relayed to the hippocampus and therefore cannot be recalled later on.

The second important factor influencing the formation of explicit memory contents is the motivational significance of consciously experienced perceptions. This significance is reflected in the emotions that accompany such perceptions. Therefore, two additional circuits are involved in the process. Perceptions are consciously experienced to the degree that attention is devoted to them. In the process of perception, initially only the AMPA and GABA receptors (see previous) of the receiving neurons are activated. This activation occurs at high speed and allows for rapidly changing perceptions. However, these receptors do not contribute to the formation of memory contents. Only when the receptors involved in a neural circuit are repeatedly activated—when the perception is continuous or enduring—will the NMDA receptors also be activated. Via the within-cell second messenger process that is thereby triggered, the transmission at the synapses within this neural circuit is altered for a somewhat longer term (i.e., for seconds or minutes). This, in turn, triggers a positive feedback loop. The signals that are received as the perception continues are increasingly easily transmitted and the additional within-cell second messenger cascade described previously (cAMP-PKA-CREB, see previous) is initiated. This cascade massively strengthens synaptic connections between the neurons involved in this circuit, and it does so via the transcription of genes, the formation of additional synapses and via retrograde second messengers to the presynaptic neuron. This last cascade, which is the most important one for the formation of enduring memory content, does not only involve the NMDA receptors but also ligand-governed receptors, which are the kinds of receptors specializing in the binding of neuromodulators such as dopamine, epinephrine, acetylcholine, and serotonin.

If a large amount of dopamine is available in the synaptic gap while a particular perception is being made, this dopamine becomes bound to the dopamine receptors located at the postsynaptic cell, thereby contributing to the depolarization of the membrane, and ultimately contributing to the triggering of an action potential. Simultaneously, the dopamine ion channels independently trigger a second messenger process of the kind I just discussed previously. The memory formation that is thus triggered by such NMDA receptors can be massively strengthened by such second messenger cascades. Dopamine is released by dopaminergic neurons. The dopamine system plays a central role for motivational cues and for reward experiences (Berridge & Robinson, 1998; Rolls, 2000; see chapter 4 for more detail). Therefore, the dopamine system is sometimes referred to as the *reward system*. The hippocampus, just as the prefrontal cortex and the entire limbic system, is well endowed with dopaminergic neurons. If the dopamine system is activated at the time when the perceptions that are to be remembered occur (this is the case, e.g., when strong motivational goals are activated) then the formation of new memory content is strongly facilitated via the dopamine receptors that are involved in this neural circuit. This is the reason why it is so much easier to learn things that we actually desire to learn. If somebody has just fallen in love and hears the telephone number of the beloved person, it tends to take only this single learning trial to remember these numbers relatively permanently. If one wants to be sure, however, one should intermittently attempt to memorize the numbers; that is, one should recall them into the working memory, because the duration of information in here tends to correlate highly with the later accuracy of the memory. Why this is so should hopefully be clear after the preceding discussions.

For the following, let us assume that attention as well as motivational significance are in place. The event that is to be acquired into memory is very important to us; it triggers strong emotions and fills up our entire conscious awareness.

Now the signals that have been transformed multiple times are being relayed via parahippocampal regions to the input regions of the hippocampus, the gyrus dentatus, and from there via two further regions—CA3 and CA1, in Brodmann's terms (see Figure 2–12)—to an output area, the subiculum. From there, projection pathways lead back to the areas in the parahippocampus and to the various cortical areas from which the signals originated. This is why we speak of neural *circuits*. The connections between sense-specific cortical areas and the hippocampus are almost entirely reciprocal; that is, the connections tend to feed back and feed forward.

The feedback connections between the hippocampus and other regions—mostly cortical regions—leads to a strengthening of the synaptic connections via these pathways. This is why, in time, explicit memory contents are increasingly stored in those regions that were involved in the aforementioned neural circuit. The hippocampus directs this storage process, as it were, but it is not in itself the sole location at which memory is stored (and the longer the duration, the more this tends to be true).

The first time an event occurs, all that changes are a few specific connection strengths within the hippocampus. When the event, or parts of that event, occurs again—either in the form of a recurring perception or in the form of a memory—then the parts of the hippocampus that were originally involved in the perception again take part in the reestablishment of the cortical activation pattern; the same pattern that was originally formed is established again. Every recurrence of this pattern changes the cortical synapses to some degree, and these changes are concentrated in those areas that participated in the original perception. These processes can be documented even during sleep (Louie & Wilson, 2001; McNaughton, 1998; Nadasdy, Hirase, Czurko, Csicsvari, & Buzsaki, 1999; Poe, Nitz, McNaughton, & Barnes, 2000; Wilson & McNaughton, 1994).

Wilson and McNaughton (1994) used a very sophisticated technique to measure the precise pattern of neural activity in the hippocampi of rats while the rats explored new territory. Later, while the rats were sleeping, the same activation patterns could be observed in the rats' hippocampi—as if they were dreaming of the newly explored territories. It appears that a further processing of that which has just been experienced occurs during sleep, and this processing specifically includes a consolidation of memory traces. Consolidation means in this context that more and more of the brain areas that were involved in the original experience become involved in the memory of the experience, and this occurs by the process of altering the strengths of synaptic connections within those areas. Bontempi, Laurent-Demir, Destrade, & Jaffard (1999) trained mice in a spatial-orienting task, and it is known that the hippocampus is significantly involved in such tasks. They measured brain activity in the hippocampus and other areas at specified intervals throughout the learning process, while the mice were repeatedly confronted with the same learning task. They noted that the activity in the hippocampus was initially very high, and the strength of this neural activity

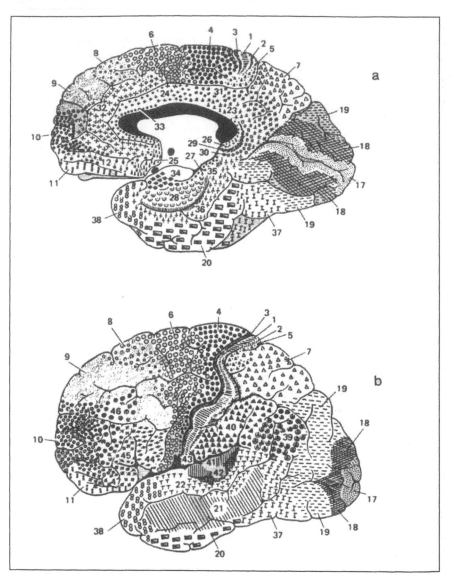

Figure 2–12. Overview of the "cyto-architectural" brain map that was formulated already in 1909 by Korbinian Brodmann. Top: (a) the interior view of one half of the cortex; bottom: (b) cortex view from the side. The foundation of this partition was the characteristic structure of the neurons in this region. The partition has nothing to do with the function of these regions. Nevertheless, the map is still being used by brain scientists to communicate about the various areas within the brain. The statements in Figure 2–8 about the brain regions involved in the perception of a chair, for instance, refer to this partition by Brodmann (Adapted from Roth, 1995; p. 50).

correlated with memory performance (how quickly and accurately the mice found the correct way). In time, though, hippocampus activity diminished and no longer correlated with memory performance. Instead, an increase in cortical activity was observed, and this ultimately correlated with memory performance. Thus, what was documented was a shifting of the location that is responsible for memory performance. This shifting process takes a relatively long time. The earlier the activity in the hippocampus diminishes during this shifting or consolidation process, the less the event is already stored in other brain regions. This is why, when the hippocampus suddenly fails to function, memory for events that occurred a long time ago tends to be better than memory for recent events.

According to Nadel and Moscovitch (1997), these findings should not be taken to mean that the hippocampus is later no longer involved in long-term memory at all. Instead, memory becomes increasingly well supported by the involvement of other brain regions, so that the failing of the hippocampus can eventually be entirely compensated by other regions. The notion that the same neural circuit that was originally involved when an experience was perceived is also involved in the later storage of that memory, is also supported by the well-established finding that the electrical stimulation of neurons in the temporal lobe can trigger lively memories of events that occurred a long time ago. Apparently, the entire originally activated neural circuit is reactivated in this process. However, this circuit can also be activated by the hippocampus, and this can only happen when the hippocampus is still involved to some degree in the memory trace.

Many such observations were reported by Penfield and Perot (1963). They examined patients who suffered from epileptic seizures during their preparation for surgery in which the brain areas that appeared to trigger the seizures were to be removed. To ensure that no functionally indispensable brain regions would be removed, they stimulated the adjacent brain regions with small electrodes. This helped them determine the function of each specific area. This process can be endured by the patients while they are fully conscious because the brain does not have any pain receptors. Thus, the patients were able to tell the surgeon very specifically what they experienced while a given area was being stimulated.

The report by Penfield and Perot continues to be one of the most impressive and illustrative documents to support the idea that our experience and behavior are based on neural processes. For example, the surgeons were able to trigger hand movements by stimulating specific parts of the motor cortex. When the patients were asked why they moved their hands, some of them were surprised to find that they had in fact moved their hand at all. Others experienced the movement as externally controlled, not as their own movement, and yet others did not really have an answer. Interestingly, some patients also stated that they had intentionally moved their hand. The experience of doing something intentionally, then, can apparently also be triggered by the stimulation of a specific area in the brain.

This holds true for the entire spectrum of subjective experiences. While the patients were lying on the operating table with clear conscious awareness of the surrounding situation, they suddenly experienced changes in their experience. As reported by Deneke (2001),

> the noises in the environment suddenly seemed louder or quieter, closer or farther away; objects within their visual field changed their appearance, became clearer or more blurred, appeared to come closer or move away; ongoing experience suddenly seemed already familiar, as in a deja-vu or—the opposite—familiar events appeared strange and unreal. Furthermore, patients might suddenly be overcome by feelings of fear, loneliness, sadness, or disgust. (p. 90)

In the context of memory, however, what is of primary interest here are the memories that were triggered via the electrical stimulation. In Deneke's (1999) words,

> approximately 8% of the patients felt as if the electrical stimulation of the temporal lobe instantaneously transported them back into their own past. They heard the voices of familiar persons, could hear a familiar tune, which they also were able to hum when prompted later on, or they experienced complex scenes, which—although they happened years ago—suddenly seemed to have come back to life. One patient, for example, had an intense memory in response to a song that was triggered by the stimulation; she remembered a Christmas church visit in Amsterdam, which she had actually experienced years earlier. This had happened during the war; in her memory, she saw many Canadian soldiers (incidentally, she later married a Canadian soldier herself and emigrated with him to Montreal, where she then had the surgery during which these memories were produced). It seemed to her as if she was once again in the church in Amsterdam. She could hear the choir and felt, as she had done years ago, intensely happy. This occurred despite the fact that she knew consciously that she was lying on an operation table in Montreal. (p. 90)

It is evident that a very complex neural circuit is involved in such memories; a circuit that encompasses many different cortical brain areas and apparently also the hippocampus, from which the stimulation originated. Thus, it appears that the hippocampus continues to be involved in some way in the storage of long-term memory content.

When we consider the complex neural circuits involved in the storage of memory, it is clear that the recollections produced by patients in psychotherapy cannot be regarded merely as objective reports of events as they actually happened. Even the original experience of an event is influenced by subjective factors, such as previous experiences, expectations, attitudes, and motivational tendencies, all of which in turn are influenced by the person's social context. This holds true all the more when we consider the factors that influence what the brain produces as memories. The subjective veracity and liveliness of memories has nothing to do with their objective validity. At the moment when the brain produces the memories, other neural circuits are also activated, and they interact with those circuits that produce the memories.

The neuroscientic skepticism about the objective "accuracy" of memories is very much consistent with recent studies that specifically examined the validity of memories. The studies by Loftus (1993) about the validity of eye-witness testimonies have attracted considerable public attention in this area because of their legal relevance. By now, however, much has also been learned about the neural bases of "false" memories (Schacter & Curran, 2000; Schacter, Norman, & Koutstaal, 1998; Schacter et al., 1996; Schacter, Verfaellie, & Pradere, 1996).

Although therapists might sometimes assume a patient's life history is reconstructed in session, it would be more accurate to say that new histories are being constructed in the therapeutic dialogue. This is all the more true to the degree that the reconstruction is based on memories produced by the patient and therapist interpretations rather than objectively validated facts. The new history is substantially influenced by the current determinants of neural experience, and such factors are usually very different from those that affected the original experience a long time ago. When a patient today reports specific early childhood memories, then this is important because these ideas, which he or she perceives as memories, are evidently meaningful in today's context. They convey information about the patient today, and this can be critically important for the therapy. However, such reports do not provide reliable information about what the patient objectively experienced in earlier years. Indeed, it is clear that valid memories prior to age four are hardly possible because the hippocampus is typically not sufficiently matured at this point to form explicit memory contents.

I will further elaborate on the implications for psychotherapy in the following section, which focuses on explicit and implicit mental processes. Before getting to that, though, two additional types of memory must be distinguished. But first, let us consider two less-pervasive circuits, which correspond to very specific experiential states. The discussion of these is intended to illustrate with clarity that all mental processes are based on specific neural activation patterns.

2.9.2   Being Captivated by Music

I opened this chapter by noting how remarkable it is that a binary sequence of magnetic states on a CD can produce something as beautiful as music. In reality, though, several intermediate transformation steps lie between the signals on the CD and the experience of music: The binary signals on the CD must be back-translated to analog signals in the form of waves of certain frequency, waves, which in turn have to be fed to the brain via the auditory system (i.e., our ears). The auditory cortex holds neural "maps" for tones of various frequencies. These maps are indeed similar to geographical maps; they have a spatial order, which is used to provide topographical order to the world of sounds. Small groups of cells respond to tones of a particular frequency. These cells lie within immediate vicinity of one another. To their right and left are the areas for higher and lower frequencies, respectively. Figure 2–13 shows the location of this "tone map" in the human cortex; the topographical order according to frequencies is shown in enlarged format here.

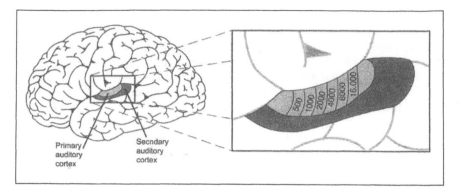

Figure 2–13. Location of the auditory cortex and tonal map in the human brain. In the area of the temporal lobe, three partially overlapping bulges—known as gyri—can be differentiated (gyrus temporalis, medius, inferior). The primary auditory center (light gray) and the secondary auditory center (dark gray) are located in the upper temporal lobe. The magnified image on the right shows that the frequencies (data in Hz units) are represented in the form of a map in the primary hearing cortex (Adapted from Spitzer, 2002; p. 185).

There are many such function-specific neural maps within the brain. Spitzer (2003) estimates their number to be approximately 700. The most well-known among these is the map of the somatosensory cortex for the body surface area, which is shown in Figure 2–14.

How such neural maps are malleable by experiences is apparent, for example, when we consider that, among violin players, the map of violin sounds is more strongly represented than the map of trumpet sounds, and vice versa, among trumpet players, trumpet sounds are neurally more strongly represented. This only holds true, however, among players who have intensively used their instrument for years. Figure 2–15 illustrates this relationship with data from a study by Pantev, Roberts, Schulz, Engelien, & Ross (2001).

The earlier one begins with the intensive practice of an instrument, the larger the area in the brain becomes that is reserved for these tones. The correlations are quite impressive ($r = .43$), and among musicians with perfect pitch, this correlation is even higher ($r = .63$; Pantev et al., 1998). Thus, it is not so important how many years one has been playing the instrument; rather, it is critical whether the practice began at an early age. Musicians who began playing prior to the age of nine show a particularly pronounced enlargement of these music-specific areas. In general, piano tones activate the auditory cortex among musicians by 25% more than what is achieved by pure sine tones of the same frequency. This difference is not observed among nonmusicians. It stems from the fact that musicians have been exposed to such tones for much longer periods than nonmusicians.

However, the auditory cortex is only one of many brain areas that has functional significance for the act of producing music and for the experience of music. Depending

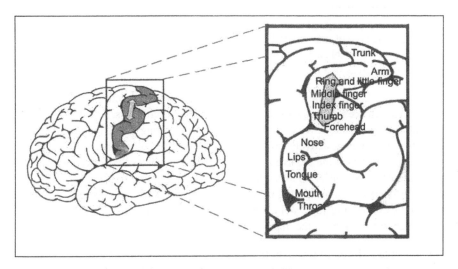

Figure 2–14. Map of the body surface (so-called somatotopic map) on the postcentral gyrus of the left cortex (gray-shaded on the left). Not only are the areas of the body surfaces represented in a map format—the somatotopic map—but there is even a direct relationship between the significance of the represented area and its location on the map. We have a fine sense of touch with our fingers as well as with the lips and the tongue. Correspondingly, there is a relatively large amount of space on the surface map for the hands, lips, and tongue. The light gray arrow refers to the enlargement of the neural representation of specific fingers, in the way that this can be accomplished via intensive training. More detail on this is provided later in the section on neural plasticity (Adapted from Spitzer, 2002; p. 184).

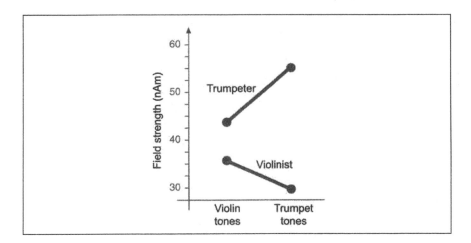

Figure 2–15. Location of violin and trumpet tones among violinists and trumpeters (data from Pantev, Roberts, Schulz, Engelien, &Ross, 2001). The figure shows the auditory evoked field strength resulting from neural activity, measured and averaged across both cortical hemispheres. Violinists have higher field strength for violin tones; trumpeters for trumpet tones. The absolute strength of the activation is less important here, but it could be related to the fact that the tones trumpeters have encountered over many years tend to be much louder than the violin tones encountered by the violinists (Adapted from Spitzer, 2002; p. 190).

on the specific type of musical activity, up to 12 specific functions of specialized areas that are involved in this activity are included in the neural circuits. Tramo (2001) has grouped these areas in a conceptual model, which is shown in Figure 2–16.

Based on all this, it is clearly justifiable to say that musical activity and intensive musical experience involves large areas of the brain. Nevertheless, this does not mean that all brain areas are activated in this process. Conducting music requires different circuits than dancing; actively producing music involves other ones than passively receiving it. But even the passive reception of music involves the activation of complex circuits, which differ, in turn, based on how this listening transpires. Among lay music listeners (i.e., nonexperts), the right brain hemisphere is more strongly activated; among experts, this tends to be the left hemisphere. Spitzer (2002) has related this difference to variations in the depth of processing.

It is possible to listen to music in the background while pursuing all kinds of other activities. Alternatively, one can concentrate intently on this act. To the degree that one does this, special types of experiences can be induced by the music. Sloboda (1991) conducted a survey among 83 persons—professional musicians, amateur musicians, and nonmusicians—about bodily reactions that were triggered in them by music. Ninety percent of the participants reported that particular music pieces could trigger "goosebumps, cold chills running down one's back." More specific questioning revealed that these reactions did not diminish with repeated listening to the music. Most of the participants were able to name a music piece as well as the precise moment that triggered the goosebumps. These findings inspired follow-up research by Blood and Zatorre (2001), who further examined the goosebumps phenomenon. Ten musicians participated in their study, and each of them named a classical music piece that effectively triggered goosebumps in them. The researchers selected a 90-second-long goosebump-eliciting piece that was to be played for them. Because each of them selected a different music piece, the sections of one musician's could be used as control stimuli for another musician. This control allowed the researchers to rule out the possibility that other aspects of the music, such as tempo, volume, dynamic, and so forth, were not responsible for the observed differences between goosebump and nongoosebump passages.

Positron emission tomography (PET)—an imaging technology that shows the relative activation at any given time point—was used to compare brain activity during the playing of the goosebump passages with those that did not elicit the goosebumps. In certain brain areas, a significant increase could be documented, whereas other areas showed a significant decrease in activation. The activated and inhibited brain areas are highlighted in gray and black shading in Figure 2–17.

The observed differences can be summarized as follows: An increase in activity occurred in the left ventral striatum, which is responsible for positive appraisals and is also active, for instance, when an addictive substance such as cocaine is being consumed or when one devours a piece of chocolate or makes eye contact with an attractive person (Breiter et al., 1997; Kampe, Frith, Dolan, & Frith, 2001; Small, Zatorre,

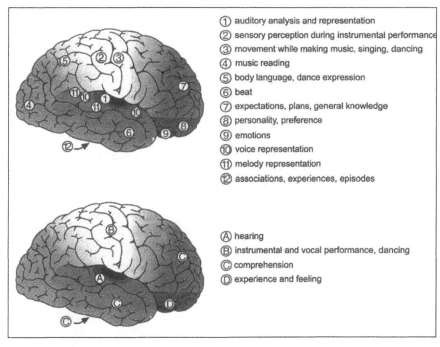

1. auditory analysis and representation
2. sensory perception during instrumental performance
3. movement while making music, singing, dancing
4. music reading
5. body language, dance expression
6. beat
7. expectations, plans, general knowledge
8. personality, preference
9. emotions
10. voice representation
11. melody representation
12. associations, experiences, episodes

A. hearing
B. instrumental and vocal performance, dancing
C. comprehension
D. experience and feeling

Figure 2–16. The entire brain creates music. Different music-related functions are indicated in the cerebral cortex (viewed from the right side; based on Tramo, 2001, p. 55); the upper image shows somewhat finer differentiations, the lower images shows only crude differentiations. (A) The analysis and representation of acoustic input transpires in the primary and secondary auditory cortex as well as surrounding areas, particularly in the right temporal lobe. Nearby are the representations of melody, harmony, dynamic, tonal colour, voice, tonal scales, intervals, key, and bars. (B) The motor and sensory areas are responsible for complex movements during the making of music and during the act of dancing; also for the corresponding perceptions involving the touch sense, and for the movement during singing and during the tapping of a rhythm. The motor areas gradually merge toward the front with areas that are responsible for the planning and understanding of music. (C) These frontal areas govern the recognition and programming of repetitions of phrases or entire music pieces; these areas also store general knowledge and, therefore, recognize unexpected musical events (key changes, syncopes, etc.). (D) Also in the frontal lobe, but further down, are areas that are responsible for private music taste, for values, but also for cultural idiosyncrasies and music styles. They are in close proximity to areas in which emotional reactions and events are encoded. These areas are closely connected with additional emotion-processing structures deeper within the brain (among others, the so-called limbic system). Therefore, they are also responsible for the physical accompaniments associated with music, such as goose bumps, crying, or the release of opiate-like substances—the endorphins—across large areas of the brain (Adapted from Spitzer, 2002; p. 209).

Dagher, Evans, & Jones-Gotman, 2001). Brain activity also increased in the dorsomedial midbrain, the orbitofronatal cortex, and the insula, which is also involved in appraisal and emotional processes. Further, increased activation was shown in the anterior cingulate gyrus, which is involved in attentional processes. A reduction in activity (i.e., inhibition), by contrast, was observed in the amygdala, which is involved in fear activation, and in the ventromedial prefrontal cortex, which is generally linked with negative emotions.

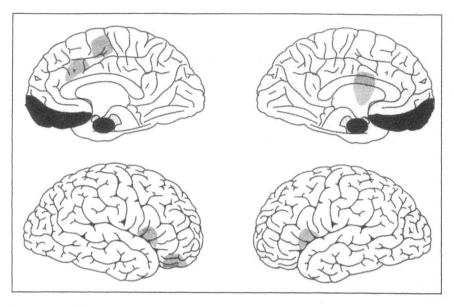

Figure 2–17. Schematic construction of some brain regions reported by Blood and Zatorre (2001) showing activation (light-gray) and deactivation (black) during the experience of goosebumps evoked by music. Increased activation was observed, among others, in the left ventral striatum (top right) bilateral insula (bottom), right orbito-frontal cortex (bottom left), the anterior cingulum, the supplementary motor area (top left) and the cerebellum (not shown). In contrast, activation of the right amygdala, the left amygdala-hippocampus complex and the ventral medium cortex (top) was reduced (Adapted from Spitzer, 2002, p. 397).

These results support very well the point that I emphasized earlier: Neural activity occurs in specific patterns of activation and inhibition. These patterns determine what we experience and which actions we take. In this case, we have a pattern that is associated with the intensive experience of positive emotions and the inhibition of negative ones. Similar experiences can also be triggered by experiences other than music. The fact that this particular experience was triggered by music could have been inferred from inspection of the full activity pattern, as revealed by PET scan. for example, as this would have shown that brain areas normally activated during music listening would also have been active in this situation. These activation patterns were not highlighted in this study, however, because the control conditions also involved listening to music.

The authors of this study concluded their report with some remarkable speculations, which seem to be valid not only in the context of music. In the following passages from the research report, the term "music" could also be replaced by other complex mental, artistic, or social activities or experiences.

> We have hereby demonstrated that music activates neural systems for reward and emotions, which correspond to the systems activated in response to specific biologically relevant stimuli, such as food or sex, and which can also be activated artificially by drugs.

This is remarkable because, strictly speaking, music is important neither for survival nor for reproduction. Similarly, music is not a substance in a pharmacological sense. The activation of these brain systems by a complex stimulus as abstract as music may be viewed as an emergent property of the complexity of human cognitive ability. Conceivably, the formation of increasingly strong anatomic and functional connections between phylogenetically older systems, which were essential for survival, and comparatively recently evolved cognitive systems might have strengthened our ability to assign meaning to abstract stimuli, and thereby might have improved our ability to infer happiness from such stimuli. The fact that music has the property to trigger intense feelings of happiness and stimulate the body's reward systems suggests that music—even though not absolutely essential for the survival of the human species—may make a significant contribution to our mental and physical well-being. (Blood & Zatorre, as cited in Spitzer, 2002, pp. 397–398)

This is in a sense a neuroscientific formulation of that which Freud meant by the term *sublimation*. To complete the picture, it should also be noted that music—a certain kind of music—can also trigger illness, which will be discussed more fully in chapter 5. Music is one of the many life experiences that have an impact on the brain. From now on, we should no longer be surprised when we encounter additional evidence for the fact that various life experiences can have a specific influence that changes the brain and that can be documented with biological methods. These influences can increase happiness, but they can also trigger sickness. Psychotherapy's task is to work with people whose life experiences have had a noxious effect on their brain, and to restore health by exposing them to life experiences that have a salutary effect. That is the essence of what I mean when I speak of "neuropsychotherapy."

2.9.3   Love-related Feelings

Before returning to questions that are more directly relevant for psychotherapy than the feeling of being captivated by music, I would like to elaborate on a second, shorter example to illustrate that even the most intimate feelings have a very specific neural foundation. After this, I hope, the reader will not require any more convincing to accept the idea that all the facets of the mental life that we encounter in psychotherapy have a neural basis and that psychotherapy, if it aims to be maximally effective, could profit from a detailed knowledge of neural processes.

In their study, "The Neural Basis of Romantic Love," Bartels and Zeki (2000) started by recruiting 70 people that happened to be in love at the time. From that pool, they selected those 17 persons that were the most deeply in love. They asked these 17 people to provide pictures of their romantic partners and of three close, same-sex friends. The experimental task was that the participants had to view the picture of either their romantic partner or their friends—in random order—for 20 seconds each, during which they were also asked to think of the persons intensively. During this period, the participants' brain activity was recorded via fMRI. In addition, their galvanic skin responses were being measured during this process, and the research participants were asked to estimate the intensity of their love feelings and their sexual arousal on

9-point scales. The emotional reactions of the participants differed significantly on all measures, depending on whether they viewed photos of their lovers or friends. The real question, however, was what differences in brain activation could be documented. Careful design of the study ensured that any observed differences could not be accounted for by factors other than the different emotional reactions during exposure to the pictures.

It became clear that acute feelings of love were based on a very specific pattern of activation and inhibition of certain brain areas, and this pattern appeared to be clearly distinguishable from those that are linked to positive emotions or sexual arousal. The pattern includes increased activation of the medial insula, of the anterior cingular cortex, and subcortically of the nucleus caudatus and the putamen. All of these regions were activated bilaterally. The pattern also included relative deactivation of the posterior gyrus cinguli and the amygdala (bilaterally), as well as parts of the prefrontal, parietal, and medial temporal lobe (right side only). The deactivated areas corresponded rather exactly to those that are activated during emotions such as grief, depression, and anxiety.

The activated areas overlapped only partially with those that in previous studies were found to be activated when participants viewed attractive but unknown faces, or with areas that are activated during positive social interactions. The closest similarity was found between the pattern elicited in this study and states of euphoria, such as the ones elicited by cocaine.

Overall, however, the observed pattern of activation and inhibition during states of love was not identical to any previously found patterns, not even those that are elicited during other types of positive emotion. The stimuli that elicited these specific patterns were not distinguishable (for outside observers) from other stimuli that did not trigger these patterns. In each case, the photo showed a friendly, sympathetic-looking person. Thus, the meaning of these stimuli was not defined from the outside. The pictures had a very idiosyncratic meaning, which existed only in the subjective world of each research participant; a very private feeling, which now became the object of a scientific investigation. The study showed that a neuroscientific approach allows us to study even such very private experiences in a scientific way. It also shows that subjective experiences, which have meaning only in the private world of one individual, can nevertheless be correlated objectively with events at the neural level of experience.

In light of the fact that in psychotherapy, we are constantly faced with such highly individual subjective meaning, it would be almost foolish to ignore the finding that such individual meanings have a very specific neural basis. If our aim is to change certain individual meanings—especially, of course, problematic or problem-producing meanings—then we must change the neural bases of these meanings. This logical consequence seems increasingly inevitable, given the mounting neuroscientific evidence in support of such an argument. Whenever we aim to influence the neural foundations of anything, we would be well advised to study the principles that govern the development and change of neural structures and processes. That is the reason why I think psychotherapy will inevitably move in the direction of neuropsychotherapy. The progress in the neurosciences cannot be halted. This progress will undoubtedly also influence psychotherapy as we know it today. Psychotherapists must now decide whether they already want to start to actively participate and shape this development or instead prefer to wait until they must eventually acknowledge that the scientific framework surrounding psychotherapy has changed fundamentally.

By including a neuroscientific perspective, psychotherapy can only win. Conventional interventions that are effective have a neural basis for their effectiveness. By explicitly addressing this neural basis, we will eventually be in a much better position to understand the mechanisms of action underlying our clinical interventions. Psychotherapy research and practice has already partially revealed *what* is effective. The neurosciences can help us understand better *why* these methods are effective. Beyond this, a neuroscientific perspective of mental activity, of the development and maintenance of mental disorders, and of the psychotherapeutic process can also provide hints about yet unexplored potentials of psychotherapy and about blind alleys that do not promise to yield effective interventions. Thus, this perspective can help us abandon unpromising paths and develop new strategies, which then must be explored and tested in order to improve therapeutic practice, one step at a time. The first steps can already be taken today, which is what I aim to do with this book. But the potential to take this further is getting better every day, in part because the neurosciences are yielding increasingly relevant findings, and in part because psychotherapists, psychotherapy researchers, and neuroscientists are now in a better position than ever to work directly together. Such collaborations are likely to become increasingly common in the not-too-distant future.

A particularly good example of how neuroscientific findings can help us to better understand the mechanisms of action of our specific clinical interventions (and the potential pathways to improving these interventions) is the emotion of anxiety.

2.10 ANXIETY

2.10.1 The Amygdala as the Anxiety Center

When I speak of *anxiety* in the following passages, I should emphasize that I don't mean anxiety as a subjectively experienced feeling in the same sense as I wrote about musical delight or love in the previous examples. *Anxiety* here also does not refer to a clinical phenomenon or disorder. Instead, what is relevant here is anxiety as the central alarm and defense system of the organism, which is activated in response to threats of any kind. At the center of this defense-against-danger system is the amygdala, which, along with the hippocampus, is localized in the temporal lobe. The central role of the amygdala in the context of reactions to threat signals is clarified in Figure 2–18, which is explained in the text following.

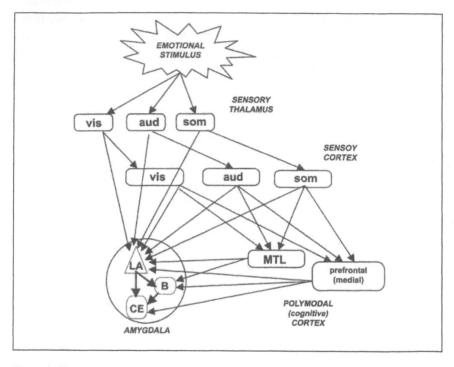

Figure 2–18. Information flow to the amygdala. The amygdala receives low-level information about objects and events from sensory-processing regions in the thalamus, and more complex information from sensory-processing areas in the cortex. Vis = visual; aud = auditory; som = somatosensory; MTL = medial temporal lobe memory system; LA = lateral amygdala; B = basal amygdala; CE = central amygdala (Adapted from LeDoux, 2002; p. 207).

When we are confronted with a sudden danger signal—a loud bang, something hurtling toward us, something that suddenly stings us—this signal (referred to as an *emotional stimulus* in Figure 2–18) is sent via the sense organs to the sensory thalamus (which is the central relay station that initially receives any form of input into the brain), and then is not passed on to the sensory cortex (which normally is responsible for the further processing of received signals) but instead is relayed instantaneously to the lateral aspect of the amygdala. All of this unfolds more rapidly than we can think. Before we even know exactly what is going on, we react with a defensive reflex or freeze momentarily. This is also how quickly our autonomic nervous system responds, releases adrenalin, increases blood pressure, and mobilizes our entire system for protective or attacking reactions. These lightning-fast, reflex-like alarm reactions have been initiated primarily by the amygdala (Figure 2–19).

One second later we already begin to relax again. By now we have noticed a young boy who has lit a firecracker. In this second, the areas of the sensory cortex have done their work and have analyzed the situation more fully. These areas also had received the signals from the thalamus. In their analysis, the cortex areas have also involved the hippocampus (medial temporal lobe), which specializes in the processing of

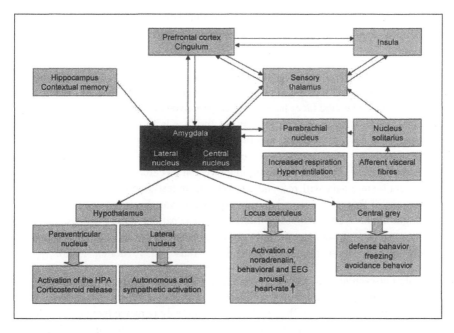

Figure 2–19. Overview of the most important connections of the amygdala. This shows especially the connections originating from the central nucleus. Signals that indicate potential danger trigger behavioral reactions such as shock-freezing, defensive- and avoidance behavior, and autonomic and endocrine reactions. Each of these reactions is being controlled from a different output of the central nucleus of the amygdala. Lesions of the entire nucleus block the triggering of all these reactions, whereas lesions of the specific output pathways only block the corresponding specific reactions (Adapted from Förstl, 2002, p. 251).

contextual information and which activates relevant memory traces, such as the memory that this is New Year's Eve, a day that is popular for fireworks.

This is the "slow way," which produces an exact representation of the situation. The result is a dewarning process—inihibitory signals are being sent to the lateral and basal aspects of the amygdala, which then is promptly deactivated. From the sensory cortex the signals have now also been sent to the prefrontal cortex, which is linked with attention and conscious action planning, and which now monitors the situation and decides that no action is required. This entire process has only occupied a few seconds.

All of this would have continued very differently, of course, if the detailed analysis had indicated that the bang was an exchange of gunshots. The amygdala would not have been deactivated in that case but would have continued to direct current activity and quickly initiate defensive or escape behavior. It would have been increasingly supported by the results of the detailed analysis, of conscious attention, and conscious decision making. But these functions require time and by now one could have been already dead. The functions of the amygdala are essential for survival. This should not be forgotten when one considers the many undesired effects that the amygdala can have for mental functioning. We will encounter many of these unpleasant effects later on.

Among the stimuli that the amygdala is predisposed to respond to particularly strongly are fearful, irritated, and angry faces (Breiter et al., 1996; Morris et al., 1996). This holds true even when the faces have not been perceived consciously (Morris, Öhman, & Dolan, 1999; Whalen et al., 1998). When the face is shown only for a very brief moment, for a few milliseconds, and then one presents a significantly longer stimulus, the face is not consciously perceived. This is referred to as *masking* of the first, preceding stimulus. Based on these findings we can be certain that in psychotherapy, the patient's amygdala will respond to even the tiniest sign of anger in the facial expressions of therapist, even when this expression was displayed only very briefly and might not have been noticed consciously by the patient. Such nonconscious perceptions that nevertheless significantly influence behavior are an essential component of nonverbal interaction regulation in therapy (Krause, 1997; Merten, 1996).

Faces do not even have to look angry in order for the amygdala to perceive them as a threat. White Americans react to faces of unfamiliar African Americans with increased amygdala activity; that is, they unconsciously perceive them to be a threat (Hart et al., 2000). Phelps et al. (2000) showed that the strength of the amygdala reaction is significantly linked with the racial prejudice that the White participants had against Black individuals. In this case, the prejudice had been measured not by questionnaire but with an implicit, more subtle measure.

If we generalize from this finding, there is a tendency to project one's own negative attitudes or hostile impulses onto other persons. When this other person is then encountered, the amygdala reacts as if this person intends to aggress against us. These processes tend to occur unconsciously because the activity of the amygdala is not always accompanied by conscious awareness. These unconscious processes have an effect not only on the relationship with the other person, but they also have an intrapsychic (i.e., mental) effect. An enduring activation of the amygdala due to stimuli that are perceived as threatening can lead to a chronic stress reaction, which in turn has rather noxious consequences. More about this at a later point.

In light of such undesirable effects we must remain mindful of the great benefit conferred by the sensitivity of the amygdala to facial expressions. People with a damaged amygdala are no longer able to recognize the emotions that others are attempting to express with their face or their voice (Scott et al., 1997). This is a catastrophe for interpersonal relationships. Such people trust everybody equally, even when mistrust and prudence would be indicated, and this naïvete tends to create great difficulties for them (Adolphs, Tranel, & Damasio, 1998).

The amygdala does not become active only when something surprising and potentially dangerous happens suddenly, or when stimuli appear that humans by nature tend to perceive as dangerous, such as angry faces, snakes, and so forth. As soon as such naturally amygdala-activating stimuli are repeatedly paired with other stimuli, the amygdala will react to these previously neutral stimuli as if they constitute a threat. This is linked to the mechanism of classical conditioning. Over the years, LeDoux (2002, 2004) has examined the role of the amygdala in the formation of conditioned fear reactions so thoroughly that only a few questions remain unanswered.

2.10.2 The Conditioning of Fear Reactions

Based on our earlier considerations, it should be easy to envision by now what transpires in classical conditioning at the level of the synapses. I will reiterate these points briefly, using Figure 2–20 to illustrate my points.

Let us assume that a moderately loud tone is impacting upon the brain. It activates Neuron A in the auditory thalamus (Input A in Figure 2–20). This neuron is not able by itself to trigger a reaction in a neuron of the amygdala (C in Figure 2–20). Now, following the tone, a painful electric shock occurs (Neuron B in the somatosensory thalamus). The activation of Neuron B immediately triggers the activation of Neuron A in the lateral part of the amygdala, which receives its input from the sensory thalamus. Now the conditions of the so-called Hebbian plasticity are fulfilled: When A and C fire together, their connections are strengthened. After a few joint activations of Neurons A and B, the synaptic connection has become so strong that Neuron A alone can activate C. Hearing the sound alone, without the electric shock, is now sufficient to activate the amygdala.

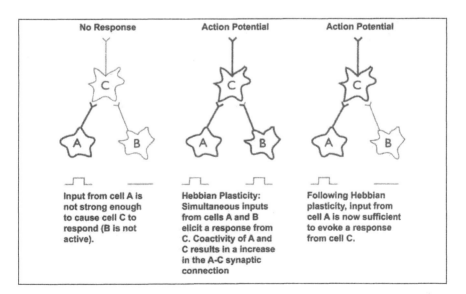

Figure 2–20. Development of classically conditioned reactions based on the model of Hebbian plasticity. Explanations in the text (Adapted from LeDoux, 2004, p.215).

Let's take another look at what happens at the synapse between Neurons A and C. While Neuron A is activated by a sound, it releases glutamate at its axon terminal. However, this is not sufficient to activate the AMPA receptors on the dendrites of Neuron C. The GABA inhibition prevents this from happening. Now Neuron C is being activated by the shock from Neuron B. This eliminates the magnesium that blocks the NMDA receptor of Neuron C and in addition opens the ion channels of these receptors in addition to the AMPA receptors. Now calcium can flow into the postsynaptic cell and can trigger a second messenger process (cAMP-PKA-CREB) within Cell C (see also Figures 2–4 and 2–5 and the corresponding explanations following). This cascade of chemical reactions within the cell results in the activation of genes in the cell nucleus by transcription factors, and these genes then produce certain proteins, which in turn strengthen the connection between the presynaptic and postsynaptic neurons; that is, between A and C. The sound from now on triggers the same reaction as the electric shock did previously.

Of course in real life it is not electric shocks that activate the amygdala; instead, angry faces, a surprise attack, a verbal threat, or a humiliating experience are potent triggers. Furthermore, it is never an isolated stimulus that indicates the threat, such as a pistol or an unpleasant remark. Such events always occur in a context. A rat that was always conditioned in one particular cage by receiving an electric shock after a brief tone will respond with fear not only to the tone, but also to the cage. It will avoid spending time in that area as much as possible. The context has been conditioned as well. A woman that was attacked in a parking garage will most likely not only fear attacks in the future but also feel considerable discomfort in parking garages, which she will thus seek to avoid.

Such context-conditioning always transpires via the hippocampus. The hippocampus specializes in the processing of relations and configurations in space and time. It provides context information for the amygdala in emotional learning situations. It relays its signals not to the lateral but to the basal nucleus of the amygdala. The basal nucleus communicates with the central nucleus of the amygdala and thereby modulates the reaction on the input from the lateral part. A tiger in the wild will trigger the wild firing of the central nucleus; a tiger in the zoo will at best trigger a much weaker activation of the central nucleus, which might be experienced as a feeling of fascination.

We have learned a lot about the relations between the hippocampus and the amygdala from studies on patients with certain forms of brain damage. Bechara et al. (1995) reported on a classical conditioning experiment with a patient whose amygdala was malfunctioning, although the hippocampus was intact. They also studied a second patient who had the opposite kind of brain damage. Both of them were presented with series of pictures and words. One of these stimuli was followed immediately with a loud fog horn sound, as an unconditioned stimulus to trigger a reaction of the amygdala. The amygdala's output, in turn, was measured as a change in skin resistance. After a few learning trials, the patients began to react markedly differently when the series was presented without the sound of the fog horn: The patient without the amygdala but with intact hippocampus was able to specify exactly when the fog horn could be heard before, even though he had no change in galvanic skin response (GSR) at this moment. With this patient, no conditioning had taken place. The other patient could not say when the fog horn could be heard (even though both patients were able to hear the sound equally well). Nevertheless, at the relevant moment, he had a strong GSR. His amygdala "knew" with precision at which moment the fog horn had been introduced. This rare combination of two cases with isolated lesions in the hippocampus and amygdala showed very clearly that fear conditioning takes place in the amygdala. This has important consequences for psychotherapy: Fear reactions can be acquired without the involvement of conscious awareness. The fact that fear reactions and fear conditioning significantly involve the amygdala has also been confirmed by brain imaging studies (fMRI; LaBar, Gatenby, Gore, LeDoux, & Phelps, 1998; Morris, Öhman, & Dolan, 1998). In the study by Morris et al. (1998), the conditioned stimulus was even presented in a masked fashion, so that it could not possibly be perceived at a conscious level. Nevertheless, the conditioned reaction developed in full form.

The amygdala, then, responds not only to threatening stimuli (even when they are not consciously perceived), but it also is involved in the formation of new fear reactions through classical conditioning and context conditioning, regardless of whether the person experiences this learning on a conscious level. The person does not know why he or she responds in a specific situation with anxiety; he or she does not know what it is that elicits the fear. It is indeed likely that the person would be unable to tell us that he or she is anxious at all because that would be a consciously experienced feeling. The reaction of the amygdala could be evident simply in an increased activation of the autonomic nervous system and an increased release of hormones, resulting in noxious physiological effects and psychosomatic symptoms—even though the person does not consciously experience anxiety at all.

Anxiety reactions can therefore develop even when conscious attention is turned away from the trigger of the anxiety. The triggers of fear in real life are typically of an entirely different kind than those in conditioning experiments. Strong anxiety is primarily elicited when important motivational goals are threatened; for example, when an important loss is looming, when something unpleasant is heading our way—something that we would like to avoid at all costs. It is entirely unclear whether our conscious attention turns toward these sources of potential anguish and uses its power of conscious decision making in order to avoid the unwanted outcome or minimize potential harm. From a rational point of view, this would probably be the best option. However, we are often highly irrational. In such situations, we often avoid all the more focusing directly on the threat because this would then consciously trigger the feeling of strong anxiety.

The neurons of the amygdala will fire, however, regardless of where we focus our attention. In such situations it is much more likely, compared to conditioning experiments, that the threat to our goal attainment will continue over longer time periods, which then will be accompanied by chronic firing of the amygdala, ultimately resulting in harmful consequences. The health-undermining effects of a chronically activated amygdala will elaborated in more detail later.

It is apparent that the processes involving the amygdala and its links with other neural–mental processes are highly relevant for psychotherapy. The amygdala is being activated by many events, and not only by external threats. More important, the amygdala tends to be chronically hyperactivated by the meanings we perceive with regard to our life situations—whether we perceive important motivational goals to be threatened. Neither these meanings nor the motivational goals themselves have to be consciously accessible; the amygdala fires regardless of our awareness. Such situations are much more common in psychotherapy than situations in which the amygdala would be activated in response to thieves, evil faces, or other types of impending doom from the external world. In chapter 4 I will elaborate on this motivational aspect of neural functioning.

The possibility that the amygdala can be activated without the presence of conscious awareness should not lead us to forget, however, that anxiety is also experienced very much consciously as a feeling state. The amygdala is of course also principally involved in this process, but the phenomenon is generally linked to a broader neural circuit.

2.10.3 Anxiety as a Feeling

Let us assume that a woman was attacked some time ago in a parking garage. Now, after considerable time has passed, she has once again developed the courage to revisit the parking garage. It is much emptier than she expected; she is all alone. Suddenly she can hear steps coming toward her. At this point, the current sensory experiences, the explicit memory content that was formed earlier via the hippocampus, and the conditioned fear reactions that are stored in the amygdala are all activated simultaneously. They are being integrated in working memory, the location of our conscious experience, and thus form an integrated experience. The result is an intense feeling of fear or anxiety. The fact that the woman experiences anxiety at this moment is closely linked with the focus of her current attention.

Another woman without the traumatizing history might enter the same parking garage, and the fact that it is so deserted might trigger no more than pleasant surprise. Her working memory is entirely occupied by her thoughts about what she might buy and cook for the guests that she expects to join her for dinner tonight. She is not even consciously aware of the steps moving toward her, even though her auditory system is encoding them. Her conscious attention is directed toward other things.

This example shows that explicit memories as well as emotional reactions involve intensive processing by the amygdala, which can conquer working memory and thereby determine the contents of our conscious experience. This then determines the focus of our attention. This system makes quite a bit of sense. Systems that would recurrently enter the very situations that are linked with negative prior experiences would not be able to survive. The amygdala and the hippocampus tend to work well together in situations requiring defense against danger, and it is necessary that they are able to conquer conscious awareness in moments when previous experience—encoded in the different memory systems—suggests that danger is imminent. Consciously experienced anxiety, therefore, is part of a biologically sensible danger defense system. It enables us to mobilize all the different capacities that are linked with conscious awareness, such as focused attention, deliberate decision making, planning, and rational action, in order to defend against danger.

Objectively speaking, both women described previously are faced with the same degree of danger versus safety. However, the woman whose amygdala alarmed her and whose hippocampus oriented her would be prepared in case the steps really signaled the approach of a man about to attack. She might perhaps flee to her car and, with trembling hands, activate the safety locks of the doors. The other woman would indeed be completely overwhelmed by a sudden attack.

If, objectively, there were no danger, if the steps came from a completely harmless person, then the second woman would be in a better position. She would happily finish her chores and would have wasted no time and energy for an objectively unnecessary danger defense. Let us not judge here which of the two women acted more rationally or reasonably—this could undoubtedly lead to lengthy discussions. What is experienced by someone as a feeling in such a situation is not dependent on rational deliberations but instead is influenced by prior experiences stored in the amygdala. The amygdala, in turn, cannot be easily influenced by rational argument because it cannot be controlled as easily by working memory, unlike motor behavior, for example.

This phenomenon can be explained by the imbalance in the projection paths between the amygdala and other brain regions. There are far more projecting paths away from the amygdala to other brain regions rather than the other way around (Amaral, Price, Pitkänen, & Carmichael, 1992). The amygdala can therefore influence other brain systems by whatever happens to trigger its activation. It can direct conscious attention to emotionally relevant stimuli by concentrating the short-term object-buffers in the sensory memory to such stimuli. It can also activate long-term memory, and it can conquer working memory via the strongly developed projection paths via the cingular and the orbital cortex (see Figure 2–21).

One important component, however, is still missing from the explanation of how a perception, memory, or imagination can be felt as a feeling. This missing piece is arousal. Whenever we have a strong feeling, the cortex as a whole is strongly aroused. The amygdala has reciprocal connections with the arousal center in the brainstem. A particularly strong connection, however, is the one with the neighboring nucleus basalis, whose neurons release primarily the transmitter acetylcholine. When the basal nucleus is activated, it releases acetylcholine throughout the entire brain, which triggers a nonspecific arousal increase. Everything that happens in the brain now transpires "hotter" than it did before. The cortical centers that process the current situations are oversensitized.

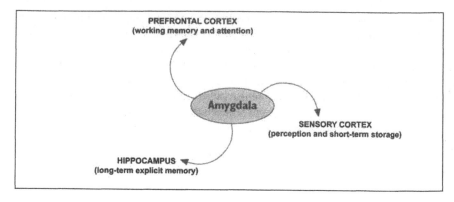

Figure 2–21. Some cortical outputs of the amygdala and their function. Areas of the amygdala project to a wide variety of cortical areas. Included are projections to all stages of cortical sensory processing, to prefrontal cortex, and to the hippocampus and related cortical areas. Through these projections, the amygdala can influence ongoing perceptions, mental imagery, attention, short-term memory, working memory, and long-term memory, as well as the various higher-order thought processes that these make possible (Adapted from LeDoux, 2004; p. 287).

The arousal systems, in turn, have connections of their own with the amygdala and thereby contribute to the maintenance of their arousal. A positive feedback loop is created; a self-sustaining emotional state. That is also the reason why it takes some time for an emotional arousal to calm down after a change of the situation. The arousal itself, however, is nonspecific. It depends on the contents of working memory how this arousal is interpreted; that is, which feeling is being experienced at that moment. Schachter and Singer (1962) postulated such a multidetermination of feelings a long time ago, but they were not able to link their model with such an elaborate neural foundation as is possible today.

Even though things are already complicated enough by now, we should still consider additional circuits relevant to the feeling of anxiety (see Figure 2–19). When the amygdala is activated, bodily reactions also follow. The motor system responds with freezing, a fearful facial expression, trembling, fight or flight. The autonomic nervous system responds with an increase in blood pressure, increased heart rate, and sweat release. Moreover, hormonal chain reactions are triggered. Adrenalin, cortisol, and several peptides are released into the bloodstream. All these bodily reactions, in turn, feed back into the brain. Damasio (1999) termed these feedback signals of bodily states *somatic markers*. According to his research, it is primarily the orbitofrontal cortex that is involved in the transformation of these body signals into somatic markers. According to Damasio, somatic markers give affective meaning to objects, situations, and events. The orbitofrontal cortex associates memory traces of past events that have high affective meaning with representations of bodily states that were elicited originally by the events. The event is thus labeled with this somatic state or marker. This is what enables us to make intuitive decisions—as it were, "from the gut"—based on previous experiences. Indeed, several research areas suggest that somatic markers influence our subjectively experienced feelings.

Ekman (1993), for example, instructed his participants to move certain facial muscles. They were unaware that these were precisely the muscles that are activated during the expression of particular feelings. Then, he asked them to rate their mood. They estimated their feelings to be much more positive when they previously moved the muscles that are characteristic of a positive emotion, compared to the condition in which their faces made the expression of a negative emotion. This suggests that it might be helpful to "put on a happy face" if one aims to improve one's mood.

Valins (1966) provided false feedback about participants' heart rate while he confronted them with anxiety-eliciting stimuli. At first he allowed them to listen to their actual heartbeats, which were rather rapid, as would be expected when people are confronted with anxiety-eliciting situations. Then he manipulated the heart rate feedback in such a manner that the participants heard how their heart rate became calmer. As a result, they estimated their anxiety to be lower.

According to Damasio's results, the influence of these feedback signals transpires via the orbitofrontal cortex and thereby leads to intuitive decisions that are often associated with a feeling of safety. In fact, people with lesions in the orbitofrontal cortex have difficulties with decision making, especially in situations that are too complex and bewildering to allow for a detailed logical analysis.

Indeed, the feeling of anxiety makes it especially difficult for us to imagine that the feeling without the bodily reactions and their resulting feedback signals would be quite the same. The amygdala is of central importance for these effects on the subjective feeling of anxiety. If the amygdala malfunctions, the next link in the chain will fail as well. Thus, the patient in the example that was described earlier in the study by Bechara et al. (1995) failed to develop a conditioned autonomic reaction when he was confronted with a loud fog horn as a shock-stimulus—for the reason of his bilateral amygdala damage,

The feeling of anxiety, then, requires at least four components:

- Working memory must be involved. This is not anxiety-specific. Without working memory, conscious experience is not possible.
- The amygdala must be involved.
- An activation of the arousal system must have occurred.
- A feedback of fear-specific bodily reactions must occur.

We see, then, that feelings of anxiety have a very broad basis in the brain. Anxiety feelings are similar to all other neural events: The more often they transpire, the more firmly they become facilitated or ingrained. What about anxiety feelings that have been facilitated so well that they can be triggered extremely easily or can even occur spontaneously? This question leads us toward the domain of mental disorders. When a person recurrently experiences anxiety, especially in situations that are not linked with anxiety for most people, and when this person suffers because of this, anxiety can become a clinical problem. The central question for psychotherapy becomes, then

2.10.4 Can Anxiety Be Extinguished?

Because of the imbalance of the projection paths between the amygdala and the prefrontal cortex (PFC), our emotions influence the contents of our consciousness more
than we can consciously exert control over our emotions. Many people would be
thrilled if they were suddenly able to control feelings of anxiety more effectively.
That does not work as well as perhaps one might wish because in its developmental
history, the brain has not yet developed prominent projections from the areas in the
PFC that are linked with conscious decision making to the amygdala. To be sure,
there are some paths from the PFC, especially the orbitofrontal aspect, to the amygdala, and these paths do allow for a certain degree of conscious emotion control
(Jackson, Malmstadt, Larson, & Davidson, 2000). However, in order to achieve an
effective control of strong anxiety reactions, a detour becomes necessary. This also
transpires via the frontal lobes, but not in the sense of a conscious-decision act, in the
same way in which we can choose to lift our arm, for instance, if we wish to do so.

The fact that we cannot control everything by sheer willpower is part of our common
human life experience. A person who has not engaged in sports for a long time and
is also obese will not be able to run several kilometers, regardless of how much he
consciously decides that he wants to do so. After a short time, he will have to give up
in utter exhaustion. Nevertheless, a detour would be possible and allow him to reach
his goal via a conscious-decision act. He would have to begin to train regularly, several times per week, and to eat less and more healthily. After a period as short as 10
weeks, the short distance that he almost could not run at the beginning would become
"a walk in the park," as it were. He would have reached indirectly the goal that he
couldn't reach via a more direct path. He was able to exert cortical control over something that originally did not appear to be consciously controllable. The difference was
that he used targeted and repeatedly orchestrated learning experiences.

This detour, in principle, also allows us to achieve a deliberately initiated change in
the experience of anxiety reactions. Let us assume that a study participant has been
classically conditioned, such that a specific tone was repeatedly paired with an electric shock. After many such pairings, she responds to the tone alone with an increase
in GSR. She cannot consciously suppress this reaction of her autonomic nervous system. Even if we told her that the tone will now not be followed by an electric shock,
she would continue for several more trials with a GSR. However, these responses
would become increasingly weaker, until they cease entirely. This we would term
extinction.

Because the unpleasant anxiety reaction initially remains in place, the study participant, given the choice, would probably not volunteer for any additional trials but
would choose to avoid the situation. This behavior results from the effort to gain control over the situation. The effort, to do anything possible to gain control over the fear
plays an important role in the formation and maintenance of anxiety disorders. The
anxiety reaction would be maintained, in this case, by the avoidance behavior. In real
life, outside of the laboratory, an extinction of anxiety reactions can therefore only

occur when the person voluntarily exposes him or herself repeatedly to the anxiety-eliciting situation and consciously endures the anxiety until it gradually weakens. However, in real life, one typically has no guarantee that nothing terrible will happen when we expose ourselves to feared situations. The anxiety itself can be experienced as something so terrible that no additional aversive stimulus is necessary in order to further facilitate the link between the situation and the fear. Thus, one never knows for sure, as it were, whether the electricity has really been switched off. The woman in the parking garage can never be completely certain whether she might be attacked once again. If the unconditioned aversive stimulus occurs even just a single additional time (intermittent reinforcement), then it would take substantially longer for the anxiety reaction to become extinguished. This is why anxiety reactions, under natural conditions, tend to remain in place quite stubbornly over long periods. Extinction would require a forced and repeated confrontation with the anxiety-eliciting situation, without efforts to control the anxiety, and without repeated confrontation with the feared consequences. Orchestrating such corrective experiences often requires the help of therapeutic intervention, although in principle, such an extinction of anxiety reactions can also be implemented by the person him- or herself.

What happens on a neural level during such an extinction of anxiety? What happens with the well-facilitated synaptic connection potentials in the amygdala? Are they simply obliterated, as it were, or do they fall apart like the memory traces of a telephone number that we haven't used in a long time? The time periods that are required to achieve extinction are clearly too short for ordinary forgetting to take place. And how should one imagine the process of "obliteration," based on that which we already know about the synaptic transmission of action potentials? Both of these processes also seem implausible because the anxiety reaction, once extinguished, can recur in full strength once the unconditioned stimulus appears again, or once the person is placed back into the context in which the conditioning originally took place. The activation pattern, once formed, can be activated again very quickly, suggesting that it must still be present in deactivated form.

These considerations suggest that the process of extinction is based on the development of active inhibition. Once they are formed, the connections within the circuits that are involved in the anxiety reaction remain in place, but they are actively inhibited. This inhibition originates from the medial prefrontal cortex (anterior cingular cortex and orbital region). In rats, lesions in certain parts of this region lead to strongly exaggerated anxiety reactions to conditioned stimuli; lesions in other regions can either delay or completely prevent the extinction from occurring (Garcia, Vuimba, Bandry, & Thompson, 1999; Morgan & LeDoux, 1995; Quirk, Russo, Barron, & Lebron, 2000). The medial PFC and the amygdala seem to have a reciprocal relationship with one another. Strong activation of the amygdala switches off the PFC, but vice versa, the activation of other brain regions originating from the amygdala can be inhibited via the PFC. The respective parts of the PFC are critically involved in the process of adjusting behavior in changing contexts, not only by inhibiting anxiety responses when they are no longer needed, but also in the process of regulating other emotions and in the modulation of other actions in response to changing situational demands. This aspect of social behavior can be described as *comportment* and deficits in this aspect of social functioning have been found among patients with orbito-frontal lesions or damage to the head of the caudate nucleus (Mesalum, 2000). Some PFC lesions specifically lead to strongly perseverative behavior (Damasio, 1999; Petrides, 1994).

LeDoux (2002, 2004) strongly argued that in the process of extinction, it is not the implicit memory stored in the amygdala that is being extinguished. Instead, the effects triggered by this emotional memory are being inhibited by the PFC. "Unconscious fear memories that are formed by the amygdala appear to be etched permanently into the brain" (LeDoux, 2004). This view is supported not only by LeDoux's own findings but also by a study conducted by Gutberlet and Miltner (1999). They found that after a successful therapy of spider phobia, the behavior, sub-jective feeling, and autonomic nervous system reactions (skin conductance and heart rate) did not show any fear reaction; however, the EEG (electroencephalography)-observed brain activity (P3 amplitude) continued to be as different in the patient, compared to a normal control person without the phobia, as had been the case prior to the therapy. The authors conclude from this that previously mentioned quick-route amygdala activation, without the cortical processing of the stimulus, can still take place. However, the slower route now contains a cortical inhibition of the subsequent reaction chain. A successful behavior therapy builds new structures in that location, which ultimately inhibit the anxiety reaction effectively. One could indeed say that the fear reaction has been removed because it can no longer be detected in the per-son's experience and behavior; it can only be measured now via a recording of brain activity. Without such measurement, we would not even know that traces of the old anxiety readiness remain in place.

LeDoux was the most thorough about the question of what happens at the neural level during the extinction of fear reactions. In this research, he made a very interesting observation that appears to have general relevance, which is why I would like to let him describe this in his own words:

> Recently I had a scientific 'ah ha' experience, one of those rare, wonderful moments when a new set of findings from the lab suddenly makes you see something puzzling in a new, crystal clear way. The studies involved recordings of electrical activity of the amygdala before and after conditioning by Greg Quirk, Chris Repa and me. We found dramatic increases in electrical response elicited by the tone CS after conditioning, and these increases were inversed by extinction. However, because we were recording from multiple individual neurons at the same time, we were also able to look at the activity relationships between the cells. Conditioning increased the functional interaction between neurons so that the likelihood that two cells could fire at the same time dramat-ically increased. These interactions were seen both in response to the stimulus and in the spontaneous firing of cells when nothing in particular was going on. What was most interesting was that in some of the cells, these functional interactions were not reversed by extinction. Conditioning appears to have created what Donald Hebb calls "cell assemblies," and some of them seemed to be resistant to extinction. Although the tone was no longer causing the cell to fire (they had extinguished), the functional interaction

between the cells, as seen in their spontaneous firing, remained. It is as if these functional couplings are holding the memory even at a time when the external triggers of the memory (e.g., phobic stimuli) are no longer effective in activating the memory and its associated behaviors (e.g., phobic response). Although highly speculative at that point, the observations suggest clues as to how memories can live in the brain at a time when they are not accessible by external stimuli. All that it would take to reactivate those memories would be a change in the strength of the input to the cell assembly. This might be something that stress can accomplish. (LeDoux, 2004, p. 251)

At another point, LeDoux reports on studies in which he and his colleagues found that stress and stress hormones can strengthen conditioned fear reactions. He follows up on these findings with these speculations:

The finding that stress hormones can amplify conditioned fear responses has an important implication for our understanding of anxiety disorders, and in particular for understanding why these sometimes seem to occur or get worse after unrelated stressful events. During stress, weak conditioned fear responses may become stronger. The responses could be weak either because they were weakly conditioned, or because they were previously extinguished or were otherwise treated into remission. Either way, their strength might be increased by stress. For example, a snake phobic might be in remission for years but upon the death of his spouse the phobia returns. Alternatively, a mild fear of heights, one that causes few problems in everyday life, might be converted into a pathological fear and the amplifying of stress. The stress is unrelated to the disorder that develops and is instead a condition that lowers the threshold for an anxiety disorder, making the individual vulnerable to anxiety, but not dictating the nature of the disorder that will emerge. The latter is probably determined by the kinds of fears and other vulnerabilities that the person has lurking inside. (LeDoux, 2004, p. 247).

There is no obvious reason for why the ideas that LeDoux articulates here, about the bases of mental disorders in the form of "latent" cell assemblies and about the relations between stress and mental disorders, should be valid only for the anxiety disorders. This logic appears at least equally plausible for depression and perhaps several other disorders. Latent cell assemblies can be "reawakened," as it were, by experiences of chronic stress, which can make it difficult to see a direct link between the type of stressor and the specific kind of disorder. These situations are very common in clinical practice. LeDoux has arrived at these explanatory possibilities via the path of basic scientific research. It could be that basic neuroscientific research can help us in this domain to solve one of the great puzzles facing clinical psychology and psychiatry—a puzzle that has inspired all kinds of wild theories over the years. In his years of studying the acquisition and extinction of anxiety reactions, LeDoux might have encountered something that is very broadly applicable but at the same time is very difficult to see and substantiate scientifically.

In chapter 4, I will establish the link between the concept of stress and the person's motivational system, and I will define stress as a form of motivational incongruence. Many studies suggest that there is a strong correlational association between motivational incongruence and the occurrence of mental disorders, which will be explained in detail in chapter 4. One of the most important types of stress occurs when important motivational goals are threatened or prevented from being attained, and the previous passages by LeDoux point to a neuroscientifically plausible pathway that shows how the link between motivational incongruence and mental disorders might be established.

The question remains, however, how the cell assemblies that are the basis of mental disorders are formed in the first place. This was not an issue in LeDoux's studies because he was dealing with experimentally conditioned fear reactions. But how are such cell assemblies formed under naturalistic conditions? We cannot assume that all mental disorders are created via the process of classical conditioning. During classical conditioning, reactions that are in principle already available within the repertoire of the individual are linked with new signals. However, many mental disorders include forms of experience and behavior that are not part of people's normal repertoire. They have an emergent quality. We have to explain, then, how entirely new cell assemblies can be formed. This is a question that addresses not only mental disorders but also the formation of all other new forms of experience and behavior. What is the basis for such new formations? This question is discussed in detail in chapter 4 (sections 4.8 and 4.9).

Before closing this section about the neural circuits of fear and anxiety, I should note that my discussions here do not really address the full spectrum of clinical anxiety disorders. What was discussed here is applicable in the clinical domain primarily to phobias. According to Panksepp (1998), there is convincing evidence to show that panic disorder, generalized anxiety disorder, and obsessive–compulsive disorder involve neural circuits that partially overlap with the ones I described here, but that also include other neural structures. The general conclusions we can formulate for psychotherapy, however, are not affected by this caveat.

2.11 CONCLUSIONS FOR PSYCHOTHERAPY

If we want to arrive at some therapeutic conclusions based on that which was discussed here about the changeability of anxiety reactions, then we can initially note that the term *extinction* is a misleading expression. It should be consistently replaced by the term *inhibition*. In an effective anxiety treatment, the anxiety is not extinguished; it is inhibited. For the practice of therapy, it makes a considerable difference whether I as the therapist aim for extinction or inhibition.

If my goal is to extinguish an anxiety reaction, then I have to trigger it repeatedly while at the same time no truly awful event (e.g., the appearance of an unconditioned aversive event) occurs. What matters primarily is that the patient is exposed sufficiently often to the fear-eliciting situation until the anxiety reaction becomes increasingly weak. Behavioral habituation training corresponds the most closely to this rationale. However, because we know by now that the amygdala reaction doesn't really habituate, we can no longer assume that habituation is the mechanism of action that explains how this therapeutic intervention operates. Because we know that this intervention tends to work quite effectively, we can infer that it leads to a cortical inhibition of the anxiety even though it is not specifically intended to work via this pathway. This could happen, for instance, when the patient, based on the instructions he receives and the experiences he has, arrives at conclusions that alter his understanding of the situation in such a way that a cortical inhibition of the anxiety is formed. His competency expectancies might slowly increase, and his worries of what might happen turn out to be unfounded. Cognitions are formed (as interpretations of that which is really happening) that inhibit the occurrence of the anxiety.

If we conceive of therapy from the very start as a process of inhibition, then we would approach this situation very differently. We would consider how we might shift the brain of the patient toward a state that is maximally incompatible with the fear before we then expose him to the anxiety-triggering situations. This might include

- A secure attachment relationship with a therapist who conveys competence, understanding, and personal engagement.
- Providing the patient with a maximum of perceptions that have a positive value for his most important motivational goals.
- The explicit activation of important positive motivational goals of the patient, which shift him toward a maximally strong approach orientation.
- Structuring of the therapeutic situation in such a manner that the positive abilities and resources that the patient brings to therapy can be utilized.
- Triggering of a maximum number of positive emotions.
- Increase of competency expectations regarding the impending anxiety confrontation, via appropriate instructions (cognitive rehearsal—What might happen? What can I do in that case?); clarification of expectations and concerns; provision of maximally convincing models who have already succeeded; or personal modeling with guided participation (Bandura, 1977).
- Practicing of maximally positive self-statements and anxiety-mastery thoughts, which the patient can easily activate in the feared situation.

If a maximum number of these conditions are simultaneously activated, the brain of the patient will be in a state of low anxiety readiness. In this state, then, the repeated exposure to pertinent experiences that have previously triggered the patient's anxiety should be initiated. The goal is the formation of a maximum number of well-facilitated anxiety-inhibiting synapses. This requires a sufficient number and sufficient intensity of learning trials over an extended time period (the formation of new synapses via gene transcription requires time, as was discussed earlier), during which this combination of conditions is repeatedly realized.

If therapy is conceptualized from the beginning as the formation of anxiety inhibition, then the structuring of the context in which the confrontation with the feared situation occurs becomes the most important therapeutic task. The formation of an optimal cortical context in which the anxiety circuit then is triggered via confrontation with an anxiety-eliciting experience is one of the most demanding therapeutic tasks because it must be individually tailored for each patient. If one views therapy as "getting rid of something," then one faces an easier task than would be the case if one wants to replace something that currently exists (the anxiety reaction) with something new. "Putting something else in place" is more difficult than "getting rid of something," but it is also more effective.

The conclusions that emerged here from the neuroscientific fear research, of course, go well beyond the territory that was covered by the studies I cited earlier. These citations had nothing to do, in fact, with therapy as such. It is also clear that the anxiety therapists often face during a course of therapy cannot always be so neatly tied to specific situations as was the case in the previously mentioned studies. The anxiety states that come up in therapy often have to do with the threatening of important motivational goals held by the patient. This aspect is completely ignored in the research that I mentioned earlier. The motivational perspective, which is still missing here, will be added in chapter 4. After that, it will make more sense to present broader summary conclusions for clinical practice.

The example of anxiety provided a good opportunity, however, to show here that specific areas of neuroscientific research can lead to clinically relevant conclusions for psychotherapy. Based on that which I have discussed here for the example of anxiety, we can already generalize: The task of psychotherapy is very often to inhibit something problematic because "getting rid of" or "eradicating" simply doesn't work. This leads to a shift of the focus from that which is problematic to that which should be put in its place instead. Inhibition means, after all, to put another neural process in the place of the to-be-inhibited process. This other neural process should trigger the inhibition. However, that which is to trigger the inhibition must first be actively created. It must be activated and facilitated with maximum frequency, so that it becomes more easily activated than the to-be-inhibited neural processes, whose neural circuits—as we have seen—remain intact. Because they remain in place, they must be kept in check via maximally facilitated other processes that have a positive function and effect. On a neural level, to get rid of something means to replace it with something. That which is to accomplish the inhibition must initially also be activated and facilitated. The main task of the therapist, then, consists of activation and facilitation—and in order to fulfill this task, the therapist must have a good idea of what it is that should be activated and facilitated. These considerations ultimately lead us to a resource-oriented perspective. The therapist must spend more time contemplating what he or she wishes to facilitate and how this might be done, and less time contemplating that which he or she aims to remove. In chapter 5 (section 5.6), I will address these conclusions again in more detail.

2.12 INTENTIONAL ACTION

Until now, we have discussed circuits that transpire without conscious goal-setting and volitional control. Such circuits are very important for psychotherapy because in therapy we are often dealing with processes that are not consciously controllable. But one of the key characteristics of humans is, of course, their capacity for conscious thought and action. This holds unconditionally true just as well for psychotherapy patients. Conscious reflection and communication, as well as intentional action, are the most important foundations of the therapeutic change process, even though unconscious processes also play an important role in this context. However, many people also enter psychotherapy because they haven't managed to solve their problems by their own self-initiated, goal-oriented action. Therapies whose primary strategy consists of instructing patients to solve their problems by their own independent, goal-directed action—the so-called problem-solving therapies—are among the most effective interventions of all (Grawe, Donati, & Bernauer, 1994, p. 436). Goal-oriented action, then, is simultaneously a precondition and a resource for therapy. Deficits in this area, however, could also be viewed as the main problem to be targeted in therapy.

There are well-established psychological models to conceptualize goal-oriented action. G. A. Miller, Galanter, and Pribram (1960) introduced the concept of goal hierarchies, but they initially did not specify the precise levels of such hierarchies. Powers (1973) then provided a specific model of distinct hierarchy levels in the form of hierarchically organized feedback loops. Figure 2–22 shows such a goal hierarchy, targeted here to the goal-oriented action of "hosting guests."

Powers attempted as early as 1973 to specify the neural basis of these hierarchical feedback loops, but the state of neuroscientific research at that time allowed him to do this only for the lowest four levels of the hierarchy. However, he was firmly convinced that it should be possible at some point to specify in detail how the higher regulatory levels might be represented on a neural level. Today, 30 years later, we are much closer to this goal.

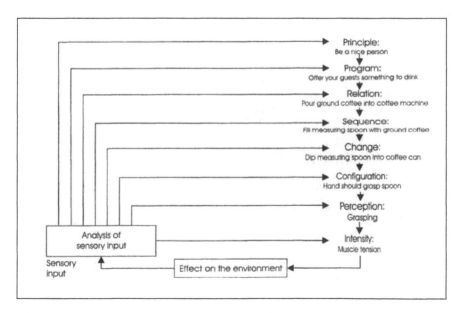

Figure 2–22. Example of a control theory view of a routine activity, which can be interpreted differently on each regulatory level. (Adapted from Grawe, 1998; p. 156).

2.12.1 The Neural Representation of Goal Hierarchies

Imagine for a moment that you were to wander aimlessly through your home and take a look around at whatever catches your eye. You might see a few unread books, begging to be picked up; you see unopened mail, which you could read and respond to; there might be a piano, waiting for you; a TV, which could be switched on; some dirty laundry here; a coffeemaker there; over here a bed, to take a nap; a telephone over there. ... The domain of all that you could do at any moment is vast. All of this has an element of appeal; otherwise you would not have placed it where it is. If you had to follow all of the appeals simultaneously, either chaos or paralysis would result in the brain. There must be something within the brain that chooses among the nearly infinite possibilities and that allows us to stick with a course of action once it has been selected. Brains that would not have such structures would surely be doomed.

For most animals, this problem never arises in this form. Their behavior is primarily determined by instincts. Their internal state (e.g., hunger) and the external conditions jointly determine their action. How they behave, in turn, is determined within relatively narrow boundaries by their species-specific behavioral repertoire.

The far greater flexibility conferred by our brains has only become possible because flexibility-promoting structures have emerged that bring order into our mental activity by allowing us to formulate and pursue goals. It is not surprising, in turn, that the part of the brain that most distinguishes us from other mammals is the area that governs goal-direction action: the prefrontal cortex (PFC). Figure 2–23 provides a rough overview of the frontal lobe.

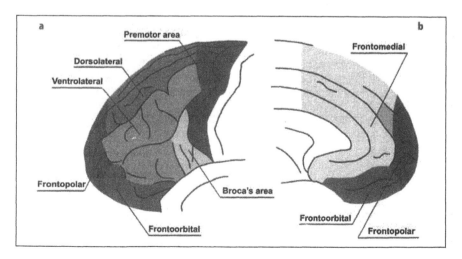

Figure 2–23. Schematic construction of subregions in frontal lobe. a. Lateral view b. medial view. Remark: The classification is strongly simplified and does not reflect our recent knowledge of the fine cytoarchitectural and functional differences of frontal subregions. (Adapted from Förstl, 2002; p. 29).

Among humans this part of the brain comprises 29% of total brain mass; by comparison, this value is 3.4% in cats, 6.9% in dogs, and 17% in chimpanzees. Based on this alone, we can suspect that this area of the brain has to do with specifically human qualities. The PFC performs this function in close association with other brain areas. It is extremely well-connected. Figure 2–24 shows the most important pathways within the PFC and between specific areas of the PFC and other parts of the brain.

The function of the PFC in the context of goal-oriented action has been the topic of E. K. Miller and Cohen's (2001) integrative theory of prefrontal cortex functioning. The following exposition is based primarily on their conceptualization.

The PFC consists of several areas that are closely interconnected. They send and receive projections from practically all of the sensory and motor systems and the subcortical areas, such as the hippocampus and the amygdala. The PFC exclusively receives strongly preprocessed signals about the internal state of the organism and the external world. They come from the various areas within the associative cortex. Basal information about surrounding contexts, such as color, contours, size, and movement is processed by the primary sensory cortex. Before such signals reach the PFC, they have already passed through different convergence zones. In the PFC—especially in the lateral and medial-dorsal aspect—multimodal information is being processed, that

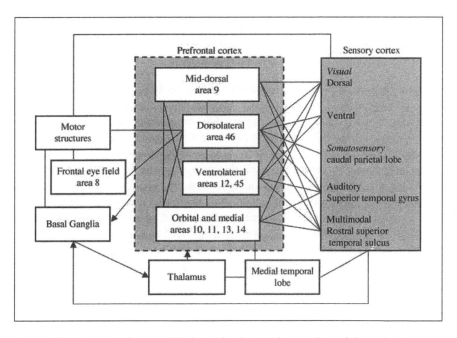

Figure 2–24. Schematic diagram of the internal and external connections of the prefrontal cortex. The convergence of input from many other brain areas and the close reciprocal connections within the PFC enable this area to play a central role in the synthesis of the input from many brain areas, which is particularly important for the generation of complex behavior (Adapted from E. K. Miller & Cohen, 2001).

is, situations with visuals objects, sounds, smells, and so forth. The neurons in this area do not respond to signals that are only visual, auditory, or somatosensory; they respond only when multiple sensory modalities stimulate the neurons simultaneously. They are specialized, then, for multimodal processing.

Similar processes hold true for the output of the PFC. It is closely connected with multiple systems that play a role in the control of voluntary motor behavior, but it is not linked directly with the primary motor cortex. The PFC controls behavior via hierarchical intermediate stations, in the same way that it receives its input via such hierarchical intermediate stations. The dorsolateral aspect, which contains the frontal eyefield, is important for the volitional control of eye movements, which is closely linked with the control of attention.

The orbital and medial PFC is closely connected with the hippocampus, the amygdala, and the hypothalamus. This enables a functional collaboration with long-term memory on the one hand and the affective and motivational system on the other. The orbitofrontal cortex, in close connection with the amygdala, plays a key role in the emotional evaluation of that which is currently being processed by the brain, and it determines the motivational direction of the processes in the sense of selecting an approach versus avoidance orientation.

Even more pronounced, however, are the reciprocal links within the PFC. They allow for a long-term maintenance of representations within the PFC, regardless of whether they currently receive input from the sensory modalities. This quality predisposes this area to the effective representation of long-term goals.

In order for goals to have a top-down influence upon behavior, the organism must have information about which behavior, under which conditions, leads to which consequences. It must, therefore, have knowledge about rules. This appears to be the exact specialty of the PFC. Watanabe (1990, 1992) trained monkeys in such a way that they received a reward, a drop of sweet juice on the tongue, when certain optical or acoustic signals appeared. He found that many neurons in the lateral PFC fired only when a sound signaled an impending reward. The sound alone was not able to activate the neuron. An analogue relationship was found with visual stimuli. Some neurons responded only when the reward was signaled by optical cues; others only responded to acoustic signals; and yet others responded to both types of information. Similar results have been obtained by Quintana and Fuster (1992), Bichot, Schall, and Thompson (1996), as well as Asaad, Rainer, and Miller (1998).

The researchers even found neurons that selectively responded only to even more complex if-then rules. In a study by White and Wise (1999), monkeys had to find out, in order to receive a reward, where a specific visual target object would appear. They briefly showed the monkey one of four stimulus patterns at one of the four possible locations. The stimulus patterns signaled where the target object would ultimately appear. However, this happened according to two different rules: In one condition, the *position* of the pattern signaled where the target object would appear; in another condition, the *type* of stimulus pattern conveyed this information. About half of the neurons in the lateral PFC whose signals were analyzed in this study responded differently based on which rule was in place (Asaad, Rainer, & Miller, 2000; Hoshi, Shima, & Tanji, 1998; Wallis, Anderson, & Miller, 2000). These neurons responded in a rule-specific manner, of course, because they were embedded within a larger cell assembly; they don't respond differentially as independent, isolated neurons. The rules are represented by cell assemblies, not by single neurons.

It might seem somewhat strange to use studies of monkeys when one investigates a brain area such as the PFC, which is supposedly linked with specifically human qualities. However, for ethical and technical reasons there are no studies about the response characteristics of single neurons in humans. The logic of these studies in monkey brains is that if we can show that, even in the monkey brain with its much less developed PFC, specific neurons exist for the representation of rules, then we can infer that in the human brain it is even more likely that neurons exist that specialize on such complex tasks. After all, much of our knowledge about the activation-transfer at the synapses comes from studies of even far less developed organisms, such as sea snails. The path to reach analogue conclusions about the human situation is even longer in that case. The logic behind these scientific studies is that the solutions that evolution has devised to solve adaptive problems, if they tend to work well, are likely to be passed on and form the basis for the emergence of even more complex structures. We will not be able to find out much about language, consciousness, or rational planning from the study of monkey brains. The studies summarized here about rule learning are in the boundary area where the generalizability of findings from animal studies to humans appears to be warranted. Beyond this boundary, however, we have to rely on other sources of information, such as observations among patients with organic brain disease or studies with EEG and neuroimaging technology. Without these newer research methods we would still not be able to say much about the function of the PFC or the associative cortex. We can have even more faith in conclusions based on analogue animal studies to the degree that their findings are supported by studies using other methods. In the case of the PFC and its importance for rule learning, such studies with other methods have indeed been conducted. Petrides (1990) was able to show that patients with lesions in the PFC had great difficulties with the learning of rules.

These findings as a whole show that the PFC is, among other things, specialized in the learning of rules. When a rule is activated, incoming perceptions that are of particular relevance for that rule are preferentially attended to. Similarly, relevant memories are preferentially activated, and relevant behavioral strategies are generated. The rule, therefore, governs attention, memory, and behavior. If one could obtain a PET image of the brain during a game of cards, one should—based on these results—find considerable activity within the PFC of the players.

Rules are typically applied over a longer time period. The neural representations of rules, therefore, must remain activated over a longer period, even when outside disturbances interfere or other actions become necessary at some point during this period. The fact that the rule-representing neural activity is maintained over a longer time period has by now been supported by many studies, including recent studies using functional neuroimaging in humans (J. D. Cohen et al., 1997; Courtney, Ungerleider, Keil, & Haxby, 1997; Prabhakaran, Narayanan, Zhao, & Gabrieli, 2000).

Rules are important to allow for the achievement of goals, but they alone are not sufficient for this purpose. The acting individual must also know the conditions in which a rule leads to goal attainment and those in which it does not. This feedback function is accomplished in the PFC by dopaminergic projections that have their origin in the ventral tegmentum and the nucleus accumbens. In the lateral as well as the ventromedial part of the PFC, neurons exist that become active when rewards are anticipated. Their activity correlates with the magnitude and expectancy of the anticipated reward (Leon & Shadlen, 1999; Tremblay, Hollerman, & Schultz, 1998; Tremblay & Schultz, 1999; Watanabe, 1996). The dopamine neurons fire especially when pleasant consequences appear suddenly; for example, earlier or more intensely than originally expected (Mirenowicz & Schultz, 1994, 1996). With increasing learning the dopamine neurons are activated ever earlier by events or stimuli that announce the reward or, in the case of goal-oriented action, by events that announce the attainment of the goal. Gradually, as the reward/goal attainment occurs as anticipated, the neurons become less activated (Schultz, Apicella, & Ljungberg, 1993). If the reward does not occur at the point when it is expected, the dopamine neurons are even actively inhibited (Hollerman & Schultz, 1998). The dopamine neurons, then, encode deviations from expectancies, in positive as well as negative directions (Montague, Dayan, & Sejnowski, 1996; Schultz, 1998; Schultz, Dayan, & Montague, 1997; Schultz & Dickinson, 2000; O'Doherty et al., 2000). The discrepancy signal, encoded by the differential activity of the dopamine neurons, provides feedback about which behavior leads to goal attainment and which does not. Therefore, it plays an important role in the formation of situation-specific strategies for goal attainment.

Once a goal hierarchy is well-established, it becomes increasingly automatized and requires less and less conscious attention. This tends to work well as long as the automatized strategies lead to goal attainment; that is, as long as the expected feedback signals are perceived. If the goals are not attained, however, this is fed back within the hierarchical feedback loops, as Powers conceptualized them, by a so-called incongruence signal. The differential feedback of goal attainment, as it was just described here, is one way "motivational incongruence"—the nonattainment of important goals, can be represented on a neural level.

Goal-oriented action is motivated action. There will be much more to say about the neural representation of the motivational aspects of mental functioning. This topic will be discussed in detail in chapter 4. At this point I would like to note, however, that there are neural circuits in the human brain that fulfill all the conditions to allow us to engage in complex goal-oriented action. These circuits are primarily based on structures in the PFC. The afferent as well as the efferent circuits are organized hierarchically. The PFC provides the precondition for the longer-term representation of goals and rules. The circuits contain a feedback component that signals whether the goal is being attained. E. K. Miller and Cohen (2001) summarize the functioning of these circuits as follows:

> When a goal that is already represented in the PFC reaches working memory, whose contents is characterized by the quality of conscious awareness, then selective attention leads us to prefer those perceptions, among all the potential perceptions, that are of particular relevance for the goal. Other perceptions are actively inhibited. The same holds true for the recall of memory contents and for the activation and generation of cognitive and behavioral strategies that serve the purpose of goal-attainment. This selective preference remains in place for as long as the goal, or parts of the goal hierarchy, remain represented in working memory. This preferential treatment means that the goal is being pursued with the special quality of conscious, deliberate action; that is, with conscious attention, conscious reflection, decision-making, volitional action-control, and so forth. Goal-oriented action occurs over a longer period and can be divided into a sequence of different phases. Psychology provides well-established models for this process, such the action-phase model by Heckhausen, Gollwitzer, and Weinert (1987); the so-called "Rubicon model."

2.12.2 The Neural Circuits for the Implementation of Actions

Figure 2–25 shows the different phases of the action-psychological phase model by Heckhausen, Gollwitzer, and Weinert (1987).

Intentional action begins, according to this model, with the motivation phase, in which the person contemplates various goals. This process is influenced by wishes and fears. This phase of contemplation is completed by the formulation of an intention. This is the shift from choosing to wanting. From here on, everything is directed toward the implementation of the decision, toward the attainment of a particular goal. What follows are planning activities for how the goal might best be reached and ultimately the execution of action, which entails the screening out of other, competing

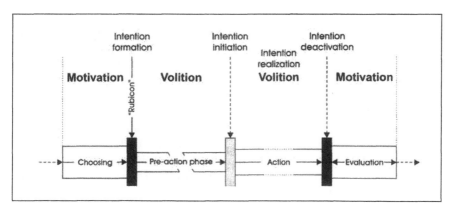

Figure 2–25. The action phase model (Rubicon model) by Heckhausen. (Adapted from Grawe, 1998; p. 50).

intentions. Finally, the action ends with an evaluation of the action consequences with reference to the pursued goal.

The PFC is also critically involved in the neural processes underlying the different action phases. This becomes obvious when we consider that many of these action components are impaired among patients with lesions in the frontal lobe.

- It is difficult for them to switch flexibly between different action phases, which is important during the planning phase.
- They are easily distracted and externally controlled; they find it difficult to maintain their focus of attention, which is important for internal intentional control.
- PFC patients are often confused; they leave out important parts of an action sequence. Their behavior appears thoughtless and unplanned.
- Intention and action among these patients often don't seem to correspond; they may know what they ought to do, but they don't follow through or simply do something else.
- Although their general memory appears unaffected, their prospective memory —the memory for that which one intends to do, for intentions and plans— is massively impaired.

Some authors assume that a central executive system, which plans, prepares, controls, and evaluates actions, exists in the PFC. Goschke (1995, 2001), however, argued persuasively that the optimizing of an action sequence involves several control systems that cooperate and compete with each other. They compete in the sense that the subsystems fight for the occupation of the most limited resource within the brain, working memory, with its very limited capacity. Because of this "bottleneck," it is impossible for several actions to transpire simultaneously with the advantages of conscious action planning and control. At any point in time, many goals tend to be activated simultaneously (see chapter 4 for more detail on this). Because of this, the goals that do not currently occupy working memory are pursued in the implicit mode of functioning (see following).

Conscious actions should not be placed at the top level of Powers' hierarchical model (see Figure 2–23); they correspond to the second-highest level, the program level. This means that the things that we have just consciously reflected on, that we have decided, planned, or performed, are always in turn influenced by goals that are even higher up in the hierarchy, and these superordinate goals are typically beyond the conscious awareness of the person. Our subjectively experienced consciousness—or something like a conscious, acting ego—is not part of the highest control levels of mental activity. Which parts of mental activity transpire in the conscious control mode is not determined by the activity itself but is the result of previously occurring complex evaluative processes that in turn transpire without conscious awareness. This will be elaborated in more detail later.

In concrete terms, how should one envision the neural processes involved in volitional action? Until recently, it was assumed that the prefrontal and orbital cortex, along with the parietal cortex, prepare for specific actions and then send the respective orders to the premotor and motor cortex areas. It was thought that these signals then initiate actual motion via the pyramidal pathway (the main exit of the motor cortex), the medulla oblongata, and the spinal motor nuclei. By now, however, it is clear that these processes are far more complex. The original assumptions had not considered that the PFC is massively connected with the basal ganglia and the limbic system.

According to Roth (2001), the action-initiating signals that originate from the PFC run through additional complex circuits before the action is executed. Roth refers to these circuits as the *dorsal loop* and the *ventral loop*. I will not describe all the specific stations that are involved in the processing of the action-initiating signals while they run through these loops because this would require knowledge of many brain areas that have not yet been introduced here and that are not particularly important for our subsequent discussions. Specific details about these loops can be found in Roth (2001, pp. 397–405 and pp. 415–426).

In greatly simplified terms one could say that within the dorsal loop, the basal ganglia have the function of selecting among a vast memory for all the various actions those actions that in the current moment, under these specific conditions, will best fulfill the current intentions. The signals are sent to the motor cortex for execution only after this detailed evaluation and action selection.

The decision of which action disinhibits the basal ganglia is influenced by the ventral loop, which is composed primarily of parts of the limbic system. Emotions and motivations influence final action initiation via the amygdala and the mesolimbic system. In this loop, our wishes, fears, and motivations are being evaluated with reference to the current situational conditions before they then are either confirmed or abandoned. This evaluative process also involves the hippocampus with its access to explicit memory contents. We will encounter the dorsal and ventral loop once again in chapter 3 in the section on obsessive–compulsive disorder. In a neural model of obsessive–compulsive disorder by Baxter et al. (1996), an imbalance between the ventral and dorsal loop plays an important role; however, these are referred to in that model as the *direct* and *indirect* paths.

2.13 CONSCIOUSNESS FROM A NEURAL PERSPECTIVE

Of all these complex neural processes, we experience consciously only those that occupy working memory for at least a few seconds. This is a very limited portion of the neural processes that are produced by our experience and behavior, because working memory only has a very limited capacity. More than a handful of distinguishable content simply does not fit simultaneously in working memory. According to Baddeley (1986), working memory consists of a visual–spatial and phonological buffer, which can hold visual and acoustic information for a few seconds, and a central component, which performs the processing as such. This part, in turn, is organized in a modular fashion

and has access to the remaining associative cortex, to preprocessed sensory information of sufficient intensity and duration, to episodic and semantic memory, and to the
systems involved in action control. Working memory is localized in the PFC. It works
in close collaboration with the anterior cingulate cortex, which plays a key role in the
internal control of attention. The content of working memory is that which we call the
"stream of consciousness." The stream of consciousness is filled not only by the components of conscious action but also by perceptions, thoughts, images, sentiments,
feelings, and memories without action intentions.

The stream of consciousness is characterized temporally by continuity and simultaneously by consistency. Both of these characteristics emerge primarily because the
current content of working memory largely determines what will enter into working
memory next. Content that does not fit at all does not enter. It does not become conscious. This is how cognitive dissonance (Festinger, 1957) is avoided. Section 4.8
will discuss in more detail this and other mechanisms that ensure consistency in mental functioning.

What enters next into working memory is determined not only by its current contents
but also by external influences. Important in this context is whatever enters the focus
of attention. The attentional system is in itself a complex system, involving various
brain areas. Via the activation of these other components, working memory can
become occupied by external influences, for example, by orienting reactions; by
unexpected perceptions that are judged as important by the evaluative system; by
somatosensory signals such as pain, hunger, shortness of breath and so forth; and primarily by processes of high motivational significance that have until then transpired
unconsciously.

All forms of consciousness are linked to the associative cortex. A key characteristic
of the associative cortex is the fact that its internal connections far outweigh those
that it has to, and from, the outside. One afferent or efferent thread is matched by
about 5 million intracortical threads (Roth, 2001, p. 214). The number of connections
in the associative cortex alone is estimated by Roth to be 500 billion. He follows up
on this with these considerations:

> Let us assume for a moment that superintelligent beings that are far superior to us, with
> brains that are radically different from our own, are visiting us on earth and want to test
> our brains for performance. These super-beings would come to the conclusion that this
> system, because of its high degree of internal interconnectedness, is occupied primarily
> with itself. Stimuli, or information, certainly do enter the system and activation also
> originates from it—as these beings observe—but these effects are negligible compared
> to the internal processes. The alien brain scientists would conclude that a high degree of
> autonomous regulation characterizes this system. They would be able to predict that an
> independent representational world emerges within this system, a world that for the
> external observer is somehow loosely connected to the events transpiring outside of the
> cortex. For the elements inside of the cortical representational world, though, this repre-
> sented world would be the only one that exists. While the observer would regard all cor-
> tical events as virtual, the interior elements—that is, the states of the cortex—would

regard these virtual processes, and thereby itself, as the perception of real events, or as the origin of real activities" (Roth, 2001, p. 214–215).

Roth regards the close interconnectedness within the associative cortex as the main neuroanatomic basis for our subjective experience in all its facets. It allows us to generate and maintain internal states that rely very little on external input. Just consider Beethoven or Mozart while they were composing, or Einstein, when he contemplated the foundations of his relativity theory. In these cases, the associative cortex interacts with itself, not with the external environment. Most of the things that humans devised to change the world were produced originally as internal brain states. It was a genius-like move of evolution to allow beings to experience such brain states in the form of thoughts and images, and to allow them to be communicated via the language system. This facilitates our ability to translate these brain states into concrete reality, which in turn has often increased our fitness in an evolutionary-biological sense. The fact that we can use these abilities for things that do not confer any obvious biological-evolutionary advantage, such as certain artistic endeavors, or even for things that constitute great risks for the development of humanity, such as the construction of hydrogen bombs, is in a sense an evolutionary by-product.

The emergence of consciousness as we know it in ourselves was surely not an end in itself for evolution; instead, it had concrete evolutionary advantages, as demonstrated every day by the dominance of humans on our planet. The specific abilities that go along with consciousness were selected by evolution, not the conscious experiencing itself. Subjective experiencing is not as important as it pretends to be, as it were. It is a product, not the producer, of the processes that constitute our selves or our personalities, and these processes transpire primarily unconsciously.

In the passages cited previously, Roth addresses the great discrepancy that exists between our subjective experience and the image that results when we examine the brain, as a part of the material world, with objective methods. Subjectively, there is nothing that feels more real than the stream of consciousness, even when know at the same time that these experiences are products of processes that cannot subjectively be known. One of the characteristic qualities of our consciousness is the tendency of our conscious "I" to experience itself as the author of that which we feel, think, and consciously do. But it is precisely this feature that, from an objective perspective, is an illusion. What we think, decide, and do is not determined by our conscious processes but by processes that have transpired previously without conscious awareness.

We know about these processes that do not become conscious primarily from studies that analyze electrical activity in specific brain areas. The advantage of such studies is their very high temporal resolution. Visual activation, from stimulus initiation to the first demonstrable reaction in the primary visual cortex, requires about 60 milliseconds. The negative N1-wave, which appears after 100 milliseconds in evoked potential studies, already indicates an initial evaluation of the importance of the stimulus, as the magnitude of this wave is proportional to the importance of the stimulus for the respective person.

Conscious perception appears no earlier than about 300 milliseconds after stimulus presentation. With very complex or meaningful stimuli, it can even last 1 second or longer until the presentation of the stimulus is consciously perceived. In the period from initial stimulus presentation to conscious perception, the stimulus properties are processed in a complex manner; in particular, the current significance of the stimulus for the organism is being evaluated. What we perceive is the final product resulting from this process, not the process itself. Meanings, therefore, are constituted unconsciously. They are supplied for consciousness but not produced by consciousness. However, we typically experience this the other way around because we are normally not aware of the preceding evaluative process.

Stimuli that reach the brain via the sense organs—a vast number of them per time unit—are continually being evaluated in terms of their importance and novelty. The evaluation can be accomplished only "in consultation" with the various memory systems and the brain areas that are involved in the representation of motivational goals and emotional valences. If the stimulus is deemed novel and/or important, a conscious perception results, and an elaborative process ensues that provides further clarification and is increasingly involved in forming a response to the stimulus. If no important or novel information is received from the outside, the brain generates the contents of consciousness "from within," either by accessing memory or by engaging in exploratory behavior. The question as to which rules govern this internal generation of consciousness is very interesting but has hardly been addressed by research so far. Parts of this content appear to be determined volitionally, but we also have spontaneous ideas, which take us by surprise.

This internal generation of consciousness states is also determined motivationally. The motivational determinants of mental activity are relentlessly active; they continually influence the contents of consciousness, either by evaluating external stimuli as important or unimportant or by contributing to the internal generation of consciousness. We experience only the result of this continuous influence of motivational determinants on our consciousness, but we do not "sense" the determining process. Because we do not experience the generative process, we regard the conscious thought itself as the primary cause of that which we experience as volitionally controllable. We experience causal relationships within our consciousness, and we do so with the quality of absolute certainty. The most important determinants of that which we think and do, however, are located outside of consciousness. They have already exerted their influence before we ever notice any of this. This is a provocative insight. What is most provocative and contradictory to our subjective experiencing is the finding that our conscious acts of will are in truth not consciously executed decisions at all.

2.14 ACTS OF WILL FROM A NEURAL PERSPECTIVE

Take a little break from reading at this point and rest your hands next to the book. Now I would like to ask you to make a decision about which of your fingers you would like to lift at which point within the next half a minute. You are completely free to decide as you wish.

If you have performed this little exercise, you will have the feeling that it was completely up to you to decide which finger you have lifted, and you will feel sure that it was your choice that led to the lifting of the finger. You feel that you are the originator of the finger movement.

And, in a sense, you certainly are. However, this is true in a different sense than how you have experienced it. The decision was indeed made in your brain, almost half a second before you had the conscious sense of making the decision. If one had just now measured your electrical brain activity above the premotor area, then already 350 milliseconds before you subjectively made the choice to lift your finger one could have detected neural activity indicating that the action had already been initiated.

Experiments of this kind were first explored by Benjamin Libet (Libet, 1978; Libet, Wright, & Pearl, 1983). Libet originally wanted to show that the subjective decision precedes the physiological process. In his planning of the experiment, he originally assumed a dualistic position. His participants were instructed to observe a small point that was rotating quickly around a clock-like device. They were asked to remember the position of the point at the precise moment when they made the decision as to which finger they wanted to lift. In this way it was possible to determine the time-point of the decision very precisely. In addition, the action potential above the premotor cortex was being measured. Contrary to his expectations, Libet observed that the neural activation preceded the decision in all of his study participants, with a range of between 150 and 1025 milliseconds. The temporal order of cortical activation and volitional choice demonstrates that it could certainly not be the conscious choice that triggers the neural activity. Libet interpreted the observed order by noting that our volitional choices appear to be preceded by unconscious processes that set the stage for the subjectively experienced decision to lift the finger. Our consciousness, then, suggests to us a causality that in fact is at odds with the true cause–effect relationships.

These results have understandably led to heated debates in the field. Various methodological concerns were voiced that initially cast doubt on the findings. Recently, however, Haggard and Eimer (1999) repeated the experiments with various modifications that in each case addressed the different concerns. This time the participants were also free to decide whether they wanted to react with the right or the left hand. The result, however, remained exactly the same, even with this new methodologically improved study. One has to assume, then, that our conscious acts of will are preceded by unconscious processes that are beyond our conscious control.

It seems to be the case that our subjective experiences, be they perceptions or acts of will, are immediately preceded by unconscious processing events in the brain, and this process appears to take at least several hundreds of milliseconds. Subjective experience appears to lag behind by a few hundred milliseconds. The act of will (decision) ends with the subjective experience instead of beginning with it. This is irritating for our understanding of ourselves in as far as it implies that all that we think, feel, and do is influenced and set up by unconscious processes that are beyond our control. At second glance, however, this finding is less irritating or provocative. If I regard the preparatory, unconscious processes as a part of myself, just as much as my conscious experience, then it is still me that makes decisions. My self—my personality—consists of implicit (unconscious) and explicit (conscious) aspects. My volitional decisions are not forced upon me by some external agent. The determinants of my behavior are my own determinants, even if they may not always be conscious. One characteristic of well-reflected volitional choices is their tendency to be experienced as consistent or congruent with the self. I can stand behind such decisions. The experiential quality of a choice that is made "wholeheartedly" is, in a sense, the confirmation of the decision's consistency with my most cherished values. In those cases, I can feel that I am the autonomous author of that which I volitionally decide and do. This authorship includes my implicit aspects. The idea that acts of will are made completely freely in the sense that they are not at all predictable might indeed be an illusion. It would indeed be quite problematic if we had to always freely decide in such a manner. The unconscious determinants that influence my volitional choices—primarily my most dearly held values and motivational goals—fortunately do not change from one day to the next. They are of an enduring nature and ensure continuity and consistency in my mental life. As soon as I accept that my self encompasses more than the parts I am aware of, I can accept that my decisions, in the moment that I subjectively make them, were already predetermined by the implicit parts of my self.

2.15 EXPLICIT AND IMPLICIT MENTAL PROCESSES

The distinction between *implicit* and *explicit* mental processes is still relatively new. This terminology was introduced by Schacter (Graf & Schacter, 1985; Schacter, 1987) for the difference between two entirely different memory systems, which had previously been called *declarative* and *nondeclarative*. Figure 2–26 shows the various forms of long-term memory that are distinguished in the literature today, as well as their relations to the explicit and implicit mode of mental functioning.

The differentiation between an explicit and an implicit mode of functioning applies not only for memory but also for perceptions, learning, emotions, action control, motivation, emotion regulation, and interpersonal behavior. All of these functions can transpire explicitly; that is, accompanied by conscious awareness, focused attention, volitional control, and the sense that one's self is truly personally involved in the activities. In that case, the processes have a different quality and effect compared to situations in which they transpire in the implicit mode of functioning. Implicit processes have advantages and disadvantages:

- They are independent of the capacity limitations of working memory; that is, many implicit processes can transpire simultaneously without interfering with one another

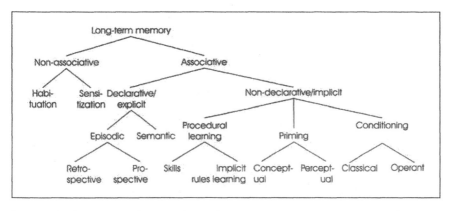

FIGURE 2–26. Taxonomy of different forms of long-term memory (Adapted from Goschke, 1996a; p. 367).

- They typically transpire quickly and without effort.
- They are not error-prone.
- They do not require attention and conscious awareness, which can even interfere with their smooth execution.
- They are typically linked to a specific sensory modality. Content that has been learned in the acoustic modality is not easily transferable to the visual mode, whereas it is not so important for semantic memory contents whether the information was originally stored acoustically or visually.
- They cannot easily, or not all, be controlled volitionally.
- They are typically learned more slowly than explicit memory contents, which can be learned in single trials. Under certain circumstances, they need many trials to be learned. This is why it is difficult to change them once they are well ingrained.
- If they become conscious at all, this tends to be in a fuzzy and unfocused manner. It is often impossible to report about them in verbal detail (this is why they were originally termed *nondeclarative processes*).

Some implicit processes could also transpire consciously, but they don't because they have become automatized or because the eliciting stimuli are too brief or too weak to cross the threshold of conscious awareness. They could relatively easily be moved into the explicit mode of functioning by a targeted redirecting of attention or by increasing stimulus intensity. Other implicit processes, by contrast, remain unconscious because the brain areas that govern their execution do not have connections to the associative cortex. Yet other processes transpire in the implicit mode of functioning because they are categorically denied access to working memory. This is the case with processes whose content is incompatible with the current contents of working memory, and which therefore would trigger a sharp dissonance or inconsistency in the working memory. By "repression" of such material from consciousness, this inconsistency is avoided. Repression is one of the processes that function to ensure mental consistency, which will be discussed in more detail in section 4.8.2.

Implicit processes involve the activity of brain areas other than those that are involved in explicit processes. If implicit processes are to be transferred into the explicit mode, therefore, it is important to establish a connection to those brain areas that are linked with conscious awareness. The main strategies for this are increasing stimulus intensity, as well as duration, and targeted redirecting of attentional focus. This is relatively easy to accomplish in situations involving automatized processes or insufficiently strong stimulus intensity. It is more difficult, however, with processes that transpire in the implicit mode of functioning because their entry into working memory has been actively inhibited. This would require establishing a state of considerable inconsistency, which is something that the brain tries to avoid at all costs (see section 4.8).

2.16 CONCLUSIONS FOR PSYCHOTHERAPY

If we apply the distinction between explicit and implicit mental functioning for the process of psychotherapy, we arrive at the following image: The mode that is clearly better for relearning or new learning is the explicit mode of functioning. Clearly, the patient can also acquire new experiences in the implicit mode of functioning, but this requires a large number of trials to achieve a change, and the new experiences will be limited to the concrete situation at hand. It is difficult to generalize them, compared to that which is acquired in the explicit mode of functioning. The processes in the implicit mode of functioning are typically very well-rehearsed. If the processes are problematic, then one would have to massively interrupt and inhibit these well-rehearsed processes in order to replace them with new behavior. This massive interruption, however, would inevitably direct attention to the disruption, which would thereby establish a link to the explicit mode of functioning. It would be practically impossible to achieve the large number of necessary trials of the new and corrective experiences, unless one somehow communicates about this process with the patient, which then would again return us to the explicit mode. The utility of the implicit mode for the direct achievement of therapeutic change, in the sense of learning new behavior, is therefore rather limited.

The explicit mode of functioning, then, is difficult to bypass if one aims to facilitate new learning. According to Roth (2001), the most important function of conscious awareness is, indeed, the facilitation of new learning. What this entails, according to Roth, is the changing of the connection patterns among neurons, particularly in those areas that govern the function that is currently being targeted. Roth argues that working memory plays a key role in the connecting and new facilitation of neural groupings. Once the groupings become well-facilitated and consolidated, they require less and less working memory capacity and ultimately become automatized. Their activation then increasingly involves brain areas other than working memory. Indeed, neuroimaging studies show that large areas of the associative cortex are activated in the process of new learning of complex cognitive or motor tasks. When these tasks are increasingly practiced and become familiar, these areas diminish in their activation, until finally no activity can be measured anymore. At this point, the execution of these tasks no longer requires working memory capacity.

The qualities that are linked with the explicit mode of functioning, such as conscious reflection, intention formation, planning, volitional control, and verbal communication, can all be viewed as resources in the context of the therapeutic change process. These are all instruments that can be used to devise experiences that, in turn, facilitate new neural groups to replace or inhibit the previously present problematic groups. Because the facilitation of neural groups requires longer time periods until the synaptic connections have achieved sufficient strength, it is necessary for the patient to hold strong intentions and to maintain goals over a long time, so that the required experiences can be repeated as often as necessary. The formation, facilitation, and maintenance of appropriate goals and intentions in the patient is one of the most important therapeutic tasks. These goals and intentions are the necessary foundation based on which patients volitionally expose themselves to new, planfully devised experiences that, although unpleasant at first, promise to have positive effects on their problems. All of this requires the qualities associated with the explicit mode of functioning.

The problematic processes that are to be changed, however, transpire primarily in the implicit mode of functioning. Anxiety, depression, and so forth are characteristically not experienced as volitional and controllable. Originating from the implicit mode of functioning, they conquer working memory against the patient's wishes and become manifested in consciously experienced negative thoughts and feelings. Their points of origin, however, are processes in the implicit mode; that is, in brain areas outside of the associative cortex, such as the amygdala. The changes must occur primarily in those areas, then, or certainly must somehow affect those areas. Therefore, one could say that a critical part of psychotherapy consists of utilizing the explicit mode of functioning to attain change in the implicit system.

The types of experiences necessary to change the neural structures and processes in the implicit systems that underlie specific mental disorders and other problems must be determined by disorder- or problem-specific research. A reasonably good fund of knowledge about this topic already exists in the field of psychotherapy, but there is certainly still room for improvement, especially once the neural foundations of the specific disorders become increasingly well understood (see chapter 3 for more detail). We saw earlier that situation-specific phobias, if they are to be changed, require a repeated confrontation with the feared situations while the brain is in a state of low fear-readiness. This constellation leads to the gradual inhibition of phobic reactions. We learned earlier that the therapist's most difficult and important task entails establishing an appropriate context for the confrontation with the feared situation. One of the key issues in the treatment of phobias is, therefore: How can I, as the therapist, structure the situation in such a way that the patient's brain is in a state of minimal fear-readiness? This question arises in slightly more generalized form in any type of therapy: How can I, as the therapist, establish a maximally optimal context in the implicit mode of functioning that will enable us to work effectively in the explicit mode?

We saw earlier that intentional action in the explicit mode of functioning corresponds to the second-highest level within Powers' goal hierarchy (see Figure 2–22). One level above this is that of situation-independent motivational goals, which influence mental activity in the implicit mode of functioning. The foundations of these goals emerge very early in life, and they differentiate and change over the course of development, although these changes tend to occur as a consequence of life experiences rather than conscious decisions. The goal hierarchies that are influenced by these general motivational goals are highly automatized and exert a pervasive influence on mental life in the implicit mode of functioning. They determine to a large degree the content with which people occupy themselves in the explicit mode; that is, they influence the contents of conscious awareness.

The goals that determine the contents of consciousness, however, are themselves not typically consciously known. Many of them would, in principle, be consciously knowable, but they are simply not represented in consciousness because our consciousness is already occupied with other things, such as the complex tasks of daily life. Other motivational goals, however, would trigger a dissonance if they entered into conscious awareness because they are not consistent with the content that is currently represented in consciousness. This is predictably the case when motivational goals are in conflict with one another. If one of the conflict components is represented in consciousness, then the appearance of the second component would trigger a consciously experienced dissonance, and we know from cognitive dissonance research that this aversive state tends to be avoided. Therefore, it is common that both conflict components are not consciously represented at the same time, given that this is a certain way of avoiding conscious dissonance. In that case, the motivational goals influence unconsciously what the person thinks, sets out to do, actually does, and so forth.

If I as the therapist interact with the patient in the explicit mode of functioning, then this always takes place in the context of the patient's implicit mode. From a therapeutic perspective, the most important parts of this context are the patient's motivational goals and implicit emotional evaluations that continuously influence what happens and what is possible in the explicit mode of functioning. Therefore, if my aim is to facilitate certain goals in the explicit mode, if I want the patient to expose himself or herself to new, previously avoided situations, if I want him or her to personally engage with the treatment goals and direct his or her conscious abilities toward the attainment of these goals, then I have to establish a motivational and emotional context in the implicit mode of functioning that is maximally supportive of these aims.

In order to achieve this, I would have to activate already existing approach goals in the patient, that "energize" that which I intend to elicit and facilitate in his or her explicit mode of functioning. I would have to ensure that the patient, as frequently as possible, experiences perceptions that have a positive meaning for his or her motivational goals. This can happen in the explicit as well as the implicit mode of functioning. For example, I could (explicitly) address issues that cast a positive light on the patient and increase his or her self-esteem. I could simultaneously (implicitly) use my nonverbal behavior to facilitate perceptions that help meet the patient's attachment and control needs. Such need-satisfying perceptions trigger positive feelings in the patient and thereby facilitate his or her approach potentials. Embedded in an approach-oriented emotional and motivational context, the problems that are simultaneously addressed in the explicit mode will appear less threatening and more likely to be mastered, and approaching these problems will seem more consistent with the patient's own goals. Problematic contents in the explicit mode of functioning should be addressed by a therapist only once the brain of the patient has been shifted, with implicit and explicit means, toward an approach-oriented and emotionally positive state.

These considerations about the motivational and emotional context of problem processing apply regardless of the type of problems that are being addressed and irrespective of the type of therapeutic strategy used in the explicit mode of functioning. In light of all the different processes that occur within the patient's brain during a course of therapy, it would be extraordinarily naïve to explicitly plan only those processes that transpire on a conscious level. Because most neural processes take place implicitly—and this applies especially to the crucially important motivational and emotional processes—such a focus on explicit strategies would be akin to viewing a vast landscape through a tiny peephole. What the patient experiences consciously is only a small proportion of the totality he or she experiences. Moreover, consciously transpiring processes are massively influenced by those that simultaneously take place unconsciously. As a therapist, therefore, I have to work continuously in the explicit as well as the implicit mode of functioning. My own implicit processes are not under my conscious control either, but by planning the course of therapy explicitly, I can arrive at an understanding of the implicit processes in the patient, which then enables me to consciously regulate the volitionally controllable parts of my own behavior in such a manner that a context is established that is emotionally and motivationally maximally constructive for working on the patient's problems.

Because these processes hold true across clinical problems and therapeutic methods, training therapists in their ability to generate these change-conducive contexts should be regarded as one of the most important aspects of therapy training. From the perspective I took here, based on neural brain processes, two core components of therapeutic work can be distinguished: *Context generation* and *problem processing*. These two tasks are intertwined and must be addressed simultaneously. Within a single therapy session, as well as over the longer period of a course of therapy, it is perfectly reasonable that sometimes one and sometimes the other of these components take center stage. What is important, however, is that the therapist is consistently aware of both aspects.

I have included these therapeutic considerations here because they follow directly from the prior sections on the structural and functional architecture of the brain. In chapters 4 and 5 I will provide a more substantial foundation for the points I only briefly alluded to here, and I will elaborate on the therapeutic conclusions in more specific detail.

2.17 COVARIATION OF NEURAL AND MENTAL ACTIVITY

Up to this point, we have discussed the general functioning of the brain, as it applies broadly to all people. However, the brain is a biological system, and all biological systems are characterized by variability. People differ in their shoe sizes, their muscle strength, their intelligence, their physiological responsiveness, and their different blood parameters. Thus, it seems plausible that they should also differ from each other in their neural structures and processes. And, of course, they do. The question then becomes whether differences in specific neural activity in a certain brain area covary with differences in specific aspects of experience and behavior. If the neural processes really directly influence experience and behavior, then there should be a strong correlation between neural and mental activity. Kosslyn et al. (2002) addressed this question in their excellent general overview, in which they compiled empirical evidence about the topic. Studies in this area can be described as follows:

In one study by Kosslyn, Thompson, Kim, Rauch, and Alpert (1996), participants were told certain letters, such as "A," for example. They were asked to close their eyes and to envision the letter for 4 seconds, in its capitalized form, as intensively as possible. Then they were told a specific quality that might characterize the letter, such as "bent lines," and they were asked to indicate whether the quality applied to the letter or not. In the case of "A," the answer would be "no"; in the case of "B," it would be "yes." The accuracy and speed of the response were measured, as well as blood flow as an index of the intensity of the neural activation in those areas that are known to contribute to visual imagination. Before correlating neural activity and performance, the blood-flow levels of the 16 participants were statistically normalized (z-transformation) to adjust for individual differences in overall levels. A strong correlational association was observed between the strength of the blood flow in Brodmann's area 17 and 18 (primary and secondary visual cortex) and response speed. Multiple correlations were used to establish that it was, in fact, the intensity of the activation in this particular area that correlated with performance. The correlation had a value of $r = .93$! That means that knowledge about the activation within a brain area that "governs" a specific mental function would allow us to predict the corresponding behavior with remarkable precision.

Similar relationships were also found for other brain regions and other tasks. For example, activation in the occipital cortex and thalamus correlated with the accuracy of face classifications (Alexander et al., 1999); activation in the temporal lobe correlated with performance in a word-recognition task (Nyberg, McIntosh, Houle, Nilsson, & Tulving, 1996), and activation of the right amygdala during the viewing of an emotional movie correlated with later memory performance about the viewed content (Cahill et al., 1996).

Of particular clinical relevance are the studies by Davidson and his colleagues in the area of "affective style" (Davidson, 2000). For example, when participants were shown negative emotional pictures—for example, images that elicit fear or disgust—the right prefrontal region was more strongly activated, as measured by EEG, whereas the left region was more activated when positive emotional pictures were used.

The baseline level of brain activity in resting mode in these two brain areas can be regarded as a stable personality trait (internal consistency above .90; test–retest reliability between .65 and .75; Tomarken, Davidson, Wheeler, & Kinney, 1992). People with stronger left-sided activation experience more positive emotions; people with more right-sided activation tend to experience more negative ones. This holds true for their responses on emotion questionnaires as well as for their reactions to emotional movies (J. Henriques & Davidson, 1990; Wheeler, Davidson, & Tomarken, 1993). Even among 10-month-old infants, those with stronger right-sided prefrontal brain activity were more distressed by the short-term absence of their mother than those with stronger left-sided activation (Davidson & Fox, 1982).

In later studies, Davidson, Jackson, and Kalin (2000) found associations between left- versus right-sided prefrontal activation and emotional reactions even among monkeys. Shelton and Davidson (2000) found that monkeys with dispositionally stronger right-sided prefrontal activation had a higher level of the stress hormone cortisol, compared to those with stronger left-sided activation. In a study by Davidson, Coe, Dolski, and Donzella (1999), the individual differences in prefrontal activation were even able to predict the activity of natural killer cells in resting state and during the reaction to a provoked challenge of the immune system. Among depressed patients, Abercrombie et al. (1998) found a correlation between .40 and .55 between activation of the right amygdala and self-reported negative emotions. In a study by Irwin et al. (as cited in Kosslyn et al., 2002), the viewing of emotionally negative pictures led to a higher group mean activation level in the right amygdala, compared to the condition in which neutral pictures were shown. Within the group, the degree of activation in the right amygdala also correlated(.61) with self-reported negative affect.

These studies suggest, then, not only that certain brain regions—practically always in collaboration with other areas—are responsible for specific aspects of experience and behavior, but also that people with stronger activation in these areas show the corresponding experience and behavior to a stronger degree. This suggests the following conclusion: If we could manage to increase the activation in certain brain regions or neural circuits, then this should have the effect of simultaneously strengthening the types of experience and behavior corresponding to these regions. The reverse is true as well: Inhibiting neural activity in these regions should lead to a reduction in the corresponding experience and behavior. This immediately suggests the question, of course, whether and to what degree it might be possible to alter the brain with targeted interventions in such a manner that these specific effects on experience and behavior can be achieved. The correlations reviewed previously do not, after all, inform us about the degree of changeability of the neural processes involved. The crucial question for psychotherapy is the one following from these correlational associations: To what extent can we achieve enduring change in the neural structures and processes that underlie mental disorders and clinical problems?

2.18 NEURAL PLASTICITY

I briefly mentioned the question about the degree of brain plasticity at several points earlier in this chapter, and I already assumed a high degree of plasticity in the discussions in my opening chapter. Nevertheless, I would once again like to review at this point the relevant facts and findings and specifically explore their implications for psychotherapy. Ultimately, the question of how much the brain can be altered via psychotherapeutic means will only be settled in the future, once the practice of what I term *neuropsychotherapy* has been broadly applied and the effects on the brain have been assessed with appropriate methods.

Studies that specifically examine the degree to which the brain can be changed by psychotherapeutic interventions are still far and few between. I mentioned earlier in the section on fear a study by Gutberlet and Miltner (1999) in which changes in EEG activity was measured among patients whose spider phobia had been treated successfully. They observed that the amygdala continued to respond to the phobic stimulus even after the treatment, but additional posttreatment activation could be documented in the prefrontal cortex that had not been present prior to treatment. This new activation suggested the formation of new neural circuits that effectively prevented the further transmission of amygdala activation.

Van der Kolk (1987) was one of the first to assess aspects of brain activation before and after psychotherapy as an outcome criterion. Among patients suffering from posttraumatic stress disorder who had been treated with eye movement desensitization and reprocessing, he found increased metabolic activity in the prefrontal cortex and reduced activity in the limbic system. Using magnetic resonance imaging, Baxter (1992; see also J. M. Schwartz, Toessel, Baxter, Martin, & Phelps, 1996) found that behavior therapy in patients with obsessive–compulsive disorder led to changes in brain activity equal to those obtained for pharmacological treatment. Similar findings were reported by Brody, Saxena, and Stoessel (2001) and Martin, Martin, Rai, Santoch, and Richardson (2001) in depressed patients who were treated either with medication or interpersonal psychotherapy. In a single case study of a depressive bipolar patient who had been treated with psychodynamic therapy, Viinamäki, Kuikka, and Tiihonen (1998) were able to document a normalization of her serotonergic metabolism that was not observed in a comparison group of untreated patients with the same disorder. None of these studies, however, was designed in a manner that would permit us to conclude that the changed brain activity was specifically caused by the therapy. The symptoms among depressed patients typically change over time, even without therapeutic intervention, and these changes are most likely accompanied by activation changes in the brain areas that control symptom intensity. Using neuroimaging technology in itself does not automatically guarantee the conclusiveness of a study. Better-designed studies would have to examine which specific interventions influence which specific aspects of brain activity or result in which lasting changes. To show that psychotherapy produces changes in experience and behavior via its effects on neural structures and processes, one would also have to show that the degree of change in neural functioning is linked with the extent of symptom changes.

The first study that came close to meeting these requirements was a controlled therapy study conducted in Sweden, which was quite sophisticated in its design and statistical analyses and therefore allows us to draw clear conclusions. Because this study can be regarded as a milestone on the road to the kind of neuropsychotherapeutic research I envision, I will describe it here in some detail.

Furmark et al. (2002) examined 18 participants with intense fear of public speaking who had responded to a newspaper advertisement to join the study. The selected participants met diagnostic criteria for social phobia. Therefore, they are referred to as patients in the study, which I will also do here. The researchers formed triplets of patients who were matched on a maximum number of characteristics, and they then randomly assigned the members of these triplets to one of three treatment conditions:

1. Cognitive–behavioral group therapy: This involved eight weekly therapy sessions of 3 hours each. The therapy consisted primarily of exposure exercises conducted in the group as well as cognitive restructuring with directions to attempt appropriate exercises in real-life circumstances (homework exercises). The sessions were led by two experienced clinical psychologists who were also part of the experimenter team.
2. Pharmacotherapy: The study participants received individually tailored dosages of the serotonin reuptake inhibitor Citalopram, administered by an experienced psychiatrist, with an average dosage of 40 mg. The psychiatrists attended to issues of compliance and side effects but otherwise did not provide specific therapeutic instructions, such as encouraging exposure to feared situations.
3. Wait-list group: The participants assigned to this group received no treatment at all over the course of 8 weeks.

Assessments were conducted immediately prior to and after the therapy as well as at a 1-year follow-up. Four scales were used to directly assess fear of public speaking: One rating for fear and one for distress during public speaking, on a scale from 0 to 100, the Spielberger State Anxiety Inventory, as well as heart rate while patients gave a two and a half minute long speech. In addition, they completed a battery of questionnaires assessing various aspects of social phobia: The Social Phobia Scale, the Social Interaction Anxiety Scale, the Personal Report on Confidence as Speaker, and the Global Assessment of Functioning Scale in a self-rating format. The posttreatment assessment also included an interview in which patients were able to express their subjective experience of the treatment experiences and resulting changes.

To measure brain activity during the confrontation with the feared situation, the researchers used imaged obtained by a PET scanner. Twenty minutes prior to the scanning session, patients were instructed to prepare for a two and a half minute long speech about a holiday trip. Their brains were scanned while they gave the speeches.

During this period, a group of six to eight people stood near the PET scanner and silently observed the patient, who in turn was instructed to look at the audience while giving the speech. To further increase the fear of being observed, the speeches were recorded by a camera from a very close distance. During each speech, fifteen 10-second-long PET scans were obtained. To increase the "signal-to-noise ratio," each patient had to give two speeches with PET scanning prior to and two after the treatment.

Prior to the therapy, the groups did not differ in any of the obtained measures, which ensured that the starting conditions for the groups could be regarded as equal. Figure 2–27 shows the changes in seven of the measures described previously, which were observed during the pre-to-post period of 9 weeks.

Cognitive–behavioral therapy resulted overall in the strongest pre-to-post changes. These changes are significant for five of the seven measures, using the 5% level of significance. In the Citalopram treatment condition, significant improvements were observed on two of the measures; in the control group, no significant differences were observed. There were no statistically significant differences between the psychotherapeutic and the medication conditions, which can be explained, however, by the low statistical power resulting from an N of six in each group. When the two active treatment conditions were pooled and compared with the control group, significant or nearly significant differences were observed on four measures.

Based on their results on the various measures, the patients were classified as responders versus nonresponders. Those who achieved an improvement of more than one standard deviation in seven to nine measures were labeled as "much improved" and those who achieved such improvement in four to six measures were labeled as "moderately improved." The other patients were labeled as "minimally improved." In both

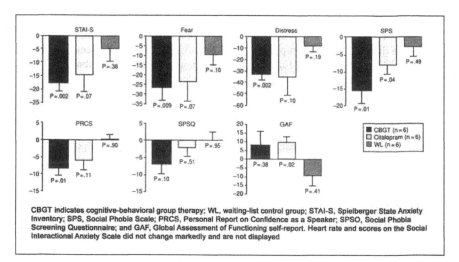

CBGT indicates cognitive-behavioral group therapy; WL, waiting-list control group; STAI-S, Spielberger State Anxiety Inventory; SPS, Social Phobia Scale; PRCS, Personal Report on Confidence as a Speaker; SPSO, Social Phobia Screening Questionnaire; and GAF, Global Assessment of Functioning self-report. Heart rate and scores on the Social Interactional Anxiety Scale did not change markedly and are not displayed

Figure 2–27. Change scores ($M \pm SE$) reflecting social phobia symptom changes with 9 seeks of treatment or waiting time (Adapted from Furmark et al., 2002).

treatment conditions, two patients could be grouped into each of these categories. In the control condition, one patient could be assigned to the "moderately improved" category. The results that are most relevant here concern the comparison in brain activity between the responders and nonresponders, either separately by treatment condition or combined for all nine of the responders (much or moderately improved) versus the nine nonresponders across the treatment groups.

After the exposure test, successfully treated patients showed reduced activation in the amygdala, the hippocampus, and surrounding areas in the temporal lobe (rhinal, parahippocampal, and periamygdaloid cortex). In both of the active treatment conditions the activity in these areas was significantly reduced compared to the control group, especially in the right brain hemisphere. One year after the treatment, the patients who had improved after the treatment were still characterized by significantly lower activity, compared to the control group, in the amygdala, central gray areas, and the left thalamus. The brain activation had, therefore, changed permanently as a result of the treatment, and the degree of brain activity changes correlated with the changes observed in other anxiety measures.

The reduction in brain activity from pretest to posttest could be used to significantly predict the clinical status (responder vs. nonresponder) 1 year after the therapy. Figure 2–28 shows the reduction in brain activity in four areas during the treatment, separately for patients who were still improved versus not improved after one year.

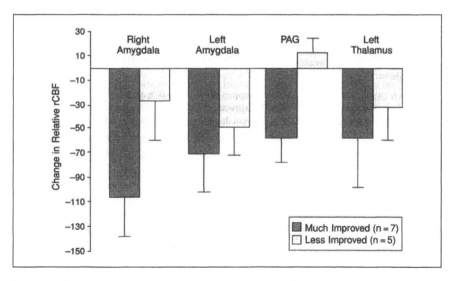

Figure 2–28. Change scores ($M \pm SE$) reflecting social phobia symptom changes with 9 weeks of treatment or waiting time. CBGT indicates cognitive-behavioral group therapy; WL = waiting-list control group; STAI-S = Spielberger State Anxiety Inventory; SPS = Social Phobia Scale; PRCS = Personal Report on Confidence as a Speaker; SPSQ = Social Phobia Screening Questionnaire, and GAF = Global Assessment of functioning self-report. Heart rate and scores on the Social Interactional Anxiety Scale did not change markedly ($p > .10$) and are not displayed (Adapted from Furmark, 2002; p. 432).

This is the first convincing evidence that successful psychotherapy can achieve enduring structural changes in the brain, and that clinical improvement goes hand in hand with changes in brain activation. When considering that these changes in brain activation, compared to the control group, were statistically significant despite the small sample sizes, one has to conclude that the effects were large in magnitude. If one converts the values observed in this study into measures of effect size, it is clear that the resulting effect sizes are substantial. Because of the small number of participants in this study, at this early stage of the research, it does not seem sensible to compute specific effect-size values, especially because effect sizes tend to magically develop a "life of their own" once they are removed from the context of the studies in which they were found. It is important to note, however, that the effects observed in this study were significant at least at the 1% level.

This study also yielded two additional interesting observations. In another, previous study, the same group of researchers (Tillfors et al., 2001) had been able to document differences in substantially more brain areas between socially phobic versus control patients who had undergone a similar exposure test, compared to their later study. The brain areas overlapped only partially. This suggests the conclusion that therapy changes only a part of the brain activity that differentiates socially phobic individuals from normal comparison individuals.

The differences also overlapped only partially between the two therapy groups. Whereas the changes in the temporal lobe were largely the same, a significant reduction in central gray areas—those areas that, according to Behbahani (1995) are involved in avoidance behavior—was observed only among patients who had been treated with behavior therapy. However, only the Citalopram group showed a significant reduction in the left thalamus. Furmark et al. suspect that the heightened serotonin level dampens activity in the thalamus and thereby reduces the excitatory input from the thalamus and other cortical areas to the amygdala. The authors further suspect that psychotherapy and medication might achieve similar effects via differing routes. The similar changes in the temporal lobe in both treatment conditions could be a shared final pathway, as it were, that must be reached in order to attain clinically meaningful effects. The route toward this goal, however, could be different in these two interventions.

The results of a study by Goldapple and coworkers (2004) suggest that clinical improvements observed after psychotherapy or medication not only follow different routes that finally end in changes of neural activity in the same cortical sites; they showed that these changes could have inverse directions depending on the applied treatments. The authors used PET to compare changes in metabolic brain activity underlying the response of depressed patients to cognitive–behavioral therapy with changes occurring in pharmacotherapy (paroxetine). A significant symptom reduction was observed in almost all patients who received cognitive–behavioral therapy. This reduction was associated with an increase in metabolic activity in the hippocampus and the dorsal cingulate cortex as well as a decrease in metabolic activity in the dorsal, ventral and medial prefrontal cortex. In contrast clinical improvement in pharmacotherapy was associated with increased activity in the prefrontal cortex, whereas activity in the hippocampus and the subgenual cingulate was decreased. These differential changes in activation patterns support the assumption that psychotherapy and medication target partly different primary sites, both resulting in activity changes in the critical prefrontal-hippocampal network. These modulations might be the most critical neural changes associated with the clinical improvement in depression.

Studies by Paquette et al. (2003), as well as Straube, Glauer, Dilger, Mentzel, & Miltner (2006), aimed at further clarifying the effect of psychotherapy on cortical activity in specific phobias. Remarkably in their studies, partly different sites from those identified by Furmark et al. showed phobia-related activation. This difference might be attributed to the different experimental paradigms being used. However these studies consistently showed a normalization of the cortical activity after psychotherapy in regions that had shown significantly increased or decreased activity before treatment.

These studies provide a preview of the future of psychotherapy research. It is practically certain that more studies of this kind will appear in the literature, and soon the number of treated cases will also increase. These studies will provide further clues about how—that is, by affecting which brain processes—therapeutic strategies might achieve their effects. Based on such results we will be able to further optimize the therapeutic strategies by concentrating more on the critical elements that are responsible for the effects. These investigations also show that this future has already begun. Study results like the ones just reviewed here can inspire us psychotherapists. It is not just an assumption that we can change the brain with our methods; it is a fact. The study by Furmark et al. will have a unique place in the history of psychotherapy research because it first transformed this assumption into fact.

But even if these studies had not been conducted, there would be reason for optimism with regard to the possibility of changing brains via psychotherapeutic means. My optimism rests upon the many replicated facts about the close causal relationships between neural and mental processes, of which I reviewed a few earlier that seemed particularly relevant in the context of psychotherapy. My confidence is especially supported by the convincing evidence showing that the brain responds to stimuli of all types with enormous adaptability.

I already mentioned earlier that, among violin and guitar players, the projection areas of the sensomotoric cortex for the second to fifth finger in the left hand are hyperdeveloped, compared to "normal people." The hyperdevelopment recedes once the musician stops playing. The neural structure, then, is dependent on its actual utilization. Violin players have better-developed projection fields for violin tones in the primary auditory cortex; trumpet players have better-developed fields for trumpet sounds. The difference is apparently produced by more or less intensive stimulation.

As I briefly mentioned before, the hippocampus is important not only for the formation of new memory contents but also for spatial orientation (Hampson, Simeral, & Deadwyler, 1999). This is well established for rats and seems to work similarly for humans. Maguire, Frackowiak, and Frith (1997) and Maguire et al. (1998) showed among London taxi drivers that especially the right hippocampus is highly activated during the successful navigation of taxis in traffic. In a follow-up study, Maguire et al. (2000) even documented increased volumes in the dorsal hippocampus among taxi drivers with longer work experience.

Instead of musicians and taxi drivers, one could probably also use other specialists such as wine tasters, top chefs, entertainers, chess players, and so forth, and one would observe that the intensive utilization and stimulation of specific brain areas manifests in structural changes within these areas. We can assume that coming years will produce many more similar studies to document this. In all likelihood, one will find over time that every intensively practiced activity manifests in the form of structural neural characteristics.

There is quite a bit of additional evidence for the usage-dependent over- and underrepresentation of projection areas in the brain. One of the first studies was conducted by Merzenich et al. (1984). They found that a reorganization of cortical representation areas took place in the brains of monkeys after the amputation of one of their fingers. The adjacent fingers used the resulting gap, as it were, and usurped it for themselves. The shifting of the adjacent areas into the newly vacated area required a period of a few weeks. Very similar cortical reorganizations were found by Weiss and colleagues (2000) in a man whose ring and middle fingers of the right hand had to be amputated. They found a clear shift of the projection fields only 10 days after the amputation.

However, cortical representations change not only after injuries but can also be achieved by targeted training. In a follow-up study to the previously described amputation study, Jenkins, Merzenich, Ochs, Allard, and Guic-Robles (1990) were able to show that the cortical representation zones of the fingers could also be enlarged by an intensive discrimination training of single fingers.

The studies on the training-induced cortical reorganization among patients with stroke-related paralysis are particularly clinically impressive. Liepert, Bauder, Miltner, Taub, and Weiller (2000) studied 15 patients who had suffered a unilateral paralysis of arm and hand as a consequence of a prior stroke. Six months after the stroke—that is, after a period when normally no further spontaneous improvements would be expected—the researchers conducted a "constraint-induced movement therapy" (also known in Germany as Taubian movement therapy). Over a period of 12 days, the healthy arm was immobilized by a sling for 90% of the waking period. For at least 6 hours per day, the patients were intensively trained with the paralyzed arm in specific movement tasks that resembled daily activities. A precondition for the training was that a minimum of movement capacity remained in the paralyzed arm and hand. Substantial improvements in the usability of arm and hand could be demonstrated in all patients, and these improvements were maintained half a year after the conclusion of the training. The cortical representations of the involved muscles were measured via transcranial magnetic stimulation. Prior to the therapy, these regions that represent arm and hand were substantially larger in the healthy compared to the injured brain hemisphere. The reduction of the representation in the injured hemisphere was probably caused by the nonuse of the paralyzed arm. Immediately following the treatment, the representation in the injured hemisphere had grown beyond its normal size, which corresponded to an increased excitability and clear clinical improvement. At the same time, the analogous region in the healthy hemisphere slightly shrunk in size. During the course of half a year, however, the area that had grown substantially because of the training slowly normalized in size. The right and left hemisphere were once again brought in line in terms of their size. Furthermore, the excitability was reduced to its normal level. These changes, combined with a maintained high functional capacity, could be linked to an "improvement in the effective connectivity of the neural connections." The effort that the brain has to master if performance is to be kept at the same level becomes increasingly less with advancing automatization. This has also been found with many other cognitive abilities.

Amputations often result in phantom pain. Flor (2003) reported on studies showing that the impairment due to phantom pain correlates with the extent of subsequent cortical reorganization. If arm and hand are amputated, the reorganization often involves the shifting of the adjacent representation fields of the face and mouth into the newly vacated areas. Phantom perceptions can subsequently occasionally be triggered even by slight touches to the face. The cortical representation of the face is asymmetrical after such a reorganization. This asymmetry can be alleviated, however, by administering targeted sensory stimulation in the representation zone of the amputated limb. By using prostheses that electrically stimulate the muscles at the stump (myoelectrical prosthesis), this kind of electrical sensory stimulation can be achieved. Compared to a merely cosmetic prosthesis, this led to a reestablishment of symmetry in the cortical representation areas and a discontinuation of the phantom pain. In a controlled intervention study, patients with phantom pain received nonpainful electrical stimulation in the amputated limb via eight electrodes that had been fixed to the stump of the prosthesis. The stimulation occurred in variable locations and frequencies. The patients had to attempt to recognize the location and the frequency and were rewarded for accurate recognition. In comparison to an attention-control group, the discrimination ability of the patients improved significantly and the phantom pain was reduced by more than 60%. The originally strongly shifted cortical representation of the mouth shifted back toward its originally symmetrical position. Discrimination ability, cortical reorganization, and pain were significantly intercorrelated. This supports the idea that the cortical reorganization was achieved by the training of the discrimination ability, and that the cortical reorganization in turn led to the reduction in pain.

2.19 CONCLUSIONS FOR PSYCHOTHERAPHY

The studies reported in the previous section are only a small selection of the many findings that have become available in recent years that show that the brain responds to injuries, as well as facilitative influences, with a great deal of adaptability. In the case of complete nonactivation, existing neural connections are weakened at remarkable speed. The newly available brain areas are quickly used up for other purposes. Conversely, intensive and repeated experiences that recurrently activate the same synapses over time lead not only to functional but also to structural changes in the corresponding brain areas. These structural changes, achieved by facilitation of neural connections, ensure a permanently heightened synaptic transmission potential. The facilitated processes transpire more and more easily and ultimately become automatized. The increased transmission strength is only maintained, however, for as long as the corresponding function is being utilized. Very well developed brain structures that are not used over a long period begin to atrophy, which triggers decrements in the ease and quality of the processes associated with these structures.

These findings about the great malleability of brain areas by intensive utilization/stimulation and about the deactivation of certain areas—collectively known as *neural plasticity*—arc good news for psychotherapy. One can change the brain via sufficiently intensive influences in such a manner that self-sustaining new structures emerge, which then become the foundation for enduring changes in experience and behavior. The stimulation by which such changes can be achieved, however, must be sufficiently intensive and strong. This is reflected in practically all of the available studies. In an fMRI study by Karni et al. (1995) with normal participants, it was evident that practicing a simple motor task for 10 to 20 minutes per day triggered changes in the corresponding brain areas only after 3 weeks. In a study by Merzenich et al. (1996) with dyslexic children, a daily 2-hour-long training over a period of 4 weeks was required to achieve a significant improvement in phonetic discrimination. Sporadic and relatively brief activation of brain processes does not result in long-term learning. Only intensive facilitation of the desired actions will produce a structural basis for these actions in the brain. Only this establishes the foundation for the self-sustaining long-term maintenance of therapeutic changes. It is possible that psychotherapists will have to adjust their ideas about the intensity and dosing of their therapeutic interventions. In the treatment of stroke-related paralyses, it became clear that truly meaningful changes could only be achieved once the "dosage" of therapeutic interventions was massively increased. Stroke patients almost always receive physiotherapy in order to reduce the paralysis as much as possible. However, this tends to be about 1 to 2 hours per day, on average. This dosage is apparently not enough to create truly new, self-sustaining structures in the brain. By using a dosage of at least 6 hours per day, combined with the inhibition of movement in the healthy arm, it became possible to achieve much better results with the constraint-induced movement therapy.

However, it is at this point still unclear whether we can generalize from these findings to psychotherapeutic interventions. The possibility exists, however, that many of the current therapeutic attempts to change intractable problem behavior do not lead to better results because the interventions—although correct in principle—are not administered in sufficiently intensive dosages and frequencies. Conceivably, some behavioral therapeutic techniques, such as prolonged exposure therapy, are so successful because they facilitate and activate neural circuits in a very targeted, structured, intensive, and enduring manner.

The intensive and repeated facilitation of new processes requires the conscious and volitional collaboration of the patient. It is frequently necessary to overcome an initially dominating avoidance tendency in these situations. With an increasing number of repetitions, however, this avoidance will typically recede. In order to elicit conscious, self-motivated engagement, the patient must be informed in detail about the goals to be attained by the repeated exercises and facilitations; he or she should be told why they will result in effects that are personally important; and he or she should know what the key principles in doing the exercises are. This requires, on the one hand, a high degree of transparency about the therapeutic goals and the strategies for attaining the goals. On the other hand, it requires that a considerable volitional strength has been established in the patient—a desire fueled by the patient's own motivational goals that prompts the patient to repeatedly expose him or herself to the new, facilitative experiences. In order to be able to accomplish these intensive and repeated exercises, then, there needs to a preexisting motivational context to support the problem-processing aspect of therapy. Establishing and maintaining this context is just as important for the overall goal of attaining the desired changes as the execution of appropriate, problem-specific interventions itself. If the goal is to create new ways of experiencing and thinking, to form new reactions to emotional situations, to facilitate new coping abilities to replace or inhibit old, problematic ones, then it is critical that all this transpires embedded in a positive motivational and emotional context. In our own detailed process analyses of psychotherapy, we had to acknowledge that this context was not always present. If it is not possible to establish this positive context for problem processing, then the problem processing itself practically never leads to the desired outcome (Gassmann & Grawe, 2006; E. Smith & Grawe, 1999, 2003). Intense new learning of the kind that is associated with structural changes in the brain cannot be accomplished in opposition to the patient's volition or resistance. This type of learning requires conscious, volitional, and self-motivated patient collaboration. Therefore, it can transpire only in the explicit mode of functioning. In parallel, however, the implicit mode of functioning is of key importance for the creation of this facilitative motivational and emotional context.

The studies on neural plasticity suggest yet another important conclusion for psychotherapy: Structural changes in the brain occur only as the result of very intensive, long-lasting influences. As I discussed in the introduction, and as I will further elaborate in the next chapter, mental disorders are often—probably even typically—accompanied by structural changes within the brain. The formation of these structures in the brain requires a relatively long time period. The facilitation of the brain structures must have occurred as a consequence of real-life experiences that have occurred over a longer period, and/or as a consequence of frequently repeated internal processes that transpired over a longer period. Such processes must have had an important function in the intrapsychic system of the respective person. Both—the emotional meaning of life experiences and the function of frequently repeated mental processes—are determined by currently activated motivational goals of the person. Mental disorders, in this view, originate in a motivational context that plays a causal role in the facilitation of pathways underlying the disorder. The motivational aspect of neural functioning is one of the three topics that would have fit well into this chapter, but because of their special significance, I have not yet fully discussed these processes. Instead, I have devoted a separate chapter to this topic—the motivational aspect takes center stage in chapter 4.

A second question that I have not yet fully addressed concerns how new neural activation patterns emerge in the first place. When new mental disorders are formed, new neural activation patterns emerge in tandem. But how exactly does this new onset of mental disorders occur? What determines whether such new neural activation patterns establish themselves in the brain? This basic question about the origin of mental disorders is also addressed in chapter 4.

I have also postponed for the moment the detailed discussion about the specific neural circuits underlying the various mental disorders. The neural foundations of mental disorders are undoubtedly within the domain of knowledge that psychotherapists ought to possess. Because of their special significance for psychotherapy, I have devoted a separate chapter to this topic, which follows next.

CHAPTER THREE

NEURAL CORRELATES
OF MENTAL DISORDERS

3.1 WHAT CAN BE SAID TODAY ABOUT THE NEURAL CORRELATES OF MENTAL DISORDERS?

Based on the findings reviewed in the chapter 2, we can assume that the differentiated facets of our experience and behavior correspond to correlated, equally differentiated facets on the neural level. Experience and behavior differ substantially between individuals. This also holds true for individuals who have developed mental disorders, and it holds true for the mental disorders themselves as well. Two depressed individuals may not have much in common, even if they score identically on, say, the Beck Depression Inventory and both meet the DSM-IV diagnostic criteria for "major depressive disorder." Their depression could appear completely differently. In one patient, it might be accompanied by various other disorders, whereas in the other case, the symptoms are limited strictly to depression. One of them might be bossy and irritable; the other one patient and gentle. How likely would it be, if we used neuroimaging scanners to measure the "resting" brain activity of these two individuals, that we would obtain identical images? It would be extremely improbable. The common label of depression and the fact the both brains were scanned without external interference are rather weak commonalities. If we obtained fMRI or PET scans from two depressed individuals we should not expect, based on these considerations, that they would have much in common. This would be all the more true if one of the scans had been obtained in New York and the other one in Bern. There are many factors that would influence the researchers' reports: The different scanner types, the recording methods; the temporal, geographical, interpersonal, and meaning-related conditions under which the scans were obtained; the competency and experience of the researchers themselves—all of these would potentially differ between the locations and would influence what would be found about the brain activation of the two depressed individuals. Therefore, we should currently not expect that the different research groups would arrive at identical results in their investigations of neural processes in patients with particular mental disorders. We should be quite satisfied when replicated and mutually confirming findings emerge at all in such studies. This is all the more true because the study samples tend to be very small because of the enormous effort required to conduct such studies. Because of these small samples, idiosyncratic patient characteristics would influence the results more than would be the case with larger groups.

An additional complication is that the diagnostic criteria of the DSM and ICD are strictly descriptive. They are based on considerable expert knowledge, but ultimately they are mere conventions. We should not assume that these conventions correspond to disorders with identical etiology. If we defined disorders based on common etiology, we might arrive at a different classification of mental disorders. This path had been abandoned, for the time being, because the etiological models (psychoanalytic, behavioral, etc.) were so different that it was impossible to find a common denominator.

We might see in the future, however, that disorders can be classified based on their common neural underpinnings. Over time, this might lead to functional definitions and, ultimately, perhaps even to a truly etiological classification. The observable symptoms in experience and behavior would be the starting point for the diagnostic process, but the more important basis for diagnosis would be underlying neural processes and structures.

The findings I review in this chapter about the neural correlates of five mental disorders (depression, posttraumatic stress disorder [PTSD], generalized anxiety disorder, panic disorder, and obsessive–compulsive disorder) provide some clues suggesting that a reclassification of subgroups within disorder categories, based on neural criteria, might be quite useful and achievable in the foreseeable future. Such new diagnostic boundaries would complement—and over time perhaps even replace—the crude descriptive groupings we have today. This new classification would probably also suggest that certain disorder-specific neural characteristics inform the process of case conceptualization and treatment planning. However, we are not yet at that point. For the time being, researchers (and clinicians) must use the currently available disorder categories as their starting point. The following overview is based, therefore, on the conventional disorder categories.

I opened the chapter with these preliminary considerations to elicit realistic expectations about what we can expect when we attempt to clarify the neural correlates of mental disorders based on today's classification systems. The research in this area took off no more than about 10 years ago. Only a few studies exist so far for most of the disorders that are frequently encountered in clinical practice. Given the complications I noted earlier, the results from these studies should be regarded as preliminary clues, not as definitive facts. There are many basic scientific findings about anxiety as an alarm system, as was discussed in the second chapter. However, there are far fewer studies specifically about the clinical anxiety disorders. The anxiety disorder that has received the greatest amount of attention in this domain is PTSD. The research on this disorder already allows for relatively clear-cut conclusions. Considerably less progress has been made with generalized anxiety disorder, panic disorder, and obsessive–compulsive disorder, although it is still worth reporting these preliminary findings. The research on phobias and eating disorders, unfortunately, is still so undeveloped or contradictory that more time is needed until preliminary conclusions can be drawn. A greater number of studies have been conducted on pain disorders and schizophrenia, and the findings discovered in these areas would suffice to produce regular review articles. However, because these disorders do not constitute the core domain of psychotherapy, I did not consider it sensible to include them in the context of this book. By far the greatest number of findings are available, however, on the topic of depression. As mentioned before, it is not surprising that not all of these findings are consistent. However, a large number of studies have been completed and clearly replicated findings are emerging, which can by now be regarded as relatively well substantiated.

The research on all mental disorders will dramatically improve in coming years because so many studies are currently in progress. The summaries detailed following, therefore, are preliminary in character. Even for the most common mental disorders, many questions remain. Because my aim in this chapter is to provide a positive perspective on the potential of neuroscientific research and its capacity to inform our understanding of mental disorders, I open with the disorder that is currently most intensively investigated—depression.

The brain regions that are particularly important for each disorder will not be specifically introduced again if they have already been discussed in the preceding chapter. New brain regions that we have not yet encountered in the previous chapter, however, will be described in as much detail as necessary to comprehend their most important, relevant functions.

3.2 NEURAL CORRELATES OF DEPRESSION

Among the many researchers who have studied the brains of depressed patients, one stands out among the others in terms of the continuity and breadth of his research program: The psychologist Richard Davidson of the University of Wisconsin. He embeds his approach to depression in a long series of studies on "affective style" (Davidson, 2000, p. 890), which he regards as a basic neural and psychological disposition and which he began to study as early as 1982 (Davidson & Fox, 1982). This dimension, on the face of it, has little to do with depression per se but is related more closely to two basic neural circuits that have to do with approach and avoidance. I will revisit Davidson's approach once again in chapter 4, in the context of discussing the motivational aspects of neural functioning. Since 1990 (J. Henriques & Davidson, 1990), Davidson has conducted an impressive series of studies in which he systematically evaluated the relevance of this dimension for depression. In Davidson, Pizzagalli, Nitschke, and Putnam (2002), he discusses his own research findings in relation to other studies on the neural correlates of depression. Other good overviews can be found in Drevets (1998, 2001) and Mayberg (1997).

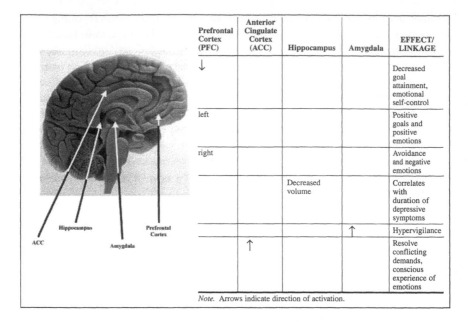

	Prefrontal Cortex (PFC)	Anterior Cingulate Cortex (ACC)	Hippocampus	Amygdala	EFFECT/ LINKAGE
	↓				Decreased goal attainment, emotional self-control
	left				Positive goals and positive emotions
	right				Avoidance and negative emotions
			Decreased volume		Correlates with duration of depressive symptoms
				↑	Hypervigilance
		↑			Resolve conflicting demands, conscious experience of emotions

Note. Arrows indicate direction of activation.

Figure 3–1. Brain structures and linkages to emotional experience. This figure summarizes major research findings, described in chapter 3, related to the association between brain structures and specific emotional states. Note: The general location of the hippocampus and amygdala are provided above, as specific structures are lateral of centerline. The anterior cingulate cortex is a collar-like structure that surrounds upper portions of the corpus callosum.

Four large brain areas have been studied particularly intensively with regard to their involvement in depressive disorders (see Figure 3–1):

- The prefrontal cortex (PFC)
- The anterior cingulate cortex (ACC)
- The hippocampus
- The amygdala

In the following section, I will first review the function of each of these brain areas and then discuss how each of them is involved in depressive functioning.

3.2.1　What is the Role of the Prefrontal Cortex in Depression?

According to E. K. Miller and Cohen's (2001) integrative theory of the PFC, which we first encountered in chapter 2, one of its most important functions is the representation of goals and the means for their realization. Whenever one pursues an important positive long-term goal, such that the emotional satisfaction of goal attainment can occur only much later, the goal and its positive value must be continuously represented on a neural level, and these representations must be defended against distracting situational influences and competing goals. This is the case, for example, if one orients one's behavior to be consistent with some moral value while resisting other types of temptations. We know from Damasio's (2000) studies of individuals with lesions in the ventromedial PFC that especially this function of orienting one's behavior according to valued goals is severely impaired among them, even though intelligence tends to remain intact. Such individuals are, as it were, turned into completely different persons.

The case of Phineas Gage, who suffered a terrible accident as a 25-year-old in the year 1848, has become one of the most important cases in all of neurology because it showed, for the first time, the devastating consequences of severe damage to the PFC. While working with explosive material, an iron rod almost 2 m long and more than 3 cm thick, blasted through the young man's left cheek, traveled diagonally through his head, exited through the top of his skull, and then flew another 30 m through the air. Astonishingly, Phineas Gage survived the accident. After only a short time, he stood up, was able to speak sensibly and sit in a chair, and was brought to a physician, who then documented the case in detail. Phineas Gage was known prior to the accident as a very reflective, determined, and goal-oriented man who was generally well liked; despite his young age, he had already been promoted to foreman. After the accident, his behavior changed entirely. As his physician reported, the balance between his mental and animal faculties (Damasio, 2000) had been disturbed. His manner of speaking had become so rude and vile that women were advised, in consideration of their sensitivities, not to remain in his presence for too long. He had become erratic, inconsiderate, cursed frequently in the most repulsive manner (which had not been among his old habits), did not show respect for others, responded impatiently to external limits and advice that ran contrary to his own wishes, behaved stubbornly at times but at the same time was capricious and emotionally labile, and constantly came up with new goals, which—once set—were immediately abandoned again (based on Damasio, 2000). Phineas Gage went on to live an unsettled life, drifting further and further down in society, until he died 13 years later from complications of epileptic seizures.

Much of what determines our personality takes place in the PFC. In particular, this includes the goals and values that guide our behavior. It is clearly established by now that the two hemispheres of the PFC specialize in different functions in this context. The left PFC "hosts" positive goals and generates positive emotions, whereas the right specializes in avoidance goals and negative emotions. Such a lateralization and its specific behavioral consequences can be documented as early as the 10th month of life. In adult life, differences in the level of spontaneous activity and situational responsiveness between the left and right PFC can be regarded as stable personality features (Davidson, 2000). In some people, the left PFC tends to be more strongly activated, and these individuals experience more positive emotions, whereas in others, right PFC activity predominates, which tends to be accompanied by negative feelings. Among the latter individuals, negative feelings are also easily elicited by situational context, which is related to their greater responsiveness of the right PFC. This does not mean, of course, that in people whose one side is dominant that the other side cannot be activated. Inhibitory, avoidant behavior tends to be accompanied by activation in the right PFC. This was shown, for example, in a study using event-correlated fMRI (Garavan, Ross, & Stein, 1999).

Among depressed individuals, the left PFC is underactive, both in terms of absolute level and in comparison to the right PFC. This corresponds to the shortage of positive feelings and the low level of behavior oriented toward the attainment of positive goals. The hypoactivity of the left PFC makes it difficult for depressed patients to replace their automatized negative emotions and behaviors such as negative ruminations with more positive goal-oriented behavior. The lateral asymmetry among depressed patients reported by Davidson and his colleagues has by now been replicated by several teams of independent researchers (Debener et al., 2000; Gotlib, Ranganath, & Rosenfeld, 1998; Reid, Duke, & Allen, 1998).

The orbital and ventral PFC areas are also involved in the representation of reward and punishment. The left medial region of the PFC responds more strongly to rewards, and the right one to punishment (O'Doherty, Kringelbach, Rollo, Hornak, & Andrews, 2001). Kawasaki et al. (2001) documented with depth electrodes during preparation for brain surgery that single neurons in the right ventral PFC fired specifically in response to the presentation of aversive images. J. B. Henriques, Glowacki, and Davidson (1994) found a differential responsiveness to reward and punishment among subclinically dysphoric students, and J. B. Henriques and Davidson (2000) found the same pattern among clinically depressed individuals. In a task in which the aim was the accurate recognition of objects and words, normal participants adjusted their behavior regardless of whether they were rewarded with an increase of money or punished with a loss of money, whereas depressed individuals responded only to the punishments, not the rewards. Therefore, it was not possible among the depressed persons to establish a positive goal that actively influenced their behavior. After successful pharmacological antidepressant treatment, the activity in the left dorsolateral PFC—which, according to Davidson, represents approach goals—once again increased significantly (Kennedy et al., 2001). The reduction in neural activation therefore appears to be reversible.

A particularly strong habitual hyperactivation of the right PFC, which is linked with avoidance and negative emotions, appears to be a negative prognostic indicator for the pharmacological treatment of depressive disorders. Bruder and colleagues (2001) measured the extent of lateral asymmetry among depressed patients before they were treated with a serotonin reuptake inhibitor (SSRI). The patients who responded well to the treatment had less of a right-sided dominance prior to the treatment. In this study, however, the right-sided hyperactivity was not limited to the PFC but extended to right-lateralized cortex areas that are located further toward the back of the brain. The activation in these posterior regions correlates with negative arousal and anxiety, according to findings reported by Heller and Nitschke (1998). It is well established that depression tends to go hand in hand with anxiety. The studies by Bruder et al. could mean, therefore, that patients with *global* right-sided activation tend to have this common form of comorbidity, and that these comorbid patients tend to respond more poorly to treatment with SSRIs.

The low level of activity in the PFC of depressed patients is not limited to the left side but can also be documented in other areas. The fact that the right ventromedial and dorsolateral parts have repeatedly been found to be more active than their left counterparts does not conflict with these other findings. What we are dealing with here is a *relative* hyperactivation. On the whole, the PFC in depressed individuals could be better characterized as hypoactive, which is linked with an overall reduction in the qualities the PFC specializes in (conscious action planning, control of planned behavior, active thinking, and problem solving, etc.).

The prefrontal hypoactivation has been associated with a reduction in volume of the gray matter in the PFC. This was documented in neuroimaging as well as neuropathology studies. Coffey et al. (1993) found a reduction of PFC volume by 7% in 48 depressed inpatients compared to 76 normal individuals. Rajkowska (2000) studied the neuropathology of brains of depressed patients after they died. She found, in comparison to normal brains, an even more dramatic reduction in volume, which was linked to the fact that the density of large, well-developed neurons was lower, but the density of small, poorly developed neurons was higher. The cell density varied based on which of six cortex layers was being studied. The reduction in neuronal density ranged from 17% to 30%. The density of the glia cells was also reduced by 19%. Rajkowska's results suggest that the volume reduction is a consequence of nonusage and does not indicate a total loss of cells. This provides hope that the volume reduction can be reversed via the activation of the withered neurons.

In samples of unipolar as well as bipolar depressed individuals who also had depressed relatives Drevets (1997), as well as Öngyr, Drevets, and Price (1998), found that the PFC volumes were reduced by values between 39% and 48%, and they showed a reduction in glia cells on the order of 24% to 41%.

These substantial volume reductions clearly show that one cannot expect normal performance from the brains of severely depressed persons, as far as deliberate, rational, active behavior and problem solving are concerned. They lack the structural requirements. Their difficulty to feel joy in experiences and to motivate themselves for activity also has a concrete neural basis.

The currently available findings are all correlational in nature. It is still unresolved whether a reduced PFC volume constitutes a diathesis for the development of depression, whether it is better viewed as a consequence of prolonged deactivation, or whether both possibilities are true. The findings by Drevets, that depressives with a family history of depression had a particularly pronounced volume reduction, could support the argument that this is a familial risk factor. But even if this were true, it would still require formative life experiences to trigger the genes that ultimately lead to such a reduced PFC volume. The results reported by Rajkowska would seem to indicate that the neurons in the PFC begin to wither if they are not used. A dispositional lateral asymmetry in the direction toward right-sided dominance might also constitute a risk factor for the development of depression. However, it could equally well be a consequence of negative life experiences. I will address these questions about cause and effect in more detail in chapter 4, when I revisit the issue of the formation of mental disorders. Answering these questions requires a longitudinal research design. Considering the dynamic nature of depression research today, we should expect that such prospective longitudinal studies will be available soon, which then can further illuminate the questions raised here.

3.2.2   The Role of the Anterior Cingulate Cortex (ACC) in Depression

The ACC is not a functionally homogeneous region. One can differentiate at least two subregions, each of which has its own connections with other brain areas that are in turn specialized on certain tasks. The *affective subregion* has strong connections to the limbic system and is critically involved in the regulation of autonomic and visceral reactions to stressful experiences, as well as in the regulation of emotional expression and social behavior. The *cognitive subregion* is closely connected to the dorsolateral PFC and to motor areas. It is particularly involved in the processing of cognitively demanding information and in determining how one should respond to a particular situation (Davidson et al., 2002; Devinsky, Morrell, & Vogt, 1995; Thayer & Lane, 2000). For our purposes, however, it is not necessary to discuss the two subregions separately, especially in light of the fact that they tend to strongly interact with each other.

The ACC is practically always activated when one is faced with equivocal, uncertain situations or confronted with conflicting demands that allow for multiple possible interpretations. Such situations occur in cognitive (Carter, Botwinick, & Cohen, 1999; MacDonald, Cohen, Stenger, & Carter, 2000) and motivational conflicts (Rogers et al., 1999), in tasks that simultaneously elicit mutually incompatible response tendencies (Pardo, Pardo, Janer, & Raichle, 1999), or in situations with uncertain decisional conditions that require a high degree of risk-taking (Critchley, Mathias, & Dolan, 2001). All of these situations require an increased level of attention and the availability of executive functions. Some authors regard the monitoring of conflicts as the central function of the ACC (Botvinick, Braver, Barch, Carter, & Cohen, 2001). *Conflict*, in this context, has a very broad meaning. The term refers to all situations that entail an incompatibility or interference of simultaneously transpiring perceptions, action tendencies, emotions, and so forth. For example, it includes situations in which one notices that a mistake has just been made; that is, when during the pursuit of a goal, an *incongruence signal*, to use Powers's (1973) term, has been generated (W. Miltner, Braun, & Coles, 1997). One could also say that the ACC always becomes active when some form of inconsistency appears in simultaneously transpiring processes. In those cases, the ACC ensures that processing resources will be made available by involving other brain regions in the processing of difficult situations.

The ACC, therefore, is a sort of monitoring system that always becomes active when some inconsistency appears in neural functioning. Its input consists of impending or already occurring inconsistencies among reactions; incompatibility or incongruence of perceptions, expectations, or goals; activation of mutually competing or mutually exclusive motivational tendencies; and similar phenomena. Its output entails the mobilization of additional resources. This includes the executive functions of working memory and volitional effort. The ACC always becomes active when something might go wrong, when something does not go according to one's wishes, when one encounters a situation that one would rather have avoided, and so forth. It mobilizes all available powers by recruiting other brain areas, such as the PFC, which is "responsible" for the representation and maintenance of goals in the process of resolving the situation.

Among depressed individuals, the ACC is chronically underactivated (Beauregard et al., 1998; Drevets et al., 1997; George et al., 1997). This is why they do not respond alertly when some situation has gone awry or when something is about to go wrong, and why they do not mobilize their available resources, such as volitional effort, to change something about the situation. The no longer activated state of their ACC signals that they have resigned, that they are no longer involved in actively coping with the demands of their environment. It could be that the hypoactivation of the ACC results from active inhibition by the hyperactive right dorsolateral PFC (avoidance goals) because these two brain regions have close reciprocal connections with each other.

When depression is reduced, ACC activity once again increases (Bench, Frackowiak, & Dolan, 1995; Buchsbaum et al., 1997; Mayberg et al., 1999). ACC activity also increases after pharmacological antidepressant treatment; however, it does so only among treatment responders (Mayberg et al., 1997; Pizzagalli et al., 2001). Several authors (Drevets et al., 2001; Ebert & Ebmeier, 1996; Mayberg, 1997) have proposed that the ACC plays an important role in the pathogenesis and symptomatology of depression. The dorsal region of the ACC tends to be linked to the impairment in attention and executive dysfunction, whereas the ventral region seems more involved in the emotional blunting, anhedonia, reduced responsiveness, and lack of coping potential.

There is still much that is unknown, however, about the transactions among the various subregions of the ACC, as well as between the ACC and PFC in depressed individuals. It seems clear, however, that these two brain regions play a central role in the pathogenesis of depression. Davidson et al. (2002) even suspect that two subtypes of depression might be plausible:

• An ACC subtype—These would be patients who have resigned, who do not perceive conflicts between their state and the demands posed by the environment, and who have lost the will to change.
• And a PFC subtype—These would be patients who very much experience the discrepancy between their state and the demands placed upon them, who suffer because of this but nevertheless are unable to activate the PFC-related goal-oriented functions of their behavior in such a manner that their situation is effectively changed.

Based on currently available findings, the PFC subtype, which is characterized by a stronger activation and responsiveness of the ACC, would have a better treatment prognosis than the ACC subtype (Mayberg et al., 1997; Pizzagalli et al., 2001; Wu et al., 1999). However, this is only established for pharmacological treatments. On a phenomenological level, one could say that the PFC subtype entails a greater ability and readiness to face existing demands despite the depressive symptoms. This should actually be a positive feature, especially in the context of psychological interventions. Taking a look toward the future, we might imagine that a part of routine diagnostic procedures would be to test how the ACC responds to inconsistency-eliciting demands, and to base differential treatment decisions on the results of such tests. Functional neural characteristics could then replace or complement the previously descriptive diagnostic criteria. It is still a long way until then, but is seems that the path would be passable in principle. One could experimentally test which psychological or pharmacological interventions have the potential to reactivate or disinhibit an extinguished or inhibited ACC activation. Similar studies could be conducted among patients with the PFC subtype. The proximal outcome criterion for such studies would be a specific change on the neural level. One could test whether, and in whom, these changes actually occur and to what degree the neural changes go along with ultimately targeted changes in experience and behavior.

3.2.3   What About the Hippocampus in Depressed Individuals?

In patients with major depression, the volume of the hippocampus is reduced by 8% to 19%. This is a well-established finding. Davidson et al. (2002) provided an overview of studies in this domain. According to a study by Sheline, Sanghari, Mintun, and Gado (1999), the reduction in volume is correlated with the total duration of depressive phases over the lifetime but not with the patient's age. The reduction in hippocampal volume is, however, not a unique feature of major depression. It is also evident in bipolar disorder (Noga, Vladar, & Torrey, 2001), in PTSD (Bremner et al., 1995, 1997; Stein, Koverola, Hanna Torchia, & McClarty, 1997) and in patients with borderline personality disorder (Driessen et al., 2000). It is a plausible assumption that in all of these disorders, the reduction in hippocampus volume might be a consequence of enduring stress (Sapolsky, 2000; Sheline, 2000). The hippocampus contains many glucocorticoid receptors, which render it particularly sensitive to a chronically increased level of the stress hormone cortisol. Normally, the glucocorticoid receptors of the hippocampus are part of a regulatory mechanism that downregulates the cortisol level if it gets too high. This negative feedback loop, however, does not function among depressed individuals (Pariante & Miller, 2001), leading to a chronically elevated cortisol level. The fact that the cortisol level spins upward instead of being downregulated might be related to chronic stress that cannot be resolved (see chapter 4, section 4.5.2.2 for more detail), or it could be linked to the hippocampus's impaired ability to fulfill this regulatory function. It could also be that both possibilities simultaneously hold true.

The volume reduction in the hippocampus can result from a loss in neurons or glia cells as a consequence of elevated cortisol levels. It could also result from stress-related reductions of neurotrophic factors as well as a lack of formation of new neurons. The exact mechanism of the volume reduction still remains unclear. A chronically elevated cortisol level does, however, correlate with hippocampal atrophy (Lupien et al., 1998). On the whole, there is quite a bit of evidence suggesting that the reduced hippocampus volume among depressed individuals ultimately can be linked to psychosocial stress.

This assumption is all the more plausible given that neural plasticity also holds true in reverse direction. As a consequence of antidepressant pharmacological intervention, new neurons are formed in the hippocampus (G. Chen, Raikoska, Du, Sraji-Bozorgrad, & Manji, 2000; Malberg, Eisch, Nestler, & Duman, 2000). New neurons in the hippocampus are also formed when animals are exposed to enriched environments with many positive forms of stimulation (Gould, Tanapat, Rydel, & Hastingo, 2000; Kempermann, Kuhn, & Gage, 1997). Studies with humans have also shown that new neurons can be formed in the hippocampus—even among adults (Eriksson et al., 1998).

These findings do not preclude the possibility, however, that the hippocampus of depressed individuals might already have been smaller than normal prior to the onset of the depression, and that this might potentially have contributed to the onset. It could be, for example, that a hippocampus that is naturally "weaker" is less effective at downregulating cortisol levels under conditions of stress, which then renders the person more vulnerable to the adverse effects of stress. Another pathway by which a malfunctioning hippocampus might contribute to the formation of mental disorders concerns the poor context-regulation of emotions. One of the most important functions of the hippocampus is the adjustment of behavior to fit with current contexts. This requires in each case the ability to quickly and efficiently establish a link to previously stored contextual information, which is precisely the function of the hippocampus. Many mental disorders, such as depression, anxiety, and borderline personality disorder are characterized by the tendency to respond to certain contexts with "wrong"—that is, not situation-appropriate—emotions. This could be explained by the fact that a poorly developed hippocampus does not fulfill its distinctive function. An underdeveloped hippocampus could then be regarded as a risk factor for the development of depression and other mental disorders. A twin study by Gilbertson et al. (2002) among patients with PTSD suggests the viability of this idea. In this study, one of two monozygotic twins had developed PTSD as a consequence of combat experiences in the Vietnam war, whereas the other one had not experienced this trauma and therefore also had not developed PTSD. A reduced hippocampal volume was evident in both twins, not just the one with PTSD. This suggested that a reduced hippocampus existed already prior the onset of the disorder and might have facilitated the development of the disorder. This could also function similarly in depressed individuals, but this possibility has not yet been examined because of lacking longitudinal studies. The causal function of the hippocampus in the context of the development of depressive and other disorders can only be clarified convincingly by prospective longitudinal studies. A reduced hippocampal volume could contribute to the cause as much as it could be a consequence of depressive disorders. These causal pathways are not mutually exclusive. What can be said with confidence at this point is that the hippocampus tends to be in poor shape in depressed individuals, and that there are clear signs that its condition can improve over the course of successful treatment. The specific functional relationships will have to be clarified by further research.

3.2.4   What Role Does the Amygdala Play in Depression?

We learned about the amygdala primarily as the anxiety center in chapter 2. This function of the amygdala can be viewed as a special case of a more general function: the continuous evaluation of all incoming stimuli in terms of their importance for one's motivational goals. The amygdala directs attention toward stimuli that have high emotional–motivational importance, and it ensures the thorough processing of these stimuli via an increase in cortical arousal and vigilant monitoring of the environment. It is specifically activated by surprising, ambiguous, and uncertain situations. Because uncertainty can imply danger, the activation of the amygdala tends to be accompanied by negative emotions (Davis & Whalen, 2001, Holland & Gallagher, 1999).

Among depressives, the amygdala is frequently enlarged, as a consequence of its continuous hyperactivation. Research has documented increased metabolic activity in the amygdala of depressed individuals, during waking states (Drevets, 2001; Drevets et al., 1992) as well as during sleep (Ho et al., 1996; Nofzinger et al., 1999), which supports the idea that it is chronically hyperactivated. This hyperactivation also manifests in neuropathological changes of the kind that Hrdina, Demeter, Vu, Stonyi, and Palkovits (1993) found in the brains of depressed individuals who had committed suicide. The amygdala in depressed persons is not only chronically hyperactive, it is also particularly responsive to situational influences, such as when confronted with fearful faces (Yurgelun-Todd et al., 2000).

Drevets et al. (1992) and Abercrombie et al. (1998) also found a significant correlation between the degree of amygdala activation and the severity of the depression—that is, the extent of negative emotions—among depressed patients. After successful pharmacological treatment of the depression, the activation level of the amygdala returned to a normal level. Among patients with a family history of depression, however, the hyperactivation of the amygdala remained intact (Drevets et al., 1992). This could be regarded as a sign that in these patients, a heightened responsiveness of the amygdala is one of the risk factors that contributes to the development of depressive episodes. Bremner et al. (1997) showed that patients who had previously been pharmacologically treated were more likely to experience a relapse after discontinuation of the medication if they already previously had a higher amygdala activation level than those who did not relapse. Based on these findings, the amygdala appears to play an important functional role in the onset of depression. We know from epidemiological longitudinal studies that the high comorbidity of depression and anxiety results primarily from the fact that most of the patients initially develop an anxiety disorder and then form an additional depressive disorder on top of this (Alloy, Kelly, Mineka, & Clements, 1990; Nutt, Ballenger, Sheehan, & Wittchen, 2002; Schulte, 2000).

We already know from chapter 2 that the amygdala plays an important role as the anxiety center. It appears that it does not lose this function when an additional depressive disorder is formed. The chronic hyperactivation of the amygdala tends to be accompanied by heightened anxiety-readiness and expectations of negative events. Therefore, primarily negative content is being stored in memory, which in turn can easily be recalled in the form of worried ruminations. The amygdala, then, could play a key role in explaining why depressed individuals are so preoccupied with negative thoughts.

3.2.5   Evaluation and Conclusions for Psychotherapy

We have seen that the brain is massively impacted by depression. If one regards the brain as just another human organ, similar to the heart, the kidneys, and so forth, then it would make sense to view depression as an organic disease. Compared to its constitution among normal individuals, the brain among depressives is structurally and functionally clearly impaired. Biological psychiatrists, who tend to regard mental disorders as diseases of the brain anyway, might interpret the findings reviewed here as grist for their mills. In light of that which I discussed in chapter 2, however, these findings ought to be viewed differently. They confirm once again what we already knew: All mental processes correspond to underlying neural structures and processes. In this sense, there is no difference between the healthy and the diseased. If these studies had not found massive differences between the brains of depressed versus normal individuals, this would have been completely incompatible with the previously reported findings about the functioning of the brain and the parallel nature of mental and neural processes.

However, the question of how these differences originate and if, or whether, they can be repaired is an entirely different matter. There is little evidence to suggest that depression is an endogenous disease that can only be influenced by medications and medical procedures. The brain specializes in the processing of sensory information from its environment and in the positioning of the organism within its context in such a manner that its integrity, its survival, and the passing on of its genes are ensured. All brain areas that are negatively affected in depression have a central function in the brain's transactions with the environment. Because they are so strongly impacted, it is appropriate to regard depression as a severe disorder. However, especially because the respective brain regions have such a central function in the conduct of brain–environment transactions, their states cannot be analyzed separately from the interactions of the specific brain with its specific contextual environment.

The discussions in chapter 2 have shown that brain development is critically influenced by concrete life experiences. There is no reason to assume that this works any differently in the case of depression. The brain of a depressed patient, at the time that symptoms emerge, has become what it is under the prior influences of concrete life experiences. This development is certainly also influenced by the individual's genetic predisposition, but specific life experiences are required to activate the genes that render the brain prepared to develop depression. The reported findings include many clues that stress—that is, the exceeding of current adaptation and coping potentials—plays a critical role in the pathogenesis of depression.

A study by Heim et al. (2000) is particularly informative as to the significance of stressful life experience for the pathogenesis of depression. In this study, four different groups of women between the ages of 18 and 45 were compared:

- Women *with* a current diagnosis of major depression *with* a history of sexual and/or physical abuse during childhood.
- Women *without* depression *with* a history of sexual and/or physical abuse during childhood.
- Women *with* a current diagnosis of major depression *without* a history of sexual and/or physical abuse during childhood
- Women *without* a current diagnosis of major depression *without* a history of sexual and/or physical abuse during childhood

The researchers studied how the women responded as adults—years after the abuse experiences—to a stress test. This test consisted of a speech and mental arithmetic in front of an audience with a preparation period of 10 minutes. Among the measures were heart rate and the two stress hormones ACTH (adrenocorticotropic hormone) and cortisol over a period of 90 minutes, starting 15 minutes before the test.

The depressed women with abuse experiences had by far the strongest stress reactions on all of the stress measures. The ACTH level was six times as high among them, compared to the nondepressed women without abuse experiences. The second highest ACTH increase was observed among nondepressed women with abuse experiences. Intense abuse experiences during childhood undoubtedly constitute intense stress for the person. These early stress experiences seem responsible for the fact that the respective individuals respond with greater stress than persons without these stress-sensitizing traumatizing early experiences. With these predispositions, the later risk for developing a mental disorder increases substantially when additional stressful experiences are encountered later in life. In a study by Bremner, Southwick, Johnson, Yehuda, and Charney (1993), it was shown that especially those Vietnam veterans who had prior experiences of abuse during childhood were likely to develop PTSD in response to their war experiences.

Remarkably, the depressed women in Heim et al.'s (2000) study who did not have childhood abuse experiences showed normal stress responses. This shows once again how important it is to differentiate subgroups of depressed patients according to functional criteria. There is also another path to depression, apart from the one via heightened stress intolerance due to traumatic life experiences. However, one of the main pathways into depression is clearly the one from traumatizing life experiences via increased cortisol levels to a damaged hippocampus. This and the acquired stress intolerance could be viewed as diathesis factors for the development of depressive disorders. Other diathesis factors might be related to a hyperactive amygdala and a lateralized right-hemispheric dominance of the PFC.

Such predisposing characteristics are always the result of previous gene expression processes, which in turn are influenced by life experiences. The findings reviewed here about neural plasticity among depressives as well as the findings emerging from psychotherapy research about the malleability of depression via psychological factors suggest strongly that specific life experiences are the most potent means for shifting the brain of a depressed patient from its impaired state toward a more healthy state.

It is true for the development into a depression as well as out of a depression that the influences impacting upon the brain must be maintained over longer periods of time in order to achieve the structural changes—in negative as well as positive directions—that exist among depressed patients. Neither getting depressed nor becoming healthy again will happen overnight. These processes rest upon the expression of genes, which takes time to transpire.

For both processes—the gene expression that leads toward the development of depression and the gene expression that leads the patient out of the depression—no studies are yet available that specifically examine the interactions between life experiences and neural changes. To study the process of individuals' descent into depression, one would need prospective longitudinal studies with high-risk groups; to study the ascent out of depression, one requires prospective, controlled therapy studies. Fortunately, we can be quite certain that such studies will shortly become available because the research that has already been accomplished in such a short time must be regarded as impressive.

Studies that clarify the transactions among the involved brain areas, the PFC, ACC, hippocampus, and amygdala, will also be quite promising for the further development of psychotherapy for depression. However, the research about the roles of these areas in depression has tended to focus on only one of these areas. It is well known, though, that these areas work together in close collaboration. Very little is known at this point about the negative and positive feedback processes among the involved neural circuits that lead into and out of depression. Such knowledge would be required, however, in order to target our interventions more specifically. For instance, where would one have the greatest leverage to reactivate a deactivated ACC? If the ACC is being inhibited by another brain region, such as the PFC, then one would have to target a different process, such as the excessive avoidance, compared to situations in which the inactivity of the ACC is caused by other processes. Because of the close interconnections of the neural processes involved in these brain regions, it is likely that an intervention that effectively targets one of the circuits will also significantly impact the other brain areas. One could imagine that, based on the presenting circumstances, one would begin with different interventions for different functional subtypes, always targeting the region that promises to offer the greatest leverage for the particular subtype. However, the effect on the total system might ultimately not differ so much because a change in one location also alters the change-input in another, connected circuit.

This could also clarify current findings about the various interventions for depression. Largely equivalent final results can be achieved with Beck's cognitive–behavioral therapy, with Klerman and Weissman's interpersonal therapy, or with Lewinsohn's therapy that aims to increase the frequency of reinforcing activities (Grawe, Donati, & Bernauer, 1994). However, these therapies begin by targeting different processes. Moreover, it appears that the therapies tend to work better with patients who come better equipped for the respective therapy: Cognitive therapy tends to work better for patients with fewer cognitive distortions, and interpersonal therapy tends to be more effective for people with relatively intact relationships. The therapies appear to utilize the functions that have remained relatively intact in order to provide a change-facilitative input to the interconnected total system. This perspective is also supported by findings from a study by Bruder et al. (1997). In a group of depressed patients, these researchers simultaneously played two different syllables for both ears (dichotic listening); the syllables consisted in each case of one consonant and one vowel. Participants differ in such tests in the degree to which they can more accurately discriminate with either the left or the right ear. Normal right-handed individuals can discriminate more accurately with the right ear, indicating that they process the stimulus more accurately with the left hemisphere. Depressed patients who successfully completed a course of cognitive therapy were more than twice as accurate with their right ear, prior to the therapy, than patients who did not improve in therapy. Good left-hemispheric processing of the tones was the best predictor of successful therapy outcome in this study. The therapy used in the study relied heavily on verbal processing; for example, in the evaluation and refutation of irrational assumptions and so on. No differences were observed in this study in a nonverbal discrimination task.

Jacobson, Martell, and Dimidjian (2001) regard as the common mechanism of action in successful psychological therapies for depression their ability to reestablish positive behavioral activities. On a neural level, the activation of the dormant left PFC would thus be the common target of all psychological interventions that have been shown to be effective in the treatment of depression. In absolute form, it is perhaps not possible maintain this interpretation of all available study results. It could be, however, that the activation of positive approach goals and positive emotions in the left PFC is one of the most easily achieved change-facilitative inputs into the complex network maintaining the depression. Further, it could be that this input plays de facto a role in all forms of psychological therapy, regardless of whether the treatment's rationale regards this as a mechanism of action at all.

In all therapies for depression, however, there are some patients who do not profit from the treatment. We know very little about the extent to which these "nonresponders" are treatment-specific nonresponders who might benefit more from another therapy, or whether they tend to be patients that are generally difficult to treat. It would be highly desirable to obtain more systematic knowledge about the neural characteristics of these nonresponders. Some studies exist about the characteristics of patients who did not profit from pharmacological treatment. I reviewed those characteristics in the previous section. However, such information is not yet available for nonresponders in psychological therapies, except perhaps for the study by Bruder et al. (1997) that was just reviewed. Such information would be invaluable for the purpose of differential case conceptualization and treatment planning. This information could help us design entirely new therapeutic strategies that on the one hand would utilize the neural characteristics of depressed patients as resources, and on the other hand would avoid targeting those areas that hold little promise for treatment success.

In terms of future of treatments for depressive disorders, there is probably reason to be mildly optimistic. On the one hand, the neural findings show that this is a truly serious disorder with structural brain changes. On the other hand, our understanding of the neural bases of depression can be expected to improve rapidly, and it also seems clear that this domain is characterized by considerable neural plasticity. It can be easily imagined that in 20 years time the routine diagnosis of depressed patients will look completely different from today, such that neural characteristics that inform a patient's likely treatment response will be routinely assessed.

Based on currently available neuroscientific and psychological research findings, I would argue that it is already possible today to conceptualize depression differently from how it is typically done. I will not specify this in more detail here because further foundations for this argument will be built in the next chapter. The neuroscientific research on depression still has a considerable weakness: It neglects the motivational dimension of neural functioning. In my view, an appropriate understanding of depression cannot be reached unless one explicitly relates the depressive changes to motivational functioning, and this may also explain why interventions' effectiveness remains below the level of what could actually be achieved. I will revisit this question in chapter 4, in the section on the development of mental disorders. I will use depression as my preferred example in that section in order to elaborate my view on the emergence, maintenance, and treatment of mental disorders in maximum detail.

3.3 NEURAL CORRELATES OF POSTTRAUMATIC STRESS DISORDER (PTSD)

Current overviews of the neural correlates of PTSD can be found in Wessa and Flor (2002), Flatten (2003), and Fujiwara and Markowitsch (2003). Similar to depression, research has identified specific structural and functional changes in the brain.

Bremner et al. (1995) found among Vietnam veterans with PTSD that right hippocampal volume was reduced by 8%, compared to a matched control group. No differences were found in other brain regions. In another sample of psychiatric patients who had been selected based on whether they reported physical or sexual abuse in their childhood, Bremner et al. (1997) found a reduction in left hippocampal volume by 12%, compared to a matched control group. All patients who had been selected based on this criterion also fulfilled criteria for PTSD. This study also showed no significant differences in other brain regions. Both samples were characterized by massive comorbidity with other mental disorders. It is unclear, however, why it was the right hippocampus in the one but the left hippocampus in the other study that was reduced in volume. Gurvits et al. (1996) found a reduction of 20% to 25% in the hippocampus among patients who had developed chronic PTSD after traumatizing childhood experiences. Even though these studies are inconsistent in the exact percentage of the volume reduction, and the lateral differences still need to be clarified, it seems highly probable that PTSD is accompanied by a severe impairment in the hippocampus.

Several studies have confirmed by now that sexual and physical abuse during childhood can lead to a reduction in hippocampal volume (e.g., Stein, Koverola, Hanna, Torchia, & McClarty, 1997). Until recently, it was regarded as nearly certain, based on the available studies, that the obtained impairments in the hippocampus among PTSD patients, similar to the situation among depressives, could be linked to a temporally or chronically elevated cortisol level (Bremner, 1999a; 2001; see also the study reviewed in the section on depression by Heim et al., 2000). However, the significance of stressful experiences during childhood for the later development of mental disorders holds true not only for PTSD and depression but appears to apply to mental disorders more generally. I will revisit this point in detail in chapter 4, where I elaborate my ideas on the development of mental disorders.

Traumatic experiences suffered during childhood predispose generally to the later emergence of mental disorders. Individuals with such histories often have several mental disorders that are more at the center of the clinical presentation than PTSD itself, even though PTSD criteria are also met. Traumatic events can of course also be encountered later in life and can still lead to PTSD then. In such cases, the PTSD diagnosis is more likely to take center stage and will be assigned as the principal diagnosis, even though it is still common that the criteria for one or several other mental disorders might also be met.

Compared to other mental disorders, PTSD that is elicited by a specific trauma in adulthood has the advantage that the proximal trigger of the disorder is clearly identified. This is why adult PTSD is a particularly good model for studying the pathogenic process. If one studies large groups of people who have experienced the same traumatic event, some go on to develop PTSD whereas other do not. What would be more obvious than studying these groups in terms of all conceivable differences?

We know from such studies that individuals who develop PTSD as adults are more likely to already have a history of traumatic experiences during childhood (Bremner et al., 1993) and that they have a much stronger hormonal response to stressors than adults without such predispositions (Heim et al., 2000). Until recently, it was assumed that this increased excretion of stress hormones leads to the atrophy of the hippocampus that is regularly found in PTSD patients. However, this assumption had to be revised in light of new research results.

Gilbertson et al. (2002) conducted a methodologically very convincing study with 70 monozygotic twin pairs. In 24 of the pairs, one of the twins had developed chronic and severe PTSD as a consequence of combat experiences in the Vietnam war. Among the other 46 twin pairs, one twin in each case also had combat experience but had not developed PTSD. In the comparison between the twin groups with versus without PTSD, the expected small difference in hippocampal volume in the twins with PTSD was indeed observed (measured by fMRI). Surprisingly, however, within each pair, the other twin, who did not have the traumatic experience, had an equally small hippocampus. In twin pairs in which one of them had PTSD, both twins equally had a smaller hippocampus than the twins in the pairs in which neither of them had PTSD. Moreover, the PTSD severity in the groups with the PTSD twin correlated significantly and inversely not only with the size of the hippocampus of the diagnosed twin ($r = -.64$) but also with the size of the hippocampus of the undiagnosed twin ($r = -.70$).

The results of this study suggest that a dispositionally smaller hippocampus constitutes a genetic risk factor for the development of PTSD. Gilbertson et al. support their argument with studies in mice and monkeys that showed that animals with a dispositionally smaller hippocampus showed stronger fear reactions and a heightened cortisol level in conditional experiments. This is consistent with the idea that humans with a less effective (smaller) hippocampus acquire conditioned fear responses particularly easily and may be more likely to form prolonged anxiety reactions when exposed to traumatizing events. The fact that the twin with PTSD did not have a hippocampus that was smaller than that of his brother is inconsistent with the notion that the toxic influence of increased stress hormones leads to a reduction in hippocampal size in PTSD patients. If the heightened level of stress hormones had damaged the hippocampus above and beyond its already small size, then one should have found an even smaller hippocampus in the PTSD twins, compared to their brothers. That was, however, not the case.

These results cast doubt on the previously common assumption that PTSD patients' smaller hippocampus are a result of the disorder. Their smaller hippocampuses are apparently not a consequence of traumatic experience but instead the cause that the stressful experiences cannot be processed more effectively. Therefore, it appears to be the cause responsible for the traumatizing effect and its adverse long-term consequences. It is certainly premature to draw such far-reaching conclusions based on a single study. However, Gilbertson et al. took great care to consider whether their findings could also be interpreted differently, more consistent with previous assumptions. All of the other interpretive possibilities are inconsistent with their data. It could well be, then, that further studies will force us to change our previous assumptions. It may not always be the case that adverse life experiences lead to structural changes in the brain; sometimes, dispositional structural brain characteristics may make it more difficult for people to cope with difficult life situations. This raises the question of whether similar principles might also hold true for disorders other than PTSD. This would further differentiate the research in this area. The fact that uncontrollable stress has very unfavorable effects on the brain is so well established that even new results could not easily undermine its validity (see chapter 4). However, new findings suggesting that this association might also hold true the other way around suggest that the relations might turn out to be even more complex than previously assumed.

Apart from this not-yet-clarified etiological question, the actual development of PTSD depends primarily on how the traumatic event is being stored in memory. If one attempts to draw a conclusion from a great number of studies, it seems that the trauma tends to be stored exceedingly well in implicit memory but insufficiently in explicit memory, and that the symptoms of PTSD emerge, in part, as a result of a dissociation between implicit and explicit memory (van der Kolk, Burbridge, & Suzuki, 1997; van der Kolk, Fisler, & Bloom, 1996; von Hinckeldey & Fischer, 2002; Wessa & Flor, 2002). Because the hippocampus plays a crucial role in the formation of explicit memory contents, the assumption that the formation of PTSD is linked to an impairment in explicit memory storage is very much consistent with Gilbertson et al.'s finding that individuals with a dispositionally smaller hippocampus were at increased risk for the development of PTSD.

A traumatic event is an extremely potent fear stimulus that triggers an immediate, very strong reaction in the amygdala, which then leads to the sort of conditioning of visual, acoustic, and other sensory cues associated with the trauma that later become very difficult to extinguish. These memory traces become engraved in the amygdala and can subsequently be easily triggered in full strength by corresponding sounds, smells, or images—without such exposure leading to extinction. Based on the discussions on fear and anxiety in chapter 2, we already know that the anxiety reaction in the amygdala generally cannot be extinguished but, at best, can only be inhibited. Individuals who develop PTSD are distinct from others because they do not develop this inhibition.

We also know from chapter 2 that such an inhibition is possible primarily via the orbitofrontal cortex, which is closely connected with the hippocampus. The hippocampus is responsible, among other things, for the processing of and storage of contextual information. It is part of the symptomatology of PTSD that the people affected by it consciously avoid recalling anything having to do with the trauma. However, this is not only a question of motivation but also has to do with an inability to do so. For them, recalling explicit trauma-relevant memory contents is objectively more difficult. Not only do they wish to avoid the memory of what they experienced, they also cannot recall this as well as people who do not develop PTSD after such experiences. Based on currently available research, we can infer that in persons who develop PTSD, an impaired encoding of the experiences into episodic memory occurs in the time immediately following the event. The reasons for this might include:

- An already impaired hippocampus.
- The impairment of current memory formation because of an excessively high level of stress hormones.
- The avoidance of processing of that which has occurred.

The normal processing of terrifying experiences usually includes the retelling of the experiences to persons we feel close to, the reenvisioning of what exactly has taken place—reliving it, as it were, but this time no longer in an acutely threatening context. During this processing, the hippocampus tends to be highly activated. Even later, during sleep, and even after a period of several weeks, a further encoding of the experiences takes place, and an increasing number of the brain regions originally involved in the trauma are included in this processing. This process of forming explicit memory content for the traumatic experience apparently does not occur to a sufficient degree among people who do not develop PTSD.

Simultaneously, however, implicit memory traces have been formed, primarily in the amygdala, and these can be frequently and easily triggered by sense-specific cues, bodily states, and so forth. Normally, these cues would be "defused" when they appear by the context that has been formed in explicit memory during the normal processing of the traumatic events. A dewarning takes place, based on the context that has by now been formed, that removes the threatening meaning from the cues. If this context were not quickly coactivated each time trauma cues appear—if the exact localization of these traumatic past experiences in their current, nonthreatening context did not take place immediately, at lightning speed—then the easily triggered amygdala activation would not be inhibited. A well-functioning hippocampus is indispensable for the relearning of context that normally happens when a person processes that a terrifying event has occurred. If the hippocampus malfunctions because of a previous impairment or a current functional inhibition (see Fujiwara & Markowitsch, 2003), the formation of easily recalled explicit context memory is undermined. After "normal" processing, the activation of memory traces in the amygdala immediately triggers the coactivation of this context memory, such that the new situation is viewed in its current, nonthreatening context and the emerging anxiety reaction is inhibited just as it is about to take off. The inhibition of the central nucleus of the amygdala actually originates from the PFC, which evaluates the total situation based on current perceptions and currently activated memory contents. These processes transpire so quickly, of course, that not they, but only their result, is experienced consciously.

Empirical findings appear to be consistent with this conceptualization. Not only is the hippocampus structurally and functionally clearly impaired among PTSD patients, but dysfunctions in the medial and prefrontal cortex were also documented when these patients were confronted with trauma-relevant cues (Bremner, 1999b).

Interestingly, Shin et al. (1997) found in such a study that Broca's area, which is important for language processing, was simultaneously deactivated as the patients were confronted with trauma-related cues. The same finding was reported by van der Kolk (1997a). Based on these results, it makes sense that patients would experience a welcome, therapeutic effect when they are encouraged to write in detail about their distressing memories. Pennebaker (1993) reported in such studies remarkable effects not only on individuals' psychological well-being but also on their immune system and physical health.

Based on these findings, one should expect a symptom reduction among PTSD patients to the degree that the dissociation between explicit and implicit memory contents can be effectively removed. By using therapeutic procedures that entail the reexperiencing and repeated reprocessing of trauma-relevant cues in a safe context, a fully formed explicit trauma memory is created, which then becomes the basis for the increasingly effective inhibition of implicit memory traces and anxiety reactions in the amygdala. This requires repeated confrontations with sense-specific triggering cues after a well-elaborated explicit trauma memory has been created. The formation of inhibition is not simply a question of insight but rather a question of the frequency with which the anxiety-triggering cues and the anxiety-inhibiting context are simultaneously coactivated. It almost goes without saying, based on what was discussed in chapter 2, that this anxiety-inhibiting context also entails a safety-promoting environment, a trusting therapeutic relationship, the activation of positive motivational goals, and so forth.

The formation of a detailed explicit trauma memory is a distinctive feature of PTSD treatment, compared to interventions for other disorders. This can be regarded as the disorder-specific component of effective PTSD treatment, whereas other mechanisms of therapeutic effectiveness—such as the therapeutic relationship, a resource-oriented context, and the confrontation with the problematic situation in a problem reaction inhibiting context—are as important in other disorders as they are in PTSD.

The perspective that emerges from the neuroscientific research on PTSD appears remarkably consistent with the view that has been articulated by psychological and therapy researchers without specific reference to neuroscientific findings (Ehlers, 1999; Ehlers & Clark, 1999). Ehlers (1999) noted in this context:

> It is now understood that the trauma memory in PTSD is insufficiently elaborated (i.e., processed in its meaning) as well as insufficiently integrated in its context of time, location, preceding and subsequent information, and other autobiographical memories. Therefore, the semantic recall route is relatively weak; the resulting memories do not have temporal context ("here-and-now" quality); they are not connected with later information (e.g., "I have not died"); and they are easily triggered by relevant cues." (p. 17)

The intrusive reexperiencing has precisely those qualities that characterize implicit memory: Sensory impressions are being experienced as if they happened right now, not as if they were memories of an earlier experience. The reexperiencing can be triggered by all kinds of cues, which are more or less accidentally linked with the trauma (have been conditioned by the trauma), and which often do not have a truly meaningful relationship with the trauma. From a neural perspective, all of these qualities emerge because neural circuits in the amygdala and surrounding areas are being activated directly, without participation of the hippocampus. According to Ehlers, these undesired sensation-like memories contrast starkly with the difficulties PTSD patients report about their inability to fully recall the trauma volitionally. In light of that which I reported earlier about the reduction and dysfunction of the hippocampus, the deficits in autobiographical trauma-related memory among PTSD patients should not at all surprise us but, rather, they seem perfectly consistent with expectations.

PTSD can, therefore, serve as a good example of how the combination of psychological and neuroscientric research can yield a comprehensive understanding of the factors contributing to the etiology and maintenance of the disorder, which then enables therapists to tailor their treatment to the unique needs of each patient, based on their understanding of the processes creating the disorder. The conceptualization of the disorder determines which elements of the intervention must necessarily be realized in order to achieve an effective improvement in the disorder. In the concrete treatment process, the therapist then combines this disorder-specific knowledge with his or her general knowledge of therapeutic change processes, which also considers other characteristics of the patients that should be taken into account in treatment planning. The treatment of two patients with PTSD can look quite different on a concrete level, even though the interventions might be based on the same understanding of disorder etiology and therapeutic change processes. Beyond clarifying our understanding of PTSD, the neuroscientific studies of this disorder can also inform models of the pathogenesis of mental disorder more generally. Most of the PTSD samples included in these studies are characterized by massive comorbidity. Nearly all of the patients tend to have either one or several additional clinical disorders. This is particularly true for patients who have been selected based on the criterion of having experienced abuse during childhood. This massive comorbidity makes one point crystal clear: Traumatizing events—that is, life experiences that cause uncontrollable stress—play a central role in the development of mental disorders, including disorders other than PTSD. The example of PTSD shows clearly that it is not the event itself that causes the trauma. If we exclude extreme traumatic events such as horrific torture or having to witness family members being gruesomely murdered (up to about 50% of individuals will develop PTSD after such extreme events), it appears that about 7% to 15% of individuals will typically respond with PTSD to a catastrophic event such as the attack on the World Trade Centers. About 8 or 9 out of every 10 persons are able to cope with such events without developing PTSD. Therefore, the individual personality characteristics of the individuals largely decide whether PTSD emerges or not. The terms *trauma, traumatic, traumatizing*, and *stress* must be used very carefully, then. A terrifying event typically becomes a trauma only for a minority of the individuals affected by it, by triggering an unrelenting neural–physiological stress reaction with enduring harmful consequences.

One can truly comprehend the development of this mental disorder only when one relates the event to the individual person with his or her idiosyncratic motivational goals, capabilities, and coping repertoires. The disorder remains a part of the same neural system that has created and continues to create it. From a neural perspective, the structures of the brain that create and maintain a disorder are so closely inter-linked with many other neural circuits that it seems short-sighted to target only the disorder in an intervention while disregarding the neural environment in which it is embedded. I will more fully discuss these issues, which are of particular importance for psychotherapies of mental disorders, in chapter 5.

3.4 NEURAL CORRELATES OF GENERALIZED ANXIETY DISORDER

Anxiety disorders can be conceptualized as unsuccessful forms of emotion regula-tion. Strong emotions emerge in situations in which they are generally deemed inap-propriate. Anxiety is linked with heightened arousal and elevated activity of the sym-pathetic nervous system. From a neural perspective, normal, well-adjusted behavior results from a smooth coordination of arousal and inhibition. From this perspective, excessive anxiety can also be construed as a deficit in inhibitory activity of the parasympathetic nervous system. The appearance of excessive anxiety can be related to the malfunctioning of the inhibitory processes that normally suppress the emer-gence of anxiety in this context. This model of anxiety disorders is elaborated by Thayer and Lane in their model of neurovisceral integration in emotion regulation and dysregulation (2000). Friedman and Thayer (1998) applied this view to panic dis-order, and Thayer, Friedman, and Borkovec (1996) applied it to generalized anxiety disorder. These authors regard the variability of heart rate as an indicator of success-ful and failed anxiety regulation. This variability is controlled by a complex neural circuit, which the authors term the *central autonomic network* (CAN). Other authors, such as Devinski, Morrell, and Vogt (1995) called it the *anterior executive region* (AER) or *rostral limbic system*. According to Thayer and Lane, this system plays a crucial role in emotion regulation. The system encompasses the anterior cingulate cortex (ACC), the insular and orbitofrontal cortex, as well as the amygdala, the cen-tral gray, the ventral striatum, and nuclei in the autonomic brainstem. PET studies have repeatedly found that specific areas in the AER are particularly activated during the experience of normal and pathological emotions (Reiman, 1997).

We already encountered some of the regions involved in this system, along with their functions, in previous sections. This is especially true for the amygdala, the orbitofrontal cortex, and the ACC. Regarding the functions of the ACC, however, a few notes to add to the information provided in the section on depression are in order. We previously learned about the ACC as an "inconsistency motor," as it where, that directs attention and the resources of the volitional control mode toward areas where they are needed in order to resolve interferences in the neural system. However, the ACC also plays an important role for the conscious experiencing of emotions (Reiman, 1997). Lane et al. (1998) triggered different emotions in 12 healthy women by showing them films or asking them to recall memories. The ability for the differentiated, conscious experiencing of complex emotions was measured with a Level of Emotional Awareness Scale. The values on this scale correlated significantly with the activation of the ACC (Brodmann area 24) during the experiencing of the emotions. It seems, then, that the ACC is crucially involved in the processing of emotional information and the conscious experiencing of emotions. However, it is clear that the ACC is not the only area involved but functions as a part of the CAN or the AER. For a different part of the emotion-regulating system—tne insular cortex—Lane and Thayer (2000) found a significant correlation between its activation and inhibited heart rate variability.

According to Devinsky, Morrell, and Vogt (1995), the AER evaluates the motivational meaning of external and internal stimuli. It generates context-relevant emotions and corresponding behavior by triggering some and inhibiting other neural activation. The process of inhibition is at least as important as that of activation in this context. The various regions of the CAN or the AER are reciprocally connected with each other, which enables rapid negative and positive feedback processes. Negative feedback loops are particularly important in order to maintain the emotional reactions within a certain area of variability. If this inhibition via negative feedback does not function, a positive feedback loop is set in motion and leads to emotional dysregulation. According to Thayer and Lane, anxiety disorders can be viewed as such forms of dysregulation.

The coordination between sympathetic and parasympathetic arousal is an important output of the CAN. It can flexibly increase or reduce heart rate via the vagus nerve and the stellate ganglia in order to match current situational demands. A high degree of heart rate flexibility is a marker of good adaptability. A heart rate that can vary flexibly in response to situational demands tends to go along with flexible attention, good emotion regulation, and flexibility in psychophysiological as well as behavioral responsiveness. Normally, the vagus nerve exerts a continuous inhibitory control on heart rate. This tonic inhibition prevents the heart from consistently overreacting.

Thayer, Friedman, and Borkovec (1996) found in an experimental comparison between patients with generalized anxiety disorder (GAD) versus normal participants that the GAD patients had a tonic inhibition in heart rate variability. A second finding was that both study groups had a phasic reduction in heart rate variability during an experimental task in which they were asked to ruminate for 10 minutes about an individualized worrisome topic. Worried rumination, then, typically tends to go along with reductions in heart rate variability. Thayer and Lane interpreted their findings as consequences of positive feedback processes in the CAN and breakdown of the regulation via negative feedback loops that normally happen in that location to prevent a dysregulation in the form of chronic hyperarousal. They argued that GAD is characterized by a dysregulation in the CAN, resulting in a breakdown of tonic heart rate inhibition via the vagus nerve, which then leads to continuous hyperstimulation of the heart. Friedman and Thayer (1998) also use this principle to explain the heart symptoms in panic disorder. However, I would like to maintain the focus on GAD for the moment. In GAD, some of the functional characteristics—and especially the core symptom of "worrying,"—can also be related to this tonic hyperarousal and the associated reduction in heart rate variability.

GAD patients respond to novel, nonthreatening stimuli with a reduced orienting response. A well-functioning orienting response is important in order to react quickly and appropriately to novel situations. However, the orienting response habituates more slowly in GAD patients. Both processes limit these patients' situational adaptability. GAD patients are continuously hypervigilant, scanning their environment for potential danger. Their avoidance system is practically chronically activated, even though their avoidance goals can never really be attained (see section 4.7.5. on this point). These goals require constant vigilance and tend to occupy attention. This, in turn, limits GAD patients' ability to turn their attention to other stimuli with more positive valences. Being continuously preoccupied with impending dangers for which no active defense is possible—because these dangers exist primarily in the patients' imagination—leads to considerable restrictions in actual behavior. This also limits the possibilities for experiencing events that would clearly expose the fears as unfounded. All activity, then, tends to center on this "worrying," on being ruminatively preoccupied with personal concerns.

GAD patients are permanently oriented to the perception of danger; however, when they are actually confronted with threatening stimuli, they respond with cognitive avoidance. Deeper elaborative processing of the threatening information does not occur. When threatening events are anticipated, their heart rate at first slows down, then increases when the events occur, and is slow to habituate when the events are encountered repeatedly. This also indicates the kind of defensive reaction that has been repeatedly found among individuals with intense anticipatory anxiety. From this perspective, the worrying—the core symptom of GAD—has an avoidance function. It ties up attention and thus prevents the person from engaging with those aspects of real-life situations that are truly important for his or her motivational goals. The GAD patient does not fully engage with emotional experiences and thereby avoids the occurrence of corrective emotional experiences.

In summary, GAD patients have a reduced orienting response to novel stimuli; they habituate poorly even to innocuous stimuli; they respond with a conditioned reduction in heart rate when threat is anticipated and with increased heart rate when such events actually occur. In all of these aspects, they differ from individuals without anxiety disorders. GAD patients are characterized by impairments in attentional regulation and emotion regulation. The disorder consists of an inability to turn off the threat and avoidance system when no threat is objectively present. This leads to chronic hyperarousal and anxiety. They cannot experience safety in situations when they are actually safe. Their emotional responsiveness is restricted; their emotional sensitivity blunted (Sollers, Mueller, & Thayer, 1997).

In a relatively small sample of patients with generalized anxiety disorder, Thayer and Lane (2000) reported that cognitive–behavioral treatment was associated with an increase in heart rate variability that went along with symptom reductions and was present even 2 years after the termination of treatment. If one regards heart rate as an output of the CAN, then this supports the idea that the treatment achieved enduring changes in this neural circuit, which is critically involved in emotion regulation, and it appears that these effects were attained via the facilitation of anxiety-inhibiting processes. In their model, Thayer and Lane generally emphasize the importance of inhibitory processes for effective, context-appropriate emotion regulation.

3.5 NEURAL CORRELATES OF PANIC DISORDER

We already briefly mentioned that heart rate variability is reduced in panic disorder as well (see Friedman & Thayer, 1998). The heart-related symptoms that often are part of the presenting problem in panic disorder can be regarded as the consequence of CAN dysregulation, in a similar way to what was described previously for generalized anxiety disorder. However, panic disorder encompasses additional neural characteristics.

In an experimental study, Wiedemann et al. (1999) compared levels of brain electrical activity among 23 patients with panic disorder but no concurrent depression versus a healthy control group. They measured electrical activity in various brain regions, during resting states as well as while participants viewed several types of pictures that differed in their emotional appeal: Emotionally neutral pictures (mushrooms), erotic pictures, pictures of spiders, and pictures of a horrific accident. The accident picture was regarded as particularly relevant for the triggering of panic-specific reactions. All patients were asked to complete anxiety and depression questionnaires and to semantically rate the meaning content of the pictures. Indeed, the panic disorder patients rated the accident images—and only these images—as significantly more negative than did individuals in the control group.

In the panic disorder patients, but not among control patients, the researchers found a significantly stronger activation in the right than in the left frontal lobe during the resting phase as well as during the viewing of the pictures, with the exception of the emotionally neutral mushroom pictures. We already know, based on the previously described findings by Davidson and his group, that right-sided lateral asymmetry is regularly found among depressive individuals and that this can be interpreted as a hyperactivation of the avoidance system. In this study, the researchers had specifically ensured that the panic-disordered patients were not also depressed at the same time. The stronger right-sided activation in the patients of this study shows, then, that negative emotions and avoidance tendencies are particularly easily triggered among patients with panic disorder. The experimental situation itself, which was rated as more aversive by the patients than by the control participants, was enough to trigger these tendencies. But also all the pictures with emotional contents—even those with erotic rather than anxiety-specific content—led to a stronger right-sided activation. It was only the emotionally neutral picture that was not associated with a difference between the panic disorder versus control individuals. It is also noteworthy that lateral asymmetry among the patients, but not the controls, correlated significantly with the anxiety level as measured by the SCL-90-R and STAI-S but not with the depression level as measured by the BDI. The lateral asymmetry found in this study was, therefore, not confounded by differences in depression but appears to be a specific marker of patients with panic disorder as well. Clinically, it is of course highly plausible that the neural circuits that are linked with negative emotions and avoidance tendencies would be particularly well established and easily activated in individuals with panic disorder.

It should also be noted that the assessments in this study were not obtained during a panic attack. The findings, therefore, do not inform us about brain activity during acute panic attacks. It would be expected that the amygdala is massively involved in such attacks. This feature, however, makes the study even more interesting, if anything. Patients with panic disorder differ on a neural level from others, even independently of the occurrence of their core symptom of panic. They are characterized by more easily triggered negative emotions and avoidance reactions, which might be a preliminary step on the path leading to panic attacks. The actual occurrence of a panic attack then probably also involves the dysregulation of the CAN, as discussed in the previous section. From this perspective, the lateral asymmetry in the frontal cortex could be viewed as a potential target point for therapeutic interventions. The idea would be to use appropriate therapeutic strategies to inhibit the tendency to respond with negative emotions and avoidance responses to all types of emotional situations. Such strategies would initially require the activation of neural circuits that could produce the desired inhibitory effects. These circuits would likely be the ones having to do with approach tendencies and positive emotions.

We arrived at similar therapeutic conclusions several times in previous sections: In order to inhibit an undesired reaction, it appears necessary to first establish a positive context from which the inhibition can originate. Once this has been achieved, the facilitation of the inhibition requires that the patient is repeatedly confronted, in this positive context, with situations that activate the undesired reaction, until finally the inhibition of this undesired activation predominates. This might be a basic therapeutic principle that can be applied to many specific disorders but that, in itself, is valid across various disorders.

A very different perspective on panic disorder has been described by Panksepp (1998, chapter 14). He argues that a close connection exists between the neural system underlying panic and the "separation distress system" (p. 262). This separation system, which is activated in many young mammals when they are left alone, has been examined in great detail in animals. It overlaps considerably with the system known as the attachment system. The neural correlates of this system will be discussed in chapter 4.

The activation of the separation system initially manifests behaviorally in the form of isolation calls or distress vocalizations (DV); that is, plaintive utterances that resemble crying. The natural trigger for these DVs is the sudden occurrence of alone-ness. Being alone for longer periods is often accompanied by behavior resembling depression. The neural circuit that underlies the plaintive utterances and corresponding emotions can be activated in various brain areas via electrical stimulation. This has been examined in detail in different mammalian species, including primates. Panksepp explicitly refers to this neural circuit as the *PANIC system* (1998, P. 262). He specifically describes the locations in the brain that can activate this circuit via stimulation by electrodes. These areas tend to be near other areas whose activation triggers pain, which leads Panksepp to suspect that the panic system might have emerged phylogenetically from the pain system.

The PANIC system can be activated neurochemically primarily by the neurotransmitter glutamate and the neuropeptide corticotropin releasing factor (CRF)s, even in situations when the animal is not socially isolated. The system can be dampened or deactivated by opiates as well as by the neuropeptides oxytocin and prolactin. Once the PANIC system is activated, various other transmitters further modulate the process, but they are not able to activate the system by themselves.

Panksepp clearly differentiates the PANIC system from the fear system, which is characterized by anticipatory and conditioned fear reactions. An important point in support of this argument is that the two systems respond to different chemical agents. The fear system responds to benzodiazepines, but the PANIC system does not. With opiates, the opposite holds true. The trycyclic antidepressant imipramine reduces symptoms not only in depression but also in panic disorder. It is also highly effective in the reduction of distress vocalizations in animals. Librium and valium, by contrast, exert effects on the fear system, but they have been shown to be ineffective for the treatment of panic disorder. Similar findings have been shown for selective serotonin reuptake inhibitors (SSRIs). Tricyclic antidepressants also have a positive effect in the treatment of "school phobias," which are closely linked with the separation system.

Additional points that Panksepp uses to document the difference between these two anxiety-related systems are the symptomatic similarities between panic attacks and acute separation anxiety: feelings of weakness; "lump in the throat" sensations; breathing difficulties; and the feelings of suddenly being abandoned, alone, and helpless. He also referred to a study by Torgersen (1986), which showed that patients with panic disorder often have a history of separation anxiety in childhood.

Panksepp defends his argument in response to the obvious objection that panic disorder has much in common with other anxiety disorders, and is often accompanied by other anxiety disorders, by noting that these circuits partially overlap and interact with each other. For example, he argues that it is only natural to respond with anticipatory anxiety to the possibility of additional panic attacks (fear of fear). In the case of this anxiety about anxiety attacks, the fear system rather than the panic system is activated.

The findings and differing perspectives on panic disorder reviewed here demonstrate that a great deal of additional research will be necessary to clarify the neural bases of the various anxiety disorders. The question of to what extent they are based on common and/or separate neural circuits is so far only insufficiently resolved. The fact that different symptoms are related to different neural correlates appears obvious, based on all that we know about the brain's functioning. The high degree of comorbidity among anxiety disorders, however, supports the idea that there are important commonalities in the neural foundations of the pathogenesis of these disorders. An important goal for future research will be, then, to clarify the extent of the neural commonalities and differences in the various anxiety disorders. If they are indeed related to very different brain regions and neural circuits, then it would perhaps be advantageous to take this into consideration in the diagnostic categorization of these disorders. It could also be that this results in novel suggestions for therapeutic interventions in the anxiety disorders. Another important, yet unanswered, question concerns the neural differences between patients who do not respond well to current therapeutic interventions versus those who do profit from them. Are the neural causes of treatment failure disorder-specific or are there neural commonalities uniting all nonresponders?

3.6 NEURAL CORRELATES OF OBSESSIVE–COMPULSIVE DISORDER

A good number of studies have been published about the neural correlates of obsessive–compulsive disorder, and most of these come from the group led by Baxter at the University of California, Los Angeles. Since their first publications on the neural correlates of obsessive–compulsive disorder (Baxter, Phelps, Mazziotta, & Guze, 1987), this group has increased their number of cases—which was originally rather small—but many of the publications are apparently based on overlapping patient samples, which means that the various publications by members of this group cannot be regarded as independent replications. When considering this caveat, it is clear that the empirical basis for the neural correlates of obsessive–compulsive disorder is still relatively narrow. Overviews can be found in Baxter et al. (1996, 2000) and Saxena et al. (2001).

All authors writing about the neural correlates of obsessive–compulsive disorder converge on the idea that a certain neural circuit—involving the orbitofrontal cortex, the basal ganglia, and the thalamus—is hyperactivated among patients with this disorder (Baxter et al., 1996; Insel, 1988; Modell, Mountz, Curtis, & Greden, 1989; Rapoport & Wise, 1988; Rauch et al., 1994; Swerdlow, 1995). Based on these findings, Baxter has developed a neural model of obsessive–compulsive disorder, which is reproduced in Figure 3–2.

When the strength of the lines in Figure 3–2 is ignored, it can be regarded as a general model of how behavioral programs that are already present in an organism's repertoire can be triggered. We already learned in previous sections about the dorsolateral part of the prefrontal cortex (PFC), which is more closely connected with the associative cortex, and the ventral part of the PFC, which is more closely connected with the limbic system. Impulses originating from these parts of the PFC are relayed via various basal ganglia nuclei to the thalamus and at that location have the effect of either releasing or inhibiting behavior programs. However, because we have not yet learned about the

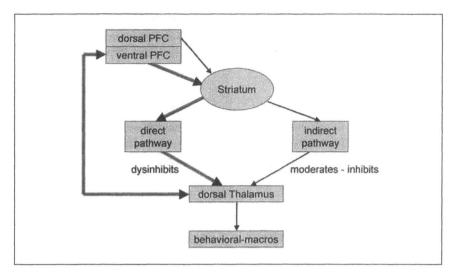

Figure 3–2. Neural model of obsessive–compulsive disorder, based on Baxter et al. (2000, p. 594). Thin lines indicate inhibition, thick lines indicate activation or disinhibition. The thickness of the lines represents the strength of the synaptic activation transmission in obsessive–compulsive disorder. (Adapted from Baxter et al., 2000; p. 594).

basal ganglia, which are deemed very important in obsessive–compulsive disorder, I will now devote more specific attention to these brain structures and their functions. This discussion is based primarily on Gall, Kerschreiter, and Mojzwisch (2002).

The basal ganglia make up a heterogeneous region that consists of various nuclei. The most important ones are:

• The *striatum*, consisting of the caudate nucleus, the putamen, and a ventral part. The putamen receives its input primarily from the somatosensory and motor cortices; the caudate nucleus receives its input from the prefrontal, temporal, parietal, and cingulate cortices; and the ventral striatum receives its input from the limbic system.
• The *pallidum*, which is differentiated into dorsal (globus pallidus) and ventral parts. The pallidum receives inhibitory input from the striatum and activating input from the subthalamic nucleus, as well as activating and inhibitory input from the thalamus. The external globus pallidus has many efferent, primarily inhibitory, projections to the subthalamic nucleus, the internal globus pallidus, the striatum, and the thalamic reticular nucleus.
• The *subthalamic nucleus*, which receives its input primarily from the motor cortex; its output is primarily activating. The subthalamic nucleus is part of the so-called indirect path from the PFC to the thalamus, which will be discussed in more detail later. It connects the striatum via the external globus pallidus with the internal globus pallidus and the substantia nigra.
• The *substantia nigra*. Two areas of the substantia nigra can also be differentiated: the reticular substantia nigra and the compact substantia nigra. The reticular substantia nigra receives input that is similar to the internal globus pallidus, and it

projects via the thalamus to the premotor and prefrontal cortices. The compact substantia nigra has many dopaminergic neurons (the degeneration of these neurons
plays a key role in Parkinson's disease). It receives input from the striatum and modulates activation in the striatum and the cortex.

The neural circuit between prefrontal cortex and thalamus shown schematically in
Figure 3–2, then, is an extreme simplification. The intermediate structure of the basal
ganglia, of which only the striatum is explicitly mentioned, is itself a highly complex
network whose functions are not nearly fully understood.

Neural impulses originating from the different parts of the PFC arrive in parallel, by
two different routes, via the basal ganglia to the thalamus. These two routes correspond to the ventral and the dorsal loops, which we encountered once before in chapter 2 in the section on the implementation of action. Within these two loops, there is
in each case a direct and an indirect path. The direct path activates already available
behavior programs; the indirect path inhibits them. Normally there is a balance
between activation/disinhibition and inhibition, which is associated with the flexible,
situation-appropriate initiation and termination of behavior. However, when this balance becomes disturbed at any point within these parallel circuits, severe behavioral
consequences can emerge. For example, Huntington's disease and Tourette's syn-
drom can be viewed as the consequences of a tonic imbalance in favor of activation,
whereas the symptoms of Parkinson's disease can be regarded as linked to a dominant indirect path.

According to Baxter, in obsessive–compulsive disorder the neural tone in the direct
path—from the orbital PFC via the substructures of the basal ganglia to the thalamus—is stronger than the indirect path. Behavior "macros" are set in motion via this
direct path. The term *macros* refers to complex, situation-specific behavior programs
that are activated in a semiautomatic fashion in specific situations. In obsessive–compulsive disorder, such macros often involve territorial or social behavior patterns surrounding themes such as aggression, hygiene, or sexuality. Baxter argues that frag-
ments of these macros are activated in obsessive–compulsive disorder even when the
relevant situational conditions are not present. The hyperactivity in the direct path via
the basal ganglia leads to a disinhibition in the thalamus, which then triggers a reciprocally enhancing interaction between thalamus and orbitofrontal cortex, culminating
ultimately in impulse-governed behavior that is difficult to stop. The neural tone in
the indirect, inhibitory path is too weak to suppress or stop these impulses.

Because this direct path within the ventral loop originates in the emotion-relevant
orbitofrontal cortex, and the indirect path within the dorsal loop originates in the dorsolateral PFC, which is more relevant to the initiation of planned, reflective behavior,
Baxter regards the hyperaction of the orbitofrontal cortex as one of the main causes
of obsessive–compulsive disorder. This process involves emotions that cannot be
controlled by goal-oriented, rational, and reflective behavior that typically originates
from the dorsolateral cortex. The impulses coming from the hyperactivated

orbitofrontal cortex, then, lead to obsessive symptoms via the basal ganglia. Because of the causal role played by uncontrollable emotions, the neural model also suggests that it is sensible to group obsessive–compulsive disorder with the other anxiety disorders. However, the neural circuits involved in this disorder differ substantially from other anxiety disorders.

Several empirical findings that support Baxter's neural model have been reported in the literature. In two empirical studies with patients diagnosed with obsessive–compulsive disorder, the degree of activation in the orbitofrontal cortex correlated significantly with the activation in the caudate nucleus on the one hand and the thalamus on the other (see Baxter et al., 1997). Such a correlation was observed only among patients with obsessive–compulsive disorder but not among depressed or normal comparison groups. This association, therefore, appears to be specific for obsessive–compulsive disorder. Interestingly, such correlations were observed prior to treatment only in those obsessive–compulsive patients who later improved in response to cognitive–behavioral or pharmacological treatment. After the treatment, the activation in the respective brain areas was no longer correlated. This supports the idea that the correlations indicate a pattern of activity that is specific to obsessive–compulsive disorder and that can be altered effectively in treatment.

It is particularly interesting that this pattern was observed only among patients who later responded to treatment, not among the nonresponders. It appears that a subgroup of patients with obsessive–compulsive disorder exists for whom Baxter's assumptions about specific neural patterns do not apply. This group did not respond to either behavioral or pharmacological treatment. The neural foundations of obsessive–compulsive disorder in this prognostically unfavorable group remain a mystery for the time being. Baxter and colleagues have so far been unable to find any neural characteristics of this group other than the nonpresent brain activation pattern that differentiates them from treatment responders. It is equally unclear, at this point, how one could intervene with these patients in order to achieve treatment success.

In another study conducted by this research group (Saxena et al., 2001), a sample of patients with pure obsessive–compulsive disorder was compared with a sample of patients with only major depression and a sample of patients with obsessive–compulsive disorder as well as depression. In all depressed patients—the ones with as well as without obsessive–compulsive disorder—a highly significant hypoactivity in the hippocampus could be documented. This was not caused only by the volume reduction in the hippocampus, which has been repeatedly found among depressed patients (see the section on depression), but it was also found that—even when hippocampal volume was controlled for via standardization—activity in the hippocampus was reduced. In addition, in all patients the degree of depression correlated significantly and negatively with activity in the hippocampus. The hypoactivity in the hippocampus, then, can be regarded as a stable characteristic of the depressive state, according to this study as well as others. Surprisingly, this study showed increased activity in the patients with obsessive–compulsive disorder only in the thalamus but not, contrary to expectations, in the orbitofrontal cortex and the caudate nucleus. To complete the confusion, the comorbid patients, compared to the purely obsessive–compulsive patients, had an even more reduced metabolism in the thalamus and caudate nucleus.

The authors concluded from this unexpected and confusing pattern of findings that one probably needs to differentiate among various subtypes of obsessive–compulsive disorder, and that it is not reasonable to expect the same patterns of brain activation for entirely different obsessive–compulsive symptoms. This appears sensible, in light of the findings about the specificity between neural and mental functioning, which have been discussed and illustrated with various examples in chapter 2. According to Grice, Boardman, & Zhang (1997), at least three symptom factors must be differentiated in obsessive–compulsive symptomatology. Rauch, Dougherty, & Shin (1998) were indeed able to find different brain activation patterns for these three symptom factors. For patients with elevated scores on Factor 1 (obsessive thoughts with aggressive content and control obsessions), an elevation in striatum activity was found. For Factor 2 (obsessive orderliness and repetitions), by contrast, reductions in the activity in this region were found. Finally, in the Factor 3 (fear of contamination, compulsive cleaning), increased activation in the anterior cingulate cortex and the orbitofrontal cortex were documented. If one mixes all these patients together based only on their common diagnosis of obsessive–compulsive disorder and then examines their brain activity, it should not be surprising, based on this study, that replicable findings would be the exception, given that the samples in the different studies will probably consist of different variations of subtypes. The suggestion that different subtypes of obsessive–compulsive patients are probably characterized by different functional relationships in neural activity is in itself an important insight. This could mean that the therapeutic levers have to be applied in different locations for the various subtypes of obsessive–compulsive disorder.

The findings reviewed here on obsessive–compulsive disorder provide important clues. However, they cannot yet be regarded as established facts. Most important, they show that much more research will be required before the neural correlates of obsessive–compulsive disorder are fully understood. This research also provided clues, yet again, to suggest that studies on the neural correlates of mental disorders can lead to the insight that our current diagnostic differentiations are not yet optimal because they do not correspond to underlying functional relationships.

In terms of therapeutic implications, we can deduce from the majority of findings, and from the most detailed theoretical model—that of Baxter—that the treatment of obsessive–compulsive disorder also requires the facilitation or creation of inhibitory structures. It appears that two points would be relevant in this context:

1. Inhibition of the negative emotions that create and maintain the obsessive–compulsive symptoms from the orbitofrontal cortex via the basal ganglia and thalamus.

2 Facilitation of positive, goal-oriented activity and thereby strengthening of inhibitory impulses, which, originating from the dorsolateral PFC via the indirect

path, can prevent or balance the disinhibition of the thalamus that occurs via the direct path.

Both of these goals are probably reached to some degree in the intervention that is regarded as particularly effective: exposure with response prevention. However, these goals can also be achieved via different paths. After all, a purely cognitive form of therapy has also been shown to be very effective in the treatment of obsessive–compulsive disorder (Emmelkamp & van Oppen, 2000). However, a clear improvement in the disorder-specific interventions available for obsessive–compulsive disorder can probably be expected only once the underlying functional relationships are better understood.

3.7 WHAT PSYCHOTHERAPEUTIC CONCLUSIONS CAN BE DRAWN FROM THE NEUROSCIENTIFIC RESEARCH ON MENTAL DISORDERS?

If we ask which concrete psychotherapeutic implications arise even today, based on the research on the neural correlates of mental disorders, we must clearly acknowledge that the answer will depend on which specific disorder we are dealing with.

In the case of depression, the neuroscientific findings suggest a treatment perspective that differs substantially from the currently most established and elaborated depression theory, Beck's cognitive theory of depression (a new version of this theory can be found in D. A. Clark & Beck, 1999). Beck's cognitive theory (1967, 1979) has proven to be enormously fertile by inspiring a great—almost bewildering—number of research studies. Beck himself has recently attempted to review the many findings and relate them to core statements of his theory to examine the degree to which they are empirically substantiated (D. A. Clark & Beck, 1999). This effort, as well as the detail-orientation and sober attitude with which it was conducted, honor the scientist Beck and positively set him apart from many other psychotherapy gurus. As a scientist, one can take pleasure in noting that Beck has by now become the most frequently cited author in the social sciences, even ahead of Freud, who was first on this list for a very long time.

When I had processed the research overview by D. A. Clark and Beck, along with students in my seminar course, and once I had recuperated from the sometimes oppressive wealth of research findings, I had to acknowledge something that took me quite by surprise: Beck's original theory is confirmed only to an astonishingly small extent by available research findings. There is, in fact, very little evidence suggesting that depression-specific cognitive schemas play an important causal role in the pathogenesis of depression, and it also appears that they do not play a critical functional role in the attainment of therapeutic gains. Changes in the depressive state can be achieved equally well if one completely ignores the idea of cognitive schemas. It is true, to be sure, that Beck's cognitive therapy is one of the most effective therapies for depression (cf. Hautzinger, 1998), but the work on cognitive schemas does not appear to be as important in creating the therapeutic effects as was originally assumed. The available research results beg for a different explanatory framework, which can better integrate the findings than Beck's cognitive theory.

If we also consider the findings about the neural correlates of depression, the doubts about the explanatory power of Beck's cognitive theory grow even larger. How should one explain the atrophy and hypofunctioning of the hippocampus by referring to cognitive schemas? Why should the lateral asymmetry in the prefrontal cortex depend on cognitive schemas, given that such asymmetry can already be documented in the 10th month of life? Do we need to assume the existence of depression-specific cognitive schemas if we want to explain the association between traumatic life experiences during childhood and the excess in stress hormones among depressed persons? How can we explain the documented therapeutic effect of the neurotransmitter serotonin by using cognitive theory? It seems to me that it is time to integrate the findings from the various research areas in a new theoretical context. Beck has identified important aspects of depression, but it seems that his theory has now fulfilled its function as an inspiration of research questions. So many new findings have been added—including those in the neurosciences—that a new effort is needed to integrate them all in one coherent theoretical framework.

In light of the fact that the long-term success rates in the currently predominant treatment forms for depression are less than fully satisfactory—the relapse rates after 2 years range from 60% to 80% for both psychological and pharmacological treatments, with a slight advantage for psychotherapy (Elkin, 1994)—there is more than enough reason to search for treatments that promise better long-term success. This also entails, of course, the questioning of currently dominant perspectives on depression in order to develop models with greater explanatory power. The neuroscientific findings are among those that a good depression theory will have to account for from now on. The research findings I reviewed here suggest that an imbalance between approach and avoidance plays a crucial role in the creation and maintenance of depressive disorders. In chapter 4, I will discuss the motivational aspects involved in the pathogenesis of mental disorders in more detail. If one views depression from a motivational perspective, the contours of a new framework on depression become visible, and this perspective differs noticeably from Beck's cognitive theory of depression. A brief preview of this was already provided in the case example in the introduction. The research examples discussed in chapter 4 form the scientific foundation for the therapeutic interventions used in this case example.

In the case of depression, I highly value the utility that we, as psychotherapists, can derive from the neuroscientific research. Our own therapeutic practice provides clues that substantially better results can be achieved in the treatment of depression if the motivational aspects of approach and avoidance are explicitly considered. This will have to be further established, of course, in appropriately designed research studies.

Current neuroscientific findings also suggest useful applications for the treatment of PTSD, albeit in a different form than what holds true for depression. In the case of PTSD, the neuroscientific findings largely support the perspective that psychologists have developed in recent years on this disorder. The memory–psychological model by Ehlers and Clark (1999) can be integrated very well with current neuroscientific findings. Both research branches converge on the finding that a dissociation between explicit and implicit trauma memory constitutes the core of this disorder. In light of the fact that two research traditions that emerged in relative separation from each other come to a surprisingly consistent conclusion, we can be relatively certain that we are on right path in our conceptualization of this disorder. For the practicing therapist, it is reassuring that research provides such clear-cut suggestions with regard to the issues that should be targeted in order to maximally help patients with PTSD. Nevertheless, there is still much that remains to be clarified, even in the area of PTSD. This includes, for example, the question of how one should therapeutically deal with the massive comorbidity that often characterizes PTSD. It seems clear, however, that further progress in this domain can be expected, given that the core of this disorder appears to be relatively well understood at this point.

With regard to the other three anxiety disorders that were discussed here, I do not yet see a similar degree of practical utility that psychotherapists could derive today from available neuroscientific evidence. Psychological research has proceeded further in these areas than has neuroscientific research. There is little reason at this point, in my view, to introduce important changes to the psychological interventions that have been developed and tested in this area. The neuroscientific research in this domain is still too unresolved to warrant such changes in treatment approaches. This can, of course, change when more neuroscientific studies on these disorders become available, which will certainly be the case in years to come.

Psychotherapists would be well advised, in my view, to closely follow the neuroscientific research produced in these areas. There are many unresolved questions about these disorders, and neuroscientific research will likely contribute much to their resolution. This includes, for example, questions about the relationships among the various anxiety disorders, questions about the diagnostic classification of subgroups with common functional characteristics, questions about predictors of differential treatment response, questions about the causes of treatment nonresponse, and questions about the comorbidity among the anxiety disorders and between these and other disorders. Research has already begun to provide clues about some of these questions, but the findings remain too uncertain and unclear to deduce concrete therapeutic conclusions. It would greatly surprise me, however, if this did not change considerably within the next few years.

Across the specific disorders, however, one can still articulate an important conclusion that arises from previous neuroscientific research on anxiety and the anxiety disorders: Across these disorders there have been replicated findings to suggest that anxiety and the symptoms of anxiety disorders can be understood as the result of dysfunctions in attention and emotion regulation. These dysfuctions are accompanied by hyperactivity in the avoidance system—or they may indeed be caused by it—and they are characterized by a lack of context-specific inhibition. A lot of evidence suggests that the treatment of anxiety can be regarded, in large part, as the creation and facilitation of context-specific inhibition. This requires, among other things, the redirection of attention toward the content that has previously been avoided. However, attention is motivated to be directed toward the content that it happens to be directed toward. Its redirection, therefore, would have to occur against current motivational tendencies. This would inevitably activate the avoidance system, which is easily activated in these patients to begin with. What is needed, then, is a counterweight to the avoidance, a context that strengthens the activation of the approach system. Without such a context of approach, attention cannot be redirected, and inhibition of the anxiety reaction cannot be established.

The terms *approach* and *avoidance* suggest that we are now operating within the motivational system of the patient. His or her approach and avoidance potentials are activated via his or her idiosyncratic motivational goals. If we note that the patient's avoidance system is overly active, which has been shown directly in the case of panic disorder and depression, and which also seems to be the case in generalized anxiety disorder, then we must also ask what has created this hyperactivation. This is a question we cannot resolve unless we refer directly to the patient's individual motivational system. The question of how the disorder developed also refers to neural circuits that influenced mental functioning prior to the emergence of the disorder, and this includes approach as well as avoidance processes. The neural circuits underlying the disorder have emerged under the influence of previously existing circuits, which are connected with the circuits involved in the motivational system. These motivational circuits later interact with the newly formed circuit that underlies the disorder; they do not simply disappear. If the disorder remains in place, then the motivational circuits are part of the context that maintains the disorder. From a neural perspective, we cannot separate the disorder from the surrounding neural system with which it inevitably interacts. Viewing the disorder processes as completely separated from other neural processes would be entirely incompatible with the known interconnectedness in general neural functioning. And this general neural functioning transpires in a motivated fashion, beginning with the first breath taken by each human.

Even though mental disorders might exhibit neural characteristics that differ from normal neural functioning, they always remain embedded in the context of general neural functioning; they do not exist separately from this context. Disorder-specific research—and this holds true for neuroscientific as well as psychological research—does not allow, for methodological reasons, for a detailed consideration of individual motivational contexts. If one compares the brain activity of 20 depressed patients with that of 20 normal individuals, then the depression symptoms are at the center of the research focus. The idiosyncratic motivations and other individual differences of the participants' neural structures and processes vary at random and represent nuisance or, at best, error variance in the context of the study. The findings that emerge from studies of this kind must necessarily ignore the motivational features of the individual study participants. When hundreds of these studies accumulate, however, a sort of virtual reality is created, as if the examined disorders exist separately from the individual persons who carry the disorders. The research literature as a whole mirrors the process occurring at the level of individual studies. Only those topics that have been investigated take center stage, and these tend to be disorder-specific characteristics. It makes little sense, of course, to criticize this practice because it is perfectly understandable from a methodological point of view. We should avoid, however, losing awareness of that which we have chosen to ignore for methodological reasons. The fact that the disorders emerge, in each individual case, under the influence of the very processes that we regard, for methodological reasons, as error variance—the processes that in fact constitute the individual life of the examined person—must be added back to our considerations after reviewing such research findings.

Most researchers are undoubtedly aware that they examine in their studies not just the carriers of particular disorders but, instead, individual persons with idiosyncratic motivations, value systems, life histories, and so forth. Their treatment of the study participants will undoubtedly often reflect this awareness. However, what is documented and processed is usually the disorder characteristics and not the participants' individual features, and findings are reported only at the level of disorders. Neuroscientists who examine the brain act similarly to radiologists examining x-rays of the lung. There is nothing wrong with this approach, in principle, because it would be relatively pointless to wonder in each case what kind of person it might be whose lung is being viewed in each case. However, if one aims to use neuroscientific studies to make general statements about the nature of human depression, then one treads on somewhat different territory. It would be desirable, at least, to contemplate the role of those aspects that one is unable to consider directly in one's studies. In my review of the neuroscientific literature, I practically never encountered any passages that would indicate an explicit awareness of these issues, or that would directly address the problems arising from these considerations.

If we ask questions about the practical implications of neuroscientific studies on the mental disorders, then of course we should not simply adopt these biases. Instead, we ought to add again that which has been removed in these studies. We should do so by asking, and attempting to answer, the kinds of questions that are not being asked in these studies. Such questions go beyond specific mental disorders, but they add to our understanding of them. What is the relationship between mental disorders and general neural functioning? How can we understand the fact that the brain produces mental disorders in the first place? Under which conditions does this occur? Why is it that no mental disorders are formed in some people, whereas several ones emerge in others? Which factors influence which specific mental disorder develops? And how are mental disorders related to a person's general life on a broader level? As practicing clinicians, we tend to ask these questions not in generalized form but always in the context of a specific, concrete patient. It also tends to take a while until we even begin to ask such questions.

When we sit across from a patient in the therapy room, a different set of questions tends to emerge at first. What kind of person is it who sits across from me, over there?

What is it that moves him or her, and what does he or she want from me? How does he or she live, and what has brought him into his or her current position? We first attempt to get a general picture of this person and his or her life situation, and then we attempt to embed his or her problems in the context of this understanding, to appreciate the problems against the backdrop of this general understanding. The patient is much more for us than merely a carrier of symptoms. He or she is a human being, with wishes and fears, hopes and disappointments, successes and failures, and likeable as well as less likeable features.

Based on the findings reviewed in this chapter, how are we supposed to think, then, about individual patients? Should we ask whether his or her hippocampus might perhaps be shriveled up, which side of his or her frontal lobe might be more strongly activated, or what his or her amygdala might currently be doing? All of these are questions that might truly arise in the process of differential diagnosis, and perhaps we should indeed be asking these questions. But before this, we should surely ask ourselves, "what is it that moves this person?" The questions about the factors that move a person, both positively and negatively, about what he or she pursues and avoids, and how well he or she manages to do so, are among the most important ones in the context of psychotherapy. Why this might be the case, and what this has to do with mental disorders, is discussed in detail in the next chapter.

Because neuroscientific research has so far devoted little attention specifically to human motivation, the following sections will be less neuroscientific in flavor than the preceding sections. As much as possible, and as far as relevant findings are available, I will discuss what the neurosciences have to say about the motivational aspects of mental functioning. However, I will also draw more heavily on findings reported in purely psychological research, and I will discuss the results of our own research efforts because the questions that now take center stage have been at the focus of my—and my colleagues'—interest for many years.

CHAPTER FOUR

NEED-FULFILLMENT AND MENTAL HEALTH

4.1 BASIC HUMAN NEEDS

We know quite a bit about the brain by now, but we still know little about things that might help us understand another human being. In order to understand a person, we have to know something about the factors that move him or her, both in a positive and negative sense, and about his or her wishes, goals, plans, values, and fears and dislikes. It might be quite useful for psychotherapists to know how exactly mental functioning transpires and how it is represented on a neural level, but without some knowledge about the issues and concepts to which the person's mental processes are oriented, the very aspect that lends meaning to a human life would be missing. The world of meanings opens up when we consider motivational aspects. This is the case not only from a hermeneutic perspective but also from the perspective of the biological natural sciences.

The organismic functioning of all species has been selected in a manner that ensures each species' survival and the transmission of its genes, in the context of its unique life conditions. The constitution of each species mandates what it requires for survival and reproduction. If the life context does not match these objective requirements—we could also say: if the needs of the species are not being fulfilled—then each individual organism fails to flourish and drops out of the selection process. The human species appears to be endowed with excellent biological fitness. It proliferates at massive rates and transmits its genes abundantly. Our species is able to do this because it has created a societal and cultural environment that enables this process. This also holds true at level of the individual person: If his or her constitution and surrounding context match well, he or she tends to flourish. But what are the conditions that must be met for a member of the human species to flourish? To phrase this another way: What are the specific basic human needs that must be fulfilled in order for him or her to feel well and develop well?

Many answers to this question would seem obvious: The person needs air to breathe, sufficient nourishment to satisfy hunger and thirst, a certain amount of sleep, conditions that enable him or her to maintain a constant body temperature, and so on. Humans share these biological needs with other life forms. Of course, these are not the basic needs that we, as psychotherapists, are interested in. Our question is: Beyond these biological needs, are there specific human psychological needs that are shared by all humans and that must be met in order for an individual to feel well and to have good mental health? This is ultimately an empirical question, which we can answer by examining the conditions under which humans experience good versus poor mental health.

Evolution ensured that basic needs tend to be met regularly, with certainty. It formed organisms in such a manner that they feel thirsty when they require liquid; hungry, when they require nourishment; tired, when sleep is required, and so forth. Whenever a deficiency state emerges, the activity of an organism is automatically oriented in a way that the deficiency is removed. To quench one's thirst becomes a goal, and drinkable liquids become goal objects. Thirst, goals, and potential goal objects are represented on a physiological and/or neural level in the organism. The physiological feedback loops and neural circuits for hunger, thirst, sleep, sexuality, and so on have been researched in detail (Panksepp, 1998). Do analogue regulatory circuits, or other forms of neural representations, exist for psychological needs? At least with regard to basic needs that are of existential importance, our current knowledge about the brain would suggest that neural structures and mechanisms that are oriented to the fulfillment of these basic needs have formed over the course of evolution. However, before we address the issue of neural representation, we must first clarify specifically which needs might have been so important or basic that neural hardware for them has been selected in human brains over the course of evolution.

The questions of which basic psychological needs must be met in order to attain mental health and well-being, and how these basic needs influence human behavior, have been answered very differently from various theoretical perspectives. The founders of different therapy schools have responded to these issues and, indeed, have erected entire ideologies of human nature based on such ideas. Because these questions concern fundamental human values, the debates in this domain can become quite heated.

In my view, psychotherapists cannot really evade this question because even those who do not explicitly address this point hold an implicit view of human nature, and such views also contain implicit assumptions about the factors that determine human happiness and misery. It is clear that such assumptions influence how other people are perceived and, therefore, they also influence the implicit assumptions therapists bring to their patients. It seems preferable, then, to address this question explicitly. After all, it is not necessary that critical thinking and rational reasoning be automatically replaced by subjective faith or belief just because we are dealing with value-related questions. I would argue that the domains that are personally relevant to us deserve especially to be the focus of our scientific questioning, exploration, and clarification.

The question of whether and which basic needs must be fulfilled in order for humans to attain well-being is certainly personally relevant to each of us. Beyond this, the question is also of key importance to our profession, which has human misery as its central concern. There are many clues suggesting that a severe and enduring failure to meet basic needs is, ultimately, the most important cause in the development of mental disorders, as well as an important factor in the continued maintenance of such conditions. In this chapter, I elaborate conceptually on these processes and substantiate them with empirical findings when possible.

In order to empirically evaluate the role of human basic needs in mental functioning, a first necessary step concerns the decision of which basic needs one wishes to focus on. The choice is enormous. Empirically oriented psychologists have provided a wealth of proposals and findings, beginning with those by McDougall (1932), Maslow (1967), Heckhausen (1980), Gasiet (1980), Epstein (1990), Ford (1992), and Deci and Ryan (2000)—not to mention the various therapy school founders such as Freud, Adler, Sullivan, and Rogers. In light of this wealth of different concepts, it would not seem reasonable to assume that a definitive and valid answer to the question of basic human needs is easily available. All of these concepts are best viewed as assumptions rather than established knowledge. Assumptions that are not supported by cogent empirical evidence, however, cannot be considered seriously among the many available options.

Basic psychological needs, in my view, are needs that are present among all humans, and their violation or enduring nonfulfillment leads to impairments in mental health and well-being. This principle alone allows us to delete some of the candidate needs from our long list. Many needs and motives examined by psychologists do not have the status of basic needs. This is the case, for example, for the need for power and achievement (Heckhausen, 1980). Undoubtedly, many humans strive to attain power and to achieve highly. However, it is equally clear that many are also doing quite well despite a lack of power and achievement. Sometimes they are even so satisfied with their situation that others, who highly value power and achievement, might feel provoked or offended. Similarly, we might be tempted to conclude that a general need exists for candy and pleasant fragrances, based on the quantities of chocolate and perfume sold every day. However, there is little evidence suggesting that the nonfulfillment of these desires would have adverse health consequences.

It took me many years to get the sense that a clear answer had emerged as to which among the many concepts could be considered basic human needs. My understanding grew clearer when I became acquainted with Seymour Epstein's (1990, 1993) cognitive experiential self theory. Originally, I was particularly intrigued by aspects of Epstein's theory other than his assumptions on basic needs, such as the compatibility of his model with Powers's control theory, the assumption of two different modes of mental activity, and his conception of motivational schemas. The four basic needs he proposed appeared at first no more than yet another among the many possible options. My view on this has changed by now, particularly after I studied intensively the neural foundations of these needs and the empirical evidence on their implications for well-being and health. Neurobiological research has demonstrated by now that at least three of Epstein's four needs are very deeply rooted in the constitution of the human nervous system. In my view, it is possible to have different opinions about the exact definitions of these needs and about the boundaries among them, but it would be unreasonable to deny their existential significance for human beings. I will attempt to convey the rationale for this argument in the following section. Epstein differentiates among four basic needs:

- A need for orientation, control, and coherence
- A need for pleasure
- A need for attachment
- A need for self-esteem enhancement

The need for attachment was not included in his first version of the theory. He added it later, which was a good idea, given that it is one of the most substantiated basic needs. I would argue that Epstein's conceptualization of the basic needs could be further revised in this aspect. This point is rather essential because it is linked with a very central principle of mental functioning that cannot be simply subsumed as one of the four previously listed basic needs. It concerns that which Epstein terms *the need for coherence* and which I will refer to as *the consistency principle* in the following section.

4.2   CONSISTENCY REGULATION AS A BASIC PRINCIPLE OF MENTAL FUNCTIONING

The term *consistency* refers to a state of the organism. That is, it refers to the agreement or compatibility of simultaneously transpiring neural/mental processes. Basic needs refers to experiences that occur in the interaction between the organism and its environment. The experiences result in specific perceptions, which can have a positive or negative meaning with regard to the respective need. This process functions differently in the case of consistency. Consistency refers to the internal relations among intrapsychic processes and states. Needs are satisfied or violated by relevant sensory experiences. There are no specific experiences, however, that would satisfy a desire for consistency. Even though the human organism clearly prefers states of consistency and has developed many mechanisms to avoid inconsistency, even though the mental system functions as if it is striving toward consistency, and even though enduring states of inconsistency clearly impair well-being and health, it is not justified to simply regard consistency as one among other basic needs. It would be more appropriate to regard consistency as a basic principle of intraorganismic regulation that supersedes all specific needs. If one regards basic needs as requirements of human existence, as conditions necessary for human flourishing, then it would make sense to view a striving for consistency as a basic human need. Because of the points outlined previously, however, I prefer to regard consistency as a core principle of mental functioning. It is a necessity or condition for good mental functioning and, therefore, also a condition for the effective satisfaction of the basic needs. If this condition is severely violated, far-reaching consequences for the well-being of the respective person can be expected.

The fact that consistency regulation must play an important role in mental functioning became clear to me in the context of working on my last book, *Psychological Therapy* (Grawe, 2004). Based on available neuroscientific and psychological evidence, I already concluded in that book that many simultaneously transpiring processes underlie mental functioning, which then suggests questions about the internal coherence and compatibility of these processes. Once I had begun to ask this question, it became increasingly clear that the regulation of these simultaneously transpiring neural processes must be of basic importance for our understanding of mental functioning. This question has not let go of me since then. Compared to when I wrote the previous book, however, it is possible today to present far more empirical evidence documenting the importance of consistency regulation in the context of mental functioning, mental health, and ultimately, psychotherapy.

Consistency regulation cannot be understood outside of its context of goal-oriented activity, which in turn is largely oriented toward the fulfillment of basic needs. Thus, consistency regulation and need satisfaction are intrinsically interlinked. The link connecting the two is the construct of *congruence*, the compatibility of current motivational goals and actual perceptions. In order to clarify how I understand and distinguish among these constructs, I will discuss in the next section the relation between basic needs and consistency regulation in mental/neural functioning, and I will conceptualize the way in which they influence a person's experience and behavior. After providing this overview, I will substantiate the specific components of this conceptualization in more detail in subsequent sections.

4.3   BASIC NEEDS, CONSISTENCY REGULATION, MOTIVATIONAL SCHEMAS, AND INCONGRUENCE

The following discussion elaborates on the view of mental functioning I developed in my last book, *Psychological Therapy* (Grawe, 2004). This view is derived from findings and models that are broadly accepted in much of empirically oriented psychology. The model specifies, among other things, that mental activity is governed by goals; that schemas, which can be viewed as a sort of ordering template of mental activity, form over the course of mental development; and that goal-directed mental activity is organized hierarchically. These tenets represent "mainstream psychology," as it were. By contrast, it is a bit less of a mainstream position to argue that the goals a person forms during his or her life ultimately serve the satisfaction of distinct basic needs. This view and the delineation of specific basic needs differ somewhat from the mainstream but are nevertheless within the broader frame of contemporary psychology. Examples for such conceptualizations are Epstein's cognitive experiential self theory and Deci and Ryan's (2000) self-determination theory.

The fact that multiple processes continuously transpire simultaneously within the mental system is probably not regarded as controversial by most psychologists; however, the consequences arising from this fact for mental functioning have only rarely been addressed explicitly. An exception to this are the connectionist models that regard massively parallel distributed information processing as the basis for their theoretical and methodological approach (e.g., Caspar, 1992; Rumelhart & McLelland, 1986). The construct of consistency places the aspect of compatibility of the many simultaneously transpiring mental processes into the center and, thereby, establishes a link with connectionist core assumptions and with other models that address conflicts and dissonances in mental functioning. Such models have a long history in psychology (e.g., Festinger, 1957; N. E. Miller, 1944). However, they are currently not among the most intensely researched domains.

The specific view of consistency that I develop in the following sections extends beyond the model articulated in my previous book and, as such, can be regarded as an independent theoretical approach. When I speak in the following sections of *consistency theory* or *consistency-theoretical*, I refer to the subsequently fully developed perspective on mental functioning. Some elements of this view have quite a bit in common with other theoretical models, which in fact tend to be integrated within the theory. As a whole, however, this perspective is new. The development of this model was based, from the start, on a neuroscientific view of mental functioning (Grawe, 2004), but the neuroscientific foundation of the previous version of the model appeared to me to be lacking in detail and depth. Having laid the foundation in the preceding chapters, I will in this chapter elaborate on the consistency-theoretical view of mental functioning, with specific attention to its neural foundations. Toward the end of the chapter, I will apply this view to the pathogenesis of mental disorders, and in the next chapter, I will apply it to the problems arising in psychotherapy.

Before going into detail, I provide in this section an overview of the most important constructs of consistency theory and of their interrelationships. Figure 4–1 presents a graphical overview of the core of the consistency-theoretical model.

The experience and behavior of a person are governed directly by his or her motivational schemas. *Motivational schemas* are the means the individual develops in the course of his or her life in order to satisfy his or her basic needs and protect them from violation. Two main classes of these are *approach* and *avoidance* motivational schemas. If a person grows up in an environment that is entirely oriented to the fulfillment of his or her needs, then he or she will develop primarily approach-motivational goals and will gain great experience in achieving such goals. This includes corresponding expectancies and a differentiated behavioral repertoire for the realization of the goals under various conditions. By contrast, if a person grows up in an environment in which his or her basic needs are repeatedly violated, threatened, or disappointed, he or she will develop avoidance schemas in order to protect himself or herself from further hurt. In a truly harmful environment, however, avoidance can be regarded adaptive behavior. However, strongly developed avoidance schemas can later get in the way of positive need satisfaction—even in situations that would actually be appropriate in this context—because the situations will more easily activate the better facilitated avoidance rather than approach tendencies.

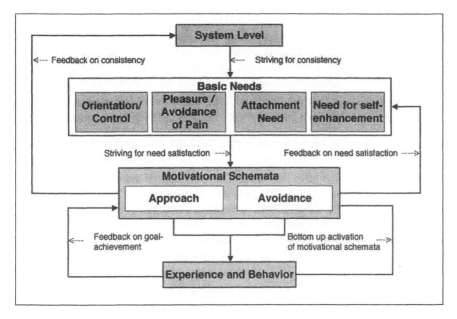

Figure 4–1. The consistency-theoretical model of mental functioning. See text for explanation.

At the lowest level of the model, which corresponds to situational experience and behavior, a continuous stream of perceptions is produced in the course of the individual's interactions with his or her environment. These perceptions inform the extent to which currently activated motivational goals have thus far been reached or avoided. Consistent with Powers's (1973) control theory, these feedback signals are termed *incongruence signals*. Goal attainment corresponds to positive emotions, whereas incongruence is linked with negative emotions.

If avoidance dominates over approach; that is, if the more established, neurally facilitated avoidance tendencies inhibit the simultaneously activated but less deeply ingrained approach tendencies, incongruence signals with regard to unfulfilled approach tendencies will emerge. In this case, *approach incongruence* is created. If the person is unable to avoid feared experiences; that is, if the undesired consequence actually occurs, *avoidance incongruence* is created.

Approach and avoidance tendencies can also be activated simultaneously and mutually inhibit each other. Such cases can be termed motivational conflicts or *motivational discordance*. Incongruence signals are created with regard to both approach and avoidance goals. Motivational conflicts, then, lead to incongruence. Discordance can also result from two simultaneously activated approach goals (approach/approach conflicts) or simultaneously activated avoidance goals (avoidance/avoidance conflicts).

Discordance and incongruence represent two particularly important forms of inconsistency in mental functioning. *Discordance* refers to the incompatibility of two or more simultaneously activated motivational tendencies, whereas *incongruence* refers to the mismatch between actual experiences and activated motivational goals. In both cases, neural patterns that are incompatible with each other are simultaneously activated. In section 4.8, we will encounter additional examples of inconsistencies in mental functioning. Discordance and incongruence are particularly important for mental health because they are linked with the activation of important motivational goals, and activation of such goals is always accompanied by strong emotions.

Goal-oriented activity, incongruence signals, and the accompanying emotions can transpire in the explicit or the implicit mode of functioning and, thereby, can occur consciously or unconsciously. The chronic or continuously repeated failure to meet approach and avoidance goals leads to an elevated incongruence level. This is accompanied by a chronically elevated level of negative emotions. To the degree that these emotions are experienced consciously, they could be felt as anxiety, disappointment, or anger, for example. We know from the discussions about anxiety reactions in chapter 2 that emotions do not manifest only as subjective feelings—their activation is linked with a cascade of physiological, hormonal, and neural reactions, regardless of whether a subjectively clearly experienced feeling is present. An elevated incongruence level can be regarded, therefore, as a highly complex stress state. We already encountered the negative consequences of a chronically elevated stress level on several occasions. I will address the relationship between incongruence and stress in detail in section 4.5.2.

If we consider the model in Figure 4–1 as a whole, we find that there are two "levers" of mental functioning: the striving for congruence and the striving for consistency. Mental functioning is consistently oriented in a way to enable perceptions that are consistent with activated motivational goals. The motivational goals, in turn, are based upon the basic needs. These needs do not influence behavior directly but do so via the motivational goals that develop around them. Goals and behavior, in contrast to basic needs, correspond to concrete situations or classes of situations. In these aspects, individuals differ from each other because of their different life experiences. They do not differ, however, in basic needs. In the same way that every human has a body, every person has the same basic needs. In the same way that arms and legs may differ between individuals, people can also differ in terms of the absolute and relative constitution of their basic needs.

In the stream of ongoing activity, while the organism is focused on the attainment of perceptions that are consistent with activated goals, it frequently happens—because of the multitude of processes involved—that constellations emerge in which mutually incompatible neural processes are simultaneously activated. Such recurring states of actual inconsistency are inevitable. Inconsistency is characterized by the fact that the simultaneously activated processes hinder each other in their smooth execution. Therefore, inconsistency impairs the effectiveness of the individual's engagement with his or her environment. Inconsistency that is maintained over a long period thus results in additional incongruence and, thereby, in additional impairments in need fulfillment.

Inconsistency is a state that an organism strives hard to avoid. The human mental system has formed all kinds of mechanisms to avoid or remove inconsistency. Like all mechanisms, these have a neural basis, which will be discussed in more detail in section 4.8. If one regards incongruence as a specific form of inconsistency—because mutually incompatible mental processes, such as goals, expectancies, and current perceptions, are simultaneously activated—then one arrives at the formulation that the striving for consistency is the ultimate moving force in neural/mental functioning. The striving for consistency, that is, the avoidance or reduction of inconsistency, is experienced only in exceptional circumstances as a conscious goal or motive. Consistency regulation transpires predominantly unconsciously and pervades mental functioning so fully that it seems appropriate to regard this as a highest or pervasive regulatory principle of mental functioning.

In this view, the basic needs define the criteria according to which consistency must be measured in mental functioning. They are the supplied standards, as it were, that evolution has furnished for the human mental system. If it fulfills these standards, the person flourishes, feels comfortable, tends to be healthy, and will most likely reproduce.

A good fulfillment of basic needs can only be attained to the degree that the individual has developed flexible, successful mechanisms of consistency regulation. From this perspective, even the motivational schemas a person has developed are means of consistency regulation. Whenever a state of incongruence emerges, the goals, means, plans, and behaviors that have previously been effective in the down-regulation of incongruence under the specific conditions are activated. Situation-specific goal-oriented behavior is not the only possibility, however, by which inconsistency can be down-regulated. Because inconsistency within the system is detrimental to need fulfillment, we can assume that mental systems over the course of their development form mechanisms to avoid states of strong inconsistency or to down-regulate them if they occur. Such consistency-securing mechanisms have been known under various labels in psychology, including, for example, defense mechanisms, coping, and emotion regulation. These mechanisms transpire predominantly in an automatized fashion, without conscious awareness. Mental disorders, in this view, also emerge in the process of consistency regulation. This assumption, which is of critical importance for the clinical significance of consistency theory, will be discussed more fully in sections 4.8 and 4.9.

Traumatic incongruence experiences have enduring detrimental consequences. They lead to a dominance of avoidance schemas, which impairs the possibilities of positive need fulfillment and can thereby lead to a permanently elevated incongruence level. This is accompanied by poor well-being and poor mental health. Traumatic incongruence experiences can also lead to structural and functional damage in the brain and, thus, can limit the individual's ability to positively cope with difficulties later in life. Section 4.9 addresses how these adverse effects of earlier incongruence experiences can later—in situations of elevated inconsistency in mental functioning—lead to the formation of mental disorders.

The following sections contain more detailed descriptions of the various components of this perspective. I begin with a detailed explanation of the importance of the previously mentioned basic needs for the content orientation of mental functioning and for mental health.

4.4   THE ATTACHMENT NEED

Hardly discussed in decades of scientific discourse, regarded by Freud as a derivative of the pleasure principle, viewed as an addition that almost did not make it into the list of basic needs in Epstein's theory, our human reliance on other people as close reference persons has had great difficulty in being recognized and accepted for its critical importance for human well-being. Today the need for attachment can be regarded as the empirically most substantiated basic need, especially with regard to its neurobiological foundation. However, in the currently dominating theories of mental disorder etiology, this need still does not nearly play the important role that the evidence suggests it deserves.

Sullivan (1953) was among the first to explicitly regard interpersonal aspects as the central causes of mental disorders, but he failed to substantiate his view with empirical evidence and did not manage to elicit his colleagues' enthusiasm for his views. Thus, another nearly 20 years passed until Bowlby (1969, 1973, 1980, 1988) erected such a visible post in the psychological landscape with his attachment theory that from then on no one could ignore the interpersonal domain. Bowlby was the first to explicitly postulate an innate need to seek and maintain the physical proximity of a primary attachment figure.

4.4.1   From Attachment Need
to Attachment Styles

Bowlby summarized the core of his attachment theory in three central postulates:

1. When an individual trusts that an attachment figure will be available whenever needed, then this individual will be less likely to experience intensive or chronic anxiety than a person who does not have this trust for whatever reasons.
2. Trust in the availability of an attachment figure—or the lack thereof—develops little by little prior to adulthood, in infancy, childhood, and adolescence, and whatever expectancies develop during these years will tend to remain relatively unchanged for the rest of life.
3. The manifold expectancies with respect to the availability and responsiveness of attachment figures that different individuals develop during the early years are relatively exact reflections of the actual experiences that were encountered by these individuals (based on Bowlby, 1973, p. 246).

Bowlby referred to the result of these experiences as an *inner working model*. In terms of consistency theory, these models correspond roughly to the motivational schemas that develop around the need for attachment. The child internalizes his or her early dyadic relationship experiences. They become encoded in implicit memory in the form of perceptual, behavioral, and emotional response potentials and motivational potentials. Young's (1994) "early maladaptive schemas" can largely be regarded as reflections of such early relationship experiences as well (p. 9). Which types of motivational schemas develop from the relationships to the primary attachment figures of early childhood will crucially depend on the availability and empathy of these first important figures. In a good attachment relationship, the attachment figures are a continuously accessible safe haven, which offers physical closeness, protection, security, and solace. According to these views, unfavorable motivational schemas develop either when no attachment figure exists at all or when the attachment figure does not possess the necessary sensitivity; that is, when it can be predicted that this person is either consistently inaccessible or inconsistent in his or her accessibility.

> Rejection leads to an emotional alienation of the child to the attachment figure; the unpredictability renders the child overly dependent on the attachment figure, which means that the attachment system is consistently activated because of the fear of losing the attachment figure. In an undisturbed attachment relationship, the child receives consolation, nurturance, and protection when this is desired, but it is also free to pursue its curiosity and its social wishes for new contacts, without being hindered in these pursuits or even punished for them by the attachment figure. (Grossmann, 1990, p. 232)

Bowlby's colleague, Mary Ainsworth (Ainsworth, Blehar, Waters, & Wall, 1978), developed a standardized observational procedure to examine how children respond to separation from their primary attachment figures. In this method, children between the ages of 11 and 20 months are typically studied. This observational procedure was later also used by many other research groups, which then produced additional empirical studies about attachment processes (for overviews, see Schmidt & Strauss, 1996, and Strauss & Schmidt, 1997).

From these studies, four attachment patterns were recurrently observed, which—based on the preceding discussions—can be regarded as a still image, as it were, of the particular constellation of motivational schemas that has developed in the children up to that point.

1. Children with *secure* attachment relationship behavior. These react with distress to separation from their mothers and immediately seek her proximity when she returns. This attachment pattern is accompanied by solid basic trust and enables the development of conflict-free approach schemas to satisfy the attachment need.
2. Children with *insecure attachment and avoidant* relationship behavior. They avoid proximity after being separated from the mother and respond even to the separation itself with no signs of the distress that is common among securely attached children. The child does not expose himself or herself to the possibility of further harm by not permitting intimate closeness. The price for this, however, is a poor positive satisfaction of the attachment need.
3. Children with *insecure attachment and ambivalent* relationship behavior. These children are very anxious during the separation from their mother and fluctuate

after her return between aggressive rejection of contact and seeking of contact. These children become completely preoccupied with the relationship after the separation and are not free to pursue other activities. This relationship pattern is characterized by conflicted motivational schemas. Closeness is accompanied by worries of losing the attachment figure, and lack of closeness is linked with fears of being alone.

4. Children with *insecure attachment and disorganized/disoriented* relationship behavior. These children respond to separation from and return of the caregiver with bizarre and stereotyped behaviors. These attachment patterns occur less commonly than the other ones. They result from severe violations of the attachment need either due to an absent or an abusive relationship with a primary caregiver, which has severe consequences for intrapsychic regulation (see following).

In the studies by Ainsworth et al. (1978), two thirds of the children exhibited a secure attachment pattern. That also means that about one third of children between 11 and 20 months in an unselected (American) population exhibit relationship patterns that are strongly marked by avoidance and/or negative emotions. Thus, already at a very early stage these children have developed rudimentary avoidance schemas, which exert unfavorable influence on the subsequent course of their relationship experiences. Main, Kaplan, and Cassidy (1985) reassessed children whose attachment patterns had been examined between 11 and 20 months about 5 years later, when the children were six years old. They showed that the attachment patterns remained stable in about 80% of the children. Waters, Merrick, Treboux, Croswell, and Albersheim (2000) studied attachment patterns of 12-month-old children and then again 20 years later. The categorization of secure attachment remained the same in 72% of these individuals. Changes in attachment styles could be explained in most cases by profound negative life events such as divorce or depression of the parents. Beckwith, Cohen, and Hamilton (1999) were able to document plausible life–historical causes for the changes in attachment styles both in negative and positive directions.

Preschool-age children with secure versus insecure attachment styles showed marked differences in their play behavior, in social relatedness, in the balance and fluency of their communication, and in autonomy and self-confidence (Erickson, Sroufe, & Egeland, 1985; Grossmann, Grossmann, Sprangler, Suess, & Unzne, 1985; Main, et al., 1985; Renken, Egeland, Marvinney, Mangelsdorf, & Sroeufe,1989). The differences were always in favor of the children with secure attachment patterns.

The relationship experiences individuals encounter even in their first months of life lay the foundation for motivational schemas, which in turn begin to influence the relationship behavior of the child at a very early stage in such a manner that schema-confirming feedback becomes more likely. A key influence on this is exerted by the responsiveness and particularly the sensitivity of the infant and toddler's primary attachment figure, whose sensitivity, in turn, is at least in part a result of his or her own relationship experiences in early childhood. Benoit and Parker (1994) examined the attachment style of mothers with the Adult Attachment Interview, a commonly used instrument for the assessment of attachment patterns among adults. They were able to use these attachment assessments of the mothers to predict in 81% of the cases which type of attachment the children had with their mothers. Even the attachment style of the grandmother was able to predict with 75% accuracy which attachment style their grandchildren had! Similar results were obtained by Main et al. (1985) and by Grossmann et al. (1989).

4.4.2 Neurobiology of the Attachment Need

The need for attachment is certainly one of the basic human needs. However, humans share this basic need with other social animals. Many studies have been conducted by now in this area, with many different species; the most detailed ones have focused on the rhesus monkey. These studies show that the attachment need is also basic for these animals, and when this need is not met, severe impairments result. These neurobiological studies cast an entirely new light on the attachment need among humans as well. They almost render Bowlby's conception in an anthropomorphic light. The need for attachment is much more deeply rooted in human neurobiology and physiology than we could have imagined based on the studies conducted only on humans. Therefore, I will discuss these animal studies in somewhat greater detail. They clarify, for the first time, why unfavorable attachment experiences have such devastating consequences on mental health, as I will describe in the following section.

When social animals such as young chicks or young rhesus monkeys are left alone in a strange environment, they show a typical pattern of reactions that is generally referred to as *protest reactions*. This is characterized primarily by plaintive utterances, similar to crying, which in the literature are known as *distress vocalizations* or *isolation calls*, as well as by increased motor activity and vigilant scanning of the environment. Heart rate is elevated, body temperature rises, and stress hormones are increasingly strongly secreted. The neural circuit underlying this reaction has been examined in detail (Panksepp, 1998, pp. 265–271). Panksepp termed it the *PANIC circuit* and also described its link to clinical panic disorder. I already described this relationship in my discussion of panic disorder in section 3.5.

Quite a lot is known about this circuit because it has been relatively easy to examine which brain regions must be stimulated with electrodes in order to elicit the typical plaintive utterances. This circuit originates in the central gray matter, close to the pain center, which is why Panksepp suspects that the separation distress system might have emerged phylogenetically from the pain system. Plaintive utterances can also be triggered by stimulation of the medial midbrain, especially the dorsolateral thalamus, and further to the front, in the anterior cingulate cortex and in specific parts of the amygdala. This circuit overlaps considerably with those that are involved in the release of stress hormones (CRF) and endorphins. Endorphins (bodily opiates) reduce the plaintive utterances, whereas CRF intensifies them. With increasing age, the easy excitability of this circuit diminishes, more so among males than females, which is apparently related to the animals' testosterone level, as testosterone reduces these plaintive voices even among young animals. Panksepp regards this as a contributory cause explaining why women cry more frequently and are more prone to develop panic disorder.

The chemical agents that modulate the activity of this circuit particularly effectively are also well known. Transmitters that stimulate the glutamate receptors, and especially the NMDA receptors, are most effective at activating this circuit. Glutamate administrations can activate distress vocalizations even when the animal is not alone. An effect that is not as strong, but still notable, is exerted by CRF.

Apart from the opiates, the neuropeptides oxytocin and prolactin also have a calming effect on the separation-anxiety system. Interestingly, oxytocin and bodily opiates also play an important role on the other side of the attachment relationship—that is, in the activation of maternal caregiving behavior. According to Panksepp, they can be viewed as "prime movers" in the creation and maintenance of social bonds among mammals (p. 256). In a well-functioning attachment relationship, both partners release equal amounts of these neuropeptides. These agents also have a memory-facilitative function. All perceptions that occur in the context of a close, physical, and affective interaction—including smells, touches, sounds, views, bodily sensations—are stored and consolidated particularly well in memory. This is why these memory traces—actually the synaptic circuits representing the implicit memory systems—can be easily reactivated by relevant perceptions. If a child with fear or pain finds solace in the arms of his or her mother, an entire neural cascade is activated that involves the release of oxytocin and opiates. This reaction pattern, which has been facilitated by a good attachment relationship, rapidly inhibits the previously activated pattern of anxiety or other negative emotions. If a positive attachment relationship is absent, these reaction readiness/memory contents are not ingrained and facilitated. Such a child would be much more difficult to console. His or her negative emotions cannot be easily down-regulated. Because this child has not been able to establish effective emotion regulation in affective transactions with his or her primary caregiver, he or she later lacks the corresponding reaction readiness in implicit memory, at a time when normally the regulation of emotions in the attachment relationship gives way to internal emotion regulation. An inability to regulate negative emotions is a marker of almost all mental disorders. Neurobiological attachment research strongly suggests that the foundations for this deficit and the risk for the later emergence of a mental disorder originate at a very early ontogenetic stage in human development, long before any memory content is formed that can later be recalled. Based on this view, it makes little sense for therapists to search in a patient's life history for events from which their misery originated. As far as the motivational schemas that formed around the attachment need are concerned, we have to assume that the interpersonal transactions underlying their formation cannot be recalled. The relationships with the primary caregivers typically extend, of course, beyond this nonrecallable period (the first 3 to 4 years of life), and later events could perhaps be remembered. Such events, however, would typically represent not the basis for the problematic development but, rather, a link in a chain that began much earlier.

The chemical agents released during positive attachment interactions, by the way, also exert an inhibitory function on aggressive behavior. Panksepp regards this as one reason why we find much lower rates of aggression in cultures that emphasize physical closeness and contact between adults and children and in which early, premarital intimacy is common, whereas higher rates are found in cultures that discourage or even punish physical proximity and early sex.

Attachment patterns are the result of recurrent relationship experiences that the child encounters with his or her primary caregivers. One result of these relationship experiences is the specific constellations of approach and avoidance motivational schemas that influence experience and behavior in close relationships. Another result—and this is particularly important in terms of the significance of attachment relationships for mental health—is the formation of neurophysiological circuits that later influence emotion regulation and the regulation of neurophysiological parameters.

Studies examining animals that are separated from their mothers for prolonged periods show just how profoundly violations of the attachment need can affect neurophysiological processes. If the separation continues for prolonged periods, the initial protest reaction I described earlier is transformed into a despair reaction, which is in many aspects the opposite of the protest phase. Motor activity, the distress vocalizations, and, more generally, any vocalizations decrease in frequency; the animal no longer seeks contact and play with other animals; limp postures and sad facial expressions can be observed; and the animal reduces its food intake. In addition, new behavior patterns are observed, such as rocking back and forth and self-soothing, self-hugging behavior. The behavior of these animals greatly resembles hospitalism syndrome as described by Rene Spitz (1945). Physiologically, the despair state during prolonged separation is characterized by reduced, irregular heartbeat, lowered body temperature, reduced oxygen use, weight loss, reduced REM phases during sleep of elevated arousal, and a reduced secretion of growth hormones as well as impaired immune system functioning (Amini et al., 1996).

Animals that are separated for longer periods early on in their life tend to show a much stronger despair reaction when they are again separated later on. Thus, a sensitization process takes place. Despair reactions occur not only among very young and nonmature animals. Even among adult female rhesus monkeys, which—unlike their male counterparts—remain in the same social group throughout life, one can observe despair reactions when the animals are separated from their mothers for prolonged periods (Suomi, Eisele, Grady, & Harlow, 1975).

Animal studies also inform our understanding of how the interorganismic affective–physical interactions between mother and child influence the later intraorganismic regulation in the child. Myron Hofer (1984) has examined the various components of the despair reaction in rats, which also occurs in this species after prolonged maternal separation. He found several specific regulatory circuits. The heart rate of the rat infants was regulated by the availability of mother's milk; their sleep–wake rhythm was regulated by the length of the feeding intervals; the secretion of growth hormone was influenced by the degree of tactile stimulation infants experienced from their mothers; and the activity level of the infants was regulated by the bodily warmth of the mother and by odor cues and tactile stimulation. Whenever a single aspect of these maternal characteristics changed, a specific effect on the corresponding physiological parameter in the infant was observed. These findings led Hofer (1987) to conclude that the despair reaction following separation is in reality a chain of mutually independent processes, each of which has its own regulatory mechanism.

Based on these and similar findings, authors like Hofer (1984), Field (1985), and Kraemer (1992) arrive at the conclusion that the nervous system of social animals after birth initially functions as a system of open homeostatic feedback loops, which require external input for the maintenance of internal system homeostasis. This input is created via the social contact and synchronization of affective and physical transactions in the attachment relationship. In this view, the effects of a separation from the attachment figure depend on the degree to which the child has already developed intraorganismic physiological feedback loops that enable homeostatic functioning without external input. This would explain why prolonged separations result in more harmful consequences than those of briefer duration. If a child grows up with a mother who does not display the characteristics of good maternal caregiving, he or she will be unable to develop optimal intraorganismic physiological and emotional feedback loops. Thus, the child will respond to later separation with a stronger despair reaction, and later stress situations will more easily shift it into a dysregulated state.

Even more extreme than the despair reaction are the consequences arising when rhesus monkeys grow up in enduring isolation, without the presence of any attachment figure. Kraemer (1992) describes this so-called isolation syndrome as follows: The animals have severe social deficits; they are unable, for example, to utilize the facial expression of other animals for their own behavior; they are unable to engage in normal interactions with other animals. This holds true for mating behavior as well as for maternal behavior. When animals that had been traumatized in this way were artificially inseminated and gave birth, they ignored or even killed their offspring. The animals tended to suck or bite themselves, beat themselves, bash their heads against hard surfaces and push their eyes out of their sockets. They were unpredictably violent toward other animals, in a way that is normally never observed in rhesus monkeys. Their feeding and drinking behavior also tended to be completely dysregulated. These impairments could be somewhat reduced when the animals were reintegrated with normal, but younger animals, but on the whole the impairments were irreversible. Even among animals who had profited from the "therapy" with younger animals, the impairments immediately resurfaced under conditions of stress or demand.

One could question the relevance of these animal studies by noting that such extreme violations of the attachment need are practically never observed in humans. This is, fortunately, true. However, the studies are nevertheless useful to show that primates—and most likely also, and especially, humans—absolutely require an attachment relationship to enable a relatively normal course of development. Humans rely on this indispensable relationship, which is why the need for attachment can rightfully be regarded as a basic human need.

In conditions that occur in normal human reality, we are not typically dealing with the complete absence of persons who could serve as attachment figures; instead, the quality of the attachment relationships is the more relevant aspect. Animal studies have also been conducted to simulate the conditions that occur in humans in normal, typical circumstances. Andrews and Rosenblum (1991; see also Rosenblum & Andrews, 1994) compared two groups of rhesus monkeys in a prospective, experimental study. In one condition, plenty of food was consistently available, which allowed the mothers to not spend their time searching for food but instead spend it with their offspring. In the other condition, periods of plentiful food alternated every 2 weeks with a condition in which the mothers had to search for food for several hours per day in order to satiate their hunger. During these periods, they were not able to take care of their children but had to leave them alone for several hours. These alternating phases began when the children were 3 months old, and they were maintained over 14 weeks. Immediately after these 14 weeks, the rhesus monkeys who had missed their mothers for several hours on a daily basis showed less exploration and play behavior than the monkeys that consistently had their mothers nearby. The necessity of having to search for food had created the effect that the youngsters were less securely attached to their mothers. This was not just a temporary effect. At the age of 3 years, the monkeys that had been intermittently separated 2-1/2 years earlier still responded significantly more strongly to the administration of (arousing) noradrenalin and significantly more weakly to the (inhibiting, calming) administration of serotonin. At the age of 4 years, these monkeys had a significantly higher concentration of corticotrophin releasing factors (CRF) in their cerebrospinal fluids (Coplan et al., 1996). The relatively mild disturbance in the attachment relationship had led to an enduring elevation in stress susceptibility.

The experimental disturbance in attachment relationships that has received the most research attention is the one in which young rhesus monkeys are reared not in the presence of their mother or another adult attachment figure but instead in the presence of other young rhesus monkeys. Suomi (1999, 2000, 2001) reviewed the results of a large number of studies. The peers had to play a dual role under these conditions—that of attachment figure as well as that of playmate. Compared to normally reared rhesus monkeys, these youngsters tend to be retarded in their play and social contact behavior; they end up at the bottom of the dominance hierarchies in mixed monkey groups; they respond more sensitively to being socially isolated, both in terms of their behavior and in terms of their stress hormone and noradrenergic neurotransmitter release; and these differences remain present over the long term, into adolescence and adulthood.

Beyond this, rhesus monkeys that have been raised alongside their peers and without their mothers tend to be more impulsive and aggressive when they reach puberty; they also drink more alcohol when it is available; they develop greater alcohol tolerance; they have a lower level of the central serotonin metabolite 5-HIAA; they require higher doses of anesthesia to be sedated; and PET scans show that they have a higher glucose metabolism throughout the entire brain. Animals with low 5-HIAA levels in the cerebrospinal fluid also tend to be the animals within monkey colonies that fall asleep last in the evening. All these are clues that suggest the presence of chronic hyperarousal. Because these are experimental studies, the hyperarousal can be causally linked to the absence of a secure attachment relationship. These effects remain stable into adulthood.

Female rhesus monkeys that are raised with their peers and without their mothers show lower levels of mutual grooming behavior than those who grew up alongside their mothers; they also tend to be bad mothers, who neglect or even abuse their offspring. This is observed primarily for their first child. With later children, their maternal behavior tends to improve markedly. The ventral contact with their children continues to be lower even with them, and it is especially this ventral contact that appears to be particularly intensive in good attachment (Suomi, 1999). The female animals, by the way, exhibited the same metabolic changes as their male counterparts. Higley, Suomi, and Linnoila (1996) assembled social groups of female rhesus monkeys that did not know each other prior the experiment. The positions that the animals attained in the emerging dominance hierarchies could be predicted with remarkable accuracy by the previously obtained levels of 5-HIAA concentration in the cerebrospinal fluid. Thus, it appears that there is a meaningful relationship on the one hand between attachment experiences and the presence of serotonin, which exerts an inhibitory and calming effect in neural functioning, and on the other hand, between available serotonin and social behavior. Animals with more available serotonin overall appear to have better social relationships.

Based on the repeated observation that rhesus monkeys growing up without maternal presence have a chronically lowered serotonin level, the research group of Suomi, who has conducted many of the experimental studies in rhesus monkeys, joined up recently with the research group of the genetic scientist Lesch, who several years ago was able to identity a serotonin-transporter gene (5-HTT) in humans (Lesch et al., 1996). Lesch and his colleagues found that a weak expression of this gene led to lowered serotonergic function. Bennett et al. (1998, see also Suomi, 2000) conducted a genotype analysis in each of the rhesus monkeys that had grown up with or without their mother in Suomi's study. It was expected that monkeys with the "short" HTT-allele, which is less efficient in terms of gene expression, would exhibit a lower serotonin concentration in their cerebrospinal fluid, compared with monkeys with the more efficient "long" HTT-allele. This was indeed what was found, but only for the rhesus monkeys who had been reared without their mothers' presence. Among monkeys that had grown up normally alongside their mothers—and which therefore had enjoyed a normal attachment relationship—the genotype played no role for their later serotonin metabolism. Among monkeys with a genetic risk for serotonergic hypofunction, the risk was alleviated by the secure attachment relationship. The genetic risk did not exert its potentially adverse effect. It was only under stress—in consistency theory we would speak of a severe incongruence with regard to the need for attachment—that the genetic disadvantage actually materialized. Among the monkeys with the longer HTT-allele, which is more advantageous for serotonin production, the unfavorable life conditions did not lead to an impairment in serotonergic functioning. They were protected by their favorable genotype against the harmful consequences of an absent secure attachment relationship.

This was probably one of the first times that a highly relevant example could be used to show concretely how a differential gene expression, via the pathway of favorable versus unfavorable life conditions, can function as the foundation for an enduring impairment in mental health. Once more studies of this kind become available, the process by which mental disorders develop—from initial genetic predisposition via a long series of protective and harmful life experiences to the actual onset of the disorder—will become increasingly comprehensible.

The studies discussed up to this point have primarily highlighted the unfavorable effects of an absent or poor attachment relationship. However, Suomi (1987, 1991) has also produced a very impressive study that illustrates the positive effects for the later functioning of a child that can be achieved via a very good, secure attachment relationship. In his decade-long studies among large monkey colonies, Suomi noticed that some rhesus monkeys appeared to be particularly easily aroused and vulnerable to impairments, based on their natural, genetically based predisposition. He then specifically bred such "high risk monkeys," as well as monkeys with normal arousal potentials, in order to conduct the study described in the following section.

Both groups of monkeys were assigned, right after birth, to a substitute mother that played the maternal role for the youngster during the next 6 months. The mothers were also divided into two groups. One group consisted of animals that had stood out previously because of their particularly nurturing, optimal care of their offspring. I will refer to these animals as *supermothers*. The other group consisted of animals that previously had shown average maternal behavior. All young rhesus monkeys in this study, then, were raised not by their biological mother but by a substitute mother. The youngsters with normal arousal potentials did not differ from others in their development depending on whether they had been raised by a normal mother or a supermother. Both of these groups showed normal development. Dramatic differences were observed, however, among the high-risk monkeys, depending on whether they had grown up alongside a supermother or an average mother. The youngsters of the normal mothers showed the kind of behavior that, based on previous long-term studies, would be expected among such high-risk groups. They showed little exploratory behavior and reacted with heightened anxiety even to minor disturbances. The high-risk monkeys reared by supermothers, by contrast, developed most excellently. They were even superior to the normal youngsters in their development. They left their mothers earlier, showed more exploratory behavior, and responded with fewer problems to the termination of maternal nursing, not only in comparison to the other high-risk monkeys but also when compared to youngsters with normal levels of arousal sensitivity. They showed all the signs of a particularly secure attachment relationship with their substitute mothers. Later in their development, after all youngsters had been separated from their mothers, they were especially able to enter alliances with other monkeys, and most of them achieved top positions in the dominance hierarchy of their respective groups. The high-risk monkeys reared by average mothers, by contrast, tended to end up toward the bottom of the dominance hierarchy.

Later, when the female animals within this study became mothers themselves, the animals that had been raised by particularly nurturing mothers also showed especially nurturing behavior toward their own offspring, regardless of whether they had originally been classified as high-risk monkeys, and regardless of the maternal style shown by their biological mothers. Maternal style, then, appears to be passed on across generations, and this transpires not via a genetic pathway but via the attachment experiences that a female youngster encounters with her "psychological" mother (Suomi & Levine, 1998). The fact that the maternal attachment behavior a mother shows toward her own children is highly significantly dependent upon her own attachment experiences with her mother, regardless of whether this is a biological or a substitute mother, has also been documented in several additional studies in rhesus monkeys (Champoux, Byrne, Delizio, & Suomi, 1992; Fairbanks, 1989). The degree of ventral contact has been shown to be a particularly stable transgenerational characteristic in this regard. These findings cohere remarkably well with the findings in humans, described earlier, which also found that attachment styles are passed on from one generation to the next.

The neurobiological studies on attachment styles so far relied almost entirely on animals, primarily on rhesus monkeys, and not on humans. The reasons for this are obvious. For ethical reasons, such experimental studies could not be conducted with humans. This shows, however, the enormous importance of attachment experiences for the subsequent well-being of humans.

How relevant are such studies in monkeys to processes in humans? The research findings can certainly not simply be directly applied to humans, primarily because humans possess an additional quality that cannot be examined in animals. For example, rhesus monkeys probably do not develop an inner working model of their relationships in the sense of Bowlby's original model. Studies about the more cognitive aspects of attachment processes can only be conducted in humans.

On the other hand, attachment processes take place at a point in life when the cognitive capacities of humans do not yet play such an important role; instead, basal neurophysiological mechanisms that overlap considerably in rhesus monkeys and humans are more relevant in this context. The studies with rhesus monkeys are probably highly relevant in terms of what is observed about the effects of attachment experiences on basal neurophysiological reaction patterns. Based on animal experimental studies on neural development in phases of particularly high neural plasticity, Braun et al. (2002) describe far-reaching conclusions for the pathogenesis and treatment of mental disorders in humans. Such conclusions seem all the more justified to the degree that more and more studies in humans and animals show similar results.

Earlier, I summarized the results of one of Suomi's studies, in which rhesus monkeys with an unfavorable, easily aroused temperament subsequently developed excellently when they were raised by a particularly nurturing, competent mother. An analogue study with a similar result has been conducted on human infants. Van den Boom and Hoeksma (1994) studied extreme groups of highly irritable and nonirritable neonates. They found that the mothers of the irritable infants stimulated their children less during the first 6 months of life. They were also less sensitive toward them and overall engaged less with them. Unfavorable temperamental characteristics of the child, then, exerted a disadvantageous influence on the emerging attachment relationship. Van den Boom (1994) explained this by noting that the degree of maternal interaction competence is negatively influenced by the problematic infant behavior. Infants that cry a lot and are difficult to console encounter fewer control experiences because their attachment figures tend to be less responsive to them. Over time, this has an effect on the inner working model that the child forms of his or her relationships. The innate temperament of the child can therefore exert a negative effect on his or her attachment experiences (for a detailed discussion of this possibility, see Pauli-Pott & Bade, 2002). Van den Boom (1994) was able to show in an experimental intervention study that the unfavorable interaction between temperament and disadvantageous life experiences could be broken by targeted interventions. This study initially also showed that highly irritable infants are more likely to develop insecure attachments. This was found, however, only for the infants in the control group. In an experimental group, targeted interventions were introduced to improve maternal sensitivity. After 9 months, the children of these mothers showed a greater level of explorative competence and overall more positive behavior. After 12 months, there were clearly more securely attached children in this group, compared to the children in the control group.

The studies by Suomi and van den Boom both suggest the conclusion that negative attachment experiences are not simply inevitable fate, a product of unfortunate genes or the result of an attachment figure who just happens to be suboptimal. The unfavorable cycle can be broken by targeted interventions. This is quite reassuring, especially when considering the high degree of concordance of attachment styles across generations, which has been documented both in monkeys and in humans.

A logical consequence of these studies would be to attempt to preventatively influence the attachment competence of mothers and fathers of high-risk children; for example, by using sensitivity training. A controlled longitudinal study on precisely this topic is currently being conducted in Ulm, Germany. Parents of very prematurely born infants receive targeted interventions such as sensitivity training, attachment-oriented supportive individual therapy, group discussions with other parents in the same position, and home visits in order to master the specific difficulties arising from the premature status of their child so that the development of insecure attachment can be avoided (Brisch, 2002). Such preventative intervention options would certainly also be useful for parents who themselves have an insecure attachment style, so that they are prepared for the task and learn the key behaviors to facilitate secure attachment in their children. Broad-based prevention programs could probably have far more beneficial consequences in this area than other preventative programs, such as those targeting smoking or other health-risk behaviors. The prevention of insecure attachment patterns via primary and secondary prevention programs would be a worthwhile domain for health psychology. It can probably be assumed that we have not yet seen more efforts in this direction because the findings from attachment research are still very new. As far as mental disorders are concerned, one could probably achieve much more with low-cost preventative interventions in the first months of life than what can be achieved with much more expensive interventions that become necessary later in life when the negative long-term consequences of early insecure attachments require treatment. The long-term consequences of violations of the attachment need are discussed in the following section.

4.4.3   Violations of the Need for Attachment and Their Consequences for Mental Health

Which mental health consequences follow from violations of the attachment need? I will answer this question in the following three steps:

1. How are insecure attachment styles formed?
2. What influence does a person's attachment style exert on his or her well-being in early and middle childhood?
3. What influence does a person's attachment style exert on the development of mental disorders in adulthood?

The number of studies on attachment styles has grown so large in recent years that it would be impossible to provide even a nearly comprehensive overview here. Good general overviews can be found in E. A. Carlson and Sroufe (1993) and in the handbook by Cassidy and Shaver (1999). Information regarding the aspects of attachment research that are particularly relevant for psychotherapy can bests be found in Strauss, Buchheim, and Kächele (2002). Assertions in the subsequent section that are not otherwise specifically substantiated by citations refer to findings summarized in these sources, which also contain the relevant references.

4.4.3.1 How are insecure attachment styles formed?

In the detailed observation of primary attachment figures—that is, typically the mothers—of insecurely attached infants and toddlers, the following findings emerged:

- They responded less empathically to the child in many daily situations, from the very beginning.
- They responded with greater delay and less overall empathy to the child.
- Mothers of children with insecure–avoidant attachment often rejected the child when he or she sought contact, even though they had the same overall degree of physical contact. They expressed their love to the child more by kissing than by

hugging or cuddling. When the child seemed sad during play, the mothers tended to withdraw from their child.
* Mothers of children with insecure–ambivalent attachment more frequently interrupted the children in their activities but then also ignored them at other times.
  Overall, they showed more inconsistent behavior toward the child. They clearly
  enjoyed hugging the child but did not respond consistently to the contact
  attempts by the child, when these were shown. They generally showed little
  response to the contact initiatives of the children.

On the whole, compared to mothers of securely attached children, the mothers of insecurely attached children were less well coordinated with the child, less emotionally
involved in the transactions with the child, responded less to the child's signals, and
often were more intrusive. These results have been confirmed repeatedly in many
other studies. A good summary of these results can be found in E. A. Carlson and
Sroufe (1993). These findings, however, are based primarily on studies of normal,
middle-class families.

Other studies have also examined lower-class families with chaotic life circumstances, in which the children were often even abused. These children were frequently emotionally and physically rejected, they were treated with hostility and aggression, verbally and emotionally threatened, and the parents tended to respond very little to the children's requests or concerns. Compared to children growing up in
matched lower-class families without such abuse, these children growing up in such
abusive conditions were much more likely to exhibit insecure attachment styles. The
numbers vary between the different studies on the order of 50% to 100%. In one of
these studies (V. Carlson, Cicchetti, Barnett, & Braunwahl, 1989), the percentage of
children developing an insecure attachment style in abusive families was 80%, whereas this percentage among children in nonabusive families was 19%.

What are the causes of maternal relationship behavior that ultimately results in insecure attachment styles of their children? One major reason concerns the relationship
experiences that the mother herself had with her own mother. In several studies (Main,
Kaplan, & Cassidy, 1985; Ricks, 1985), various methods and samples repeatedly confirmed that mothers of insecurely attached children themselves had an insecure
attachment style and experienced their own mothers as nonnurturing and noncompe-
tent. This relationship certainly appears plausible. The rejection and lack of reliability that insecurely attached children experience from their primary attachment figures
is accompanied by intensive emotions. At this early age, there is little that would be
more important for the child than the mother's care and attention. If such a thing as
motivational goals already exist at this early age, then obtaining maternal attention
and care would surely be among the most important goals of a young child. If the
child repeatedly experiences severe incongruences with regard to this goal because it
is beyond his or her control to reliably attain this goal, then intense negative emotions
will repeatedly occur. At this point, the intrapsychic emotion regulation independent
from the relationship with the mother is not yet established. However, the mother is

not available for the important task of down-regulating these emotions. Instead of responding empathically to these emotions, her rejecting, insecure, and incompetent reactions to the emotional behavior of the child accomplish little other than intensifying the negative emotions even more. For example, Grossman and Grossman (1991) found that mothers of insecure–avoidant children withdrew when their children were sad. Such a child was left alone with his or her feelings. One of the side-effects of an insecure attachment, then, is poor emotion regulation. Because she has not learned effective emotion regulation with her own mother, the mother with insecure attachment cannot help her child to cope with his or her negative emotions successfully. Good attachment behavior and good emotion regulation are apparently best learned in the context of a good attachment relationship. Individuals who did not experience this cannot pass on to their children that which they haven't learned themselves. It seems clear, then, that parents who did not have the chance to develop a secure attachment relationship with their own parents will have greater difficulty in being good parents, compared to parents who were lucky enough to grow up in a secure attachment relationship.

A further major reason for why parents often cannot be good attachment figures for their children concerns the presence of mental disorders in the parents. This is particularly true for parental depressions. In a study by Sameroff, Seifer, and Zax (1982), it was shown that even very young infants of depressive parents are more difficult to soothe and have less differentiated emotions than children of parents who have other (physical) illnesses. Subsequent negative consequences on the development of autonomy in these children were also documented. In several studies, highly significant associations were found between different types of parental depression and insecure attachments in the children (Gaensbauer, Harmon, Cytryn, & McKnew, 1984; Radke-Yarrow, 1991; Radke-Yarrow, Cummings, Kuczynski, & Chapman, 1985; Zahn-Waxler, Cummings, McKnew, Daveport, & Radke-Yarrow, 1984). The strongest relationships were found for bipolar disorder. The negative effects associated with parental psychopathology were not just short-term but tended to endure. These studies also showed that emotion regulation constitutes a central aspect of the attachment relationship. It is easily apparent that depressives do not have good emotion regulation skills and tend to be preoccupied with themselves. Therefore, they are unable to direct their attention continuously toward the child and to be emotionally responsive to the current states and needs of the child, regardless of their own feeling state. How would a child learn from an acutely depressed mother how to master negative feelings?

4.4.3.2 What influence does a person's attachment style exert on his well-being in early and middle childhood?

Three research groups—the ones of Maine and colleagues, Sroufe and colleagues, and Grossmann and Grossmann—have conducted longitudinal studies that examined how children whose insecure and secure attachment styles had been assessed at an early age (between 12 and 18 months) continued to develop over the course of child-hood. Many individual articles, which I will not all discuss in detail here, have been published on this topic. The following represents an overview of the main findings. More detailed results can be found in E. A. Carlson and Sroufe (1993); Grossmann and Grossmann (1991); Main, Kaplan, and Cassidy (1985); and Sroufe, Carlson, and Shulman (1993).

When the development of securely and insecurely attached children is compared with each other in various age groups, practically all of the examined aspects show marked advantages favoring the securely attached children. It is important to note that these other characteristics are obtained in a blind test; that is, the other raters or judges did not know whether the child whose characteristics were rated had been classified as secure or insecure. The advantages in favor of the securely attached children concern their self-confidence, self-esteem, self-efficacy expectancies, their resilience (robust-ness) under conditions of stress, and particularly their interpersonal relationship behavior and the quality of their relationships with their peers. Compared to insecure-ly attached children, securely attached children are rated by peers and teachers as more socially oriented, relationship-competent, empathic, and popular. They are bet-ter able to express their impulses, wishes, and feelings, including especially negative feelings. Their greater resilience under conditions of stress is evident not only in observable behavior but also on a physiological level. Gunnar, Brodersen, Nachmias, Buss, and Rigatuso (1996) found a stronger cortisol response after a mental stress sit-uation among anxious–inhibited 2-year old children who had been classified as of the insecure rather than the secure attachment type.

Even though, compared to the securely attached children, both insecure–avoidant and insecure–ambivalent children tend to experience worse outcomes, remarkable differ-ences between these two types have also been documented. Both have been rated as dependent by their teachers, but the insecure–avoidant children were less able to express their wishes to the teacher. They either did this indirectly or waited for a par-ticularly safe situation to arise. When some emergency occurred—such as a minor injury, a disappointment, or a stressful situation—they did not dare to turn to the teacher for help. Avoidant children elicited more punishing reactions from their teach-ers. The teachers demanded more discipline from them, tended to be more controlling toward them, expected more motivation from them, and showed less warmth and more anger in interactions with them, compared to the other children. The children, then, reproduced in these relationships with important adults similar types of relation-ships than what they already knew from home.

When they saw another child in difficulties, the insecure–avoidant children respond-ed less empathically and supportively and often behaved inappropriately and hurtful-ly; for example, by teasing a crying child or mocking a child that had hurt himself or herself. The relationships with their peers were often hostile or aggressive or very dis-tanced. When they were introduced to play partners, they often tended to take advan-tage of them, to devalue them, or to behave in a hostile manner, including even beat-ing or physically punishing them. In each case, they took the role of the perpetrator.

In this aspect, too, then, they perpetuated the pattern of rejection and isolation that they knew so well from their own experience.

Insecure–ambivalent children, by contrast, tended to seek the proximity of their teachers. They tended to wait for the teacher's initiative—for example, until the teacher would invite them to join in. Compared to other children, the teachers acted more caringly and patiently toward them; they were more tolerant, helped them out, and so forth. In play situations with other children, they tended to relinquish initiative and leadership to other children and preferred to be followers. When other children teased or exploited them, they passively accepted this. In each case, they took the role of the victim. They also perpetuated the pattern they already knew from home.

These differences between securely versus insecurely attached children, on the one hand, and insecure–avoidant versus insecure–ambivalent children, on the other, tended to remain stable at all stages: kindergarten, preschool, elementary school, secondary school, and holiday camps, all the way up to adolescence. After the 1st year of life, in which the attachment style is largely formed, then, children with different attachment styles encounter a very different set of experiences. These experiences result in different perceptual tendencies, motivational tendencies, emotional response tendencies, and a different behavioral repertoire. The securely attached children consistently benefit from more positive life experiences, which then results in additional positive memories that, in turn, further increase the likelihood of encountering further positive experiences. For them, the early positive attachment experiences set in motion a positive feedback loop that is associated with positive control experiences, self-esteem enhancing experiences, and positive emotions. Thus, as a consequence of an early satisfaction of the attachment need, they also benefit from a greater probability that their other basic needs will be satisfactorily met. The result is an overall positive development and a state of good mental health.

The opposite developmental pattern is evident among children with clearly insecure attachment patterns. They do not develop adequate social repertoires and effective emotion regulation skills, nor do they develop positive expectancies about what might be anticipated from others. They also do not benefit from the experience that their own behavior can effectively lead to positive perceptions in terms of their immediate goals and needs. The sum of these experiences is certainly not self-esteem enhancing. Therefore, not only do they experience a poor satisfaction of their attachment need, but also their need for control remains unmet. Instead of encountering self-esteem-enhancing and positive emotion facilitative experiences, they encounter the exact opposite. Their basic needs not only remain unmet; they tend to be continuously squashed and violated. Every such violation leaves memory traces, which further reduce the chances of attaining positive need satisfaction in the future. After several years, they have either developed a schema structure in which stable avoidance schemas clearly outweigh approach tendencies, or they have acquired a schema structure in which approach and avoidance tendencies are continuously in such conflict that neither approach nor avoidance can ever be successfully implemented. In terms of consistency theory, the result of both cases would be a permanently high level of incongruence.

In disorganized attachment, which has not been examined as often because of its lower frequency, the consequences of attachment need violations were even more severe in studies that specifically investigated this topic. The formation of insecure attachment patterns in childhood, then, provides an early foundation for the additional, continuous failure to meet the attachment need later on, which then almost inevitably leads to further violations of other basic needs. If a high level of incongruence is indeed the main foundation for the development of mental disorders—as is posited in consistency theory—then one would assume that individuals with insecure attachment styles in adulthood would show an increased prevalence of mental disorders. The research findings on this topic are summarized in the next section.

4.4.3.3 What influence does a person's attachment style exert on the development of mental disorders in adulthood?

The relationship between attachment patterns and adult mental disorders has been examined for about 10 years now. The method of assessing attachment styles in these studies has almost always been the Adult Attachment Interview, which is a semistructured interview targeting the person's current representations of attachment experiences in the past and the present. The interview focuses on memories of early attachment relationships, the individual's access to attachment-relevant thoughts and feelings, as well as the person's judgment with regard to the influence of early attachment experiences on subsequent development. The interview is evaluated not only based on the content produced but also based on the coherence of the narrative and in terms of minimized and underregulated affects (Buchheim & Strauss, 2002). AAI ratings result in a classification of five attachment types, the first three of which correspond to the types defined by Ainsworth et al. (1978), although they are given a different label. The five types are

1. Secure–autonomous persons (F = free to evaluate)
2. Insecure–distancing persons (Ds = dismissing)
3. Insecure–preoccupied persons (E = enmeshed/preoccupied)
4. Persons with unresolved trauma/unresolved grief (U = unresolved)
5. Nonclassifiable persons (CC = cannot classify).

Table 4–1 provides an overview of the results of 11 studies on the relationships between attachment styles, as measured by the AAI, and mental disorders.

The studies with small samples in Table 4–1 can certainly be interpreted only with great caution. Three studies were conducted with larger samples, however, which therefore allows for more confidence in drawing conclusions. In the study by Fonagy et al. (1994) in 85 psychiatric patients, 90% of the patients had an insecure attachment

Table 4–1. Studies on the Association of Specific Attachment Patterns With Mental Disorders

Study	Sample	Findings
Dozier (1990)	14 schizophrenic patients, 28 depressive patients	Increased attachment insecurity in both groups, more frequently among schizophrenics.
Dozier et al. (1992)	21 schizophrenic patients, 18 depressive patients	Among the schizophrenics, by comparison, more frequently distanced attachment.
Carnelley et al. (1994)	25 depressive patients	Attachment pattern predicts later interpersonal ability better than degree of depression.
Harris & Bifulco (1991)	Reanalysis of Walttharnstow study (225 women)	5 out of 6 attachment patterns were attributed to depressive patients, according to Bowlby (forced care = 60%).
Cole & Kobak (1991)	65 students (male and female) with eating disorders and/or depressive symptoms	Eating disorders only—distanced pattern, depression only—ambivalent pattern, eating disorder + depression—both patterns.
Fonagy et al. (1994)	85 psychiatric patients	90% insecurely attached, high proportion of unprocessed experiences of abuse in borderline patients.
Patrick et al. (1994)	12 patients with borderline disorder, 12 patients with dysthymia	12 borderline patients vs. 4 dysthymia with complicated attachment; in 6 borderline patients vs. 0 dysthymia patients, indications for unresolved trauma.
Sperling et al. (1991)	24 borderline patients	Borderline patients in comparison to student population mostly insecurely attached; dependent attachment style in borderline population correlates with less pathology.
Adam (1994)	132 psychiatrical symptomatic youths	80% insecurely attached, considerable portion suicidal.

(continued)

Table 4–1. *(Continued)*

Study	Sample	Findings
Van Ijzendoorn (as cted in Scheidt & Waller, 1996)	Meta-analysis of AAI results	Proportion of insecure attachment in clinical populations: 41% insecure— avoidant; 46.5% insecure— ambivalent; 12.5% secure; in control group of "healthy individuals": 23.2% vs. 17.6% vs. 59.2%, respectively.
De Ruiter & van Ijzendoorn (1992)	Overview on studies on agoraphobics	Proportion of ambivalent patterns very high among agoraphobics, but not specific to the disorder.

Note. From *Psychologische Therapie* (p. 331), by K. Grawe, 1998, Göttingen, Germany: Hogrefe. Copyright 1998 by Hogrefe. Adapted with permission.

pattern. In Adam's (1994) study among 132 adolescent psychiatric patients, 20% of the participants were securely attached. In a meta-analysis of more than 14 studies with a total number of 688 patients, Ijzendoorn and Bakermans-Kranenburg (1996) arrived at the following classification of patients' attachment patterns: Insecure–avoidant (dismissing), 41%; insecure–ambivalent (enmeshed/preoccupied), 46%, securely attached, 13%. In normal populations, the proportion of securely attached persons, as measured by the Adult Attachment Interview, tends to be near 60%. In a meta-analysis, the value of 59.2% was reported. In the studies by Fonagy and Ijzendoorn, the percentage of securely attached individuals in clinical samples, by comparison, was only around 10%!

Dozier, Stovall, and Albus (1999) attempted to differentiate this global picture for the different disorders. For this purpose they analyzed five studies that had not been included in the meta-analysis by van Ijzendoorn and Bakermans-Kranenburg. The resulting grouping of AAI attachment types in various clinical disorders is shown in Table 4–2.

The first three columns in the table refer to an analysis in which only the distinction among F , E, and Ds was considered. The next four columns refer to an analysis in which the category U was also considered. This table shows clearly that the frequencies of the insecure attachment styles E, Ds, and U are substantially higher than those of the secure attachment style, F. Figure 4–2 shows the results from the first type of analysis in somewhat more easily digestible, graphic format.

This figure illustrates clearly that the proportion of securely attached patients in most disorders is just above 10%. In the majority of disorders, the insecure–preoccupied attachment type outweighs the avoidant type, which is particularly pronounced in

Table 4–2. Diagnostic Groups and AAI Attachment Types

Disorder	Authors	F	E	Ds	F	E	Ds	U
Mood disorders								
Unipolar								
Depressive disorders	Cole-Detke & Kobak, 1996	4	6	4				
Mixed	Rosenstein & Horowitz, 1996	0	22	10	0	19	8	6
Major depressive disorder	Tyrell & Dozier, 1997	5	1	0	3	1	0	2
Dysthymic disorder	Patrick et al., 1994	2	4	6				
Bipolar		0	0	7	0	0	3	4
Mixed	Fonagy et al., 1996	18	41	13	9	6	5	52
Schizo-affective disorder	Tyrell & Dozier, 1997	1	1	6	1	0	5	2
Anxiety disorders	Fonagy et al., 1996	7	29	8	2	1	3	38
Comorbid samples								
Eating disorders and depressive disorders	Cole-Dethke & Kobak, 1996	4	10	5				
Conduct disorders and depressive disorders	Rosenstein & Horowitz, 1996	0	3	9	0	1	6	5
Borderline personality disorder	Fonagy et al., 1996	3	27	6	2	1	1	32
	Patrick et al., 1994	0	12	0				

Note. F = free to evaluate; E = enmeshed; Ds = dismissing; U = unresolved.
From "Attachment and Psychopathology in Adulthood," by M. Dözier, K. C. Stovall, & K. E. Albus, in J. Cassidy & P. Shaver (Eds.), *Handbook of Attachment*, 1999, New York: Guilford. Copyright 1999 by Guilford. Adapted with permission.

borderline personality disorder. In schizophrenia, by contrast, the avoidant type even more strongly outweighs the others.

Figure 4–3 shows the image that results when one takes the U category into consideration in the analysis of five studies by Dozier et al.

The frequency distribution shows that for most of the disorders, there are clear indications of severe traumatic experiences in 70% to 90% when they are asked to speak

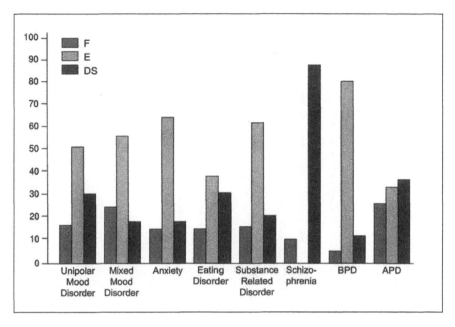

Figure 4–2. Distribution of AAI-attachment types in various disorders (in %; F = secure; E = enmeshed; DS = dismissing; BPD = Borderline Personality Disorder; APD = Antisocial Personality Disorder) (Adapted from Dozier et al., 1999).

about their relationship history. The much lower value for unipolar depression cannot be regarded as reliable. It is based on very low sample sizes, as can be seen in Table 4–2. The typically very low sample sizes are generally still a problem that hinders our efforts to understand the distribution of attachment styles across different disorders. The exact percentages in each study always depend on the particular characteristics of the studied sample, on the severity of the disorders, and so on. Such sample differences have not been taken into consideration in the meta-analyses up to this point. Once we have the results of much larger, more representative samples, it will become clearer whether associations between the type of disorder and type of attachment style indeed exist at all. It could well be, for example, that an insecure attachment pattern predisposes patients for the development of mental disorders, but that other factors determine which specific disorder ultimately emerges.

It is also conceivable that the studies in this area are still too crude in their differentiations of attachment styles, undermining our ability to conduct more specific analyses. Pilkonis (1988) and Strauss, Lobo-Drost, and Pilkonis (1999) have developed a rating system for a differentiated distinction of seven attachment prototypes, with the goal of differentiating subgroups with different attachment patterns within the different diagnostic groups. The differentiated distinctions were regarded as necessary primarily in order to test whether different attachment prototypes would respond differentially to psychotherapy on a global level and to specific forms of therapy (Meyer & Pilkonis, 2002; Meyer, Pilkonis, Proietti, Heape, & Egan, 2001).

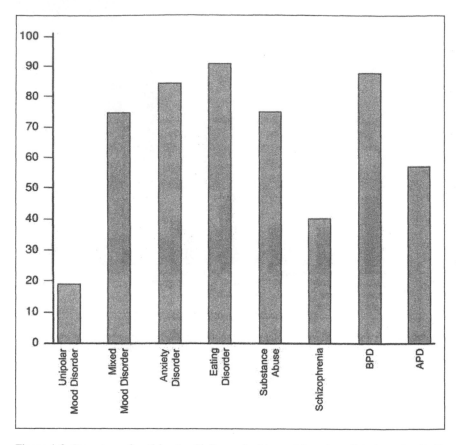

Figure 4–3. Percentage of participants with "unresolved trauma" in various clinical samples (in %; BPD = Borderline Personality Disorder; APD = Antisocial Personality Disorder) (Adapted from Dozier et al., 1999).

Even when using this more fine-grained approach of seven attachment prototypes, the main result was once again a clear predominance of insecure attachment patterns in patients with mental disorders. This time the result is based on a large sample within a multicenter study of 528 patients who were treated in inpatient psychotherapy (Schauenburg & Strauss, 2002). Figure 4–4 shows how the patients were distributed across the attachment prototypes.

These results specifically apply to a group of psychotherapy clients. They are based on the results of an attachment prototype interview conducted prior to the beginning of an inpatient psychotherapeutic treatment. The findings show that fewer than 10% of the patients encountered daily by psychotherapists in inpatient settings are securely attached. If one regards the predominant attachment pattern that a person develops as an accurate reflection of his or her actual relationship experiences, this finding also means that almost all psychotherapy patients were not able to satisfy their basic need for a secure attachment relationship in their historically most important relationships.

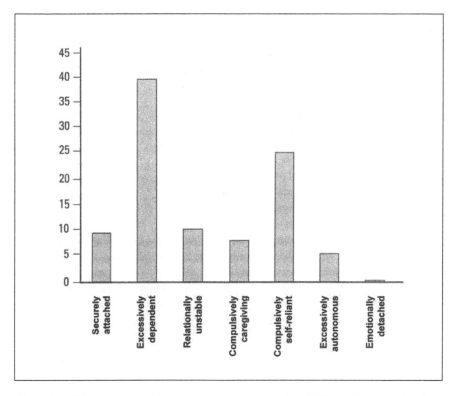

Figure 4–4. Distribution of AAI-attachment types in a sample of 528 psychotherapy inpatients (in %).

If we also consider that about 60% of individuals in normal populations have a secure attachment pattern, the conclusion seems obvious that the nonfulfillment or even violation of the attachment need might have played an important role in the process that brought them to psychotherapy.

This causal interpretation of the correlational associations found in adults between their insecure attachment styles and the presence of a mental disorder appears justified because, based on the previously summarized studies of attachment processes in the first months after birth, we can have considerable certainty that the violation of the attachment need was present prior to the mental disorder, and that this violation and its immediate consequences had a causal influence on the unfavorable development that children with insecure attachment styles experience from their 1st year onward into adulthood. If we add all those findings that I previously reported, one could not have any reasonable doubt anymore that it is the violation of the attachment need that leads, via many intermediate steps, to the development of mental disorders, and not vice versa, that the disorders somehow come first and lead to an insecure attachment style. Based on the totality of findings on the attachment need, an insecure attachment style can be viewed as the greatest single risk factor known today for the development of a mental disorder. I know of no single characteristic from the entire area of etiological research on mental disorders that would have even nearly the high prognostic significance that has been documented for insecure attachment styles.

4.4.4   Conclusions for Psychotherapy With the Example of Depression

Based on the findings I reviewed here, attachment styles are something very stable. We have no reason to assume, then, that a depressed patient who is treated with antidepressant medication or with Beck's cognitive–behavioral therapy and then recovers from the depression would change his or her attachment style in tandem with the reduction in depression. Such a finding has never been documented and also would appear very improbable if one considers what the term *attachment style* really denotes.

Attachment styles refer to memory content that has become deeply ingrained over the course of a lifetime. These styles refer to reactions to stress, to emotion regulation, automatized perceptual tendencies, the automatized aspects of relationship regulation, avoidance reactions that are stored in implicit memory and have been facilitated from the 1st year onward, and so forth. One would have to believe in magic if one assumed that all of these completely automatized components of attachment patterns, which transpire beyond the reach of consciousness, would be effectively changed by working for 16 sessions with Beck's methods on cognitive schemas or by using medication to increase the serotonin level. If one remains rational and abandons the idea of such magic change—especially because one had not even attempted to target attachment styles in these treatments—then one would come to the conclusion that insecure attachment styles that were present prior the therapy will tend to remain in place even after a successful disorder-specific treatment.

There is also no obvious reason why the attachment style, over the course of treatment, should have lost its status as a risk factor for the development of a mental disorder. If such a risk factor continues to be present, of course, one should not be surprised that the majority of the cases that initially recover after a successful disorder-specific depression therapy would later tend to relapse. According to consistency theory, if one treats an insecurely attached person but does not specifically target the avoidance and conflict schemas that have formed around his or her attachment need, then one of the most important sources of inconsistency remains in place. The activity of these motivational schemas manifests primarily in the patient's current interpersonal relationships. Patients with insecure attachment patterns typically have interpersonal problems in addition to their more narrowly defined psychopathological symptoms. One would also have to consider in treatment the direct consequences of the attachment need violations in the first years of life. Based on the previously reported results, this would primarily include reduced stress tolerance, poor emotion regulation, low self-efficacy expectancies, and low self-esteem.

The previously summarized studies suggest that 8 or 9 out of 10 psychotherapy patients, apart from their prominent disorder, also have an insecure attachment pattern, which can manifest in all of the additional problems mentioned just previously. These additional problems will differ from patient to patient. Therefore, they cannot all be considered in a disorder-specific treatment manual. If the therapy is limited strictly to the treatment of the disorder, the additional sources of inconsistency remain untreated and therefore constitute a continuously present breeding ground for the development of mental disorders. They can even undermine treatment success entirely, if the treated disorder also functions to down-regulate inconsistency, or if the patient is poorly motivated for treatment because the therapist's treatment goals insufficiently match his or her own goals.

These considerations are inspired by the findings of attachment research, and they are supported by research studies on the therapy goals of psychotherapy patients. At our psychotherapy clinic in Bern, we have, for more than 20 years, assessed the goals that patients want to pursue at the beginning of their treatment. For each of these goals— on average, patients provide about three goals—we conduct a procedure known as goal attainment scaling (Kiresuk & Lund, 1979), in order to measure the treatment success based on the attainment of these patient-generated goals. On the basis of more than 1,000 therapy goals that had been obtained with this method, we developed a taxonomy of therapy goals, the Bern Inventory of Therapy Goals (BIT; Grosse Holtforth, 2001; Grosse Holtforth & Grawe, 2002a). The taxonomy is structured hierarchically. At the highest level, five categories of therapy goals can be differentiated, including:

1. Problem-mastery/symptom-reduction (P).
2. Interpersonal goals (I).
3. Well-being (W).
4. Orientation/meaning (O).
5. Self-actualization, self-esteem (S).

Table 4–3 shows how the most important therapy goals of 383 patients in our psychotherapy clinic are distributed across these five categories.

The therapy goals are limited to problem mastery/symptom reduction in only 9% of the patients. A markedly higher proportion—15%—report only interpersonal goals. The most common combination of therapy goals is, at 18%, problem mastery along with interpersonal goals. In another 13% of patients, this constellation is combined with additional self-improvement goals. Among a full 28%, problem mastery/symptom reduction is not even listed at all as one of their own therapy goals. The most common treatment goals among psychotherapy patients concern the interpersonal domain (75%). Problem-mastery/symptom reduction goals follow in second place at 59%, although such goals tend to be named as the most important ones in the cases in which they are listed at all.

The finding that 75% of all outpatient psychotherapy clients report treatment goals in the interpersonal domain coheres very well with the previously reported finding from attachment research, according to which more than 80% of all psychotherapy patients have insecure attachment patterns. Acquired early in life, the insecure attachment pat-

Table 4–3. The Distribution of the Three Most Important Therapy Goals of 383 Psychotherapy Patients Across the Five Top Categories of the Bern Therapy Goal Inventory.

P	I	W	O	S	N	%	Total %
P	I				69	18	18
	I				57	15	33
P	I			S	51	13	46
	I			S	49	13	59
P					33	9	68
P				S	30	8	76
...	...	...	...	...	...	...	...
59%	75%	13%	11%	46%	383	100.0	100.0

Note. P = problem mastery/symptom reduction; I = interpersonal goals; W = well-being; O = orientation/meaning; S = self-actualization, self-esteem.
From "Bern Inventory of Treatment Goals (BIT), Part I: Development and First Application of a Taxonomy of Treatment Goal Themes (BIT-T)," by M. Grosse Holtforth & K. Grawe, 2002, *Psychotherapy Research, 12.* Adapted with permssion.

terns apparently lead to the kinds of difficulties in interpersonal relationships later in life that patients cannot easily cope with by themselves. Both findings show from different perspectives that a psychotherapy that limits itself to the treatment of the prominent mental disorder cannot be regarded as sufficient, neither from an objective, scientific perspective nor from the problem perspective of the patients themselves.

I attempted to illustrate this idea already in the introduction with my case example of a depressed patient. Indeed, this breadth of treatment goals seems to occur primarily among depressed patients. Figure 4–5 shows a comparison of the goals reported by patients who had been diagnosed with an anxiety disorder versus a depressive disorder, based on DSM-IV criteria.

We can see that the therapy goals of depressed patients are more spread out across the categories, compared to those of patients with anxiety disorders. From their perspective, there are more diverse problem areas beyond their narrowly defined symptoms, compared to those with anxiety disorders. Similar differences between therapy goals of patients with anxiety disorders versus depression were found in other studies with patient samples (Berking, Grosse Holtforth, Jacobi, & Kröner-Herwig, 2004; Berking, Jacobi, & Masuhr, 2001; Faller & Gossler, 1998; Grosse Holtforth, Reubi, Ruckstuhl, Berking, & Grawe, 2004; Grosse Holtforth, Schulte, Grawe, Wyss, & Michalak, 2004).

Based on these results, it seems less appropriate to treat depressed patients with a disorder-specific manualized therapy, whereas this might be more justified in patients

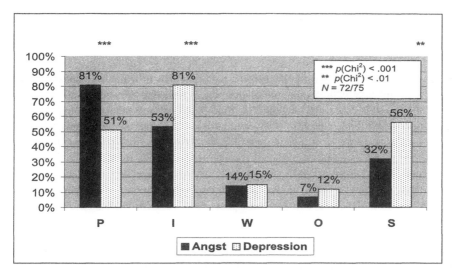

Figure 4–5. Comparison between the Therapy Goals of Patients with an Anxiety Disorder versus a Depressive Disorder (adapted from Grosse Holtforth & Grawe, 2002).

with anxiety disorders. Such narrow, manualized treatments do not do justice to the various problems that the patient himself or herself regards as important. In terms of consistency theory, a merely disorder-oriented treatment, especially of depressed patients, ignores other sources of inconsistency, including those in the interpersonal domain.

In my view, the facts I discussed here are the most important reasons explaining the rather limited long-term effectiveness of currently available depression treatments. Every practicing clinician is repeatedly confronted with the fact that depressive patients repeatedly require additional treatment, even after they have already received psychotherapeutic or pharmacological treatment. If one reviews the scientific depression literature, however, a very different impression emerges. According to that view, it would appear that depression can be treated particularly effectively with currently available methods.

The research literature on depression treatment confronts the reader with awe-inspiring effect sizes. They sometimes lie above a value of 2 or even 3, and effects of this magnitude are achieved over a treatment duration of about 10 weeks. If one keeps in mind that M. L. Smith, Glass, and Miller (1980) calculated an average effect size for psychotherapeutic treatments of .85, one would conclude that depression apparently is one of the disorders that can be treated particularly effectively. In one's efforts to explain the contradiction between the research results and the impression from clinical practice, one almost inevitably arrives at the conclusion that the depressed patients who recurrently present for treatment must not yet have been treated with the truly effective types of therapy, which is why they keep returning for treatment. If only they were finally treated with one of these truly effective depression interventions, for which such strong effect sizes were found, then they would surely be cured. It would be great if this were the case, but unfortunately the conclusion is based upon an illusion.

In our institute we have just concluded a meta-analysis in which we analyzed all of the 77 controlled depression studies that were published between 1980 and 2001 (Gallati, 2003). We considered studies that examined only pharmacological treatment ($N = 29$, with a total of 5,352 patients), studies that examined only psychological therapies ($N = 24$, with a total of 1,297 patients), and studies that compared psychological and pharmacological interventions (mixed studies, $N = 24$, with a total of 2,668 patients). The treatments lasted an average of 10 weeks, and the psychotherapies encompassed an average of 13 sessions. For each treatment and control condition we calculated pre- and posteffect sizes for each measure used in every study.

We found that the pre–post effect sizes after an average of 10 weeks of treatment were indeed very high, compared to the pre–post effect sizes that are normally found in psychotherapy. Whereas the average pre–post effect size for various forms of psychotherapy and various types of disorders has been estimated at 1.21 (Grawe, Donati, & Bernauer, 1994), this value in the pharmacological treatment of depressed patients is, on average, 2.11. The serotonin reuptake inhibitors Venlafaxin and Paroxetin even resulted in effect sizes of more than 3! Figure 4–6 provides an overview of these effect sizes, which we calculated for the various antidepressant medications.

Figure 4–6 can be expected to be warmly received by the representatives of the pharmacological industry, especially when compared with the effect sizes obtained for psychotherapeutic treatments (Figure 4–7).

The average effect sizes for the various psychotherapeutic treatment modalities are clearly lower than those observed for the pharmacological treatments, but they are still markedly above the effect sizes that one typically finds in psychotherapy studies.

These findings seem to suggest three conclusions:

1. Depression, compared to other disorders, can be treated particularly effectively.
2. Medication is the first line of treatment.
3. Psychotherapy is very effective in depressed patients, even though it is not as effective as antidepressant medication. There are no marked differences in the effectiveness of various psychotherapy modalities.

These are exactly the messages that are conveyed to visitors and to the general public at psychiatric congresses, accompanied by huge advertising campaigns by the pharmaceutical industry. They are also the messages presented as certain fact to beginning psychiatrists at the outset of their training. The real state of affairs, however, looks completely different. Each of the three statements is incorrect and would almost have to be turned into its opposite.

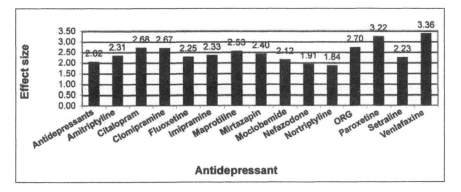

Figure 4–6. Overview of the average effect sizes for the various types of antidepressant medication (Adapted from Gallati, 2003; p.91).

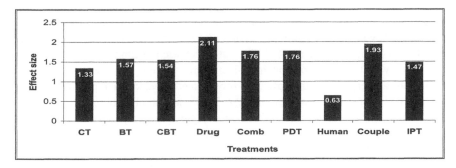

Figure 4–7. Overview of the average effect sizes for all examined psychological treatment modalities among depressed patients. Explain all abbreviations in the legend. CT = Cognitive Therapy; BT = Behavior Therapy; CBT = Cognitive-Behavioral Therapy; Drug = Medication; Comb = Combined Psychological Therapy and Medication; PDT = Psychodynamic Therapy; Human = Humanistic-Existential Therapy; Couple = Couple Therapy; IPT = Interpersonal Therapy (Adapted from Gallati, 2003).

The impression of good effectiveness is created by the fact that depressive episodes, over a period of 10 weeks, tend to improve considerably even without treatment. The degree of "spontaneous remission," which also includes placebo effects, is particularly large in depression. In studies that examined the effectiveness of medication, the changes attained across all measures in the control groups were estimated at an effect size (ES) of 1.59! Even these spontaneous improvements—occurring without any targeted intervention at all—are clearly larger than the effects that are normally found for effective forms of psychotherapy. If one subtracts these improvements, which are not accounted for by the examined medication, from the average effect of 2.11 in the treatment groups, the net effect of .52 remains. This is the true average effect of pharmacological treatment of depressed patients. When using a measure that was developed specifically for the assessment of medication effects in depressed patients—the Montgomery-Asberg Depression Rating Scale (MADRS), which is only used in pharmacological studies—then the effect size across all treatment groups was estimated to be 1.88. On the face of it, this looks like a very respectable effect size. However, the ES on the same scale in the control groups was estimated at 1.82. This yields a net effect of .06.

In addition, one needs to consider that 25% of the patients in the studies that were included in this meta-analysis dropped out of treatment, whereas this was the case for only 13% of the patients in psychological treatments. Dropouts represent treatment failures, but these are not considered in the calculation of effect sizes. High dropout rates, therefore, lead to an inflation of effect size estimates. They drive up the pre–post effect sizes because only the pre–post effect sizes of the therapy completers are considered. The therapy completers represent a positive, skewed selection of all the patients to whom the treatment was administered. The twice-as-high dropout rate works in favor of the pharmacological rather than the psychological treatments.

In the studies on psychological treatments, an ES of .97 was obtained across all studies and all measures in the control groups. The net effect size for the various psychological therapy modalities can be calculated by subtracting the value of .97 from the effect sizes displayed in Figure 4–7. The net effect size remaining for purely cognitive therapy is then .36; for cognitive–behavioral therapy, .57; for Klerman and Weissman's interpersonal therapy, .50; for present-oriented brief psychodynamic therapies, .79; and for couples therapies, .96. This is an entirely different image of the effectiveness of the various therapy forms for depression, compared to what is conveyed in much of the literature. From this perspective, the therapy modalities that devote more attention to the interpersonal problems of depressed patients tend to be somewhat more effective than the therapies that target only specific disorder components.

In order to evaluate the comparative statements about the different therapy forms, one must first examine whether this effect size comparison is really based on comparable foundations. The effect size comparison between pharmacological and psychological therapies is akin to a comparison between apples and oranges, which has been one of the main criticisms of the method of meta-analysis. This is all the more true because the two types of studies use totally different criteria for the evaluation of therapy successes.

Figures 4–8a and 4–8b compare the frequencies with which pharmacological and psychological studies employed specific psychometric measures. The differences are obvious: Pharmacological studies used almost exclusively symptom measures, and primarily those that relied on observer evaluations. These measures are the ones that also attain by far the highest effect sizes without treatment (in the control groups; see Figure 4–9). They are, therefore, particularly well suited to render treatments in a particularly positive light.

The researchers in these pharmacological studies typically are not particularly interested in the perspectives of the patients themselves. Key characteristics of the mental

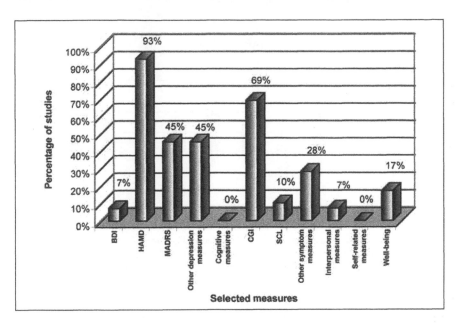

Figure 4–8a. Frequencies of the use of selected measures in studies that examined only pharmacological treatments. BDI = Beck Depression Inventory; HAMD = Hamilton Depression Scales; MADRS = Montgomery-Asberg Depression Rating Scale; CGI Clinical Global Impression; SCL = Symptom Checklist (Adapted from Gallati, 2003).

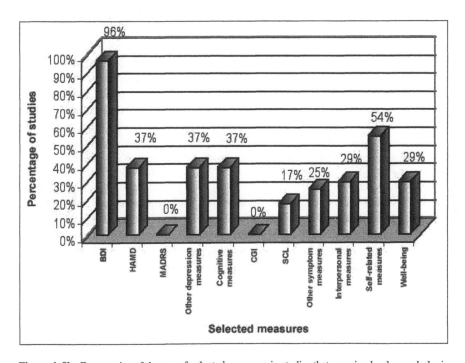

Figure 4–8b. Frequencies of the use of selected measures in studies that examined only psychological treatments. BDI = Beck Depression Inventory; HAMD = Hamilton Depression Scales; MADRS = Montgomery-Asberg Depression Rating Scale; CGI Clinical Global Impression; SCL = Symptom Checklist (Adapted from Gallati, 2003).

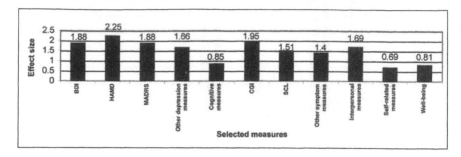

Figure 4–9. Magnitude of the average effect sizes that resulted for the various change measures across all 77 studies. BDI = Beck Depression Inventory; HAMD = Hamilton Depression Scales; MADRS = Montgomery-Asberg Depression Rating Scale; CGI Clinical Global Impression; SCL = Symptom Checklist (Adapted from Gallati, 2003).

constitution of depressed patients, such as their cognitive functioning, their self-esteem, their subjective well-being, and the entire interpersonal domain, are consistently completely ignored in these pharmacological studies. However, these are also exactly the measures for which typically only very small improvements can be observed. Figure 4–9 shows the average effect sizes that were found for the various change indexes across all treatment groups in all 77 studies. We can see that the measures differ in their sensitivity to change. Pharmacological studies are typically limited to the measures for which very generally—including in the control groups—the largest changes can be observed. These are the Hamilton scale, the MADRS, and the Scale of Clinical Global Impression (CGI). The pharmacological studies achieve such a high average effect size because they are limited in their measurement to exactly those domains in which even no targeted treatment will result in the greatest improvements. Changes in cognitive functioning and in self-esteem are much harder to attain than those in primary symptoms. The psychological therapies fare worse at first glance because—and only for this reason—they also tend to include measures of these other domains.

The limited view that is taken in the purely pharmacological studies is also reflected in the fact that none of these studies includes long-term follow-up analyses, whereas this was the case in 30 of the 48 other studies. This narrow viewpoint is apparently related to another aspect in which pharmacological and psychological studies differ remarkably: All pharmacological studies were financed by the pharmacological industry. All psychological studies, by contrast, were financed by public research funding.

Because of the drastic differences in that which is actually measured in psychological versus pharmacological studies, it is indeed prohibitive to conduct simple comparisons because otherwise one truly compares apples with oranges. However, there are some measures that do tend to be used in all three study types. This is the case for the Beck Depression Inventory (BDI), the Hamilton Rating Scale, and the SCL 90 as a measure of general psychopathological symptom intensity. For these measures, it would seem more appropriate to compare the effect sizes directly, even though the frequencies with which these measures were used differ among the treatment types. In two of these comparisons, weaker effects sizes were found for the pharmacological versus the psychological treatment groups. These measures were the BDI (1.26 vs. 1.66) and the SCL-90-R (1.11 vs. 1.43). In one case, the Hamilton Rating Scale, pharmacological treatments fared better (2.83 vs. 2.08). This observer rating measure targets primarily the physical and physiological aspects of depression. It makes some sense that medications achieve a stronger observable effect in this domain. This is not the case in the subjective experience of the patient, however, because the BDI and SCL-90-R also measure some of these physical aspects.

The design of the pharmacological studies makes it very clear that these researchers are not particularly interested in the changes occurring in the mental constitution and the life of the patient as a consequence of the medication. It is more than apparent that these researchers and their sponsors are only interested in the "scientific" proof of maximally impressive effectiveness of their respective medications. As far as science has something to do with curiosity and the wish to discover, these studies are beyond of the domain of that which one generally means when one speaks of science.

Despite the enormous financial investments, we still know hardly anything about what these antidepressant medications really accomplish within the mental system. From a neuroscientific perspective, this would actually be a rather fascinating question, especially from the perspective of a targeted pharmacological preparation and facilitation of learning processes that are created via psychological intervention methods. We will see repeatedly in this chapter that the naturally produced neurotransmitters in the brain play a crucial role in the learning process. It would seem rather obvious to facilitate learning processes via the targeted use of pharmacological agents that could facilitate or inhibit the availability of specific neurotransmitters at specific locations within the brain. One should also be able to use this for therapeutic purposes. This would require, however, a detailed knowledge of the processes in the brain as well as a detailed planning and control of the learning experiences that the patient is to be exposed to while his or her brain has been rendered maximally ready to benefit from such learning experiences. This is where I can see a truly rational utility for psychopharmacological agents, in contrast to the current, largely "blind" scattering or dumping of agents into the brain; agents whose exact effects on the mental processes remain largely unknown because these questions have not even been considered as specific research questions in the relevant studies. In this area, serious basic research is required. The undifferentiated research that dominates today, which aims only to legitimize, is not likely to yield a better understanding of the interactions between artificially created neurotransmitters and specific life experiences and learning processes.

With an effect size of .52—"beautified" by the high dropout rate of patients—and without convincing evidence of long-term effectiveness, the effectiveness of pharmacological depression treatments is clearly not nearly as established as is often suggested. We know from comparative studies, such as the NIMH depression study (Elkin, 1994) in which follow-up effects were examined, that the long-term effectiveness of antidepressant medication is very poor and is indeed below that of psychological therapies. The contrast between objective findings and the promises of the pharmaceutical industry could not be larger. Those working in clinical practice or those who have friends who have experienced recurrent depressive episodes for years will already know what to make of these promises, but it would be useful in the research literature and in the public discussion on this topic to down-adjust the effectiveness evidence of psychopharmacological antidepressant agents, including in terms of their numerical values, bringing them more in line with their actual effectiveness. An effect size of .52 is more than no effect at all, but it is clearly no reason to celebrate.

When interpreting these findings, we ought to keep in mind that the absolute value of the effect sizes in the treatment of depression initially seems truly impressive. In the short term, we can actually observe strong improvements in symptom severity. The individual patients typically do not care so much what causes this improvement as long as it actually does occur. Even if this effectiveness is explained by common mechanisms of action or by so-called spontaneous remission, the improvement is real from the perspective of the patient. We should also consider that the previous discussion focused throughout on average effect sizes across large groups of patients. Within the treatment groups, some patients might have responded very well to the treatment whereas others might have experienced very little or no improvement. The treatment might have been a blessing for one patient, whereas it was utterly disappointing for another one. It is quite possible, then, that some of the patients—and of course also some of the therapists—might gain the impression at the end of the treatment that the therapy was highly effective. The sobering evaluation of the specific effectiveness of the different therapies was based on the fact that the effects related to placebos and spontaneous remission were deducted from the raw score effects, and that the high relapse rates and dropout rates were specifically considered. The resulting net-long-term effects of the individual therapy modalities can look rather modest when compared to the subjective experience of successfully treated patients and their therapists immediately after the conclusion of a treatment course. In terms of the evaluation of the specific effect sizes associated with different treatments, however, this approach is the only appropriate one.

Such a sober evaluation of the objective findings clearly shows that it would be inaccurate to argue that pharmacological treatments are broadly more effective than psychological interventions. The opposite is closer to the truth. Once the higher dropout rates and the inferior long-term effectiveness of pharmacological treatments are taken into consideration, one arrives at the conclusion that psychological therapy is the treatment of choice in the treatment of depression, and that medications should be used only when combined with psychological therapy, and only in cases in which psychotherapy alone has not led to the desired success. One could perhaps also argue for a combined treatment in order to achieve maximally rapid symptom reduction and in order to utilize the positive interactions between these two treatment forms, even though the benefits of this approach are still not clearly substantiated.

However, the followers of the different therapy schools also do not have much reason to celebrate. The findings on the most commonly used forms of psychological therapy for depression, Beck's cognitive–behavioral therapy and Klerman and Weissmann's interpersonal therapy, when reviewed from a sober, objective perspective, can be presented as follows:

About 25% of the depressive patients who are invited to participate in a disorder-specific therapy in the context of a controlled therapy study will decline the invitation. The reasons for this remain unknown. A possible reason might be that they do not regard the offered treatment as an appropriate option. If we keep in mind the goals that they themselves bring to treatment (see previous), their decision not to participate in a treatment that focuses only on the disorder appears entirely comprehensible. Other reasons for the dropouts, which have nothing to do with the type of treatment, are of course also possible. Of the remaining 75%, our meta-analysis suggests that another 13%—and Hautzinger and de Jong-Meyer (1996) suggest that another 25%— of the patients terminate prior to the end of the treatment. The reasons for this might be similar to those of patients who do not enter treatment in the first place. However, it is also possible that it is not the content of the therapy goals that provides the reason for termination but rather the concrete therapeutic proceeding, which might not attend sufficiently to the optimal facilitation of the therapeutic relationship in the way that would be advisable among persons whose attachment needs have been severely violated. Of the remaining 64%, about 50% achieve clinically significant improvement (Gortner, Gollan, Dobson, & Jacobson, 1998). Of the remaining 32% of patients who have achieved short-term improvement, 60% experience a relapse within 2 years (Elkin, 1994).

The total equation, therefore, leads to the following result: Of 100 depressed patients who require treatment, 13 to 14 achieve enduring improvement from currently available psychological depression therapies that are regarded as particularly effective. In light of the fact that little evidence documents significant effectiveness differences among the psychological treatment modalities (Rush & Thase, 1999), the additional question arises as to whether these effects can indeed be explained by the presumed therapeutic rationale. Jacobson, Martell, and Dimidjian (2001) argued that the effects of all psychological depression treatments can be explained by the fact that these treatments all reactivate the approach system of the patient. Even though these authors do not provide neuroscientific evidence, a neuroscientific view of depression leads to a rather similar conclusion. I will revisit this point later.

As far as the treatment of depressive disorders is concerned, based on the reviewed evidence we have every reason to search for concepts that might lead to more effective treatments. Inspiration for more effective treatments can be obtained, in my view, from that which I discussed in chapter 3 on the neural correlates of depressive disorders. However, this would require the development of entirely new concepts and strategies. Other approaches for effectiveness-enhancement, which can already be implemented, concern the intensive targeting of the sources of inconsistency that are apparently present among depressed patients, above and beyond their core symptoms. These include, above all others, the domain of interpersonal relationships of the patient. Interpersonal aspects must also be considered more thoroughly in the context of treatment goals. These consequences arising from the perspective I developed here will be more fully discussed in chapter 5.

In the new version of Beck's cognitive theory of depression, which I already mentioned earlier, etiologic research on depression is discussed at length (D. A. Clark & Beck, 1999). The fact that violations of the attachment need or insecure attachment styles might have something to with the pathogenesis of depression, however, is not reflected at any point within this book. I do not mention this here to criticize the authors. I have already expressed at an earlier point how highly I value their scientific work. I mention it because it shows that an overly one-sided disorder perspective can lead one to completely ignore findings originating from another perspective. Attachment research, on the face of it, has little to do with the specific disorder of depression. However, it can probably contribute more to our understanding of the etiology of depression than the theory of the depressogenic cognitive schemas.

In light of the strong relationships between insecure attachment patterns and mental disorders, it would seem that the time has come to devote more attention to the interpersonal aspect, not only in etiological models of the various mental disorders, but also in the development of evidence-based therapy manuals. The interpersonal aspect is of greater importance than what is expressed in most therapy manuals, not just from the subjective perspective of the patient but also according to the research findings on the various disorders. Research findings outside of attachment research also suggest that the development of depression is often linked to the interpersonal relationships of the respective person. Study results in this area can be found in Anderson, Beach, and Kaslow (1999), Dill and Anderson (1999), Holahan, Moos, and Bonin (1999), as well as Roberts and Monroe (1999). It is difficult to understand why the findings reported in these studies are not applied to the treatment of depressed patients. In the commonly used depression manual by Hautzinger (1998), the interpersonal aspect is relatively neglected. More attention is devoted to it in Klerman and Weissman's interpersonal therapy (Schramm, 1996), but even this manual does not truly entail an intensive and targeted intervening on the level of the problematic inner working models, as described by Bowlby (1973), or on the level of the motivational schemas, as described in consistency theory. In order to do justice to the far-reaching consequences of a lifetime history of insecure attachment patterns, one would have to analyze in much more detail, in each individual case, which problematic constellation of approach and avoidance motivational schemas tends to be activated in the interpersonal relationships of the patient, and which additional sources of inconsistency have emerged as a consequence of the negative attachment experiences in the individual patient's case. This would include a lowered stress tolerance, poor emotion regulation, unfavorable consistency-securing mechanisms, and poor self-esteem regulation. In order to answer the question about additional sources of inconsistency, however, basic needs other than the attachment need will also have to be considered. The next section deals with the need for control, which also plays an important role in the development of mental disorders.

4.5   THE NEED FOR ORIENTATION AND CONTROL

4.5.1.   Control Processes Permeate All Mental Functioning

According to Epstein (1990), the need for orientation and control is the most fundamental of all human needs. Every person, based on this view, develops a model of reality into which other real experiences are then assimilated. The inner working model of Bowlby would be such a model in the domain of relationship experiences. This is an important part of that which Epstein terms *conception of reality*. Epstein argues that the conception of reality that a person forms, based on his life experiences, is synonymous with the self of that person, which he tries to maintain in his interactions with the environment. The most important instruments that this self possesses in its transactions with the environment are the motivational schemas that form in the process of experiencing real-life events. If behavior is continuously oriented toward the attainment of perceptions that fit currently activated goals, as described by Powers (1973), then this also means that the person continuously aims to achieve control over these perceptions. It is inevitable that real life experiences are accompanied by the experience that one either achieves that which one had aimed for or one fails to achieve it. Therefore, it is also inevitable that either positive or negative control experiences are encountered. This control aspect is intricately linked with goal-oriented behavior. Striving for perceptions that are consistent with activated goals always means striving for control. It is impossible to completely relinquish this control if one wants to survive or if one wants to fulfill any kind of need. Control is a pervasive, essential aspect of mental functioning.

Depending on the life experiences that accumulate for an individual with regard to his need for control (particularly in early childhood), a basic conviction begins to form about whether predictability and controllability is possible, about whether it tends to pay off to invest resources and engage in life, and about the extent to which life makes sense in general. Rotters' (1966) construct of self-efficacy beliefs corresponds to this basic conviction.

There is such a broad literature about the need for control and the related control and efficacy beliefs that it would be impossible to present a comprehensive overview here. Indeed, thousands rather than hundreds of research studies have been conducted on these topics, which is why I will highlight only some of the findings that are particularly relevant for psychotherapy. An excellent introductory overview with regard to the concept of control can be found in Flammer (1990), whereas Wegner and Pennebaker (1993) summarized some research trends related to the many aspects of mental control.

The vast extent of theoretical and empirical research work on the concepts of *control*, *control expectancies*, and *control beliefs* indicates that these terms correspond to a fundamental part of human existence. This supports the assumption, according to Flammer (1990, pp. 114–115), that a corresponding need exists, that is, that one can assume that individuals are spontaneously willing to engage effort in order to attain, maintain, and increase control. This need is not aimed at any particular effects as such but rather at the experience of control itself.

I differentiate between ordinary needs for control and the *basic* Need for Control. The basic Need for Control is innate but is always expressed in concrete contexts and corresponds to concrete contents and goals. These concrete goals are available to the individual because of his or her socialization. One could also say that the basic Need for Control materializes over the course of socialization in the form of specific contents and goals, that it becomes tied to particular scenarios and thereby multiplies and increases in specificity. There are many particular needs for control but there is only one basic Need for Control.

In his book *Action Regulation and Control*, Oesterreich (1981) substantiates the assumption of a basic Need for Control with a biological consideration:

> One has to assume that each organism or each species has the aim to preserve itself or its own species... The point is to regulate behavior in such a way that a variety of different action possibilities (activity options) remain possible. Because of the necessity that organisms strive for survival, we can deduce the assumption that more developed organisms have a striving for control. This striving aims to orient behavior in such a way that the ability to regulate one's environment is maintained. (pp. 130, 132)

Therefore, the need for control is not just about exercising control in the current situation; it is also about the ability to maintain a maximal sphere of options for future actions.

> The need for control is satisfied when a maximum number of options remain available for freely chosen action, in a maximum number of valued life domains. The person who strives for control, then, strives for behavioral flexibility, for activity options that would allow him to reach important goals. The need for control can also be satisfied by the current sense of controllability, not strictly by actual control experiences. (Flammer, 1990, p. 117)

People who feel compelled to save money in order to maintain a maximum sphere of potential future options clearly have a strong need for control. Their sense that one is able to use this money "just in case" can be interpreted as the creation of a generalized sphere of behavioral flexibility. Other people's relationship with money is determined, instead, by their need to enjoy a maximum number of pleasurable experiences. They might spend their money, for example, for culinary delights, for journeys, or for other pleasurable events. Yet others use their money to satisfy their need for self-esteem enhancement; for example, by driving a prestigious car, wearing expensive clothes, or owning an impressive villa. Individuals who have to use money to satisfy their need for attachment are perhaps in a comparatively worse situation but such instances are, of course, far from rare.

Positive control experiences—in other words, the experience that one's own behavior can successfully lead to the achievement of personal goals—will yield positive control perceptions, as described by Rotter (1966), or will yield positive self-efficacy perceptions, as described by Bandura (1977). This emphasis once again clarifies that the assumption of a need for control constitutes a core aspect of Powers's (1973) control theory, from which I derived the concept of incongruence. According to Powers and similar models, all behavior is oriented toward the attainment of perceptions that are congruent with specific goals. If this is successful, the need for control is satisfied; if not, incongruence results. Therefore, incongruence is always accompanied by a nonsatisfaction or violation of the need for control.

Thus, the need for control can be viewed as a need to be able to perform some action that is important for the attainment of personal goals. This corresponds to the competence aspect of mental functioning. If one is unable to control something in such a way as to bring one closer to important personal goals, then this constitutes a grave violation of the need for control. Mental disorders, which the patient naturally experiences as something beyond his or her control, are such a violation of the need for control. Whenever we provide something in therapy that allows the patient to master his or her problems in a better way or to cope with problems successfully, then this constitutes a positive control experience that heals the aversive violation of the patient's need for control.

Effective disorder-specific and problem-specific therapeutic interventions are always accompanied by a more optimal satisfaction of the need for control. This reduces the incongruence level in mental functioning and results in improved overall well-being. Even if a therapy aims merely to help the patient cope with the primary disorder, it will inevitably act via the reduction of his or her incongruence level toward improved overall well-being.

Control also has a cognitive component. The ability to exert control presupposes that one has an accurate view of the situation. This is what the need for orientation refers to. Not knowing what's going on in one's environment can be highly aversive, particularly when one's self or other personally important issues are involved. To gain clarity about one's situation, and about what can be done to improve it, therefore constitutes another positive experience with regard to the need for control. This explains why even supportive, clarification-oriented therapies can lead to a better satisfaction of one's need for control and, further, to a broader improvement in well-being. From the perspective of the need for control, of course, it would be best if the patient experiences both: that he or she understands his or her situation better and actually has gained control over his or her environment, by having mastered or successfully coped with personal problems. Both of these aspects together would have a cumulative positive effect on the patient's control perceptions, perceived self-efficacy, and general well-being.

At the beginning of our lives, the need for control is intrinsically linked with the need for attachment. When infants or toddlers experience that they can reliably use their behavior to achieve desired reactions in their mothers, that they can influence when and how their mothers attend to them, and when they experience that they can convey to their mothers their wishes, needs, and general situations, then these can all be viewed as positive control experiences that ultimately will lead to positive control and self-efficacy beliefs. An available, sensitive, and responsive attachment figure is critically important in this first phase of life, not just for the attachment need but also for the need for control. Most of the exchanges that occur between mother and child during this phase are relevant simultaneously for the attachment and the control need. This holds true as well for negative situations. Violations of the need for attachment simultaneously violate the need for control.

The reason for this is that the child is practically unable to do anything by himself or herself in this phase of his life. The child has to rely on an attachment figure for everything that is important to him or her in this phase. This principle applies even to the regulation of desire and distress. When something hurts, when he or she feels cold or hungry, the infant can control these negative states only via a sensitive attachment figure. This circumstance could explain, in part, why attachment figures are so immensely important. What is important here is not just the physical proximity of the attachment figure. The attachment relationship is also the locus at which control, powerlessness, desire, and distress are experienced and regulated during this phase of life. It is only as the child grows older that desire and distress, as well as control and lack of control, are experienced outside of the attachment relationship. The needs then become more differentiated and become partially independent in contrast to their early internal interdependency. However, this disjoining is indeed only partial: If an adult experiences marked distress, such as intense pain or fear, then the desire to alleviate the distress is activated along with the need to attain greater control. If one is in a position to personally control the pain, this is experienced as less aversive than situations in which the pain is uncontrollable. The incongruence with regard to the need for pleasure and the need for control have cumulative effects. The pain or fear is soon accompanied by a sense of helplessness, and these experiences together are even worse than the pain by itself. In such a situation, it would also be typical that the attachment need is activated alongside the other two. The intense wish for a competent person emerges; for someone who engages with one's personal distress and who is able to end the distressing state of affairs. If such efforts are successful, one simultaneously has positive experiences with regard to the activated attachment need and with regard to the need to reduce distress. Whether this also entails positive control experiences depends, foremost, on the degree to which one was personally involved in the alleviation of the distressing situation.

The need for control is always activated when important goals are involved. Events that satisfy the need for control will practically always lead to an improved state, whereas events that frustrate or violate this need will lead to state decrements. When someone has to undergo a serious heart surgery, his or her need for distress (pain, helplessness, fear) reduction, as well as his need for attachment (with regard to the medical staff), is activated. The fact that the need for control is activated can be recognized via the positive effects that are observed when patients are thoroughly informed about all aspects of the surgery and when they are given the opportunity to be actively engaged in the healing process after the surgery (Baltensperger & Grawe, 2000). A thorough preparation for surgery satisfies the need for orientation and control, and this need satisfaction (incongruence reduction) has a positive effect on the state of the patient, even though the patient is unable to contribute anything to the purely medical aspects of the surgery and aftercare. Among all forms of psychotherapy, the satisfaction of the need for control plays an important role as a mechanism of action. For example, psychological treatments for tinnitus have only little direct effect, if any, on the actual sounds within the ear. The patient learns new ways of dealing with the sounds, however, and thereby achieves control over something that had previously caused intense distress. This leads to a marked reduction in subjective impairment. This functions in a very similar way with pain therapies. The medical procedures alone often have little direct effect on back pain, for instance. Their effect size can be estimated at .20. However, if patients are also engaged in psychological interventions that facilitate better coping strategies or encourage positive activities, this effect size, as measured by multiple outcome assessments, attains a mean magnitude of about .70 (Kröner-Herwig, 2000). The actual organic causes of the pain have changed little. The greater control that has been achieved, however, has a positive effect via the satisfaction of the need for control on the patient's subjective impairment and perceived pain intensity. This is another example of the close interrelatedness among the basic needs. The satisfaction of the need for control also triggers a reduction in distress, which in turn strengthens the sense of control, and so forth.

Despite this close interrelatedness it seems sensible to conceptually differentiate various basic needs because, in later life—in contrast to infancy—control does not always involve attachment but rather often tends to involve, for instance, work achievement. In adulthood, control is also not always linked to pleasure, and pleasure is, in turn, experienced in contexts other than the attachment relationship. It is hard to be helpless, hard not to have an attachment person, and also hard to suffer from pain or fear. But these experiences are not equivalent, even though they are often linked with one another. The basic needs have their own neural circuits, which are strongly connected with each other and can reciprocally activate each other. When one is exposed to aversive experiences, the need to change something about these circumstances will inevitably arise. The need for distress reduction and the need for control are activated simultaneously. Nevertheless, the neural circuits are separate, even though both may be activated at the same time and might interact with each other. These circuits tend to reciprocally activate each other primarily when serious violations of a basic need occur. Because all of the basic needs become enveloped by avoidance schemas over the course of life, grave violations of these needs can only occur when these protective mechanisms are not effective, when control has failed, as it were. Violations of the three other basic needs, therefore, also engender violations of the need for control. When incongruence arises with regard to any one of the basic needs, the question of the controllability of these situations will also always arise. This leads us to a question that is of great importance for a consistency-theoretical conceptualization of how mental disorders originate: the question of the immediate and long-term consequences of incongruence.

4.5.2   Controllable and Uncontrollable Incongruence

The term *incongruence*, in the sense that I use it in consistency theory, refers to discrepancies between perceptions of reality on the one hand and activated goals, expectancies, and beliefs on the other. This incongruence is one possibility of how inconsistency can arise in mental functioning. *Inconsistency* is a term that refers to the incompatibility, discrepancy, or disagreement of simultaneously activated mental/neural processes. It is notable that this incongruence refers to the interaction between the individual and his or her environment. The mental processes that are involved in this incongruence always include the real perceptions that the individual experiences with his or her sense organs. These perceptions are also always linked, via goals, expectancies, and beliefs, to the person's basic needs. Congruent perceptions constitute need satisfaction, and incongruent ones the opposite. According to Powers's (1973) control theory, the functioning of the mental system is fundamentally designed in such a way as to seek perceptions that are consistent with currently activated goals. I have adopted this basic assumption in my formulation of consistency theory. As the very term *control theory* suggests, continuous control is required in order to achieve or maintain congruence between perceptions and goals. Incongruent perceptions activate the need for control. Because states of complete congruence—the simultaneous satisfaction of all basic needs—are very rare, the need for control is practically always somewhat activated. Thus, the individual continuously experiences perceptions that are somewhat incongruent with regard to his or her need for control or, rather, with regard to the motivational goals that have formed around this need. When a situation is evaluated as controllable, that is, when the person evaluates his or her options to bring the current situation in line with his or her goals in a positive way, the need for control is satisfied. However, when a situation that is incongruent with regard to one of the basic needs is experienced as uncontrollable, the need for control is not met or, in other words, is violated.

What I said earlier holds true not just for the positive satisfaction of the basic needs but also for the protection of their violation. Based on his life experiences, the individual develops approach and avoidance goals. These two types of goals differ in terms of how well they can be controlled, in principle. With approach goals, it is possible to envision a clear aim. Movements toward that aim can be controlled quite well because one knows the final purpose or destination. Avoidance goals, by contrast, involve a moving away from something. No clear aim is available because the goal is being defined in negative terms. Being able to confirm that something is not the case, however, requires constant monitoring. One can never be certain that the goal has been reached; one can never relax one's attention in this respect. With approach goals, it is much easier to judge how far away one still is from the goal and when the goal has been reached. There are also many other reasons why avoidance goals are less advantageous than approach goals, and I will explicate these reasons later on. In this section, however, the important point is the extent to which these processes can be controlled. Incongruence with regard to avoidance goals is more difficult to control than approach incongruence, and such avoidance incongruence requires a greater mental effort and a higher risk for violations of the need for control.

In order to detect and monitor the incongruence between perceived reality and goals one would need something like a comparator. Powers postulated the existence of such a comparator but he was unable to say much in detail about how the continuous comparison process between perceptions and goals unfolds over time. Gray and McNaughton (1996), in their updated revision of Gray's (1982) neuropsychological anxiety model, went into greater detail and specified the neural basis of such a comparator. This neural system involves the various regions within the limbic and associative cortex that we already encountered previously when we discussed the continuous evaluative processes that constantly occur in the mental system. The inputs of this comparator are

- Current sensory information about the present situation.
- The steps to be taken next within the activated executive program (prospective memory). This implicitly contains the currently activated goals.
- Knowledge stored in memory about rules and lawful cooccurrences within the currently activated domain of reality (contextual knowledge).
- Knowledge stored in memory about instrumental behavior–effect contingencies as a reflection of the person's autobiographical learning history.

Based on this input of sensory information, on the one hand, and information from explicit and implicit memory, on the other, the comparator continuously forms expectations about how the current situation will continue to develop. These expectancies can be felt very clearly when someone practices a familiar melody on an instrument and then suddenly stops. While listening to the melody, one's own brain constantly produces the tones that are expected to occur next, and when a wrong tone occurs, a rather painful incongruence can result. The same phenomenon can be observed in the context of driving a car or doing any routine daily activity. The brain continuously produces expected scenarios about how the situation will continue from this moment on. As long as incoming sensory information is consistent with these prospective scenarios—that is, as long as no incongruence signals are received within the comparator—the need for control remains deactivated. Everything proceeds smoothly within the expected frame. The person's mental activity is determined by well-engrained neural activation patterns that are smoothly in tune with the situational context. However, such situations are also characterized by the absence of new learning. The dopamine neurons that respond to deviations from expectancies, whose activation facilitates the strengthening of new neural connections (see section 2.3), are not being activated in such a situation.

As a father or mother, when one listens to one's own child perform an instrument at a concert, one can have a very intensive demonstration of the process by which the brain consistently produces a preview of the tones that are to follow at each moment. One has had the opportunity to listen to the child practice incessantly at home, so all the critical passages are well known to the brain. When the child manages to master these critical passages with surprising ease, the dopamine neurons fire. On the other hand, when our expectations remain unmet in a negative sense—when the child has difficulties with a normally easy passage—these neurons are actively inhibited. In section 2.14.1, I discussed this neural coding of deviations from expectancies in positive or negative directions via the activation and inhibition of dopaminergic neurons. This differential activation of dopamine neurons can be regarded as an important part of the comparator, above and beyond the neural circuits discussed in this context by Gray and McNaughton.

If one is the person who is actively acting in a given situation, then the expectancies that enter into the comparator also include one's own control or self-efficacy expectancies for the each respective situation. If a situation develops in an unexpected way—for example, when a surprising new event is encountered or when something aversive might happen or something valued might not happen or might be lost—the comparator activates the need for control. The subsequent experiences are then determined by the incongruence signal that results from the evaluative process with regard to the control goals and control expectancies If important goals are being threatened and controllability is being evaluated as low, anxiety results. Using the terminology of Gray and McNaughton, the behavioral inhibition system is activated.

Incongruence can therefore be regarded as an output of the comparator. The process that I have termed *incongruence* here is often discussed in the literature under the label of *stress*. For example, the diathesis–stress model of the genesis of mental disorders uses the term in this sense. The term *stress* has become part of common, everyday language, and experiencing stress is nothing that would necessarily threaten one's sense of self-worth. For that reason, it is perhaps wise to use the term *stress* when speaking with patients about the phenomenon of incongruence. However, *stress* is a rather vague term, and in a scientific context one ought to use more precise terminology. When someone says, "I am stressed," then it is hard to differentiate between the triggering situation, the stressor itself, and the reaction to this, the stress response. Stressor and stress response cannot be discussed separately. What constitutes intolerable noise for one person might not bother another person at all. This kind of burden or personal challenge cannot be defined separately from the person experiencing the challenge. Much research on stress has been conducted with animals. In these experiments, a stressor is introduced via the manipulation of external circumstances, such as via uncontrollable electric shocks, via enforced separation from the mother, and so forth. In the natural life of humans, however, stressors can normally not be defined by physical terms but rather by psychosocial or psychological phenomena. The stressor is the meaning that the situation has, depending on the goals, expectancies, values, and beliefs of the person.

The differential effects of controllable and uncontrollable stress on neural structures have been described by Huether (1998) in a article that contributes a great deal to our understanding of the possible effects of incongruence. His writings on the meaning of stress fit very well here in the context of my own conceptualization of incongruence:

> Most of our current knowledge on the consequences of the activation of the central stress responsive systems on the brain is derived from animal experiments using various kinds of physical stressors. The predominating stressors in the life of socially organized mammals and especially humans, however, are psychosocial conflicts. Psychosocial stress is not just another, but a totally different kind of stress. The difference between physical (or physiological) and psychosocial (or psychological) stressors is a simple but fundamental one: Physiological stressors elicit a stress response because a certain objectively existing, environmental or physical force is strong enough to disrupt the counterregulatory mechanisms available to an individual. Psychological stressors elicit a stress response because the subjective perception and interpretation of often rather subtle and ambiguous changes of the outer world come in flagrant variance with the expectations, beliefs or assumptions made by an individual on the basis of its previous experiences. Whereas physiological stress is caused by changes in the world outside the brain, *the very root of psychological stress resides inside the brain*: [italics added] Whether or not and to what extent a stress response will be elicited, is dependent on a subject's interpretation of the changes perceived from its outside world.

If the previously acquired basic beliefs and assumptions of an individual about the world are incompatible with its current perception of the world, this individual will experience a sustained, more or less controllable, stress reaction, until either its perception of the external world or its beliefs and assumptions about the world are corrected to better fit with one another." (Huether, 1998, p. 298)

In this last paragraph, Huether addresses exactly the process that I refer to as incongruence; an incompatibility of simultaneously activated mental processes, whereby one of these processes includes the perception of reality. My concept of incongruence is completely compatible with Huether's definition of *psychological stressor*. When Huether speaks about *stress* or *stressors*, he refers to the same processes that I have termed *incongruence*. Huether also says that this type of incongruence triggers a stress reaction that persists for as long as the incongruence/stress remains in place. In another passage, he also says that neural activation patterns that can effectively remove the incongruence/stress are, in fact, being established and facilitated in the process of reducing the incongruence/stress. I emphasize this so explicitly here because there are hardly any neurophysiological studies on the topic of incongruence, but there are many such studies on the topic of stress.

According to Huether, the brain is simultaneously the location where stress originates and where it exerts is effects. That is, stress that is created in the brain in turn affects the brain and changes its structure. The strength of the stress reaction and of its short-term and long-term effects depend, according to Huether, primarily on the controllability of the stressor/incongruence. In the following section on controllable and uncontrollable incongruence, I draw primarily on the already cited works by Huether, without necessarily listing each of the studies that he, in turn, cites. The precise foundation of these assertions can be found in the numerous original studies cited in Huether's book.

4.5.2.1 Consequences of controllable incongruence

Demands are typically experienced as controllable stress when one believes that one will in principle be able to cope successfully with them—although an element of doubt about the required coping behavior must remain, otherwise the situation would not elicit a stress reaction at all. Such a situation could be experienced as a challenge, for example. The beginning phase of the stress reaction is identical in controllable and uncontrollable situations. In the appraisal process, the new, unexpected, demanding, or threatening situation initially triggers unspecific arousal in large parts of the associative cortex and the limbic system. From there, descending pathways activate deeper brain areas, especially the central noradrenergic system, which originates from the locus ceruleus and, from there, sends neural projections across the entire brain (see Figure 4–10).

The released adrenalin influences almost all regions of the brain—the entire cortex, the hypothalamus, the hind brain, and the brain stem. In addition, the sympathetic nervous system is activated. The adrenalin release not only maintains but even increases the continued arousal of the associative cortex and the limbic system. When the sum of the arousal passes a certain threshold, the neurosecretory nuclei of the hypothalamus are activated. Then, the corticotrophic releasing factors (CRFs) and vasopressin are released, which in turn activate the hypothalamic–pituitary axis (HPA), which ultimately results in the release of cortisol. This prolonged stress reaction, however, occurs only in the case of uncontrollable stress. When the incongruence is experienced as controllable, the initially unspecific arousal energizes behavior. This can lead to a reduction of the incongruence, which in turn leads to a facilitation of the neural activation patterns that correspond to the incongruence-reducing behavior. The unspecific arousal is thereby reduced and the HPA is only weakly activated.

Because the adrenalin permeates the entire brain, and because it is not just neurons that have adrenergic receptors but also glia cells and endothele cells, this controlled stress reaction leads to a large number of reaction chains that ultimately have very positive effects. Specifically, the stimulation of the neural adrenergic receptors increases the brain's readiness to learn—it enables the easier facilitation of synaptic connections. The stimulation of adrenergic receptors at the blood vessels leads to an increased uptake of glucose and an increased energy metabolism. Activation of the adrenergic receptors at the astrocytes—a specific form of glia cell—also triggers the increased release of glucose. That is important because increased neural activity in this region is associated with increased glucose needs. In addition, the thus-stimulated astrocytes form various types of neurotrophic factors. All of these processes conjointly lead to a stabilization of the neural connections that were activated during the process of coping with the incongruence and, furthermore, to an improved facilita-

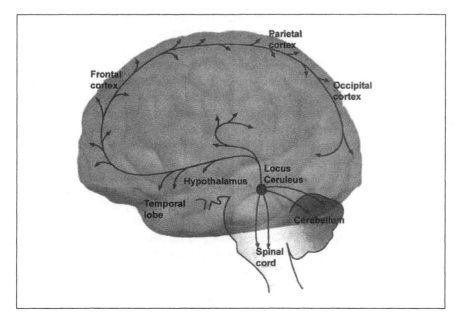

Figure 4-10. The noradrenergic system (Adapted from Andreasen, 2002; p .96)

tion of newly formed neural connections. As a consequence of these processes, repeated confrontation with the same type of incongruence situation will lead to an increasingly strong facilitation of the behavior—and its corresponding neural activation patterns—that led to a reduction of the incongruence. Eventually, this type of situation will not elicit any kind of stress reaction at all because the individual has by now formed easily activated activation patterns within his or her repertoire, and these patterns can then be easily called upon without any notable incongruence even arising.

The processes I describe here on the basis of Huether's research constitute a central neural learning mechanism. New coping possibilities are strengthened and facilitated via the reduction of incongruence at a given moment. According to Huether (1998), "as long as the activation of the central stress system can be terminated by a cognitive, emotional, or behavioral reaction, the neural circuits involved in this response become facilitated" (p. 302). Huether describes exactly those mechanisms that form the basis of Edelman's (1987) selection of neural groups. In his theory of the selection of neural groups, Gerald Edelman explained the formation of new ways of behaving and experiencing in this way: In a situation of psychological/neuronal instability, in which no already established neural activation pattern—which Edelman calls a *neural group*—is present to reduce the existing tension—and stress or incongruence constitutes a kind of tension—the neural activity will for some time fluctuate between various possible states. Because none of these states—and *states* refers here to neural constellations—is sufficiently facilitated at this point, the new constellations break apart quickly again, until finally the momentary circumstances lead to the formation of a constellation that reduces that tension. The tension reduction strengthens and facilitates this newly established neural activation pattern. If this process is repeated, the newly formed pattern occurs more and more quickly and easily and is facilitated better with each new activation until finally it becomes part of the individual's established repertoire. It is important to note here that these are entirely new neural activation patterns that correspond to qualitatively new forms of experience and behavior. This is not a case of reinforcement in the sense that a behavior that is already within the person's repertoire occurs more frequently. This is, therefore, a crucial difference from that which the behavioral learning theories mean by reinforcement in the context of instrumental conditioning.

If one is repeatedly exposed to challenging incongruence situations that can be mastered with effort, then a positive effect beyond the mere mastery of this incongruence can be observed. The efficiency of the noradrenergic system is increased by the repeated confrontation with challenging incongruence situations that the individual can potentially cope with. This occurs in multiple ways: The firing rate of the noradrenergic neurons is increased, that is, they can be more easily activated. The synthesis, storage, and release of noradrenalin in the terminals of the noradrenergic neurons become more efficient. In some brain areas, especially the cortex, a true sprouting of the axons of noradrenergic neurons has even been observed. If a person is exposed again and again to complex, multifaceted challenges, then this leads via the formation of ever more complex and differentiated neural circuits to an optimal expression of the person's genetic potential. Without the acceptance of such challenges, the genetic potential of the person would not be fully exploited.

Animal research has shown in more detail the extent to which the brain can change as a consequence of such challenges. The experimental animals were exposed to "enriched environments" and were given ever new options to explore new aspects of the environment. The exploration of the environment is always associated with increased cortical arousal and increased activity of the central noradrenergic system. Rats that grew up in such specially challenging and stimulating surroundings, compared to rats raised under normal conditions, turned out to have a thicker cortex, better formed blood vessels in the brain, an increased number of glia cells, enhanced branching of the pyramidal neurons, and an increased density of cortical synapses. In short, they had a better-formed, more efficient brain. Incongruence, then, is nothing inherently bad—as long as it remains controllable. Such incongruence drives the development of neural structures. It can be viewed as the motor of mental development; the drive to continue push and develop one's potentials beyond the currently attained level, or to utilize one's naturally provided potential by to the fullest degree possible.

The previous sentences would also fit in well in a textbook on humanistic psychology. They are directly derived, however, from neurobiological stress research. Repeated experiences of mastering limited stress can have very positive effects on later stress tolerance levels, especially when such experiences are encountered early in life. Young rats were removed from their mothers during the 3-week-long nursing phase, directly after birth, in order to be placed into another cage for 15 minutes each day (Muneoka, Mikuni, Ogawa, Kitera, & Takahashi, 1994; Ogawa et al, 1994). This treatment triggers a stress reaction among the young rats. After 15 minutes they were always reunited with the mothers, who were able to quickly calm them with their maternal behavior. This treatment continued for 21 days. The rats that had been exposed to such a limited stress were later, when they were older, less fearful in a new environment and showed a reduced hormonal stress reaction, compared to a control group of animals that had grown up without the earlier stress experiences. Among the experimental animals, the negative feedback between the attained level of glucocorticoid stress hormones and the inhibition of the hypothalamic-pituitary axis (HPA) functioned more efficiently, and this feedback is responsible for preventing the continual rising of glucocorticoid levels. The functioning of this particular feedback mechanism is typically undermined by uncontrollable, continuous stress, and this then leads to a chronically heightened glucocorticoid level, with all the negative consequences this has for the brain. The improved negative feedback via increased glucocorticoid sensitivity and the thereby weakened HPA response lasted a lifetime for the experimental animals. The treatment of the rats during their first 21 days of life had functioned in a sense as a "stress inoculation" (Meichenbaum, 1993). This shows not only how the brain is malleable, especially in the first life phase, by life experiences, but it also shows that it is wrong to keep children away from any potential stress. When this is done, they are prevented from acquiring the experience of mastering stress and they later become all the more vulnerable to stressful experiences.

The fact that this probably also applies to humans is supported even by somewhat older study results on the relations between parental child-rearing practices and mental health. G. Parker (1979a, 1979b, 1981, 1983) obtained retrospective ratings from groups of depressed and anxious patients with regard to the parents' rearing practices when the patient had been a child. The participants estimated the parents' affectionate care provision and the extent to which they typically exercised restrictive control. Patients with anxiety disorders and with major depression rated their parents as more restrictive–controlling and simultaneously as less caring–affectionate, compared to the participants in a matched control group. They had fewer opportunities, therefore, to explore their environment freely, without restrictions, and with parental support as a secure base. Thus, they had fewer opportunities to acquire positive control experiences in limited-stress situations. Among the nonclinical participants, significant correlations were found between the extent of restrictive control by the parents during childhood and lower self-esteem, increased anxiety, greater depression, and higher neuroticism in adulthood.

It is not just positive control experiences that have positive effects on later development. Mineka, Gunnar, and Champoux (1986) studied the effects of control experiences among young rhesus monkeys who handled positive things such as food, drink, and little treats. In the experimental group, the young monkeys were allowed to freely access these things. They had control with regard to when they wanted to eat, drink, or get a reward. The monkeys in the control group received the same things based on a schedule imposed by the experimenter, regardless of their own behavior. These conditions were maintained from birth to the age of 12 months. From the age of 8 months, the monkeys were observed under various test conditions. The monkeys with the positive control experiences were generally less anxious. They explored new play situations more courageously, they habituated more quickly when they were confronted with an anxiety-eliciting robot, and they showed more active coping behavior when they were temporally separated from their peers.

People who have acquired many positive control experiences during childhood will likely develop the conviction, based on this foundation, that they have control and can exercise this control in situations when this might be important. Associations between positive control beliefs and all kinds of aspects of good mental health have been studied so frequently and with such consistent results that I can summarize the findings just briefly here. More detailed descriptions on this topic can be found in Flammer (1990).

People with high internal control beliefs report greater life satisfaction, endorse more well-being, have greater self-confidence, and are more resistant to stress. If one feels that one is exposed to a situation and has no control, then any unpleasant stimuli, such as noise or pain, is experienced as far more unpleasant compared to a situation in which one would encounter the same stimuli knowing that one could avoid them, turn them off, or reduce them whenever one wishes. Thus, people who go through life with positive control expectancies, with the sense of being able to cope with difficulties or aversive states, will not only be better able to actually master such situations but will also suffer less before they even occur. Positive control expectancies, then, clearly have a protective function and are an important component of mental health.

All of these findings are consistent with the assumption that humans and more highly developed mammals have a basic need for control. When this need is well satisfied, then their development is generally better, they are more resilient, stress-resistant, and healthier. This section focused not on violations of the need for control but on the consequences of positive control experiences in early life on one's later life. These consequences are all-around positive. Early positive control experiences later on lead to even more positive life experiences.

4.5.2.2 Consequences of uncontrollable incongruence

If an individual in a challenging or threatening incongruence situation does not find the means to reduce incongruence via his or her own activity, then the increased arousal, as described before, remains in place and continues to escalate until eventually the HPA is activated and glucocorticoids are released. Glucocorticoids can easily cross the blood–brain barrier and bind to the glucocorticoid receptors of the neurons and glia cells. Normally a negative feedback loop is then triggered, which downregulates the further release of stress hormones. We have seen earlier that this feedback mechanism, which protects the brain from the effects of a continuously increased cortisol level, can be strengthened permanently via the process of stress inoculation, in which subsequent stress sensitivity is reduced.

However, if the uncontrollable stress situation remains in place—imagine, for instance, the severe physical or sexual abuse of a child—then the negative feedback mechanism is severed and an escalating glucocorticoid release is observed. Arousal levels continue to rise and, along with them, the glucocorticoid release increases. The mechanisms that are now triggered do not function as quickly as the noradrenergic initial stress reaction. Instead, they work via altered gene expression (see section 2.3 as background for the following material). The formation of cAMP is suppressed and, thereby, the activation of the second messenger cascade is prevented, which otherwise would have been activated via noradrenergic stimulation. The cerebral energy metabolism is reduced as well as the formation of neurotrophic factors and of new synapses. The excess of glucocorticoids harms the activated glutamate synapses and the pyramidal cells in the hippocampus, which have the highest density of glucocorticoid receptors. The noradrenergic axons and the neural terminals in the cortex are very sensitive to cortisol and begin to degenerate. On the behavioral level, an excessively high level of glucocorticoids can lead to the erasure of previously acquired behavioral patterns.

A prolonged uncontrollable stress situation that activates the HPA axis therefore leads to a destabilization of previously formed neural connections. The opposite of what has been observed in the case of controllable stress now takes place. The destabilization of existing neural activation patterns also affects the behaviors that correspond to these neural activation patterns. What happens, then, is a general destabilization of neural/mental functioning.

According to Huether, this can ultimately be viewed as an adaptive mechanism of the neural system. Because the behaviors that have been activated in this stress situation have not proven to be effective in terms of controlling the stress, it seems sensible to erase them so they can be replaced by other, more effective behaviors. This means that a more thorough un- and relearning takes place than what can be observed in the case of controllable stress. This reasoning makes sense in abstract terms as well, for instance, when we recognize the analogy to Piaget's (1977) principles of assimilation and accommodation, which also refer to a more limited versus a more thorough readjustment.

Huether's argument, however, targets only instrumental, mastery-oriented learning—contextual, problem-oriented coping. The learning of new instrumental behaviors in such a prolonged stress situation is already severely compromised because the formation of contextual behavior requires the hippocampus, which—as we have seen earlier—is not in a state of learning readiness at this time. This does not mean, however, that in such a situation of uncontrollable stress other forms of learning would not take place, such as classical conditioning or sensitization. Other brain areas that also play a role in the stress reaction, such as the amygdala, are very much in a state of learning readiness. The neural circuits that are activated in this stress situation are being all the more thoroughly facilitated because they are already activated, and they are being conditioned to the contextual cues that triggered the stress reaction.

If we conceptualize a prolonged situation of severe physical or sexual abuse or severe neglect of a child as a source of incongruence or stress, then it is hard to see what the adaptive value of the reactions that are observed in such situations should be—unless one regards everything as adaptive—that somehow makes the situation more bearable for the individual, such as dissociative behavior or a reduction of all instrumental behaviors and interactions with the environment, as evident in depression, for example. According to Huether's line of reasoning, it would make sense to regard such reactions, which normally are construed as disorders, as forms of adjustment to uncontrollable stress. Such newly formed neural activation patterns are reinforced and facilitated by virtue of the effect that they at least momentarily reduce an aversive, uncontrollable incongruence or render it slightly more controllable. In general, however, the behaviors formed under such conditions will not be well-organized instrumental behaviors because, as described earlier, the neural conditions for the formation of such behaviors are unfavorable. Anything that somehow down-regulates the aversive inconsistency is being reinforced. In this kind of a situation, the system is occupied with coping with itself rather than coping with the environment. If one wants to use the term *coping* at all, then this would be—in the sense of Lazarus (1984)—emotion-focused rather than problem-focused coping. However, what is being learned in such a situation is primarily the state itself that is triggered by the situation; that is, primarily, anxiety or fear.

Anxiety is a natural response to threatening, uncontrollable incongruence situations—situations in which the most important goals are threatened or violated and in which one is unable to do anything about it. The response patterns that Huether describes as a reaction to uncontrollable stress form a central part of the response system that Gray (1982; Gray & McNaughton, 1996) writes about in his neuropsychological anxiety theory. However, they are only one part—the part that corresponds to the hormonal aspects of the stress reaction. The response system that is being activated in such a situation is even more complex. Gray has termed it the *behavioral inhibition system* (BIS). The neural circuit that forms the BIS encompasses the septum, the hippocampus, and the Papez loop, which in turn includes dopaminergic input to the prefrontal cortex, descending input from the associative cortex into the septohippocampal system, ascending cholinergic input into this system, noradrenergic input to the hypothalamus, and noradrenergic connections to the locus ceruleus.

The BIS is activated when the comparator (see previous) indicates that a threatening situation is being encountered. The threat can refer to either the presence of something very aversive or the loss of something very much desired. The BIS is closely connected to the ventromedial and dorsolateral parts of the right prefrontal cortex, which play a crucial role in the control of negative emotions and the representation of avoidance goals. Avoidance goals are therefore activated and the entire system is focused on defense and avoidance. The emotional response to the threat is anxiety. *Anxiety* here does not refer to situation-specific fears or to the presence of panic; instead, it refers to anticipatory anxiety. The immediate response of the BIS to a threatening situation is a reduction of attention, an inhibition of motor behavior, an intensified scanning of the environment, and a preparation of the hypothalamic-governed motor systems to increase readiness for potentially required fight-or-flight behaviors.

The further development of the emotional reactions depends on the perceived controllability of the situation. If no control is possible or at least does not seem subjectively possible, and if even avoidance is not possible, the sequence of reactions that Seligman has described in his learned helplessness paradigm transpires (Overmier & Seligman, 1967; Seligman & Maier, 1967). Seligman regarded the responses of his experimental animals as an analogue for the development of depression because the behavior that ultimately resulted seemed to resemble human depression in many ways. Barlow (1988) and Mineka and Zinbarg (1996), however, regard this evidence more as an important contribution to our understanding of anxiety. When animals are repeatedly exposed to uncontrollable shocks, after all, they initially show behavior that seems more similar to anxiety than depression. It is only in a later phase that one can observe reactions such as resignation or depression. What is important about these studies by Seligman—which later were replicated by many others—is in this context that it was not the aversive shocks themselves that triggered the behavior analogous to anxiety or depression, but the uncontrollability of the shock. Animals that received the same amount of shock but had control over the shocks did not exhibit the same "symptoms" (Maier & Seligman, 1976). Alloy, Kelly, Mineka, and Clements (1990) conceptualize the frequently observed sequence of reactions to uncontrollable aversive stimulation as one possible pathway from anxiety to depression. When a person is uncertain about his or her capacity to exercise control over a threat, a state of high-arousal anxiety will initially result. Alloy et al. term this phase *uncertainty helplessness*. If this uncontrollable incongruence is maintained, this anxiety state is gradually transformed into a mixed anxious–depressive state (*certain helplessness*), and, ultimately, the person slips into a depressive state that is characterized by a deep hopelessness.

Chorpita and Barlow (1998) argued that anxiety and depression have a general emotional factor in common, which is largely consistent with Gray's definition of anxiety. Watson and Clark postulated such a fundamental disposition to experience negative emotional states in 1984. This perspective has been substantiated in studies by L. A. Clark and Watson (1991), as well as T. A. Brown, Chorpita, and Barlow (1998), whose factor analysis among 350 outpatients clearly found such an overarching factor cutting across the diagnostic boundaries among the various anxiety and mood disorders found in DSM-IV. The syndrome of generalized anxiety disorder had the highest loading on this negative emotionality factor. Prior to that, Arrindell, Pickersgill, Merckelbach, Ardon, and Cornet (1991) studied a sample of 432 patients with various anxiety disorders and found in a second-order confirmatory factor analysis a general factor on which all the primary factors—which corresponded closely to the subcategories of the specific anxiety disorders—loaded rather highly.

Chorpita and Barlow (1998) argued that a specific process leads from anxiety to depression. This process begins with a threat that is experienced as uncontrollable, and it escalates further as this kind of uncontrollable incongruence is maintained over longer periods. This view is supported by the fact that anxiety disorders typically precede depressive episodes, rather than the other way around (Angst, Vollrath, Merikangas, & Ernst, 1990; Di Nardo & Barlow, 1990). Chorpita and Barlow regard this disposition to experience negative emotions as a broad risk factor for the development of anxiety and mood disorders, and they ascribe central causal importance to violations of the need for control experienced in early childhood. For them, such early experiences of the uncontrollability of aversive events are among the most important vulnerability factors. They explicitly acknowledge the importance of the kinds of experiences the child encounters in the context of early attachment relationships.

The importance of violations of the need for control is most clearly apparent in the pathogenesis of PTSD. By definition, traumas are uncontrollable aversive experiences, and PTSD develops after such traumas. According to the perspective developed in the preceding section, and in the section on the attachment need, however, it would make sense to view many other disorders also as variants of PTSD, in terms of their pathogenesis, because they are often associated with violations of the need for control in early childhood—during a time when the child is most vulnerable to the effects of such violations because of his or her inherent dependency on adult caregivers. Empirically, high comorbidity rates have been documented for patients with PTSD and, even more so, for patient groups with histories of early abuse experiences. This supports the idea that early experiences with regard to the uncontrollability of incongruence predispose individuals to later develop mental disorders. For most patients, such early violations of the control need are not part of the definition of the disorder, and even psychotherapists do not always find it easy to obtain reliable information about such traumatic experiences because their patients' memories may not reach back to the time during which these violations occurred or, in some cases, because the memories are actively avoided. The scientific studies that have focused on this question have repeatedly found evidence pointing to violations of the control need. Additional findings consistent with this can be found in Chorpita and Barlow (1998) as well as in Traue (1998), among others. Ultimately, all of the studies about violations of the attachment need that were described in sections 4.4.2 and 4.4.3 can also be regarded as evidence for violations of the control need because such attachment need violations are always accompanied by the experience that the attachment figure's behavior is uncontrollable.

Violations of the control need, then, do not just trigger immediate anxiety and stress. What is more, through the processes of sensitization and conditioning, these immediate reactions leave traces that later in life can become the foundation for the development of mental disorders. In this context, I remind the reader of the study by Heim et al. (2000), which was already mentioned in section 3.2.5, in which depressive women with abuse experiences in childhood were found as adults to have a sixfold increase in the excretion of adrenocorticotropic hormone and of cortisol, compared to depressive women without similar abuse experiences. In this study, women with abuse experiences who were not depressive also had stronger stress reactions than women without abuse experiences. Abuse experiences are among the clearest examples of violations of the need for control.

The studies described here show a clear pattern that violations of the need for control are toxic for mental health. The importance of the need for control among patients with anxiety disorders is also evident because desperate efforts to regain control are often among the cardinal symptoms of the disorders. The relentless worrying in generalized anxiety disorder and the compulsions in obsessive–compulsive disorder can be regarded as dysfunctional attempts to regain control over something that has become uncontrollable. In other anxiety disorders, control in the form of avoidance plays a central role. Schulte (2000) regards these unsuccessful efforts to gain control over the anxiety as the most important cause for the persistence of anxiety disorders:

> Three mechanisms are essentially responsible for the persistence of anxiety disorder: Anxiety stabilization via the avoidance of anxiety; anxiety stabilization via excessive fighting of anxiety; and anxiety stabilization via excessive focus on the anxiety. Moreover, these processes are responsible for the development of additional symptoms. (p. 397)

The need for control, then, appears to play a central functional role in the genesis as well as the persistence of anxiety disorders.

Depression also fits in well with the ideas developed in this section and can be construed as an advanced reaction to violations of the control need. The deactivation of the anterior cingulate cortex, which plays an important role in goal-oriented transactions with the environment, points to the fact that the patient, in a sense, has "given in and given up." He or sh has relinquished his or her efforts to regain control. The patient' avoidance system is simultaneously strongly activated and transforms the patient into a generalized protective and defensive state, in which he or she is not prepared to enter into any new experiences, not even the experiencing of feelings, which also requires the ACC. Such a generalized protective state is not difficult to understand in patients who have lost all hope of regaining control. The overactivated amygdala and the equally overactivated right ventromedial PFC reflect the persistent activation of negative emotions, which accompany the path into depression. The patient's shrunken hippocampus reflects the damage resulting from the stress hormones, which have been all-too-frequently and excessively excreted throughout the course of a persistent state of incongruence, and which have by now affected the neurons and synapses in that area.

When we look at a patient's life as the history of his or her control experiences, then, we are in a better position to understand how the disorders that are currently present might have originated.

4.6   THE NEED FOR SELF-ESTEEM ENHANCEMENT AND SELF-ESTEEM PROTECTION

4.6.1   The Need for Self-Esteem Enhancement as a Specifically Human Need

> People want to feel good about themselves. They want to believe that they are compe-
> tent, worthy, and loved by others. This desire for self-enhancement is regarded as so fun-
> damental to human functioning that it was dubbed the "master sentiment" by William
> McDougall (1932) and "the basic law of human life" by the renowned anthropologist
> Ernest Becker (1971). Many other figures of historical (e.g., Allport, 1943; Cooley,
> 1902; Mead, 1934) and contemporary (e.g., Baumeister, 1991; Greenwald, 1980;
> Schlenker, 1985; Steele, 1988; Tesser, 1988) prominence have endorsed the belief that
> a drive to achieve a positive self-image is, in the words of William James (1890), a direct
> and elementary endowment of human nature. (J. D. Brown, 1993, p. 117)

We could add many additional names to the aforementioned list of authors, all of
whom regard the need for self-esteem enhancement as a superordinate human need.
In the domain of psychotherapy, one would first have to mention Alfred Adler (1920,
1927), who regarded the striving to overcome inherent inferiority feelings as the most
important source of human motivation.

The need for self-esteem enhancement and self-esteem protection occupies a special
position among the other postulated basic needs; not because it is more important
than the other basic needs but because it can be viewed as a specifically human need.
As we have already seen or will review later, even rats, monkeys, and other highly
developed mammals have an attachment need, a need for control, and a need for pleasure maximization and distress avoidance. The regulation of self-esteem, however,
requires certain qualities that are unique to humans. These include conscious awareness of oneself as an individual and the ability to think reflectively. Without these
abilities, it is not possible to have a self-image or self-esteem. The self-image develops in interaction with other people, and a critical component of these interactions is
language. The self-image—and this is even more true for the explicit rather than the
implicit self-image—is in large part the result of language-based communication and
self-reflective processes that are, in turn, also based on internalized language.

A need for self-esteem enhancement can, therefore, be present only when a conscious
self that can utilize these qualities has been formed. In the course of phylogenesis
only humans have attained this developmental stage, and in the ontogenesis of an
individual human it takes many years before these qualities are sufficiently developed
so that the child becomes aware of himself or herself; that he or she can, for example, feel shame, feel his or her dignity being violated, or experience other types of
self-esteem damaging feelings. In order for the comparison process to take place, in
which comparisons with self-image related reference values on the highest hierarchy

level result in self-esteem enhancement or self-derogation, sensory information must first be transformed multiple times. The meaning these signals have for self-esteem is generated in this transformation process. Many evaluations that originate from within the system itself and that are not directly linked to the original sensory signals also influence this multilevel processing, rendering such self-esteem reactions necessarily highly subjective. According to the hierarchical model of Powers (1973) or Epstein's (1990) cognitive–experiential self theory, self-esteem regulation involves feedback loops on the highest level of information processing. We must assume, then, that the neural circuits involved in self-esteem regulation are more complex than all of the circuits we have encountered so far. This is probably also the reason why no neuroscientific studies have yet addressed the issue of self-esteem regulation. The topic is apparently too complex and cannot be successfully tackled with our current neuroscientific methods.

What can be said today about the need for self-esteem enhancement and self-esteem protection, based on solid empirical findings, stems primarily from social psychological studies with adult human participants. This constitutes a major difference from the other basic needs, which can be studied among animals or very young children. Animal research is particularly useful for the purpose of revealing how the regulation of the respective needs corresponds to activity on the neural level. Studies with young children have shown that the satisfaction or violation of these needs in the very first life phase can have dramatic consequences for later mental health. In the case of the need for self-esteem enhancement and self-esteem protection, however, such studies are impossible, not just for methodological reasons but also because of the very nature of this need. At the age of just a few months, when the attachment and control needs can already be violated, human infants have not yet formed a self or self-representations that can be evaluated positively or negatively. However, negative experiences that the child encounters with respect to his or her attachment need or control need can strongly influence the developing sense of self and can lead to the formation of a negative self-image and, thereby, low self-esteem. Such a person will subsequently often behave in self-derogating ways, which leads us to the question of whether a general tendency to self-esteem enhancement does indeed exist, as the citation noted earlier suggests. This question is the focus of the next section.

4.6.2 Is There Really a General Tendency to Self-Esteem Enhancement and Self-Esteem Protection?

If the striving for self-esteem enhancement is really such a central human motivation, why do so many people have low self-esteem and—even more important—why do such people further derogate themselves? Why to they tend to choose interaction partners who think negatively of them (Swann, 1990, 1992)? Why do they even withdraw from relationships in which they are evaluated positively (De La Ronde & Swann, 1993)? This *puzzle of low self-regard* (Baumeister, 1993) has been the focus of a great deal of social psychological research in recent years.

How can low self-esteem originate in the first place? Sullivan (1953) offers the following response: How should a small child make sense of the fact that his or her relationship with the primary caregiver is fraught with problems, or of the fact that his or her needs are only very insufficiently met by the caregiver? In the mental world of the child, only two possibilities exist: Either I am good and mother is bad, or I am bad and mother is good. For a small child who is entirely dependent on the mother, the first option is far worse. The child is helplessly dependent on the mother, without hope that he or she can independently improve the situation. The feelings that would correspond to this are chronic disappointment, fear, or anger toward the mother as well as/or hopelessness. The alternative, by contrast, might be somewhat better: If the child attributes the behavior of the mother to his or her own "bad" behavior—that is, if the child processes the experiences in such a way that he or she concludes that he or she is not worthy of being treated any better—then this is perhaps not linked with more pleasant feelings but at least it implies the hope that perhaps things might change for the better. A certain sense of control would be maintained.

A small child whose attachment figure does not satisfy his or her needs will, therefore, tend to experience himself or herself as the cause for these experiences and will feel bad or worthless. This might indeed be expected if the child did not also receive corresponding messages from the attachment figure. However, the child will probably receive such messages because an insecure–avoidant child is in many respects quite difficult for the mother. The mother is not aware that the behavior of the child is influenced by her own lack of availability and sensitivity, and she will tend to search within the child for the reasons for the unsatisfying relationship. She might perhaps convey a message such as, "Why are you not the same as other children!" She might also verbally attack or criticize the child. The child learns this via identification, internalizes this way of self-treatment, and over time develops a stable negative self-image and low self-esteem.

Based on these considerations, the development of low self-esteem is entirely compatible with the assumption that humans generally strive to maximize their self-esteem. Fulfillment of this need also requires an appropriate environment. If we recognize that the need for self-esteem enhancement is not the only basic need to be met, we can explain even without relying on Sullivan's observations how humans sometimes seem to do all they can to maintain their low self-esteem.

Attaining low self-esteem is not in itself their goal; it is a means in the service of other needs. The reasons are not any more mysterious than would be the case with someone who, despite the presence of a basic need for an intimate relationship, continuously behaves in such a way as to prevent such relationships from succeeding. In such cases, need-violating prior experiences have led to the formation of predominantly avoidance rather than approach or intentional goals, such that one of the needs is now being sacrificed in the interest of another one. In most cases, it is not possible to understand human behavior appropriately if we neglect the fact that not just one but several needs are being serviced at any given moment. Therefore, it is important to keep in mind the total balance of the various needs.

The motivational schemas that develop in order to achieve need fulfillment act as the "executive organ" of the needs. These schemas must be compatible not just with the needs but also with the environmental context. The schemas are, in line with Epstein, schematized experiences or hypotheses about how particular needs can be satisfied under particular circumstances (e.g., with this particular mother). If none of the attempted means succeeds, it is not possible to form generalizations based on such means and experiences. Thus, appropriate approach schemas would not be formed in such situations.

However, it is not simply the case that the environment that would stimulate the formation of these schemas is insufficient; rather, the individual's attempts to attain need-satisfying reactions via his or her behavior are unsuccessful. Based on our current understanding of goal-oriented mental functioning, every failed attempt is linked with negative feelings such as disappointment. However, as we will see in more detail in the section on the need for pleasure maximization and distress minimization, humans strive to avoid negative emotions, especially when they are experienced without any possibility of personal control. Therefore, instead of developing approach schemas, the individual will develop the kinds of schemas that allow him or her to avoid the unpleasant experience of need dissatisfaction. Situations that are relevant to this need after a while elicit no recognizable approach reactions but instead activate avoidance schemas. The person appears to behave in such a way, then, as if the behavior is actively oriented toward the attainment of a goal that is incompatible with the respective basic need. The "gain" for the person can be located in other basic needs: Pain and suffering are avoided and the person exercises control over his or her own experiences.

These considerations show how important it is to differentiate the terms *goal* and *need*. Without experiences, needs don't have any goals. Goals are always context-bound (and the person's body would also be included as context here). Goals always refer to something, even if they are schematized. The schematized goal components of the motivational schemas are generalizations abstracted from concrete experiences. When one forms a goal to protect oneself from derogations and disappointments, then this does not mean that one doesn't have a goal to enhance one's self-esteem. One simply has not formed context-bound goals for the realization of this need.

Such goals could have formed as generalizations of positive, need-fulfilling experiences. If such experiences have not been encountered or if they were much less frequent or weaker than need-violating experiences, then the need-relevant situation will more quickly and strongly activate avoidance schemas than the rudimentary approach schemas that are based on only few experiences. The neural activation patterns that represent the avoidance schemas inhibit the patterns corresponding to the approach schemas. Mental activity is determined by avoidance goals. These avoidance goals are not pursued in the service of the need for self-esteem enhancement but rather in the interest of other basic needs—such as not wanting to suffer pain, wanting to have control, and wanting to have a close relationship.

The need for self-esteem enhancement, however, has not disappeared in such situations. Its fulfillment is simply being blocked by strongly developed avoidance schemas and weakly developed approach schemas. When we encounter individuals with low self-esteem who seemingly actively maintain this low self-regard, the situation is analogous to the one with patients who have insecure–avoidant attachment patterns. Self-esteem enhancing thoughts, wishes, or fantasies are blocked by the avoidance schema in much the same way that self-esteem enhancing behavior is blocked. Such impulses are prevented from entering into conscious awareness. Persons with low self-esteem behave in such a way, therefore, as if they actively strive to maintain or verify their negative self-image. Various social psychological authors have construed the "verification" of a particular kind of self-image as a source of motivation in its own right.

> "Self-verification theory assumes that as people mature, they learn that their relationships proceed most smoothly when others see them as they see themselves—even if they see themselves negatively. For example, people discover that those who develop overly positive appraisals may become disappointed and disgruntled with them. Through repeated exposure to this fact of life, people come to associate self-verifying evaluations with feelings of authenticity and nonverifying evaluations with feelings of uneasiness or bemusement. Eventually, these epistemic concerns become functionally autonomous (Allport, 1937) of the interpersonal or pragmatic concerns that originally produced them and people verify for either epistemic or pragmatic reasons. Thus for example, a man with low self-esteem may seek negative evaluations either because he fears the social consequences of being appraised in an overly positive fashion or because his past experiences have convinced him that he should expect to encounter such evaluations. Thus, both epistemic and pragmatic considerations may motivate people to seek self-verifying appraisals, even if this means displaying a preference for unfavorable evaluations. (De La Ronde & Swann, 1993, pp. 149–150)

What is described here under the rubric of self-verification theory can also be explained by the principles I elaborated previously. The *puzzle of low self-regard* (Baumeister, 1993)—the seemingly paradox effect that people with low self-esteem appear to actively maintain their negative self-images—is entirely compatible with our assumption that a basic need for self-esteem enhancement and protection exists.

For people with high self-esteem, the assumption that a striving for self-esteem enhancement exists does not conflict with the tendency to maintain this high self-esteem. What needs to be explained is the behavior of people with low self-esteem. A proper understanding of this phenomenon is of great importance for psychotherapy because many psychotherapy patients also have self-esteem problems, and improving self-esteem is probably regarded as a desirable therapy goal among most psychotherapists.

Based on these ideas, the active maintenance of low self-esteem can be viewed as the result of primarily self-protective avoidance strategies that aim to prevent pain or distress, the loss of control, and the general occurrence of inconsistency. The existence of avoidance strategies does not mean, however, that the person no longer has a need for self-esteem enhancing perceptions. There is indeed empirical evidence showing that individuals with low self-esteem nevertheless have self-esteem enhancing behavioral tendencies—as long as they don't have to openly admit to having such tendencies.

When individuals with low self-esteem are given the opportunity in research studies to evaluate (a) themselves, (b) persons with whom they share common characteristics, or (c) persons without common characteristics, then the low self-esteem participants evaluate those with whom they share characteristics particularly positively. At the same time, they do not rate themselves or strangers positively. Individuals with high self-esteem do not have this tendency—they tend to rate primarily their own efforts positively and those of others, by comparison, more negatively (J. D. Brown, Collins, & Schmidt, 1988). When individuals with low self-esteem were led to believe that they share characteristics—such as a birthday, for instance—with a very attractive person, the people with low self-esteem later rated themselves as more attractive. People with positive self-esteem did not have this "representative self-esteem enhancement." One might say they don't need to do this kind of thing (J. D. Brown, Novick, Lord, & Richards, 1992).

People generally tend to enhance their self-images by identifying with successful, attractive others (Cialdini & De Nicholas, 1989). College students are more likely to use "we" when speaking of their school's football team after victories, compared to team losses (Cialdini et al., 1976). According to J. D. Brown (1993), this tendency is particularly pronounced among people with low self-esteem. For them, national pride, identification with heroes, sports teams, and so forth, are opportunities for indirect self-esteem enhancement.

These phenomena show that the need for self-esteem enhancement has not simply vanished among people with low self-esteem. Such findings are quite important in the context of psychotherapy with people with low self-esteem. Conceivably, one ought to first find ways to enhance their self-esteem indirectly, such that they do not have to identify with the process directly, before they are more open for direct self-esteem enhancing feedback.

Epstein and Morling (1995) argued that the implicit processing system spontaneously tends to utilize self-esteem enhancing reactions, whereas the explicit system seeks to find a compromise between the need for self-esteem enhancement and the consistency principle (concordance between current perceptions and previous experience). This conceptualization is also empirically supported. Swann, Hixon, Stein-Seroussi, and Gilbert (1990), for example, interfered with the processing capacity of the explicit system by having participants choose a partner under intense time pressure. The potential partners differed in how positively they had evaluated the participants previously. Under time pressure, all participants chose the person that had previously evaluated them positively. When participants were given more time, those with low self-esteem tended to choose persons as partners that had previously evaluated them negatively. The tendency to engage in self-verifying behavior, therefore, seems to be more closely linked with the explicit processing mode, whereas self-esteem enhancing behavior is more directly associated with the implicit mode.

This difference between the implicit and explicit processing mode was confirmed by Epstein and Morling (1995) in an independent study with a different design. In that study, the gradation between the positive and negative evaluation by the potential partner was more finely tuned. Using this more fine-grained system, participants who were given enough time for explicit processing chose persons as partner who had evaluated them just slightly, but not very much, better than they had evaluated themselves. The explicit reaction was interpreted by Epstein and Morling (1995) as a compromise between the tendency to self-esteem enhancement and the verification tendency in the service of the consistency principle.

Epstein regards this as confirmation of his general conviction that behavior is typically a compromise among the satisfaction of various needs. Often, behavior that serves to fulfill one need also yields benefits for other needs. The needs might also conflict with one another, however, in which case a behavior must be found that represents a compromise among the various needs.

The different self-evaluative reactions in the implicit and explicit processing modes are probably one of the reasons explaining why behavioral tests, self-reports, and other-reports of people's self-esteem generally agree rather poorly with one another (Savin-Williams & Jacquish, 1981). That which is measured by one of the many self-esteem questionnaires probably corresponds only to the components of self-esteem stored in explicit memory. This explicit self-image may or may not be concordant with the implicit self-image, which can be derived from the behavior and emotions of the patient.

A person's implicit self-esteem could, therefore, be more important for what a person does in real situations. Weinberger and McClelland (1990) reported that motives derived from Thematic Apperception Test (TAT) stories—which can be taken to correspond to implicit motivations—predict the behavior people show in their lives much better than self-report measures of motivation.

The different self-esteem reactions in the implicit and explicit processing modes also suggest conclusions for therapeutic practice. Therapists should aim primarily, then, to modify the implicit self-evaluative reactions of their patients. This should be a better therapeutic target because, based on the findings reported previously, it is easier to activate self-esteem enhancing reactions in the implicit processing mode. Especially when working with patients with low self-esteem, then, one should initially attempt to use implicit methods to positively influence patients' self-esteem. Self-evaluations should, therefore, not be an initial explicit focus. Instead, a process-activating intervention can be used, as it were, in which the patient is initially guided to encounter self-esteem enhancing experiences, even though these experiences do not have to be explicitly identified to the self.

It is not just the tendency for self-esteem enhancement that has been empirically well supported; this also holds true for the tendency to protect one's self-esteem. Sedikides and Green (2000) have shown in a series of studies that people tend to remember self-

esteem harming statements much more poorly than they remember self-esteem supporting statements. This tendency is stronger to the degree that these statements refer
to particularly important characteristics of the person. They respond in such a way
even if the threat to the self-esteem is not real but only hypothetical. Decrements in
memory performance were observed regularly when the study participants were confronted with statements that fulfilled three conditions: They had to be detrimental to
the person's self-esteem, they had to refer to the person himself or herself, and they
had to refer to aspects of the person that were personally important. In all other experimental conditions—when the statements referred to another person, referred to unimportant self-aspects, or were positive—memory performance was not impaired or
even particularly good. Sedikides and Green interpreted these results in the sense of
an inconsistency-negativity neglect model:

> The present investigation portrays the self as a system that will reject inconsistent infor-
> mation only to the extent that this information is both central and negative. This empiri-
> cal generalization is compatible with Greenwald's (1980) view of the self as a closely
> guarded operating system that controls in a totalitarian manner inconsistent or negative
> information (i.e., the ego bias of beneffectance). In addition, the present investigation
> portrays the self system as having an exceedingly low threshold of inconsistency-nega-
> tivity potential for highly valued (i.e., central and positive) self-beliefs. (Sedikides &
> Green, 2000, p. 918)

In a large meta-analysis of studies of self-esteem enhancement and self-esteem protection Campbell and Sedikides (1999) arrived at the conclusion that the tendencies for
self-esteem enhancement (approaching) and self-esteem protection (avoiding violations) are governed by different laws. The tendency to self-esteem enhancement is limited, for example, when individuals are asked to describe themselves with objectively
verifiable characteristics, when they are asked to compare themselves to a concrete
other person, or when they interact with a person who is emotionally close to them.
The tendency for self-esteem protection is greater among people with a pronounced
performance orientation, with external control expectancies, and with high self-esteem.
It is particularly strongly present among people with narcissistic personality features.
We will see later (see section 4.7.3) that approach and avoidance generally are governed by different laws and also have different neural bases. The tendency to self-
esteem enhancement can be regarded as a part of the approach system and the tendency to self-esteem protection can be regarded as part of the avoidance system.

4.6.3   Self-Esteem-Enhancement
and Mental Health

Based on the findings reported earlier, we can assume that humans have a general tendency to enhance their self-esteem. Some realize this striving directly and openly,
whereas others do so in a more covert fashion. Because the satisfaction of basic needs
generally exerts a positive effect on mental health, one could assume that individuals
who satisfy this particular need—who actually take advantage of opportunities to
maximize their self-esteem—might be characterized by better mental health.

Empirical findings support this assumption. Self-esteem enhancing behavior correlates with mental health even when the situational context does not seem to justify such a relationship. On the one hand, many authors argued that an undistorted perception of reality is a prerequisite for healthy personality development (Haan, 1977; Jourard & Landsman, 1980; Vaillant, 1977). On the other hand, however, empirical findings have told a different story. Healthy individuals tend to exhibit self-esteem illusions as well as, by the way, control illusions. It tends to be a sign of good mental health when one can create unrealistic cognitions and perceptions that satisfy the needs for self-esteem enhancement and control. Reviews of the current research in this area can be found in Nisbett and Ross (1980), Taylor and Brown (1988), and Colvin and Block (1994). There are so many investigations that found that humans have a tendency to form self-esteem illusions that I won't describe single studies here but will only summarize some of the main findings briefly.

When people are asked to describe themselves they tend to choose many more positive than negative attributes. Most people evaluate themselves as more positive and less negative than the average. A large majority of people regard themselves, for example, as above-average drivers, which—of course—is logically impossible. In this regard, then, most people are under positive illusions. When people are asked to compare themselves to others, they tend to use their positive rather than negative characteristics as the basis for the comparison. When they are asked to rate their personalities and are simultaneously rated by someone else, it turns out that the self-ratings are far more positive than the other-ratings for most people. These positively biased evaluative tendencies are also applied, in somewhat weaker form, to friends and partners, who are also presented in a positively skewed manner.

Positive information about the self tends to be remembered better and processed more quickly than negative self-information. Successes are also more easily recalled than failures. Memories of performances tend to be reported more positively than the performances actually were. Negative self-aspects that cannot be ignored tend to be devalued in terms of their importance. Areas in which one does not excel tend to be regarded as unimportant.

These various findings hold true for the majority of participants in the respective studies. The only exceptions to this self-esteem enhancing trend are depressives and individuals with low self-esteem. These groups report a greater balance between positive and negative self-aspects, they show greater concordance between self-reports and other-reports, equally remember positive as well as negative self-esteem relevant situations, and so forth.

It is clear, then, that it is the mentally healthy have a skewed perception of reality with regard to themselves, not the ones with relatively more poor mental health. Mentally healthy people tend to enhance their self-esteem when given the opportunity. This supports the assumption of a general basic need for self-esteem enhancement. Most people attempt to satisfy this need when relevant opportunities arise. It is a sign of good mental health to view oneself excessively positively and to evaluate oneself more positively than others. One should be concerned about people who fail to do this, rather than the other way around.

As a therapist, then, one should not strive to establish an absolutely realistic self-evaluation within the patient; instead, one should support the patient's effort to self-enhance, even when this might seem objectively exaggerated. There are boundaries to this, of course. This principle would not apply, for example, to patients whose disorder consists of chronically inflated self-enhancement to the point that this tendency interferes with other basic need satisfactions. This is evident, for instance, in patients with narcissistic personality disorder.

Taylor and Brown (1988) also reported findings on the construct of unrealistic optimism. Most people believe that they are less likely than the average person to suffer from misfortunes such as, for example, being in a car accident, being a crime victim, becoming unemployed, and so on. This also applies to positive events, such that most people believe that they will be well off in the future, even when they evaluate the prospects for others more negatively. The majority of people also believe that they are happier than the average. Depressive people, once again, do not share this unrealistic optimism.

In general, one can note, then, that humans have a tendency to delude themselves to some extent when it comes to their basic needs. They imagine that their basic needs are better satisfied than is actually the case. It appears that these types of thoughts and action tendencies have a—albeit naturally limited—need-satisfying function. They lead the person to experience more positive feelings than would seem to be justified.

Because the illusions exert real positive effects, however, they have the characteristics of a self-fulfilling prophecy. They lead the person to attain a better state. Once in this state, such individuals function better, on average, than they would without these illusions, and thereby the illusions lead to better need satisfaction. Good need satisfaction, in turn, leads to improved emotional states, and so forth. This self-maintaining positive feedback process appears to be a crucial aspect of normal mental functioning. When the process no longer functions, it is a reason for concern. The fact that depressive people do not show this tendency to self-esteem enhancement can also be related to the finding that they show broader deficits in any kind of approach motivational tendencies. The approach system fails to function, and—as we discussed in chapter 3 in the section on neural correlates of depression—the left prefrontal cortex, which plays an important role in the approach system, is deactivated among depressive individuals. In the same way that depressive people no longer attempt to create pleasurable experiences, they also no longer attempt to enhance their self-esteem. Both processes are probably linked to a strong activation of the avoidance system, which actively inhibits the approach system.

It is likely that the lacking tendency for self-esteem enhancement contributes to the maintenance of the depressive state among depressed patients, but it seems improbable that this should be viewed as a cause for the depression. More generally, one can assume that violations of the self-esteem need contribute to inconsistency in mental functioning and, thereby, increase the risk for the formation of any mental disorder. Alsaker (1997) and Alsaker & Olweus (2003) were able to show that rejection and bullying by peers in adolescence correlates with self-derogation and depression. Depressive youths are more likely than nondepressed peers to report being rejected and isolated by others and to have frequent negative encounters with peers (Alsaker & Flammer, 1996). All of these constitute violations of the person's self-esteem, but these violations are not simply inflicted by the environment but also have to do with the adolescents' own behavior. Studies about the long-term consequences of negative attachment experiences in early childhood have shown repeatedly that children with insecure attachment patterns later tend to encounter experiences with peers and enter into roles that are rather detrimental to their self-esteem. Violations of self-esteem need are generally one among many links in the chain of the negative development that often originates in violations of the attachment and control needs. Especially after early childhood, when peer relations play an increasingly important role for self-esteem, violations of self-esteem need are clearly an important source of inconsistency in mental functioning. Acute violations of self-esteem need can create such high states of inconsistency tension that they become part of the triggering conditions facilitating the pathogenesis of a mental disorder, but such violations cannot generally be viewed as the original cause explaining the development of the disorder. The fact that individuals with mental disorders generally have low self-esteem can be viewed as the result of an interactive process in which the disorder, as well as other contributory causes, creates a negative sense of self, and at the same time the violations of self-esteem needs also contribute to the further development of the disorder.

4.7   THE NEED FOR PLEASURE MAXIMIZATION AND DISTRESS AVOIDANCE

4.7.1   The Good–Bad Evaluation: A Continuously Active Monitor of Mental Activity

A young child falls down on the playground and hurts himself. His attachment need is activated. He wants to run to his mother to find solace and protection. The little boy looks around, searches, but cannot find her. Now his need for orientation and control are also activated: Where is she, how can I get to her, how can I get her to notice my suffering? Influenced by his control need, the child begins to sob loudly. Negative emotions are experienced: He is hurt, he feels left alone, and he begins to feel fearful. His entire activity is now oriented toward the avoidance of distress feelings.

This example shows that in daily situations, several needs are almost always activated simultaneously. In this situation it is also evident that, in addition to the attachment and the control needs, another need system is activated, which we have not yet discussed in detail but which may well be the most obvious of all needs: The need for pleasure maximization and distress minimization. It is probably the most pervasive of all the needs postulated here and the one that is most accessible to our experience. Hardly anyone will object to the claim that we generally strive to attain pleasant states and avoid unpleasant ones. However, upon closer inspection, this seemingly obvious need turns out to be particularly complex and questions can be raised in this context that go beyond the pleasure– distress principle (see section 4.7.7., "Beyond the Pleasure Principle").

A part of this basic and complex motivational system is the automatic evaluation of all experiences with regard to their "good versus bad" quality. This evaluative dimension permeates all aspects of human experience. If one asks people to use semantic differential scales to rate or evaluate any objects of human experience with regard to their characteristics, factor analyses of these evaluations show that the first dimension of this "semantic space" can regularly be interpreted as an evaluative dimension (Ertel, 1967; Osgood & Suci, 1955). If interpersonal interactions are evaluated, the first dimension of this "interpersonal space" is always a factor that can be interpreted as "emotionally positive (love) versus emotionally negative (hate)" (Benjamin, 1974; Kiesler, 1983).

The evaluative reaction to a stimulus has a specific neural foundation. If one confronts research participants with discrimination tasks—say, for example, one plays a series of two different tones in which one tone is heard more frequently than the other one and then asks them to remember how often the less frequent tone is heard—then the highlighted event (the less frequent tone) is associated with a stronger electric potential, as measured by EEG, 300 ms after introduction of the stimulus, compared to the more frequent tone. This positive electric potential, which reaches its peak after about 300 ms, is called the *P300 component* of the event-correlated potential. It is regarded as a sign that the stimulus has been recognized and discriminated.

When analogue studies with emotionally potent stimuli are conducted—for example, by showing a series of tasty foods and interspersing these with single aversive, disgusting stimuli—then the event-correlated potentials for these other stimuli, which are inconsistent in terms of the general evaluative pattern, show after about 650 ms a so-called *late positive potential* (LPP; Cacioppo, Priester, & Berntson, 1993). This shows that the stimulus was recognized in terms of its deviating evaluative quality. The evaluative processing is, of course, not independent of the quicker identification and discrimination of the stimulus. Rather, it relies on it and builds upon it. The evaluative processing, however, takes a bit longer than the mere identification, and it requires somewhat different brain regions. Whereas the P300 component occurs symmetrically in both brain hemispheres, the LPP is more pronounced in the right hemisphere (Cacioppo, Crites, & Gardner, 1996). When certain stimuli, such as various types of vegetables and other foods, are evaluated in terms of whether they are vegetables or not, the event-correlated potentials transpired symmetrically in both brain hemispheres. However, when the same stimuli had to be evaluated in terms of whether they taste good or not, the LPP was asymmetrical and much stronger in the right hemisphere. This evaluative process also involves additional processes that take place primarily in the right brain hemisphere (Crites & Cacioppo, 1996). The lateralization of evaluative stimulus processes is independent of the question of whether positive or aversive stimuli are being encountered.

The emotional evaluation of stimuli transpires automatically and, just like the identification and discrimination of stimuli, is not a conscious process. How a stimulus is evaluated does not depend much on its objective characteristics but rather depends on the person's prior experiences and the momentary state of the evaluating individual. A cold stream of air is experienced as pleasant when one feels hot, but the same stimulus is unpleasant when the person already feels cold. People who are not used to it tend to experience chili pepper as unpleasant or even intolerable, but more than 1 billion people regard this as a culinary pleasure that they do not want to miss. The positive evaluation is acquired by learning. Even Mexican children initially do not like chili pepper. The recoding of the evaluation typically occurs among them between the ages of 4 and 7 years (Rozin & Schiller, 1980). Children also do not like the taste of coffee, beer, or wine, but adults of Western cultures regard these as preferred and highly palatable delights. It is remarkable that such a relearning of tastes occurs never or hardly ever among animals (Rozin, 1990, 1999). The relearning of taste preferences is apparently a complex process that is influenced by motives such as the wish to belong to groups, to be competent, or to feel positively about the self. The ingested substance and the reactions that are triggered in the sensory cells remain unchanged. On its route from the sensory cells to the comparator, however, the signal undergoes increasingly complex transformations, which further increase in complexity with increasing degrees of acculturation. The process is increasingly influenced by evaluations that have little to do with the hedonistic quality of the stimulus as such; instead, the role of motivational schemas becomes more dominant. The beer finally tastes good despite its bitterness because one wants to belong to those who drink it. What ultimately reaches the "evaluative comparator" is a multiply transformed and integrated signal that has little to do with the original sensory experience. However, after the relearning, the sensory experience itself triggers a positive hedonistic reaction and drinking beer is not just something that is pursued but also something that is experienced as pleasurable.

It is generally easy for us to integrate very different aspects into a single good–bad evaluation, and we often make important contingencies dependent upon such integrative judgments. This type of good–bad evaluation is often even quantified on a scale; for example, in ice skating or high diving, and decisions such as who should be awarded medals are made contingent on this quantitative judgment. In the context of degustation, wine experts also express the quality of wines on quantitative scales that might range from 10 to 20 (*Weinwisser; Wine Spectator*) or 50 to 100 (Parker, 1992; *Wine Advocate*). These evaluations can quite directly trigger approach behavior (i.e., purchase) or avoidance tendencies. When the "wine pope" Robert Parker evaluated the Chateau Montrose 1990—which normally attains between 90 and 93 points—with an unexpected 100 points, the wine was sold out within a few weeks worldwide, even though none of the buyers had had an opportunity to personally taste the wine. Positive evaluations activate the desire for pleasure maximization and lead to approach behavior. This holds true not only for processes within one person but—if the evaluation is viewed as trustworthy—also applies to interpersonal situations. In this case the trust was justified. The Chateau Montrose 1990 turned out to be one of the best wines ever produced and today is already regarded as legendary.

Wine, whose culinary delights must originally be learned, is also a good example to illustrate the qualitative differentiation of the criteria that can contribute to an overall good–bad evaluation. Let us appreciate how Robert Parker justifies his judgment about the Chateau Pichon Lalande 1982, another wine that has acquired legendary status:

> If someone wants to hold a wine-tasting with his collection of 1982 vintages, and the person is also somewhat of a player, then he should easily put all his money on the bet that the 1982 Pichon Lalande will emerge as the favorite because it appears to become ever better, and every time I taste it, it seems to go from one high to the next. My first impression was that it lacked in structure, so that I thought it would not become as great as the 1981 vintage. How wrong I was! In hindsight I had to upgrade my evaluation not only from the cask tasting but also from the initial tasting after the bottling. It is hard to believe, but it has become even better since then. Perhaps the 1986 wine will eventually overtake the 1982 vintage, but it is such a seductive Pauillac as anything I have ever tasted. Its bouquet shows unbelievable ripeness and great fruitiness—an aroma that one must have tasted oneself. On the tongue this wine is remarkably impressive with silky, rich, concentrated taste, immaculate balance, and a seductive, highly ripe charm, which it must have had within it from the very beginning. All these qualities point the fact that it can only become better over the next 4 to 8 years. It already tastes so wonderful that it would be a sin not to recommend to those readers who happen to have a few bottles to enjoy one of them today. This wine is simply enchanting—there is no other word for it. Will it become even better? I believe so, but such a fascinating wine deserves to be drunk soon. (back-translated from the German version of R. M. Parker, 1992, p. 261)

All this is ultimately integrated into an evaluation of 99 point on the Parker scale, which Parker changed a few years later to extend to 100 points. Lucky few who have a few bottles of this wine in their cellar!

It is not just bodily but also many mental pleasures that must be acquired via learning, or for which a relearning must take place before the stimuli can be evaluated positively. Many people experience free jazz as horrendous noise, but there are jazz lovers for which the same tones constitute the highest pleasure. A prerequisite for this development of pleasure experience is frequent listening. The enjoyment of music has much to do with the occurrence or nonoccurrence of expectancies that are learned implicitly (Kubovy, 1999). It is easy to form such expectancies for simple melodies, but music that contains too few surprises or whose implicit rule system is too simplistic can be experienced as boring by discriminating listeners. Inexperienced listeners tend to perceive the tones in a piece of free jazz as completely unpredictable and, therefore, they lack the basis for the experience of anticipation that results from the occurrence and nonoccurrence of expectancies. Thus, they also lack the foundation that would render the piece meaningful. An experienced free jazz listener, however, will find even the tones in a new piece not entirely unpredictable. The listener forms expectations while listening and can enjoy the play of the musicians and the excitement as these expectancies are or are not fulfilled. Listening to music requires full concentration, however, and those who want to listen to such music as if it were a background pop tune will not be enabled to access its beauty. What is said here about the beauty of listening to music also holds true analogously for the ability to appreciate beauty while viewing visual art.

Whether something is experienced as beautiful, pleasurable, or tasty does not depend only on the sensory target object, then, but to a large degree also depends on the qualities the perceiving person brings to the situation. The old adage that "beauty lies in the eye of the beholder" may exaggerate the point somewhat because there are also objective characteristics of beauty—certainly within a given culture—but the saying nevertheless has a kernel of truth.

Kubovy (1999) lists among the mental pleasures also the excitement inherent in discovery, finding out new things, and confirming expectations. The object of these activities is the unknown. The reading of crime stories, solving a jigsaw puzzle, traveling to foreign areas and cultures, and also scientific explorations and inventions can be included among these pleasures. The mental pleasures also include the joy about one's own competencies. One does not have to be a virtuoso in order to experience joy about one's own abilities. It is enough to be able to do something slightly better than before, or to have the sense that one is really good at something. Things that one can do well are normally done with pleasure. One is intrinsically motivated to do it. The pure form of the intrinsically motivated state—a form of intensive mental joy—is the state that Csikszentmihalyi (1990) has termed the *flow experience*:

> One of the main forces that affects consciousness adversely is psychic disorder (or psychic entropy)—that is, information that conflicts with existing intentions, or distracts us from carrying them out. ... The opposite state from the condition of psychic entropy is optimal experience. When the information that keeps coming into awareness is congruent with goals, psychic energy flows effortlessly. There is no need to worry, no reason to question one's adequacy. But whenever one does stop to think about oneself, the evidence is encouraging: "You are doing alright." ... We have called this state the *flow experience*, because this is the term many of the people we interviewed used in their descriptions of how it felt to be in top form." (pp. 36, 39–40)

What Csikszentmihalyi describes here is, from the perspective of consistency theory, a momentary state of complete consistency of simultaneously transpiring mental processes. Current perceptions and goals are completely congruent with one another, and the transpiring mental activity is not disturbed by any competing intentions. According to consistency theory this can be viewed as a state of optimal mental health. Such sates of complete consistency are generally attained only for very short time periods. However, consistency theory is very much compatible with

Csikszentmihalyi's flow concept because both models regard a maximal level of consistency of mental processes as the foundation of good mental health. States of "psychic disorder" or "psychic entropy", as Csikszentmihalyi calls them, clearly exert a negative influence on mental health.

The examples described here are intended to show that the evaluation of perceptions and activities in terms of pleasure versus displeasure, which transpires continuously and automatically in the mental system, does not pertain only to pleasurable bodily experiences but more generally to anything that is experienced as positive or negative. The examples are also intended to show that in most cases we are dealing with a very complex evaluative process that also involves the other basic needs and the motivational schemas that have formed to facilitate need attainment and protection. If the child that was mentioned earlier finds his mother again and is taken into her arms, then this child will experience this as a result of a good–bad evaluation as a very positive event. The satisfaction of the attachment need and the regaining of control are experienced as equally pleasant as when intense pain recedes. To be consoled by the mother and the reduction of pain are qualitatively different experiences because different goals are involved in terms of the reduction of previously existing goal incongruence. However, with regard to the evaluation of whether the experiences are positive or negative, there is no difference in principle.

It is also possible to experience discrepancies between pleasure–distress evaluations and incongruence signals with regard to goals that serve different needs. Imagine a long-distance runner who is still far away from the finish line, whose muscles are already aching intensely and who experiences shortness of breath. If the runner continues despite these intense distress feelings and perhaps even tries to step up the pace, he or she demonstrates clearly that pleasure–distress regulation is not necessarily the highest maxim of mental functioning. If it becomes the dominant maxim, this is usually associated with tragic consequences for the person and leads to intense suffering. This will be the topic of sections 4.7.4 and 4.7.5. In section 4.7.6 I will address in more detail the constellations in which the incongruence signals with regard to pleasure–distress—which were addressed in this section—are discrepant with the incongruence signals arising from other activated motivational goals.

4.7.2 FUNCTIONAL ASSOCIATIONS BETWEEN THE GOOD–BAD EVALUATION AND APPROACH–AVOIDANCE

When a stimulus or a situation is evaluated positively or negatively—and this occurs, as I mentioned, completely automatically—this triggers (also automatically) both approach and avoidance tendencies. The evaluative function is regarded by authors such as Lang (1995), Gray (1982), and Gray and McNaughton (1996) as an integral part of two motivational systems—an approach system (the behavioral approach system; BAS) and an avoidance system (the behavioral inhibition system; BIS).

There are close relationships between the emotional evaluation of a situation and the orientation of mental activity. If one has just seen an anxiety-provoking horror movie in the cinema and then, upon exiting from the theater, suddenly hears a loud bang, one is more likely to have a strong startle reaction, compared to a situation in which one would have viewed an amusing film that elicits positive emotions. In both cases, the sudden noise activates the avoidance system, but if one has previously experienced negative emotions, the activation of the avoidance system is already prefacilitated and the startle reaction will be stronger. This process has been called *motivational priming*. Negative emotional cues facilitate associations, representations, and behavioral programs in the avoidance system; positive cues facilitate them in the approach system. Motivational priming has been studied with a variety of emotional situations as the independent variable and the eye-blink reflex as the dependent variable. The results are unambiguous: Regardless of whether the emotional state is being elicited via smells, images, sounds, or imagined scenarios, it is inevitably found that negative emotional contexts lead to a stronger startle or defensive reaction, whereas positive emotional contexts weaken the activation of the avoidance system (for summaries of this research, see Ito & Cacioppo, 1999).

The fact that emotional cues do not just elicit an emotional evaluation but also trigger behavioral tendencies is also evident in analyses of facial muscles. When research participants are asked to recall and imagine positive or negative events in their lives, it is possible to demonstrate with the electromyogram that certain facial muscles are activated and inhibited that are already known to be innervated during the expression of specific positive or negative emotions. When sad emotions are being imagined, the eyebrow muscle is more strongly and the cheek muscle more weakly activated, compared to situations in which positive events are being imagined (G. E. Schwartz, Fair, Salt, Mandel, & Klerman, 1976). The same holds true when one imagines that someone holds an opinion that is contrary to one's own (Cacioppo & Petty, 1979). These reactions are not openly observable but can be detected via electromyogram. They show an automatic activation or inhibition of certain approach-related or avoidance-related behavioral programs.

In a study by M. Chen and Bargh (1999), participants were asked to evaluate words that appeared on a screen as quickly as possible as "good" or "bad" by either pulling a lever toward them or pushing it away from them. The participants were quicker in pulling the lever when they evaluated positive words. At the same time, they were quicker in pushing the lever away when they evaluated negative words. The movement "towards oneself" is thought to be linked to the approach system because it is coupled, for instance, with the act of ingesting foods. The movement "away from oneself," by contrast, is thought to be coupled with the avoidance system. Similar results were found for head movements. Förster and Strack (1996) asked participants to either nod their heads (a positive, approving movement) or to shake their heads (a negative, negating movement) while encoding positive and negative words. Participants who nodded during the encoding process were quicker to recognize the positive words in a later recall test, whereas those who shook their heads were quicker to recognize the negative words.

This close functional association between approach and avoidance on the one hand and emotional evaluation on the other hand also works the other way around. Cacioppo, Priester, and Berntson (1993) showed participants neutral Chinese language symbols, which they had not previously encountered, while their arms were either bent (an approach gesture) or while they were extended (an avoidance gesture). The symbols that were displayed while the arms were bent were later evaluated more positively by the participants than those that had been presented while the arms were extended. Neumann and Strack (2000) reported very similar results in a series of studies. Their participants were quicker to classify positive words into the correct category while their arms were bent, and quicker to classify negative words with their arms extended. The idea that arm bending can be regarded as activation of the approach system and arm extending as activation of the avoidance system is also supported by the fact that analogous results were also observed with a "toward me" and "away from me" movement. If the procedure was arranged in such a manner that the words appeared to come closer on the screen, positive words were categorized more quickly, but when the words seemed to move away, negative words were processed more quickly.

All of these findings together can be interpreted as follows: Mental processes transpire more easily and quickly when the good–bad evaluation is compatible with the behavioral approach—avoidance orientation. This could also be expressed like this: When evaluation and behavioral orientation are consistent with one another, the mental system works more efficiently; when they are inherently inconsistent, the simultaneously transpiring processes hinder or inhibit each other and the efficiency of the system for coping with various demands is compromised. This is one of the core assumptions of consistency theory. It is empirically very broadly supported. I have described here only some of the relevant studies to provide examples. There are many more studies with similar findings; such studies can be easily found in the reference sections of the articles cited previously.

Even though the interpretation of the results here is primarily of theoretical importance, the phenomenon of motivational priming—which is clearly supported by these studies—also has direct and practical relevance in the context of psychotherapy. If the avoidance system is activated one way or another, the entire mental activity is oriented toward avoidance and negative evaluations are made more quickly and easily. The same holds true—conversely—for the approach system. The more the approach system is activated for one reason or another, the more easily the processes relating to approach and positive evaluation transpire. By its very nature, psychotherapy often involves the discussion of topics that are negatively evaluated by the patient. The findings reported here suggest, however, that this leads to a motivational priming in favor of the avoidance system. This is actually highly undesirable because the goal of therapy is almost inevitably the strengthening of positive aspects of the patient; that is, to encourage an approach orientation of mental activity toward positive goals and an increase in positive emotions. The rather unfortunate activation of the avoidance system can itself hardly be avoided, however, if one aims to genuinely confront the issues in the patient's life that are evaluated negatively by him or her. If the activity of the avoidance system dominates in therapy, however, it is difficult to see how any positive goals could be attained at all. These considerations suggest the conclusion that it is very important for the therapist to activate the approach system as frequently and intensely as possible, so that positive therapy goals can be attained even while confronting the negatively evaluated patient problems. I will discuss this conclusion in greater detail in chapter 5, in which I articulate which specific consequences for therapeutic action arise from this conclusion.

Because the striving for positive goals and states and the avoidance of negative ones is of critical importance for well-being, mental health, and the attainment of beneficial effects in psychotherapy, we must clarify several additional questions with regard to approach and avoidance before we can devote attention to these concrete psychotherapeutic implications. These additional questions include, for example, those about the relationship of the two motivational systems toward one another. Are they inherently opposed to each other? Do they exclude each other, inhibit each other, or are they independent of each other? These questions are the focus of the next section.

4.7.3   Approach and Avoidance as Two Independent Motivational Systems

Psychological models of the semantic and interpersonal space (see previous) conceptualize the evaluative dimension as a bipolar dimension with two opposing poles and a neutral midpoint. This conception implies that something can only be either good or bad but not both at the same time. We have just seen that evaluative processes are closely linked with behavioral approach—avoidance tendencies. A bipolar model therefore suggests that behavior is either oriented toward approach or avoidance. This bipolar conception, however, was not supported upon closer inspection.

Neuroscientific as well as psychological authors construe the approach and avoidance systems as two separate motivational systems. The systems interact and tend to mutually inhibit each other, but they can also both be activated independently of one another, and each has its own independent neural substrates and mechanisms (Cacioppo, Gardner, & Berntson, 1997; Gray & McNaughton, 1996). This independence and the fact that good and bad are subjectively experienced as opposites seem to conflict with one another. However, we also experience hot and cold as opposite qualities, even though cold and warm perceptions have different neurophysiological underpinnings.

In section 3.2.1 on the neural correlates of depression I discussed findings pertaining to the asymmetrical activation of the left and right ventromedial and dorsolateral cortices among depressive patients. The dorsolateral PFC is critically important for the representation of approach goals (left PFC) and avoidance goals (right PFC), and the ventromedial PFC for the generation of positive (left) and negative (right) emotions.

These brain regions are important parts of that which Gray and McNaughton (1996) have termed *BAS* and *BIS*. These systems, however, do not include just anterior but also posterior brain regions, such as the amygdala, the cingulate cortex, the hypothalamus, and the sympathetic nervous system.

Evidence for the existence of two different response systems—one associated with a tendency for positive emotions and approach behavior and one associated with the tendency for negative emotions and avoidance behavior—has come not just from neuroscientific research but also from temperament research, personality research, and emotion research (Diener & Lucas, 1999; Ito & Caccioppo, 1999). There is no doubt that these differential tendencies to experience positive or negative emotions are, in large part, biologically inherent (Tellegen et al., 1988). This involves in each case several genes. In the case of the avoidance system, the genes appear to have an additive effect; that is, the more of these genes a person has, the more certain it is that one will have a pronounced tendency to experience negative emotions. The tendency to experience positive emotions, by contrast, appears to require a specific sequence of certain genes. According to twin studies, the tendency for positive emotions is apparently also influenced by environmental influences to a greater degree than is the avoidance system (Baker, Cesa, Gatz, & Grodsky, 1992).

LaGasse, Gruber, and Lipsitt (1989) were able to predict based on the sucking behavior of 2-day-old infants the extent to which they displayed inhibited behavior later on. Davidson and Fox (1989) categorized 10-month-old children after their resting EEG into either dominantly left-activated or right-activated groups. The children were later separated for 1 minute from their mother. The infants who cried during this period— for which the separation therefore more easily elicited negative feelings—tended to have a stronger right-sided EEG activation, compared to the children who remained silent. This differential response tendency appears to remain stable. Davidson (1993) observed 31-month-old children in a play situation with other small children. The children were categorized based on their behavior as either inhibited or noninhibited. A resting EEG that was conducted 5 months later showed that the inhibited children had stronger right-sided brain activity and the noninhibited children had stronger left-sided brain activity. Individuals who exhibit a stronger right-sided activation in resting EEG also have a stronger right-sided activation when they are confronted with cues that elicit negative emotions; that is, they also react more strongly to negative emotional situations.

The tendency to experience positive and negative emotions, then, appears to be a relatively stable personality characteristic. Costa and McCrae (1998) asked married partners to report on their husband's or wife's tendency to experience positive or negative emotions. Six years later, the researchers asked the participants to rate themselves in terms of these tendencies. The ratings, which were 6 years apart from another and came from different sources (self vs. partner ratings) nevertheless correlated .50 with one another. The relative stability of these differential response tendencies, therefore, appears to extend into adulthood.

According to Gray (1981), the BAS is the neurophysiological basis of the personality trait extraversion and the BIS the basis of the trait neuroticism. These two factors are construed as mutually independent personality dimensions both in Eysenck's (1981) two-dimensional personality model and in Costa and McCrae's (1992) model of the "big five" personality factors. These two personality factors, their general independence from one another, and their differential links with positive and negative emotions are empirically broadly supported (Diener & Lucas, 1999). This can be summarized in the image shown in Figure 4–11.

According to Gray's (1981) model, extraverts are more receptive to rewards and are more easily influenced to experience positive moods. People high on neuroticism, by contrast, respond more strongly to punishment and more easily slip into a negative mood after experiencing negative events. This prediction was empirically supported in studies with different designs by Headey and Wearing (1989) and Larsen and Ketelaar (1991), as well as Magnus, Diener, Fujita, and Pavot (1993). The tendency to experience negative emotions and avoidance tendencies, which were construed as the BIS by Gray, not only overlap strongly with the personality factor of neuroticism; both concepts also overlap strongly with the general factor *disposition to experience negative emotions* which—as discussed earlier (see section 4.5.2.2)—was found by L. A. Clark and Watson (1991) and T. A. Brown, Chorpita, and Barlow (1998) in their factor analytic studies on the correlations among the various DSM diagnostic criteria. According to Chorpita and Barlow (1998), this factor forms the common foundation for the later development of anxiety disorders and depression. Elliot and Thrash (2002) tested explicitly whether the different constructs could really be conceptualized as two independent factors. They used a questionnaire to measure in a sample of 167 participants the variables extraversion and introversion (measured by the NEO Five-Factor Inventory [NEO-FFI]), the tendency to experience positive and negative emotions (measured by Watson and Clark's [1993] General Temperament Scales), as well as BAS and BIS (measured by the self-report scales developed by Carver and White [1994] for that purpose). Table 4–4 shows the resulting factor matrix. Indeed, it was observed that exactly the two expected factors emerged. Elliot and Thrash term them the *approach* and *avoidance temperaments*, by which they mean exactly the constructs that I described earlier.

If we consider all these findings together, we arrive at the following picture: There is a genetic foundation, and a specific neural basis, for the disposition to experience negative emotions and to engage in behavior that is oriented to defense, inhibition, and avoidance. Newborns with this disposition are generally more vulnerable to experience disorders. They respond with strong negative emotions to challenges such as frustrations or violations of the attachment and control needs, and to distress-eliciting cues that elicit little irritation among children without this disposition. Therefore, they are more frequently and for longer durations in a negative emotional state and are more oriented toward avoidance than to exploration and approach. This increases the risk that such a child will enter into a negative interaction pattern with his or her caregiver, who feels overwhelmed and frustrated by the "difficult child." I mentioned earlier (see section 4.4.2) a study by van den Boom and Hoeksma (1994)

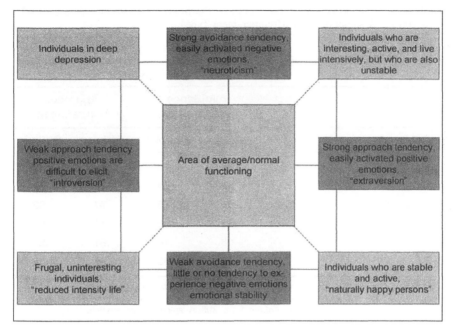

Figure 4–11. The two independent motivational systems, approach and avoidance, and their association with the personality factors extraversion and neuroticism. The four corners of the square refer to the "extreme types" resulting from the combination of the poles of the two motivational systems.

in which such high-risk children with higher responsiveness indeed showed such an unfavorable course of their attachment relationship. The consequence is that these children encounter even more emotionally negative experiences, such that an escalating feedback loop can be created in which the child experiences a sustained state of stress, along with the release of harmful stress hormones (as detailed in the previous section on "Consequences of uncontrollable incongruence").

This sequence, however, does not transpire automatically. The disposition includes only an elevated risk for such an unfavorable course, which later can result in the formation of mental disorders. The chain of events can be broken, interrupted at the beginning, or channeled into a more beneficial trajectory by positive influences from the environment. In section 4.4.2 on the "Neurobiology of the Attachment Need" I reviewed a study by Suomi (1991) in which rhesus monkeys with genetically unfavorable dispositions—similar to the disposition to experience negative emotions discussed here—were raised by a particularly caring mother with whom they developed a particularly strong attachment relationship. After half a year, these monkeys were in even better shape with regard to their development and robustness, compared to a comparison group with more favorable genetic dispositions. Their high-risk status that seemed to place them at high risk for negative development could, therefore, not just be defused but could be transformed into its opposite. Van den Boom (1994) was also able to show in an experimental intervention study that the development of an

Table 4–4. Factor Matrix With the Result of the Study by Elliot and Thrash (2002), Justifying the Assumption of Two Temperament Types

	Factor Loadings			
		Factor		
	Avoidance Temperament		Approach Temperament	
Variable	Varimax	Oblimin	Varimax	Oblimin
Extraversion	−.23	−.13	.85	.84
Neuroticism	.86	.84	−.28	−.17
Positive emotionality	−.26	−.16	.84	.83
Negative emotionality	.92	.93	−.11	.01
BAS	.08	.18	.81	.84
BIS	.80	.82	−.02	.08

Note. BAS = behavioral activation system; BIS = behavioral inhibition system.
From "Approach—Avoidance Motivation in Personality," by A. J. Elliott and T. M. Thrash, 2002, *Journal of Personality and Social Psychology, 82,* p. 808. Adapted with permission.

unfavorable attachment relationship with particularly responsive children could be prevented via a targeted sensitivity training for the mothers.

If the result of the interaction between an inherited disposition to experience negative emotions and a suboptimal environment is a more frequent occurrence of negative emotions and, via motivational priming, a dominance of the avoidance mode, then this will also influence the motivational schemas that the child develops step by step in order to satisfy and protect his or her basic needs. The greatest risk associated with an inherited tendency to experience negative emotions is the development of predominant avoidance schemas. Once these avoidance schemas are firmly in place, negative emotions will be elicited not just by unfavorable environmental situations but they are also created actively by the individual via his or her context-bound activity.

4.7.4   The Development of Motivational Goals

The basic needs have neural foundations that humans bring with them when they enter the world. These foundations allow humans to employ their behavioral repertoires from the beginning in such a way that their needs will be met. In the very limited, naturally determined environment of an infant, a very small genetically governed behavioral repertoire suffices for that purpose (e.g., crying, sucking, and wiggling of the body). The goals that these behaviors are oriented toward are also genetically determined. However, this changes rather rapidly. Whereas at the beginning any person can satisfy the infant's needs, soon very few persons or perhaps only one primary attachment person is singled out. It is not proximity to just any person but proximity to this one special person that becomes the goal of the infant's striving. Via sensory perceptions, this goal becomes associated with a face, a smell, a voice, and so forth. Proximity to this person becomes a first concrete approach goal. With this goal, the first means that are specific to this dyad also develop. DeCasper and Fifer (1980) showed that even a 3-day-old infant can use the speed of his or her sucking movements in order to get to hear the voice of his or her mother rather than that of a strange woman. This means that the infant encounters first control experiences. The seed of a motivational schema is developing, with the overarching goal of "proximity to mother," the subgoal of "hearing my mother's voice," and the means, "sucking in a particular rhythm." This schema soon becomes more differentiated. The face of the mother, her approaching steps, the way in which she turns to the child become subgoals, and also the means with which the infant can influence the mother become more differentiated. Neural activation patterns emerge that represent the goals, subgoals, and means. In chapter 2 and in the section on the attachment need we have seen how cell assemblies, which represent the single components of this approach schema, are bound ever more tightly together and form a neural network as the synapses between the respective neurons strengthen and multiply. Oxytocin and dopamine, which are more strongly released during episodes of proximity to the mother, increase the signal transmission at the synapses and enable second messenger cascades, which selectively strengthen the synapses that are activated when the mother is nearby and that form additional synapses between the simultaneously activated neurons. With each episode of proximity to the mother, the neural connections associated with this motivational schema are further strengthened, such that over time the corresponding neural circuits are more easily activated. Ultimately, these will form some of the cell assemblies with the highest excitability of all, such that they can eventually become active spontaneously, even when the schema has not been activated by the image or the voice of the mother. The child experiences a wish, a longing for the mother. The goal of the well-facilitated approach schema has been activated spontaneously. The goal will also be easily activated when a need state such as hunger or freezing is present—states that have previously often been alleviated by the mother. The child feels hunger or feels lonely, the goal "mother should come" is activated, the utterances that have previously been effective for this purpose are emitted, and attention is deployed to notice whether mother can be heard or seen anywhere. When the mother appears, her sight triggers positive emotions. This is an approach schema in action.

Context-bound goals form around the basic needs even within the first days of life. From then on, these motivational goals determine concrete behavior as well as emotional evaluations. Basic needs are realized via these goals; they do not directly influence behavior. Basic needs are nevertheless not just an abstraction. They correspond to neural mechanisms that humans bring with them to the world. These neural mechanisms render humans needy to obtain specific types of life experiences. This neediness for specific experiences, which is based on specific, inherited neural mechanisms, is what I mean when I speak of basic needs.

In the case of the need for self-enhancement, the neediness for specific life experiences emerges only after, on the basis of inherited mechanisms and under the influence of life experiences, a complex neural structure has formed, which we call self and self-image. The correspondence to innate neural mechanisms is less direct in this case. But all humans develop on the foundation of the inborn neural mechanisms a self and self-representation, in the same way that they also develop language. In the same way that language is a specifically human ability that requires a neural structure that only the human brain contains, the self-esteem need is also a specifically human need that requires a neural structure that, although not yet present at birth, develops within each human. The need for self-esteem enhancement and self-esteem protection fulfills the requirements for a basic need in the same way as the other basic needs, even though it only develops in a later phase of life. It is common to all humans, and its nonsatisfaction has negative consequences for well-being and mental health.

Life experiences are mediated via the goals that have formed under the influence of inborn neural mechanisms. Motivational goals are the mediators between the neural mechanisms that are common to all humans—which render the person needy for specific experiences—and the person's environment. Via the motivational goals, the person strives to attain specific types of transactions with his or her environment.

The extent to which the basic needs of a person are met depends in large part on the degree to which the motivational goals that he or she has formed enable the person to actually encounter need-satisfying experiences in his or her environmental context. The self-esteem need can be satisfied in different life context in very different ways. In an upper middle-class European family, the goals to perform and achieve will generally contribute to self-esteem. In a slum in which criminal activity surrounds the growing children from an early age, however, the goal to appear maximally cool might contribute more to the person's self-esteem. One determinant for good need satisfaction is, therefore, a good match between basic needs and motivational goals in the context of the person's individual environment.

Once a motivational goal has formed, need satisfaction also depends on the extent to which life experiences that are encountered actually match the goal. If they are congruent with the goal, the need is satisfied; if they are incongruent, the need remains unmet, even if the goal would be generally appropriate to satisfy the need in this particular life environment. The extent to which a person succeeds in realizing his or her motivational goals depends to a large degree on the resources he or she has available. These resources are primarily his or her abilities that can be employed as a means toward the realization of the goal, as well as the objective nature of the person's environment. A nonnurturing mother does not constitute a good environment that would allow the person to encounter positive perceptions with regard to the goal, "achieve that mother directs her love and attention toward you." Most of the time, the actual life experiences of the child will be incongruent with this goal because the mother repeatedly neglects or rejects the child in moments of need.

The nonattainment of approach goals activates negative emotions and, simultaneously, activates an avoidance-orientated behavioral tendency. In situations in which similar experiences have been encountered, mental activity will from now on be oriented toward avoidance. It will aim to avoid the repeated occurrence of negative emotions. In order for that to happen, the transactions that have previously led to the negative emotions must be avoided. This includes primarily the previously emitted approach behavior, which subsequently will be inhibited so that neglect and rejection will be effectively avoided. The pain associated with these experiences will therefore not occur; avoidance learning has taken place. What is reinforced in this situation is the inhibition of approach behavior. As a result of this avoidance learning, which is associated with an ever easier activation of avoidance goals and an increasingly well-facilitated inhibition of approach goals, the negative emotions that were previously triggered by the rejection are, on the one hand, successfully avoided, but on the other hand, the original approach goal, "achieve that mother directs her love and attention toward you" is not being realized. Via the inhibition of approach behavior, acute suffering is avoided, but the incongruence with regard to the approach goal remains in place. The more pronounced and generalized the avoidance, the more the approach behavior that is oriented toward the satisfaction of the attachment need is inhibited. The gain lies in the control that this inhibition enables, but the loss lies in the nonsatisfaction of the attachment need. If these processes are frequently repeated in the attachment relationship, the result is an avoidant attachment style. Close intimate relationships will subsequently be avoided.

All of these sequences are completely automatized because of the hundreds or thousands of times that they have been facilitated. They are determined by a powerful avoidance goal, but this fact and that which is being avoided are not within the individual's conscious awareness because, indeed, it is being avoided and therefore does not become content of the person's conscious experience. In addition to the fact that these processes are completely automatized, they become established already within the very first years of life, in which reflective consciousness for one's own goals cannot yet exist. Therefore, the person will later not be in a position to say, based on his or her subjective experience, "I avoid being rejected when I feel a need for intimacy." He or she does not know the state of "feeling the need for intimacy," at least not in the way that securely attached people experience this, and he or she also does not know the pain associated with rejection. Both were eliminated from the person's experiences early on via active inhibition. This might describe the approximate development of what, from the observer's perspective, we call an avoidant attachment style.

In the case just described we assumed that the child has predominantly encountered the experience that the mother does not attend to the child or that she rejects him or her. If this experience is not so consistent and the mother sometimes acts attentively but at other times neglectfully or rejecting, another constellation of motivational goals will develop. Now and then, in situations when the mother has her attention fully directed toward the child, the goal "achieve that mother directs her love and attention toward you" is effectively attained. It will be all the more painful for the child if the mother acts entirely differently in the next similar encounter. The child never knows what to expect. He or she does not have personal control over the result of his or her approach attempts because this depends on the state of the mother. The child can never be certain that the mother will be there when needed. The mother might be physically present but preoccupied with herself rather than being attentive to the needs of the child. In this case, the attachment need also remains unmet, but the control need will be even more severely violated. The state of intimate proximity with the mother does recur occasionally, but the recurrences are not being controlled by the child. In this constellation, the child will be oriented toward the regaining of control. He or she will continuously check whether mother is in fact there. The child's attention will be constantly preoccupied with this concern. He or she will test again and again how mother responds. If she leaves, this constitutes an alarm signal, especially with regard to the control need. The child will not form the avoidance goal, "Do not permit intimacy," or "Inhibit your wish for closeness." The opposite is true: The child will be continuously worried about whether mother's proximity is really certain. The approach goal remains continuously activated but at the same time an avoidance goal is established, "Ensure that mother's closeness will not be lost." Such a child will become clingy and dependent, with easily elicited negative feelings in response to the loss of the mother's proximity. The child is simultaneously approach- and avoidance-motivated. In this case, the approach behavior is not inhibited by the avoidance goal, as was the case in the previous example. Instead, the goals "seek intimacy" and "avoid the loss of intimacy" are simultaneously activated. The child lives in a state of constant anxious tension. He or she does not encounter the experience that he or she can control that which is of greatest importance to him or her. The child does not develop secure control expectancies but will remain preoccupied with the theme of control, constantly seeking to avoid impending loss. This constellation would be described, from the outside, as an insecure–ambivalent attachment style. The hallmark of this constellation is the frequently recurring simultaneous activation of approach and avoidance goals. This constellation will later lead the person to develop rather strong conscious wishes for intimacy, along with tendencies to consciously seek intimacy, but he or she will also be constantly worried about whether the intimacy will remain in place and will constantly seek reassurance so that the dreaded outcome will not occur.

Many factors determine which constellation of approach and avoidance goals forms within a person in early childhood around the systems of attachment, control, and distress-avoidance. Individuals who are also biologically equipped with a sensitive, anxiety-prone temperament will tend to develop more and more strongly developed avoidance goals in similar situations, compared to cases in which no genetically determined avoidance temperament is present. However, this genetic foundation starts to interact with the environment from the first day of life. There will be an increased probability that the environment will be overwhelmed by the unfavorable genetic endowment and that interaction patterns will develop that encourage the development of avoidance goals (van den Boom & Hoeksma, 1994). If the child is lucky, however, he or she will encounter a relaxed, sensitive mother in good mental health and develop a positive, close attachment relationship with her. In such a case, the genetic endowment will be defused. The unfavorable genetic potential will not be expressed and the child will develop from the basis of this "safe haven" the kinds of approach schemas that will become his or her personal foundation for continued good need satisfaction and mental health (Suomi, 1987, 1991; van den Boom, 1994).

The specific constellations of motivational schemas that develop around the basic needs are infinitely richer and more multifaceted than the differentiation of three or four attachment styles might suggest. Humans are not simply secure, avoidant, or ambivalent in their attachment. This is a crude distinction that is useful in research contexts. In clinical practice, however, individuals who have been assigned to one of these three categories may differ clearly in their individual constellation of approach-related and avoidant schemas. As children, humans also generally encounter relationship experiences with more than just one person. They develop different interaction patterns in different relationships. This means that the relationship experiences that are reflected in the motivational schemas are much richer than the examples described here might suggest. In different situations with different interaction partners, different motivational goals can be activated within one and the same person. Correspondingly, a child's emotional evaluations will differ drastically based on the specific person and situation encountered. Even among individuals with strongly formed avoidance tendencies, there are generally positive goals that one can work with if one aims to enable them to attain a generally better satisfaction of their basic needs.

4.7.5   The Functional Significance of Approach and Avoidance Goals

Based on the ideas developed so far, approach and avoidance goals are not opposing poles with positive or negative signs. Instead, they are qualitatively different types of goals. Carver and Scheier (1998) attempted to illustrate this with the image shown in Figure 4–12. Whereas approach goals have to do with the reduction of discrepancy to a positively evaluated goal, avoidance goals have do with the maximization of discrepancy from a negatively evaluated goal. Carver and Scheier therefore also use the term *antigoal* when they discuss avoidance goals.

When pursuing a positive goal it is relatively easy to determine whether one has come closer to the goal, and it is generally possible to actually attain the goal. Progress toward the goal is generally associated with positive emotions, especially when the progress is greater or transpires more quickly than the person previously expected (Carver, Lawrence, & Scheier, 1996). The route toward the goal can be chunked into smaller steps or subgoals that are more easily attainable than the final goal. Attaining these intermediate goals can also elicit positive emotions. One can plan to engage in specific actions that promise to lead to greater proximity to the goal. One can also concentrate on the goal and take action in such a way that success or effectiveness can later be determined. Approach goals can be pursued with intrinsic motivation, which facilitates the deployment of attention toward success and improvement-related information.

Avoidance goals require constant control, as well as distributed instead of focused attention. Such goals can never be completely reached. Even if a danger has been successfully averted, one can never be certain that another danger might not lurk from a different direction. Avoidance goals necessitate chronic vigilance. If a small child strives to attain closeness to the mother, he or she knows what must be done to achieve this and can relax when the goal has been reached. If the child has the goal, "Avoid the loss of mother's closeness," he or she must constantly monitor everything that is happening. It could always come true that mother leaves, attends to other things, and so forth. If the child deeply engages in a game, mother could suddenly disappear. Therefore, a part of the child's attention must constantly be devoted to the monitoring of the mother's whereabouts. Activated avoidance goals bind attention and are accompanied by anxious tension. The attention that is bound by the avoidance goal then is lacking for the successful coping with positive challenges. Positive emotions, if encountered at all, will be experienced only in the form of relief that the dreaded outcome has not yet occurred. However, this doesn't mean that it couldn't happen in the next moment. A person who has the goals, "Don't stand out unpleasantly," or "Don't do anything wrong" will endure a social event in a state of constant tension; he or she will behave in an inhibited fashion and will be occupied with self-related cognitions. If a misfortune really occurs—he or she might knock over a wine glass, for instance, or make an easily misunderstood comment—this indicates to the person that the dreaded catastrophe has occurred, and he or she will be flooded by negative emotions. A person who enters into the same situation with the goal of having a maximally good conversation will experience something entirely different and will most probably be more satisfied afterwards.

Avoidance goals, therefore, do not permit efficient goal pursuit and real goal attainment. They bind a great deal of energy and attention and, nevertheless, they never yield the satisfying feeling of really having accomplished something. They generally lead to a compromised need satisfaction because, to the degree that mental activity is determined by avoidance goals, it is not free for the pursuit of approach goals. Real need satisfaction can be attained only via the realization of approach goals.

Because of the qualitative differences between approach and avoidance goals, the conclusion arises that avoidance goals are unfavorable goals with regard to the prospects of goal attainment and need satisfaction. A natural tendency to avoid unpleasant events exists, which is common to all humans, and which constitutes an important protective mechanism when real danger actually looms. When avoidance

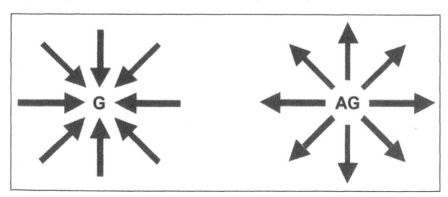

Figure 4–12. Goals and "Anti-Goals" (Adapted from Carver and Scheier, 1998).

becomes an enduring goal, however, that determines how the environment is primarily evaluated and how one interacts with one's surroundings, then one will attain and experience little in terms of positive life events. The pursuit of avoidance goals is accompanied by more frequent negative emotions because the perceptual system is oriented toward the detection of negative events in order to defend against them or avoid them. Moreover, the avoidance activity interferes with the realization of approach goals. People with strongly formed avoidance goals or with a dominance of avoidance over approach goals should therefore experience fewer positive emotions and experience less need satisfaction. As a consequence, it can be expected that their well-being is lower and their mental health generally poorer.

Over the past several years, our team has tested—and clearly supported—these assumptions, using measures that were specifically developed for this purpose. For many years, it has been common practice at the psychotherapeutic outpatient clinic in Bern, Switzerland, to use the methods of plan analysis (Caspar, 1989; Grawe & Caspar, 1984) and schema analysis (Grawe, Grawe-Gerber, Heiniger, Ambühl, & Caspar, 1996) in clinical case conceptualization in order to derive the patient's most important approach and avoidance goals from his or her observable behavior. From the starting point of the individual goals that—according to the therapist's analysis—influence the patient's behavior most strongly, we developed an Inventory of Approach and Avoidance Motivation (IAAM; Grosse Holtforth & Grawe, 2000, 2002b). This questionnaire includes 14 scales for approach goals and 9 scales for avoidance goals to assess the motivational goals that have been shown to be particularly pertinent in the context of psychotherapy. The patient evaluates each approach item (e.g., "to have my freedom," "to be taken care of") in terms of how important it is for him or her. For each avoidance item (e.g., "to embarrass myself," "to be criticized"), patients indicate how negative or awful it would be if what the item says came true. The questionnaire was developed primarily for use in clinical practice, in order to enable therapists to obtain a quick overview of the patient's most important motivational goals. Figure 4–13 shows an IAAM profile for a moderately depressed example patient.

It can be seen easily that the approach goal of an establishing an intimate relationship is particularly important for this patient. At the same time, she is quite fearful of separation and loneliness. The trigger of her depressive episode was that her husband had an extramarital affair. With this constellation of motivational goals, it is immediately apparent that such an event would trigger a maximal state of incongruence in her. Another particularly important approach goal of the patient is recognition and confirmation. Her worst fears and strongest avoidance goals, however, also pertain to this domain of social evaluation. She primarily avoids accusations and criticisms as well as situations related to personal failure. In addition, she strongly avoids tension in interpersonal relationships. This profile clarifies why the patient has entered an acute conflict situation as a consequence of her husband's infidelity. This situation has simultaneously activated her strongest approach and avoidance goals and she experiences strong simultaneous incongruence with regard to several approach and avoidance goals. Based on everything that has been said so far about need satisfaction, motivational goals, and inconsistency, it is clear that this woman should feel very negatively at this point. A high degree of inconsistency characterizes her current mental state, and consistency theory posits that such a state is an acute risk situation for the formation of mental disorders. The background for this pertains to the longer-term structure of her motivation goals, which can be assessed efficiently with the IAAM. The relations between inconstancy and the formation of mental disorders will be addressed in more detail later on (sections 4.8 and 4.9).

In addition to the different scale scores for the various motivational goals, the interpretation of the IAAM also yields sum scores for the total strength of approach and avoidance goals, as well as a value that estimates the relative strength of avoidance versus approach motivation (dominance of avoidance). These sum scores are a good measure for research examining whether a person is primarily avoidance- or approach-oriented in his or her goal-oriented behavior. We correlated these sum scores in large clinical and nonclinical samples with other measures in order to test whether the assumptions described previously could be empirically supported.

A first very important result is that approach and avoidance goals correlate only weakly with one another. This confirms once again the previously noted independence of the avoidance and approach systems. Someone who holds avoidance goals might at the same time have either strongly or weakly developed approach goals.

If someone has many avoidance goals, this has negative consequences in terms of his or her ability to realize his or her approach goals. In addition to the IAAM, we have developed a second measure, the incongruence questionnaire (INC; Grosse Holtforth & Grawe, 2003a; Grosse Holtforth, Grawe, & Tamcan, 2004). In this questionnaire, the patient is asked to rate the same items contained in the IAAM in terms of the extent to which he or she succeeds in attaining each goal. For avoidance goals, this means the respondent indicates the extent to which he or she manages to avoid the respective aversive state. This instrument also permits the computation of sum scores for approach incongruence and avoidance incongruence across all approach and avoidance scales. The degree of avoidance in the IAAM for a sample of 1,021 participants (normal participants as well as clinically disordered patients) correlated highly significantly at a magnitude of .31 with incongruence with respect to approach goals in the INC. For the association between avoidance to approach in the IAAM—the "dominance of avoidance"—this correlation was even higher, at a magnitude of .40. Avoidance goals, therefore, have a clearly negative impact on a person's ability to attain approach goals. Avoidance goals inhibit successful approach activity.

Because of this inhibitory effect on the attainment of approach goals, avoidance goals should be accompanied by reduced well-being. This assumption was also empirically supported. The empirical correlation between dominance of avoidance in the IAAM with various well-being measures is highly significant and in expected directions. In a mixed clinical sample of patients from various institutions, dominance of avoidance correlated at −.38 with the positive life orientation subscale of the Bern

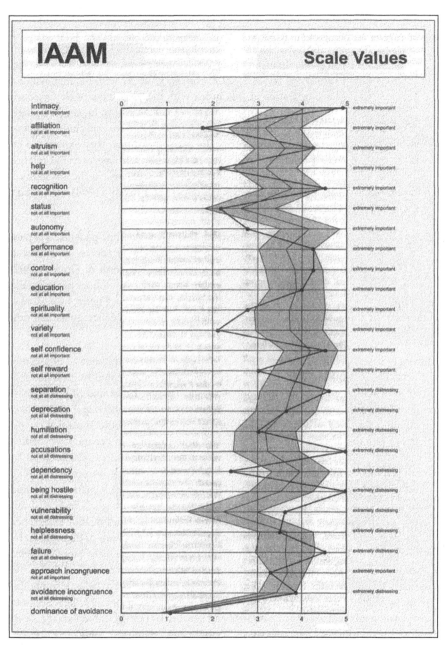

Figure 4–13. IAAM-profile. Thin line in grey area = mean of a sample of psychotherapy patients; grey area = standard deviation above and below the mean. Thick line = patient profile on this background (adapted from Grosse Holtforth & Grawe, 2002).

Well-Being Inventory (Grob, 1995), at −.37 with the self-esteem subscale, and at .26 with the depression subscale. All these correlations were highly significant with a sample of 304 participants. In a sample of 70 psychotherapy patients, the sum scale of total symptom severity of the SCL-90-R correlated at .37 with the dominance of avoidance index. The correlation of this index with the total score of the insecurity questionnaire by Ullrich de Muynck and Ullrich (1977) was even higher, at a magnitude of .55. Individuals who have primarily avoidance goals are especially characterized by a high degree of social insecurity. In social situations, they experience primarily negative feelings and have pronounced difficulties in interpersonal relationships. The sum scale for the extent of interpersonal problems in the Inventory of Interpersonal Problems (IIP) by Horowitz, Strauss, and Kordy (1994) correlated at .43 with dominance of avoidance.

Given all of these findings, it is not surprising that psychotherapy patients, compared to a normal comparison sample, attain significantly higher scores on almost all IAAM avoidance scales (for more detail, see Grosse Holtforth & Grawe, 2002b, which contains all the correlations mentioned previously, as well as more detailed descriptions of the samples and other methodological details). Figure 4–14 shows a summary of the correlations between the dominance of avoidance goals and other variables.

Strongly developed avoidance tendencies, therefore, have a multitude of unfavorable effects. The list of unfavorable effects can be further augmented by findings from studies in which goals were operationalized differently. Elliot and Sheldon (1998) examined in three studies the associations among "personal strivings" and physical symptoms. Personal strivings are construed as individual specific goals that the participants aim to pursue within the next month (one study) or in general (two studies). Participants who had freely listed a relatively larger number of avoidance goals had more physical symptoms such as headaches, coughs, sneezing, shortness of breath, chest or heart pains, stomach pains, and feelings of dizziness or weakness. In a study with a retrospective design, participants with more avoidance goals indicated relatively less progress toward the attainment of their goals. In addition, they experienced a low degree of autonomy and personal control during the pursuit of their goals. In a study by Elliot, Sheldon, and Church (1997), participants who listed more avoidance goals also reported less optimal well-being and less improvement in well-being over the past several months, compared to participants with fewer avoidance goals. In a study by Coats, Janoff-Bulman, and Alpert (1996), participants who endorsed more avoidance goals in the sense of Emmons and McAdams (1991), exhibited significantly lower scores on scales measuring self-esteem and optimism but higher scores on a depression scale.

Michalak, Püschel, Joormann, and Schulte (2006) used two different methods to measure approach and avoidance goals in a group of 53 normal participants and 65 psychotherapy patients. One method they used was a semiprojective test that was designed to measure "implicit motives." These corresponded largely to that which consistency theory means by motivational goals. The other method was a questionnaire in which participants were asked directly about their most important personal strivings. Participants were also asked to evaluate the extent to which the strivings were oriented primarily toward approach or avoidance. In contrast to the indirectly assessed implicit goals, these directly assessed goals were construed as explicit goals. In line with the consistency-theoretical model, the study was based on the assumption that different goals could be pursued in the implicit versus explicit modes of functioning. Indeed, explicit and implicit motives were only very weakly correlated in this study. The authors found very substantial correlations for both types of avoidance goals—but not for the approach goals—with various measures of psychopathologic symptom severity. The correlations with the global severity index of the SCL-90-R were around .40. The magnitude of the correlation between implicit avoidance motives and the total score of interpersonal problems on the IIP was even higher, at a magnitude of .65. This study suggests, then, that explicit as well as implicit avoidance exerts a negative influence upon mental health.

These findings show that there is strong empirical support for the idea that a habitual avoidance orientation of mental functioning is associated with decrements in well-being, lower self-esteem, and worse overall mental health. However, it should be noted that these are correlational associations. Even though it seems obvious to interpret these correlations to indicate that strong avoidance leads to these negative consequences, it seems equally plausible to conclude from these findings that someone who is in a negative state will tend to avoid more and approach less. In the study by Elliot and Sheldon (1998), participants who reported more avoidance goals also had higher neuroticism levels. The total score of avoidance goals on the IAAM also correlates

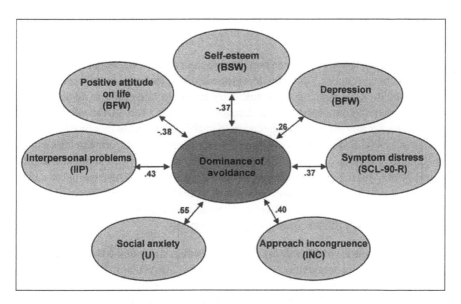

Figure 4–14. Correlations between dominance of avoidance and various measures of well-being and mental health. Detailed descriptions in text.

highly significantly with neuroticism (Grosse Holtforth & Grawe, 2002b). We know from other studies that neuroticism is part of an inherited tendency to use avoidance (see previous for detail). This suggests that it might be plausible to interpret the correlational associations in such a way that individuals with high neuroticism more easily form avoidance goals. The formation of avoidance goals, in turn, has additional unfavorable consequences.

When we note a high degree of avoidance goals in a patient in clinical practice, we have reason to assume—based on these findings—that the avoidance goals contribute to his or her problems and negative state. A reduction of the strength of avoidance goals should therefore be one of the most important therapy goals. This also holds true when we keep in mind that an easily activated avoidance system might be a stable characteristic of the human nervous system, which cannot be changed as easily as individual avoidance goals. If we want to achieve that the inconsistency level among these patients decreases and their need satisfaction increases, we will have to reduce the influence of the most dominant avoidance goals so that their inhibitory influence on the attainment of approach goals is also reduced.

In terms of the importance of avoidance goals in the context of psychotherapy, the perspective developed here is consistent with that articulated by S. C. Hayes, Wilson, Gifford, Follette, & Strosahl (1996), who proposed that the avoidance of particular kinds of experience ("experiential avoidance") can be regarded as a unique diagnostic dimension that is useful for psychotherapy. They defined experiential avoidance as

> the phenomenon that occurs when a person is unwilling to remain in contact with particular private experiences (e.g., bodily sensations, emotions, thoughts, memories, behavioral predispositions) and takes steps to alter the form or frequency of these events and the contexts that occasion them. (p. 1154)

S. C. Hayes et al. proposed that measures for the routine assessment of avoidance goals in clinical practice should be developed so that the avoidance of certain experiences could be regarded as a diagnostically relevant dimension in addition to the traditionally symptom-focused DSM diagnostic system. The IAAM was developed specifically for this purpose. It is intended to enable clinicians to measure the avoidance goals of individual patients in an efficient manner so that one can check at the beginning of therapy whether a reduction in the avoidance of certain experiences ought to be considered an independent therapy goal. The IAAM also allows clinicians to check the specific domains in which a patient avoids particularly strongly. A logical requirement for this therapeutic rationale is, of course, that avoidance goals can be changed at all in the process of psychotherapy.

In order to examine whether this requirement is met, we examined in a sample of 78 patients in our outpatient clinic in Bern the extent to which our therapists were able to weaken the avoidance goals of the patients (Grosse Holtforth, Grawe, Egger, & Berking, 2005). The patients were treated based on the integrative treatment guidelines described in the book *Psychological Therapy* (Grawe, 2004). The therapists utilized disorder-specific as well as coping-oriented and clarification-oriented interventions and particularly attended to a carefully designed relationship formation, tailored individually, based on the patient's motivational schemas. The therapies lasted an average of 26 sessions.

At the beginning of the therapies, the IAAM scores showed—as expected—that avoidance goals were highly significantly stronger than those reported by normal participants. At the end of treatment, these scores were no longer significantly different from the normal comparison sample. The changes were significant for 8 out of 9 IAAM avoidance scales. The only nonsignificant change was observed for the worry about autonomy loss subscale. The reduction of avoidance was also highly significant when the total scale was analyzed. In addition, we computed a patient-specific sum score for his or her three most pronounced avoidance goals and tested separately whether these would change over the course of therapy. The pre–post magnitude for the reduction of the individually most important avoidance goal was estimated at ES = 1.06. Given that motivational goals are much more stable characteristics than symptom measures, this can be regarded as a rather strong effect. The reduction of avoidance goals correlated at .49 with the initial strength of the avoidance goals, indicating that greater changes were observed among patients who had higher levels of initial avoidance.

These findings strongly support the proposal by S. C. Hayes et al. (1996) that therapists ought to clarify prior to the beginning of therapy the nature and strength of the avoidance goals of their patients, and that they should then try to specifically attain changes in the most important avoidance domains. The prospects for achieving such effects, based on the results described previously, are relatively good. In the study by Grosse Holtforth et al. (2005), the reduction of avoidance goals also correlated highly significantly with the attainment of patients' individual therapy goals, with a reduction in interpersonal problems, and with an increase in optimism and contentment. All this also supports the idea that a weakening of avoidance goals could constitute a distinctive goal for psychotherapy.

A reduction of avoidance goals via psychotherapeutic means is achieved not only in our own institution, in which therapists are specifically trained to achieve this. Berking, Grosse Holtforth, and Jacobi (2003a) administered the IAAM, along with other measures, in a cognitive–behaviorally oriented psychosomatic clinic, both at the beginning and at the end of an inpatient therapy. They found a roughly equivalent reduction in the strength of avoidance goals, and in this study as well the reduction of avoidance goals correlated significantly with other therapeutic changes.

In our own study we also found several additional notable associations between the reduction of avoidance goals and other therapeutic changes. That is, we calculated the respective correlations separately for various diagnostic groups and observed a highly interesting pattern.

For depressive patients, the correlations between the reduction of avoidance and other therapeutic changes was very high, whereas among anxiety patients, this correlation was not even statistically significant. Table 4–5 compared the correlations between these two diagnostic groups.

Among depressive patients, particularly high associations were found between the reduction of the total score for avoidance goals and the reduction of interpersonal problems. For them, moreover, the associations with the reduction of symptom severity and with increases in contentment and optimism are also significant. The absence of corresponding correlations among patients with anxiety disorders is remarkable because, on average, they experienced equivalent reductions in the strength of avoidance goals, compared to the depressed patients. The difference pertains to the functional relations. Whereas it is possible among anxiety patients to reduce avoidance goals without simultaneously affecting changes in symptom severity or other aspects, one can observe that, among depressives, a host of other positive changes follows when one manages to reduce their avoidance goals. For the depressed patients, then, the avoidance appears to be of greater functional significance, compared to the anxiety patients. Among anxiety patients, the disorder appears to have gained functional autonomy. It exists apart from the motivational system. One could interpret this to mean that the anxiety has become independent and is now decoupled from other areas of motivational functioning.

Consistency theory assumes that the motivational system plays an important role in the genesis of symptoms. After anxiety symptoms have been facilitated, however, their neural underpinnings—at least among the anxiety patients in our sample—become a well-facilitated part of implicit memory, and these memories can be activated even if the motivational conditions that led to their original emergence are no longer in place. This suggests that a disorder-specific approach may be needed in order to specifically inhibit the respective neural activation patterns.

Among depressive patients, by contrast, the avoidance of violations or goal-related insults appears to play an important role in mental functioning even during the course of therapy. This avoidance is an important aspect of the depressive problem constellation. The changes that have been found in the brains of depressed patients can best be interpreted in such a way that depression is ultimately due to a hyperactivation of the avoidance system. This avoidance is the origin of other neural changes. It inhibits the approach system, and the hyperactivation of the avoidance system is linked with the continuous release of stress hormones, which damages the hippocampus. The increasingly extreme avoidance tendencies finally culminate in a generalized protective or defensive stance, in which transactions with the demands posed by the environment are no longer pursued. This is accompanied by a notable deactivation of the anterior cingulate cortex (ACC).

If this perspective is correct, one might assume that a weakening of avoidance leads to a gradual reactivation of the approach system and to a revitalization of the ACC. Similarly, the negative emotions should decrease in intensity. The weakening should

Table 4–5. Correlations Between the IAAM Avoidance Scales and Other Therapy Success Measures, Separately for Anxious and Depressed Patients

	Pre—post difference in...						
	Psychopath Symptoms (BSI)		Interpersonal Problems (IIP-D)		More Optimism (VEV)		Attainment of Therapy Goals (GAS)
Avoidance motivation	.47*	.21	.64***	.02	.40*	.39*	.37* .14
Vulnerability/ loss of control	.46*	.32	.56**	.07	.54**	.28	.24 −.02
"Showing my weakness to others"	.35	.10	.78***	.14	.54**	.13	.39 .04
"Showing my own needs or desires"	.27	.23	.56**	.16	.51*	.39	.19 .32
"Being overhelmed by my emotions"	.32	.03	.11	.09	.15	.14	.15 .17

Note. IAAM = Inventory of Approach and Avoidance Motivation; BSI = Brief Symptom Inventory; III-D = Inventory of Interpersonal Problems—German version; VEV = Questionnaire to Access Changes in Experiencing and Behavior (QCEB; German VEV); GAS = Goal Attainment Scaling. From "Reducing the Dreaded: Change of Avoidance Motivation in Psychotheraphy," by M. Grosse Holtforth, K. Grawe, O. Egger, and M. Berking, 2005, Psychotherapy Research, 15, pp. 261–271. Copyright 2005 by Routledge. Adapted with permission.
*p < .05
**p < .01
***p. < .001

therefore be accompanied by positive changes in experience and behavior, in the pattern that was shown in the observed correlations.

The conceptualization that depression is a generalized protective or defensive reaction to avoid mental injuries or need-related violations is also supported by a remarkable detail within our findings. We noticed that the strongest improvements of all were found for the IAAM fear of vulnerability subscale. For this scale, the initial score also correlated highly with the change score, at a magnitude of .68. This prompted us to examine in more detail how this fear of vulnerability could be understood. Therefore, we analyzed this scale in more detail by inspecting correlations with single scale items. We found an extremely high correlation of .78 between changes in other domains and the item "Showing my weaknesses to others." To be able to show oneself to others as weak and vulnerable is apparently the critical point for many depressed individuals. This fear is, of course, particularly relevant in interpersonal contexts. It makes sense, then, that problems in interpersonal relationships improve markedly as the fear of showing one's weaknesses and thereby become vulnerable is reduced.

This pattern of findings suggests that, when dealing with depressive patients, it is particularly important to create a space in which they feel safe and protected from renewed injuries or threats. This points to the special importance of relationship formation processes. An additional primary therapy goal should be to render the patient less vulnerable to insults or injuries, so that he or she can relinquish the protective or defensive stance permanently.

Apart from the special significance of avoidance for our understanding of depression, additional implications can be derived from the functional independence of the approach and avoidance system, which was the main focus of this section. These implications are of general importance for psychotherapy. Approach goals can be strengthened and their attainment facilitated even among people who have a deeply rooted tendency to avoid. As we will see later on, improvements in approach goal attainment are very closely linked with other positive therapeutic changes—even more so than a reduction or better attainment of avoidance goals. In psychotherapy, it is of utmost importance to enable the patient to realize his or her approach goals more effectively. Because of the independence of the approach and avoidance systems, this goal can be attained in therapy even when the patient has a high level of avoidance goals.

4.7.6   Neural Mechanisms of Approach and Avoidance Learning

How do approach goals acquire their positive incentive value and avoidance goals their threatening or aversive value? How does the process from the incentive or threat to behavioral action transpire? What triggers approach and avoidance behavior and which processes inhibit such behavior? How do humans learn these processes? What is the role of reward and punishment in this context?

All these are central questions for psychology but, of course, also for psychotherapy, because the questions essentially address how we should understand the nature of motivation and learning. This section focuses on the neural mechanisms that ultimately lead to enduring changes in experience and behavior. Which types of answers to these questions are we enabled to uncover with the help of neuroscience?

The assertions made in this section are supported primarily by the works of Berridge (1999), Berridge and Robinson (1998), Hoebel, Rada, Mark, and Pothos (1999), Rolls (2000), and Shizgal (1999).

A cocaine-dependent person places his head into PET scanner and imagines that he smokes a crack pipe with his lover. Within him, a strong longing wells up for the feeling of being high. At this moment, clear activations are notable primarily in two areas: The amygdala and the nucleus accumbens (Grant et al., 1996; Volkow, Wang, & Fowler, 1997). We have previously encountered the amygdala primarily as the brain's anxiety center but we also noted several times that it also plays a role in general emotional evaluations. Brain imaging studies show that the amygdala—or, more precisely, other parts of the amygdala than those activated during anxiety states—is also activated while pleasant memories and perceptions are activated. The nucleus accumbens lies in close proximity to the amygdala. It is critically involved in the learning of behavior that leads to pleasant states and reduces fear. Therefore, it plays an important role in the processes that in behavioral learning theories are termed *positive* and *negative reinforcement*.

Animal studies have revealed quite a bit about the role of pleasure and distress feelings in the learning process. Most of these studies have been conducted with monkeys because it is necessary to implant electrodes and sensors into tightly circumscribed areas in order to study the microprocesses that transpire in these locations during the learning process. Most of these studies examine what happens in these areas in the context of eating, drinking, sexuality, or drug consumption and withdrawal. It is a clear advantage of consumption-related behavioral domains that the incentive value of stimuli can be manipulated directly and, therefore, that one can be certain that the stimulus is pleasurable or aversive for the animal in each respective situation.

Rolls (1995) studied the flow of electrical activation within single neurons in different brain areas among monkeys that were in the process of tasting sweet fruit juice. He observed an elevated rate of neural firing in the amygdala. The amygdala neurons fired even when the monkeys saw only the juice container. Firing was also observed in the neurons in brain regions that are involved in the recognition and differentiation of objects such as fruits. Particularly interesting, however, was the activity of neurons in the PFC. Some of the neurons in this area also responded while the monkeys tasted the juice, but their activity depended on the internal state of the monkeys—how hungry they were or how long ago it had been since they last tasted the juice. The PFC, then, is informed about each particular need state (i.e., the metabolic state), and this information comes from the hypothalamus, which monitors the blood sugar level, fat reserves, availability of proteins, and body temperature, and thereby generates appetite for specific nutrients. If no nutritional need is present, the juice elicits less activation in the PFC, with the result that the monkey does not reach for the juice. The decision of whether the monkey reaches for the juice or not is, therefore, at least partially made in the PFC. The neural signal that represents this decision is transmitted from the PFC to the nucleus accumbens and there triggers a process that can elicit and strengthen behavior.

The nucleus accumbens integrates incoming sensory signals, evaluations of the need state, evaluations from emotional memory (amygdala), and contents of the location memory (hippocampus) and then activates or terminates a behavioral response. It makes "go" or "stop" decisions (Hoebel et al., 1999). This process involves neurons that release the neurotransmitter glutamate.

In addition to this function of initiating and terminating behavior, the nucleus accumbens is critically involved in reinforcement processes. If the result of a behavior is evaluated positively, the synapses that were or are involved in the recently exhibited behavior are selectively reinforced. This strengthening of synapses is mediated by the neurotransmitter dopamine.

Hoebel et al. (1999) placed small sensors into the nucleus accumbens of animals in order to measure the dopamine concentration in those areas via microdialysis. They were able to document with that method that dopamine is being released when an animal eats. The dopamine release had already been observed in response to a cue that signaled the future presentation of food.

The process always transpires exactly like this: Initially the nucleus accumbens receives various input signals that lead to a go decision for a particular behavior, such as eating. If the eating is evaluated positively—that is, when the animal is hungry and then receives something that it likes to eat—dopamine is released. The dopamine binds to receptors of synapses that are involved in the perception of eating and in the motor actions constituting the eating behavior. Second messenger processes are thereby facilitated in the involved neurons. As discussed in section 2.3, this leads to an elevated synaptic transmission . If the neurons participating in this process are once again activated, the transmission among them transpires more easily because the synapses between them have by now been better facilitated. The behavior recurs with greater likelihood in the future. The process of reinforcement, then, is governed by facilitation of the neurons involved in the behavior. This facilitation is made easier when neurotransmitters are being released that lead to second messenger processes in the participating neurons. In this case, this neurotransmitter is dopamine.

Beyond eating behavior, sexual behavior also leads to a release of dopamine in the nucleus accumbens (and also in the striatum). It is not orgasm that is associated with the highest dopamine levels, but foreplay (Damsma, Wenkstern, Pfaus, Phillips, & Fibiger, 1992; Pfaus, Damsma, Wenkstern, & Fibiger, 1995).

Studies in which animals can inject themselves with dopamine into the nucleus accumbens by pushing a lever demonstrate clearly that it is really the dopamine release that leads to the reinforcement of the synapses involved in creating the instrumental behavior. The animals pushed the lever again and again, showing that it was really the dopamine injection that produced the reinforcing effect (Guerin, Goeders, Doworkin, & Smith, 1984; Hoebel et al., 1983). The process of reinforcement, then, can also occur without a natural reinforcer such as tasty food or sex. What is important for the reinforcement is not the nutrient but the dopamine release. Pleasurable stimuli such as tasty food, drinks, sex, and so forth are triggers for the release of dopamine, which explains where their reinforcing effect stems from.

The process of positive reinforcement can be differentiated from that of negative reinforcement, during which the behavior that preceded a reduction in pain or other aversive feelings is reinforced. Is this type of reinforcement also mediated by dopamine?

Rada, Mark, and Hoebel (1998) stimulated in rats a part of the hypothalamus whose activation is highly aversive. Using microdialysis, they measured dopamine levels in the nucleus accumbens and observed minor reductions in this level as a consequence of the aversive stimulation. However, when the rats were given the opportunity to reduce the aversive stimulation for five seconds by pushing a lever, the dopamine level rose once again dramatically as the lever pushing increased. The facilitation of escape and avoidance behavior, then, also appears to transpire via the dopamine route.

In other studies, in which the aversive stimulation did not originate directly in the brain but via "natural" pathways—for example, by using electric shocks or by restricting the animal's ability to move freely—these aversive situations have tended to trigger a massive dopamine release. Even normally harmless stimuli that had been paired with aversive stimuli led to dopamine releases. It is possible, then, to condition dopamine release. If the animal is given the opportunity to escape from the aversive situation, the dopamine release is even stronger. The dopamine available in the extracellular space binds to receptors of simultaneously activated synapses and thereby contributes to the facilitation of synaptic transmission . When positive stimuli are presented and approach goals are thereby activated, the approach behavior that follows this presentation is facilitated by these processes (i.e., positively reinforced); when an aversive stimulus is presented and avoidance goals are thereby activated, the avoidance or escape behavior is facilitated (i.e., negatively reinforced). If the binding of the released dopamine to dopamine receptors is pharmacologically blocked (e.g., by using Haloperidol), the avoidance learning is compromised. This means that dopamine plays an important role in both types of reinforcement learning—positive reinforcement of approach behavior and negative reinforcement of avoidance behavior (Berridge & Robinson, 1998).

By studying the activation of single neurons in specific brain regions, researchers are enabled to examine in detail what triggers the firing of single neurons. The dopamine neurons in the nucleus accumbens do not just fire when the animal already has food in its mouth. They also fire when the monkey perceives signals that food might be forthcoming (Schultz, Dayan, & Montague, 1997). Therefore, it is not just the case that instrumental behavior is being reinforced when dopamine is released. Also, the simultaneously activated neural groupings, such as the associative connections between anticipatory signals and approach behavior, are more effectively bound together into cell assemblies. In this way, previously neutral stimuli are infused with meaning with regard to positive and negative goals and acquire the ability to elicit approach and avoidance behavior.

The dopamine reinforcement system can also be activated via direct electrical stimulation of relevant brain regions. Since Olds and Milner (1954) demonstrated this effect for the first time, the phenomenon of "brain stimulation reward" has been studied in all types of animals, from goldfish to human beings (Shizgal, 1999). When electrodes are implanted into the lateral hypothalamus of rats—into the part that governs the control of appetite for food—and the rats learn that the pushing of a lever triggers self-stimulation for half a second, they will self-stimulate for up to 3,000 times per hour.

The same effect is observed when the electrode is implanted into the part of the hypothalamus that governs sexual appetite. Originally, it was believed that this indicated the discovery of "pleasure centers" within the brain. We now know that the processes are far more complicated.

If the relevant hypothalamus regions are stimulated for an entire minute instead of just half a second, the animal will immediately begin to eat or to copulate when food or, respectively, an appropriate other animal is presented. However, when the stimulation lasts for only half a second, the animal will disregard the food and the sex and do whatever it has learned to continue the self-stimulation. Of importance, this effect of self-stimulation is also influenced by dopamine. The electrical stimulation in the lateral hypothalamus triggers a massive dopamine release in the nucleus accumbens, which leads to a very potent reinforcement of the self-stimulation.

If the animal has previously been fed and the electrode is implanted into the area that controls appetite for food, the frequency of self-stimulation is notably lower. If the electrode is implanted into the sexual area, the self-stimulation reduces in frequency if the animal has just previously ejaculated. The cause for this is linked to another neurotransmitter, acetylcholine, which is, as it were, an opponent of dopamine in the process of reinforcement. When one is hungry, a normal meal begins with a great deal of appetite. With time, this appetite reduces, the eating slows down and is finally terminated. Microdialysis has shown that the level of acetylcholine increases in the nucleus accumbens during this final eating phase (Mark, Rada, Pothos, & Hoebel, 1992). If the acetylcholine level was artificially raised in earlier eating phases, the eating was terminated prematurely. This inhibitory effect is observed for moderate levels of acetylcholine. If a much higher amount of acetylcholine is released into the nucleus accumbens during the eating process, one observes the formation of an enduring taste aversion rather than just normal satiation. This shows, then, that acetylcholine plays an important role in the termination and inhibition of motivated behavior.

As far as eating is concerned, the balance between dopamine and acetylcholine—and thereby, the control over eating behavior—is governed by the appetite center in the lateral hypothalamus. The eating behavior is initiated and terminated via the release of neuropeptides and neurotransmitters in the hypothalamus. These substances also govern the appetite for very specific types of nutrients. Appetite for carbohydrates is stimulated via the neuropeptide Y along with the neurotransmitter norepinephrine and other substances that are beneficial for the processing of carbohydrates. Another peptide, galanine, produces in combination with norepinephrine appetite for preferred, especially fat-saturated nutrients. The ingestion of protein is governed by a peptide that facilitates growth (Leibowitz & Hoebel, 1998). If galanine and norepinephrine are injected into the hypothalamus, this leads not only to eating but also to a predictable change in dopamine and acetylcholine levels in the nucleus accumbens (Hoebel, 1997). Dopamine is increased and acetylcholine is reduced. The reduction in acetylcholine probably contributes to the disinhibition of dopamine-mediated approach behavior.

I have discussed these relationships in somewhat greater detail to show with this example how the specific regulatory mechanisms govern the various forms of approach behavior. These relations hold true for eating behavior. For other domains of behavior, similarly specific regulatory mechanisms probably exist, but not all of them are yet fully understood. In section 4.4.2 on the neurobiology of the attachment need, I reported some of the studies by Hofer (1984) about very specific feedback loops in which certain forms of maternal behavior regulate specific psychophysiological parameters in the infant. The availability of the mother's milk regulates the child's heart rate; the intervals by which she feeds the child regulate his sleep–wake rhythm; the degree of tactile stimulation regulates the secretion of growth hormones; and so forth. The relationships that we study as psychologists are almost always on a much more global level of analysis, and this also holds true for attachment and approach behavior. The examples from the studies by Hoebel and Hofer show that the relationships on the molar level, which is the level on which our questions are typically asked, are probably underpinned by much more specific relationships on the neurophysiological level. Many of these relationships may not be of direct importance for psychotherapy, but in some areas the research into the specific neural mechanisms that underlie observable behavior could one day also be of great practical relevance for psychotherapy. This could be true, for instance, for disorders in which approach behavior has gone awry or for the different types of addictive disorders.

Based on the findings reviewed here, dopamine plays an important role in motivated behavior and in reinforcement learning. The normal triggers for dopamine are things that satisfy the need for pleasure maximization, such as eating, drinking, sex, and so forth. These are known as primary reinforcers. However, we have already seen that cues, situations, or activities that are associated with these primary reinforcers can also lead to dopamine releases. Even pleasant mental imagery can activate dopamine neurons, as we have seen earlier in the example of the cocaine-dependent person in the PET scanner. The direct electrical stimulation in the lateral hypothalamus and the ingestion of drugs, however, are much more potent triggers of dopamine releases (see following for more detail).

Regular eating is an important means of maintaining the dopamine level at its normal set-point. When people fast and lose considerable weight, the dopamine level can sink to almost half its normal level (Pothos, Creese, & Hoebel, 1995). If we consider how important dopamine is for motivated behavior and reinforcement, a reduced dopamine level must be regarded as very unfavorable for the individual's successful engagement with the environment. Thus, it makes sense that there appears to be an innate regulatory mechanism that is oriented toward increasing the dopamine level if it is too low. When animals or humans starve to the degree that they lose substantial weight, the probability for experiencing binge-eating attacks or for seeking drugs is increased. Rats with electrodes in the hypothalamus begin to self-stimulate more vigorously. All of these are alternative means in order to increase the dopamine level. Because they are alternative methods to attain dopamine increases, they can also be substituted for one another. When someone diets and loses weight, he or she will often begin to smoke more (nicotine increases the dopamine level). Conversely, people who stop smoking often gain weight because they attempt to regain the lacking dopamine by eating more.

One of the most potent and quickest means of increasing the dopamine level is the ingestion of drugs. Almost all drugs lead to dopamine releases. The psychostimulants amphetamine and cocaine and the opiates heroin and morphine act with particular intensity because they are associated with dopamine release in various brain regions, not just in the nucleus accumbens. Nicotine achieves its effects directly on the cellular level as well as at the terminals of the dopamine neurons and, therefore, leads to a doubled dopamine dosage. Marijuana and hashish directly activate the opiate system, which in turn activates the dopamine system. Even alcohol and caffeine lead to an increase of dopamine in the nucleus accumbens.

Beyond this direct effect on the dopamine system, many drugs also lead to an intensification of the hedonistic value of food and sex and generally increase approach behavior. Therefore, they increase one's appetite and increase one's enjoyment during eating and during sex.

The findings reviewed so far about the role of dopamine in approach and avoidance learning are compatible with the assumption that humans—and not only humans— have a basic need for pleasure maximization and distress avoidance, and that this constitutes a major source of motivation. Approach behavior is oriented toward objects or goals that elicit positive feelings; things that elicit negative feelings are avoided. This seems plausible and is largely consistent with our subjective experience.

In reality, however, these processes are more complicated than the discussion so far might suggest. According to Berridge (1999; Berridge & Robinson, 1998), two neural systems underlie hedonistic evaluation and motivation. He differentiates these systems as "liking" and "wanting." In most natural situations, liking and wanting are of course closely tied, but it is possible to show experimentally that these are independent processes with distinctive neural foundations, similar to the earlier discussion in which we saw that positive and negative evaluation or, respectively, approach and avoidance behavior, have separable neural foundations.

If someone repeatedly chooses a certain dish when given the opportunity, we conclude from this behavior that he or she likes this particular type of food. In general, he or she will also confirm this verbally and we have little reason to concern ourselves further with this because actual behavior and subjective report are concordant. However, this is not always the case.

Fischman (1989; Fischman & Foltin, 1992) was able to show this in a study with drug addicts. The participants were able to push one of two buttons in order to receive two types of intravenous injections. The two injections differed in terms of the dosage of cocaine they contained. On one day, one infusion could contain a strong dose and the other a weak dose of cocaine; on another day, one infusion contained only a weak dose and the other contained none at all. The participants were able to try out the two infusions and were then free to choose which one they would inject. In addition, they always indicated how much cocaine might be included in the infusion, based on their own impressions, and how pleasurable the effect of the injection was for them.

When medium to high cocaine dosages were used, the participants reliably selected the higher dose and described the effect as pleasurable. But when a very small dose was used, their behavior and the subjective evaluation of this effect were remarkably discrepant. They were convinced that they had received only a salt solution and were not able to detect an effect. They also did not show any cardiovascular reactions to the injection, which was consistent with their subjective experience. Even though they were convinced that both solutions on that day did not contain any cocaine at all, they were significantly more likely to inject on that day with the solution that contained a small amount of cocaine. The more frequent pushing of the "cocaine button" shows that the cocaine solution had an "objectively" higher incentive value even on these days. It motivated differential behavior even though no difference could be detected in subjective evaluation. Wanting and liking, then, do not always concur and also may not always transpire with conscious awareness.

In a similar study with heroin addicts, Lamb et al. (1991) injected their participants every week from Monday to Friday with a solution that either contained morphium or a salt solution. The dosage did not vary within weeks but could vary between weeks. The participants received the injections "for free" from Monday through Wednesday. However, on Thursday and Friday there were asked to do something in order to obtain the injections. They were asked to push a lever at least 3,000 times in 45 minutes. Whenever the solution contained morphine, the participants worked hard to receive the injection and evaluated the injection positively. However, when the lowest morphine dosages were used, the subjective evaluations and actual behavior were discrepant. Even though the participants said that the solutions were useless and contained only salt solutions, they worked just as hard as they had for higher dosages in order to receive the injections, whereas they did not exert such effort when working for injections that actually contained only salt solution.

Among humans, wanting and liking can be differentiated rather easily because one simply has to ask participants about the hedonic evaluation—how pleasant or unpleasant the presented stimuli appear to them. In animal research, this operationalization is somewhat more difficult. Normally, one simply assumes that animals will like something if they exert effort to attain it. Liking is therefore viewed as equivalent to the definition of wanting.

In order to investigate whether liking and wanting are truly one and the same thing, Berridge and colleagues developed a procedure to operationalize liking independently of wanting. They filmed the expressive behavior of rats when they were given something very tasty (e.g., sweet foods) or something aversive (e.g., bitter foods). Using slow-motion analysis, the researchers were able to reliably differentiate, based on this expressive behavior, when the rats had received a pleasant or aversive stimulus. These criteria were then used in experimental studies to define liking and wanting.

Berridge and Robinson (1998) assumed that the motivation to want something corresponds to different neural substrates than those corresponding to hedonistic evaluations (liking). They argued that the neural substrate for wanting would be primarily the dopamine system with the nucleus accumbens at its core, which we already encountered in the previous sections. This also includes certain nuclei within the amygdala. Figure 4–15 shows this neural substrate schematically.

The hedonistic evaluation of foods, by contrast, was thought to involve different neurotransmitters and brain regions (see Figure 4–16). One of these regions is the ventral pallidum, a region that lies directly adjacent to the lateral hypothalamus. If this region is destroyed, sweet food that previously was very much liked by the animal suddenly elicits a strong aversive reaction, as if the animal had encountered very bitter food. The ventral pallidum is, therefore, very important for the positive evaluation of food.

The hedonic intensity of sweet food can be increased in rats when microsensors are used to release specific chemical substances into certain brain regions. A part of the nucleus accumbens responds strongly to opiates with such a reinforcement of the hedonic evaluation. There is, however, at least one additional region in the brainstem that can be stimulated with benzodiazepines or gamma amino butric acid (GABA) in order to intensify the liking of foods. This evaluation still functions even among animals without a cortex. The good–bad evaluation seems to be designed as a multilayered system. Normally the various systems work together in a hierarchical fashion. If one level fails to function, however, the evaluative function remains intact to some degree. If a high level malfunctions, such that communication with the PFC is no longer possible, these evaluations are no longer as flexible and certainly also not associated with conscious awareness. Nevertheless, rudimentary evaluations can still be made. The brain rarely relies on only one neural system for important functions. However, because the neural circuits involved in a specific function normally work together, it is difficult to realize that they are independent neural circuits with different neural substrates. The fact that the systems work independently of each other is only apparent when one of the circuits is chemically disabled or when its neural substrate is destroyed. This is exactly how Berridge and Robinson (1998) arrived at their conclusion that two separate systems exist for liking and wanting.

The dopamine system can be disabled in isolation by injecting the neurotoxin 6-hydroxidopamin directly into only the dopamine neurons while the animal is anesthetized. Other neurons remain intact in this procedure. Animals that have been robbed of their dopamine neurons in this way lose all interest in eating, drinking, and everything else that previously had incentive value for them. They starve until they die, even if they are surrounded by mountains of tasty food, as long as one does not feed them artificially. The animals whose dopamine systems were disabled in Berridge and Robinson (1998) were no longer motivated for anything. They appeared as if they had lost their enjoyment of doing anything, as if they were completely anhedonic. If sweet food or liquid was delivered to their mouths without the animals having to take any action for this, their expressive behavior showed normal, positive hedonic reactions.

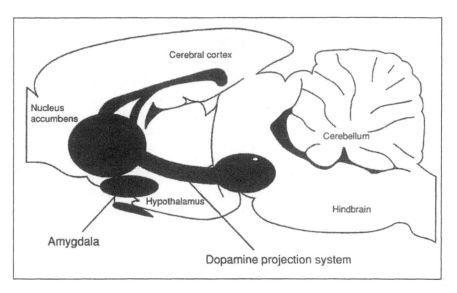

Figure 4–15. The neural substrate for the motivation to want something, based on Berridge (1999) (Adapted from Kahnemann, Diener, & Schwarz, 1999; p.539).

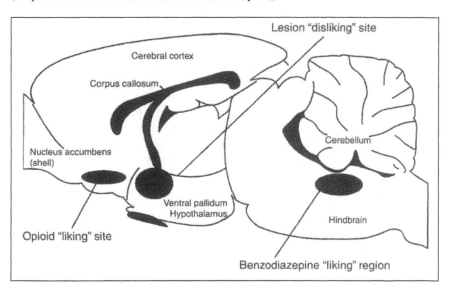

Figure 4–16. The neural substrate for hedonistic evaluations (liking), based on Berridge (1999) (Adapted from Kahnemann, Diener, & Schwarz, 1999; p.539).

When bitter food was delivered, they reacted with aversive reactions The hedonic evaluation could even be classically conditioned. When another sweet substance was delivered to an animal's mouth, it initially responded with a positive hedonic reaction. If nausea was repeatedly elicited just afterwards by injecting lithium chloride, the animal subsequently showed an aversive reaction to the sweet food. Evaluation and expectancy learning, then, function normally even without the dopamine neurons. It is only the motivation—the wanting—that breaks down entirely.

The same is observed, albeit in weakened form, when dopamine receptors are blocked pharmacologically, for instance with haloperidol. This suppresses the incentive value of things for which the animal was previously highly motivated, be it food, cocaine, or the electrical stimulation of the hypothalamus. The animals were no longer motivated to do anything in order to receive these stimuli that previously had been highly reinforcing for them. However, the dopamine blockers did not change the hedonic evaluations, which remained intact. When the animals' hypothalamus was electrically stimulated, when they were injected with cocaine, or when sweet foods were delivered to their mouths, their expressive behavior continued to show that this was all very pleasurable.

This also holds true the other way around. Berridge and Valenstein (1991) stimulated the lateral hypothalamus with electrodes while plenty of food was available in the vicinity of the rat. As we saw earlier, a somewhat longer electrical stimulation "motivates" the rat to eat. After a while the rats immediately began to eat when the electrical stimulation began, and they stopped eating when the stimulation was terminated. When this state was reached, the affective reaction of the rat was filmed while a sweet infusion was administered to the mouth. During the infusion, the electrical stimulation was activated for 15 seconds, followed by an equally long pause, followed again by the stimulation, and so on. The rat, therefore, always had the same concentration of sweet solution in its mouth, sometimes accompanied by electrical stimulation of the hypothalamus, and sometimes not. The later analysis of the video films showed that the animals never responded with increased pleasant evaluation when the electric stimulation began. On the contrary, the rats tended to display more aversive reactions to the sweet solution when the stimulation began. The same stimulus that regularly motivated rats to eat did not lead to a more positive hedonic evaluation.

Studies such as this one persuaded Berridge and Robinson that the dopamine system equips objects and events with *motivational salience*, which then leads to either approach or avoidance. In case of approach, they use the term *incentive salience*, which they define as follows:

> Incentive salience has both perceptual and motivational features. According to our hypothesis, it transforms the brain's neural representations of conditioned stimuli, converting an event or stimulus from a neutral "cold" representation (mere information) into an attractive and "wanted" incentive that can "grab attention." But incentive salience is not merely perceptual salience. It is also motivational, and is an essential component of the larger process of reward. Its attribution transforms the neural representation of a stimulus into an object of attraction .... (1998, p. 313)

When objects or events are perceived that were previously repeatedly associated with rewarding experiences (i.e., that were positively or negatively reinforced), then dopamine is released in the sense of a classic conditioning process. It is not just currently perceived objects or situations but also mental images, memories, or cognitions in general that can lead to dopamine release if they were associatively linked with rewarding experiences. Dopamine engenders a state of motivation; it orients the mental activity toward approach or avoidance. This state of motivational salience is regularly encountered when motivational goals in the sense of consistency theory are activated. By definition, motivational goals are the types of goals that influence actual behavior. They refer to that which has high motivational salience for the individual. When a motivational goal is activated, the dopamine system is also always activated. The dopamine enables the person to learn all of the experiences that were encountered under the influence of the dopamine with particular strength, because dopamine facilitates the synaptic transmission of the currently activated neurons.

This leads to the conclusion that everything that the person should learn in a therapy must have high motivational salience if the learning is to be effective. Not all therapy goals necessarily have high motivational salience. Therapy goals may not concur with motivational goals of high salience but may instead result simply from the insight that things cannot go on the same way. Some therapy goals are also suggested to the patient by the therapist or by other persons, and they may not match the concerns of the patient. Even if they promise to yield pleasant consequences, such goals do not have motivational salience and do not shift the patient into a motivated state, which is a prerequisite for effective learning. Based on this view, one of the most important tasks of a therapist is to ensure that the issues discussed and pursued in therapy have high motivational salience for the patient. This will be the case when important motivational goals that the patient brings to therapy have already been activated.

The task of the therapist is not limited, then, to the appropriate delivery of a disorder-specific intervention. In order for the interventions to be successful, the patient must be motivated to engage with them. Whether the patient is in such a state can be inferred, among other things, from the amount of effort he is willing to exert in therapy. This prerequisite is frequently not met in therapy. When it remains unmet, it does not make much sense to administer the intervention. Without activation of the dopamine system, long-term learning will not occur. According to Hoebel et al. (1999), the mental system without dopamine is like a *sailboat without a sailor*. Dopamine "energizes" the approach- or avoidance-oriented goal hierarchies. Goals that are pursued without dopamine involvement do not have much chance of being attained. They lack the motivating power to overcome obstacles, to persist over the long term, and to engage with full effort. The dopamine system is the basis for intrinsic motivation.

In section 4.8.4.1 I will revisit the question of what gives goals their motivational power. That section deals with the relationships of various goals among each other. It will show that goals are pursued with great effort when they are consistent with important motivational goals or important motives of the person. At this point we see the neural mechanisms behind this process. If goal pursuit is not accompanied by the release of dopamine, that is, if no goals are activated for which a dopamine release has already been conditioned, the probability of goal attainment is low.

4.7.7 Beyond the Pleasure Principle

The need for pleasure maximization and distress avoidance is the most obvious among all the basic needs. It is most accessible to our self-observation, which is probably also the reason why it was repeatedly construed as "the mother of all needs," from which all other needs flow. A representative quotation in this context is the following statement by the 18th century moralist Jeremy Bentham, who wrote in his book on *Principles of Morals and Legislation* (1789/1948):

> Nature has placed mankind under the governance of two sovereign masters, pain and pleasure. It is for them alone to point out what we ought to do, as well as to determine what we shall do. ... They govern us in all we do, in all we say, in all we think. (p. 1)

This conviction continues to occupy the minds of many people, even though the evidence to the contrary is incontrovertible. What about the married couple that fights bitterly until everything is broken that was dear to both of them? Is this activity accompanied by hidden pleasures or does it serve the avoidance of pain, even though everything ends in pain and suffering? What about the freedom fighter who lets himself be tortured or killed for his ideals? And what about the athlete who gives everything he has, under conditions of great pain, even though he has no chance of winning? What about our daily activities? How much pleasure is gained from a hectic working day? How well do we manage on a daily basis to avoid unpleasant feelings? Considering that this is supposed to be our most important motivational source, it seems that rather little of our activity results in truly pleasurable experiences. We simply have to turn on the evening news on TV to see that this holds true for most other people as well. Based on such daily news, the striving for pleasure and the avoidance of distress can hardly sufficiently explain what humans do, say, and think. One would have to invent abstruse common sense defying theoretical conceptualizations to explain how continuous striving for pleasure results in so much displeasure. We would have to conclude that nature has equipped us with rather poor abilities to achieve the goals that are most important to us.

The examples of individuals whose behavior is truly oriented primarily to pleasure maximization and distress avoidance—those who have become dependent on drugs, for example, or extremely obese people for whom ultimately everything is about food—demonstrate that it is tragically misguided to elevate the striving for pleasure to a highest maxim. People who strive primarily for pleasure will end up in displeasure. Fortunately, nature has not made us in such a manner.

Is that which a physician feels when he has once again saved a patient's life after an 8-hour-long surgery a feeling that is primarily hedonic in character? Is it pleasure? What the doctor feels in this situation is hardly comparable to that which we call enjoyment, even if this feeling also has a basic positive quality. Inner contentment, pride, satisfaction, compassion, to be at one with oneself—all of these are positively colored states that can result from one's own behavior but which have little to do with what we call enjoyment or pleasure. Complex feelings of this type result when people feel that they are concordant with important goals. However, it is the goal that is being pursued, not the feeling state that might or might not result when the goal is finally reached. The fact that our goal-oriented activity is permanently accompanied by emotional evaluations, which are partially experienced as conscious feelings, does not automatically lead to the conclusion that we ultimately strive only to attain the feelings, or that the goals are only a means toward the end of these feelings.

Our motivational goals have formed on the basis of more than one basic need. They do not only serve to create pleasure and avoid distress. The attachment need and the control need influence behavior from the first day of life. The need for pleasure gain and distress avoidance was not present prior to these other needs. The other needs are not just a derivative of the pleasure principle, as Freud had originally argued. They are independent needs in their own right. Their goal is not pleasure but intimacy and control. The fact that negative emotions result when these goals are missed does not mean that the point is to avoid such feelings. From an evolutionary point of view, the purpose of the attachment need is to ensure the continuity of an existentially important care provision, which human infants need so that they can later pass on their genes. The control need also serves this purpose. The need for pleasure maximization and distress avoidance also serves important survival goals such as the ingestion of palatable food, the maintenance of a certain body temperature, the avoidance of physical injuries, and so forth. In the course of evolution, specific feedback systems have formed to ensure that all these purposes are continuously fulfilled—the pain system, the disgust system, the panic system, the anxiety system, the system for temperature regulation, and so on. All these systems have their own neural underpinnings.

The formation of the good–bad evaluative system along with the behavioral orientation toward approach or avoidance as a system that overarches these specific subsystems is an evolutionary step that probably became necessary to ensure the integration of these various subsystems. The good–bad evaluation with regard to proximity to the attachment person, however, is not equivalent to the good–bad evaluation of food tastes or the good–bad evaluation of having competence of not (having control or not). Attachment and control are not about hedonic quality. The good–bad evaluation that also occurs with regard to these needs constitutes an aid of nature to further ensure the satisfaction of these needs. However, the point is not to create a sense of well-being but to create intimacy and control. In the case of food and sexuality, the creation of pleasure directly serves survival and the generational transmission of genes, as well as the avoidance of pain. In the case of eating, pleasant taste and food ingestion cannot be separated from one another. Therefore, the creation of pleasant tastes and sexual pleasure as well as the avoidance of pain—which is the heart of pleasure–displeasure—are much closer to their evolutionary purpose than is the case for the attachment and control needs. When one strives to create pleasure during eating and in the context of sexuality, then this serves an evolutionary purpose in a direct, unmediated way. The fact that the neural mechanisms that were developed for this function could be disconnected from their evolutionary purpose—such as would be the case with drug additions or electrical stimulations of the appetite center in the hypothalamus—was not anticipated by evolution because evolution never anticipates; its optimization is always achieved via post hoc selection.

From this perspective, it is not really logically consistent that I discussed the approach and avoidance systems here in the section on the need for pleasure maximization and distress avoidance. Approach and avoidance serve all the basic needs and not primarily the creation of pleasure and the avoidance of distress. Perhaps I should have discussed the topic of approach and avoidance separately from each basic need. However, it is hardly possible to discuss the tendency to create pleasurable feelings and to avoid distress without also discussing the good–bad evaluation and approach–avoidance processes. The neuroscientific authors that have studied pleasure–distress typically focus only on the pleasure principle and not on other needs, which explains why they discuss approach and avoidance only from that perspective. It would have been rather difficult to explain at the beginning of this section that approach and avoidance do not just serve the regulation of pleasure and distress. Therefore, I pursued this logically incongruent presentation in full awareness of its consequences.

The need for enhancing self-esteem joins the concert of the basic needs only at a later stage, after the self and self-representation have formed as the basis of self-esteem feelings. This need also cannot be simply subsumed under the need for pleasure gain and distress avoidance. How many people deny themselves pleasurable feelings while eating in order to maintain a slim appearance, even though they later do not in fact feel beautiful and, therefore, are not reimbursed for the sacrifice of not having ingested the tasty food by attaining positive self-esteem with regard to their appearance? In this example as well, the motivational goals that have formed around the self-esteem need are the ultimate purpose and not just a means for the attainment of positive hedonic states.

Our subjective well-being is not the result of a striving for pleasure, or if it is, then this is true only for short periods. Our more enduring well-being is not a goal that has been attained; it is dependent upon our ability to satisfy all our basic needs simultaneously. When we attempt to fulfill one need at the expense of another, we will specifically not attain subjective well-being. A person who doesn't stop eating, becomes increasingly fat, and soon cannot help but continue to eat, experiences a high degree of incongruence with respect to his or her control need, his or her self-esteem need, and probably also his or her need for intimacy. This is similarly true for others who sacrifice their basic needs for the short-term attainment of drug-induced hedonic states. According to Berridge (1999), the point in drug addiction is not so much the striving for pleasure; instead, the "wanting" has become autonomous and runs amok, as it were. According to his incentive-sensitization theory of drug addiction, the repeated ingestion of drugs such as heroin, cocaine, and amphetamine leads to a neural sensitization, which in an escalating vicious cycle causes anything that is associated with the drug consumption to acquire ever stronger motivational salience. Under the influence of dopamine, the drugs are eventually excessively and obsessively wanted, regardless of their hedonic quality. Drug addicts who have arrived at this point behave like the rat that stimulates its hypothalamus obsessively, until total exhaustion, because the dopamine released in the process functions like a straitjacket that constrains the individual within these tight behavioral parameters.

When certain basic needs or motivational goals are excessively valued, at the expense of others, the inevitable consequence is that these other basic needs are not being satisfied. On the level of motivational schemas, discrepancies emerge between perceptions and goals that serve these basic needs. The consequence, in turn, is poor well-being. Good well-being is the result of a balanced attainment of all basic needs.

Because motivational schemas develop around every basic need, many motivational goals come into existence and compete for the ability to dominate mental activity. It is unavoidable in this process that some goals are mutually incompatible and interfere with one another. This can mean that both goals are not being attained or that one is being attained at the expense of another; for example, an avoidance goal at the expense of an approach goal. Thus, it is possible to experience different forms of inconsistency in motivated mental functioning. The result is always that certain motivational goals are not being attained, which simultaneously means that the needs they serve are not adequately met.

Given the multitude of simultaneously transpiring processes, a certain degree of inconsistency in mental functioning cannot be avoided. If these inconsistencies are only temporary and can later be resolved, need satisfaction is not seriously threatened. However, if higher degrees of inconsistency endure for longer periods of time, the need satisfaction overall is compromised, which will be reflected in a generally poorer state of well-being.

Subjective well-being, then, is the result of complex mental processes that cannot be simultaneously volitionally controlled by the individual. Therefore, subjective well-being cannot be pursued in the same way that one strives for other goals. To feel maximally good is not a goal that can be pursued directly. If a person suffers from poor subjective well-being, then it is not only his need for pleasure that has been violated. It will be necessary to understand more than just this single need and to analyze the totality of his motivated mental processes in order to uncover the reasons for his poor well-being. This global view will now be the focus in the subsequent section on the consistency principle.

4. 8 CONSISTENCY AND CONSISTENCY REGULATION

4.8.1 Forms of Inconsistency in Mental Functioning

Inconsistency refers to the incompatibility of simultaneously transpiring mental processes. Inconsistency can appear in many different forms. One of them is the *interference* of two or more processes. We have already encountered several examples of this. If one is asked to nod while encoding negatively valenced words, the words will be encoded more poorly than if one is asked to shake one's head (Förster & Strack, 1996). The affirmation of negative content appears to "go against the grain," and this has a detrimental effect on subsequent recognition ability. The perception of negative content and simultaneous affirmation are mutually inconsistent. This also holds true more generally: Approach of objects with negative valence and avoidance of objects with positive valance interferes with one another, which is reflected in slowed reaction times (Neumann & Strack, 2000; see section 4.7.2).

One form of interference that has been studied extensively in psychology is the Stroop test, which simultaneously elicits two opposing reaction tendencies. For example, the word *blue* might be presented in red letters and the word *red* in blue letters. If one is asked to respond maximally quickly to either one (red) or the other (blue), then the reaction time will be significantly slower than would be the case if the word *red* were presented in red color and the word *blue* in blue color. Similar tests have been conducted in many different variations; the findings are regularly that this form of interference of simultaneously activated reaction tendencies compromises performance speed.

The Ravensburger game publishing house (popular in German-speaking countries) has published a game—"Confusion"—that takes such interferences to an even higher level. Similar to the Stroop principle, the game material elicits several reaction tendencies simultaneously, but the rules always allow for only a single reaction. The player who shows the correct reaction the most quickly, wins the game. This will be the person who can best suppress the simultaneously activated competing reaction tendencies. Such people will have a particularly smoothly functioning anterior cingulate cortex (ACC), because according to Botvinick, Braver, Barch, Carter, and Cohen (2001), this brain region functions as a conflict monitor. The ACC is regularly activated when the individual deals with challenges of this type, and it mobilizes resources to enable successful coping with such conflict. Conscious awareness is directed toward the sources of the interference, such that the qualities of the conscious awareness mode can be utilized for the resolution of the conflict. Chronic interferences of this kind are a form of stress for the brain. The steady activation of the ACC, with subsequent conscious suppression of the disallowed reaction tendencies, is experienced as effort, so that one feels exhausted or "satiated" much more quickly when playing this game, compared to other, really more intellectually demanding games.

Experimental psychologists have also examined many other forms of interference among simultaneously elicited processes. These studies show consistently that interference undermines performance. Long before this type of inconsistency had become a research focus, the Gestalt psychologists had already discovered one particular kind of striving for consistency. They noticed that the human perceptual apparatus has an automatic tendency to create concise images (*Gestalten*) from the offered stimulus material—images that are associated with a minimal degree of tension and that are experienced as pleasant (Barth, 1987). Gestalt psychology, however, did not examine what happens when this striving for clarity or unambiguity is blocked.

The blocking of unambiguously interpretable perceptions is experienced as very torturous. This was first shown by Pavlov (1927) in his studies with dogs. In one of his experiments a dog was given food when a circle was shown but none when an ellipse appeared. After the dog had learned to differentiate the circle from the ellipse and to respond differently to them, the circle and the ellipse were both changed to become increasingly similar in appearance, until the dog finally could no longer tell the difference between the two. Even though no punishments were administered in this experiment, the dog developed what Pavlov called an experimental neurosis. The state of ambiguity, nonclarity, not knowing how to act is apparently highly unpleasant for a dog. The perceptual pattern simultaneously triggers different memory content and reaction tendencies, even though it is not possible to make a decision between the two, such that the conflict between the two tendencies cannot be resolved. It is not the stimulus that triggers the suffering behavior of the dog but rather the state of inconsistency that is triggered by the simultaneous activation of mutually incompatible reaction potentials.

Heider (1982) applied the Gestalt approach to social perceptions and developed his balance theory from this starting point. This theory is about the balance or coherence among cognitive units. What is the relationship between the perceiving I and two objects of perception, and what is the relationship between these two objects? Such triads can be unbalanced, in which case a tension exists between the three elements, or it can be balanced, which can be called *consistent*. Heider postulated that humans strive to experience consistent perceptions. This assumption was supported in several experiments. Balanced triads are also evaluated as more pleasant than unbalanced ones (Lindzey & Aronson, 1968; Witte, 1989).

The form of inconsistency that has been studied most frequently in psychology is known as *cognitive dissonance*. Cognitive dissonance is present when two cognitions are relevant for one another but at the same time incompatible. An example: A woman detests a certain political party. She falls in love with man. Soon she finds out that the man is active in just that party. Festinger (1957) considers dissonance an unpleasant, harmful state that elicits motivation to end the state as quickly as possible. Four possibilities for reducing such dissonance exist, according to Harmon-Jones and Mills (1999). These possibilities refer to the reduction of just cognitive and not other types of dissonance, but most studies in this area have in fact only addressed these cognitive aspects. The possibilities are

- Removal (avoidance, suppression) of dissonant cognitions.
- Addition of new consonant cognitions.
- Reduction of the importance of dissonant cognitions.
- Increasing the importance of consonant cognitions.

In more than one thousand studies on dissonance theory, it was tested whether the participants really, as expected, reduced the dissonance and which of these possibilities they chose under which conditions (Cooper & Fazio, 1984). The research paradigms with which dissonance is created in these studies can be divided into four groups (Harmon-Jones & Mills, 1999):

1. The *free-choice paradigm:* In this paradigm, participants are forced to make a choice. Such a choice automatically creates dissonance if the available options seem equally important (good, valuable). In this case, every positive aspect of the nonse-

lected choice and every negative aspect of the selected option is a source of dissonance. Conversely, every positive aspect of the selected option and every negative
aspect of the nonselected option is consonant with the decision. The more equal the
two alternatives are, the more difficult the decision and the greater the dissonance. In
order to reduce the dissonance, the dissonance-producing aspects of both alternatives
are removed, the consistent aspects are evaluated more positively, or other consistent
aspects are introduced.

2. The *belief-disconfirmation paradigm:* The point here is to confront the person
with information that is inconsistent with his or her belief. This will lead to a change
in the belief, or to false perceptions or false interpretations, or the person will reject
information, search for information from others, or will try to persuade others of his
or her beliefs.

3. The *effort-justification paradigm:* Dissonance is created by asking the person to
do something unpleasant in order to achieve something desirable. The more unpleasant the activity, the greater the dissonance. The dissonance is reduced by the extremely high evaluation of the goal state, in the service of which the activity is performed.

4. The *induced-compliance paradigm (forced-compliance paradigm):* In this
case, dissonance is created by having the person do something that is counter to his
or her belief or attitude. The person does this either because of a promised reward or
a punishment. This, in turn, is consonant with the behavior and provides a justification for it. The stronger the justification, the weaker the dissonance and the smaller
the attitude change. The smaller the reward or the weaker the threat, however, the
greater the dissonance and the stronger the attitude change.

In general, the striving for cognitive dissonance reduction can be regarded as one
of the most broadly substantiated phenomena within all of psychology. There is
also broad agreement in the literature on the point that the attitude changes resulting from dissonance reduction are real and often enduring changes of persuasion.
What is not yet completely resolved, however, is how exactly the mental distress in
dissonance situations should be understood, even though it is clear that this distress
has a motivating function for attitude change, and that the removal of dissonance
reinforces the cognitive activity that caused the dissonance reduction. Two experimental studies by Elliot and Devine (1994) specifically examined and supported
this effect.

Cognitive dissonance refers to an inconsistency between simultaneously activated
contents of working memory; that is, inconsistency among contents that are simultaneously conscious. It is plausible that consciousness is particularly strongly protected from inconsistency because mutually inconsistent contents would compromise
conscious action planning and action direction and thereby undermine the person's
ability to have successful transactions with the environment.

Inconsistency in consciousness is also avoided by yet another mechanism: suppression. This does not refer only to the cognitive content of consciousness but also to
perceptions (perceptual defense) as well as goals and action impulses. Content that is

not compatible with that which is currently in consciousness is not allowed to enter or has great difficulty entering into working memory. Suppression as a mechanism for consistency continuity will be discussed in more detail in the next section.

Supression can lead to another form of mental inconsistency; that of *dissociation*. We have already encountered this form of inconsistency in the form of dissociation between explicit and implicit memory content in the discussion on the neural foundations of PTSD, where such inconsistency was construed as one of the causal factors in the disorder's pathogenesis. Beyond that, dissociation is also a notable feature of several other mental disorders (Fiedler, 2001). Inconsistency, then, can be a core aspect of various forms of psychopathology.

Other forms of inconsistency that have been well researched are the *motivational conflicts*. The heyday of experimental conflict research may be in the past, but it has provided us with a well substantiated but now somewhat forgotten knowledge base about the effects of various kinds of conflicts. The most famous model here is perhaps the conflict model by N. E. Miller (1944; Dollard & Miller, 1950). Good summaries can also be found in Heilizer (1977, 1978). Seymour Epstein, on whose personality theory consistency theory builds in important aspects, also devoted much of his early scientific career to the experimental study of conflicts (Epstein, 1962, 1967; for a summary, 1982). Experimental conflict research has differentiated and examined five types of conflict:

• *Approach–approach conflicts.* These are conflicts between two positive alternatives. An example is the story of the donkey who starves to death between two equally distant and equally large haystacks because it is paralyzed by the two equally strong approach tendencies. In line with N. E. Miller (1944), however, one could also construe this as a double approach–avoidance conflict because the approach of one haystack necessitates the loss of the other.
• *Approach–avoidance conflicts.* This conflict type has been examined the most frequently, in part because of its clinical relevance. The conflict here is that approach and avoidance tendencies are simultaneously activated. N. E. Miller (1944) conceptualized these types of conflicts in the way shown in Figure 4–17. One simultaneously desires and fears an object. The closer one gets to the object, the stronger the desire but the stronger also the fear. The increase in desire and fear does not have to be equal; instead, the gradients that show the strength of desire and fear as a function of the distance from the object can differ in steepness, such that they cross over at some point. This would be the moment at which desire and fear are exactly equally strong and mutually inhibit each other. It is the moment at which the person experiences the greatest conflict tension and fear because from this point on he or she will no longer approach the object because the avoidance gradient is now stronger. Figure 4–17 shows this situation for two differentially strong approach gradients. The figure clarifies that the point of maximal conflict and, therefore, the greatest inconsistency is reached later in the case of a strong approach motive, compared to a weaker approach motive. With this type of conflict, the person is, as it were, a prisoner of his or her own goals: The person can neither reach nor abandon the goal. He or she is caught in the conflict, and the inconsistency remains in place.

- *Avoidance–avoidance conflicts.* In these cases, the person is confronted with two aversive alternatives. Wherever possible, the person will "leave the field entirely."
- *Double approach–avoidance conflicts.* The person is not confronted with just one goal object that is simultaneously desired and feared but with two such goals simultaneously.
- *Avoidance–approach conflicts.* These are the types of conflicts in which the avoidance component clearly outweighs the approach component. For a long time these situations were not recognized as conflicts because the person typically simply avoids the conflicted goal. However, if by accident or due to external pressure the person is suddenly brought into close proximity with the goal, he or she will often suddenly approach this on his or her own effort even though the person had previously always avoided this. Studies about this conflict type have somewhat illuminated phenomena that previously had been hard to understand (Epstein, 1978).

After a creative pause, conflict research has experienced a resurgence of interest in recent years. However, this newer research tends to examine not experimentally induced conflict situations but rather conflicts among cognitive representations of personally relevant content. Emmons (1989, 1997; Emmons & King, 1988) focused on conflicts among personally important goals (personal strivings). The degree of conflict among their most important personal strivings is evaluated by the participants themselves. Based on Heider's balance theory, Lauterbach (1996) has developed a computerized method for the assessment of the degree of conflict among personally important life issues. This method was used in several studies on the associations among conflicts, well-being, mental health, and therapy motivation (Hoyer, Fecht, Lauterbach, & Schneider, 2001; Hoyer, Frank, & Lauterbach, 1994; Michalak, Heidenreich, & Hoyer, 2001; Michalak & Schulte, 2002; Renner & Leibetseder, 2000). Grosse Holtforth and Grawe (2001) use this method to estimate the extent to which the most important motivational goals measured by the IAAM can be construed as conflicted.

In his self-discrepancy theory, Higgins (1987) studied the discrepancies between various types of self-representations, such as the actual-self, ideal-self, and ought-self. Higgins and colleagues' hypothesis that greater discrepancies are associated with worse well-being has been confirmed in several empirical studies (Higgins, 1987, 1997; Higgins, Shah, & Friedman, 1997). In their self-concordance model, Sheldon and Elliot (1999) examined the concordance between desired goals and central aspects of the self, such as important values, interests, needs, and the compatibility of behavioral goals with higher-level goals (vertical coherence). They found that concordance and coherence, as defined in their model, led to better goal-attainment and better well-being. That which Sheldon and Elliot termed *concordance of the currently pursued goals with basic needs and values* also characterizes the state of intrinsic motivation and the state of "flow" (Csikszentmihalyi, 1990), both of which can be viewed as forms of a particularly good consistency in mental activity.

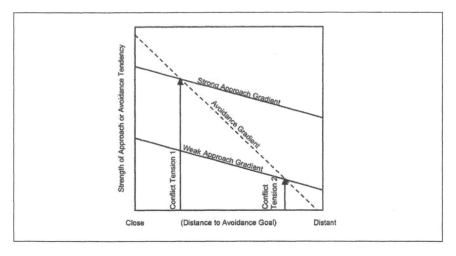

Figure 4–17. Depiction of two approach-avoidance conflicts with the same avoidance gradient and differentially strong approach gradients. In the case of the conflict with the weaker approach gradient, the strongest conflict tension and a preponderance of avoidance will be reached further away from the goal, compared to the case of the stronger approach gradient. In this case, the conflict tension and fear are clearly higher because they only occur closer to the goal. The person abandons the approach behavior at only a very short distance from the goal.

In his studies on the compatibility of explicit goals and implicit motives, and the implications of such compatibility for goal-attainment and well-being, Brunstein (Brunstein, Schultheiss, & Grässmann, 1998; Brunstein, Schultheiss & Maier, 1999) has taken his research in a similar direction as Sheldon and Elliot.

Finally I should mention what is perhaps the most important form of inconsistency— the inconsistency between the goals a person holds and the extent to which these goals have been attained; that is, the *incongruence* between motivational goals and actual perceptions. If these are markedly discrepant, it indicates that the needs of the person are not being satisfied, which, in turn, has negative effects on his or her well-being and mental health. Incongruence has been studied as one among several constructs in several of the previously mentioned studies, but it has perhaps received the most targeted attention in our own studies (Grosse Holtforth & Grawe, 2003a; Grosse Holtforth, Grawe, & Tamcan, 2004).

I have limited myself here to merely listing the various types of inconsistency that have so far been defined and studied. The associations between the various forms of inconsistency with the effectiveness of mental functioning and with well-being and mental health will be reviewed further (see section 4.8.4). I can already say here, however, that inconsistency has been found to be generally detrimental, regardless of which form of it has been correlated with which criterion. These findings suggest that chronic inconsistency is harmful for the mental system, that it is experienced as unpleasant, and that it creates a motivation toward the reduction of the inconsistency.

If we step back from the research studies for a moment and once again reflect on the terms that we used to define inconsistency, it should be already clear that inconsistency is infused with negative meaning. We used terms such as *conflict, dissonance, discrepancy, incompatibility, disharmony, dissociation, discordance, incongruence, ambivalence, ambiguity*, and *lack of clarity*. The state of consistency can be less easily described, perhaps because it is reached so rarely that just a few terms suffice for this purpose. It is characterized by harmony, being at one with oneself, consonance, concordance, agreement, and clarity. If we could choose between two worlds, one in which consistency dominates and one with predominantly inconsistency, these descriptions leave no doubt about which of the two most people would likely choose. However, this also means that it is better to live in a psyche characterized by consistency rather than inconsistency. The core assumption of consistency theory, that our mental system is furnished in such a way that it strives for consistency and avoids inconsistency, is broadly supported, not just by scientific research findings but also by the implicit preunderstanding that is contained in our language.

4.8.2   Neural Mechanisms for Ensuring Consistency Continuity

If inconsistency is as detrimental for the conduct of optimal transactions with the environment as the previously mentioned findings suggest, it makes sense to assume that over the course of phylogenetic development structures would have emerged and been selected in the human nervous system that ensure the consistency of mental activity. Given the complexity that marks higher-developed nervous systems, a good coordination among the transpiring processes is a prerequisite for the prevention of chaos. Whenever a new developmental step occurred in the course of evolution that enabled new mental processes and qualities, an important criterion for its enduring inclusion into the structure of the nervous system was that it would be compatible with what already existed. With ever increased differentiation and complexity, higher structures had to form, which would refer not just to separate functions such as hearing, seeing, and tasting but also to the compatibility and coordination of all the potential neural–mental processes.

This necessity existed at all stages of our phylogenetic development. We have to assume, therefore, that the criterion of being in tune, compatible, or coordinated with the already existing was present during all previous evolutionary developmental steps, and that our nervous system, in its present form, is also the result of a consistency-ensuring selection. In this sense, all of the regulatory processes that exist on all levels among the separate subsystems and circuits in the brain also contribute to the continuity of consistency. There are some structures and functional principles in the neural system, however, whose function for consistency continuity is particularly obvious. If I highlight some of these consistency-assurance mechanisms in the following, then this should be interpreted with the understanding that consistency regulation is a very common functional principle of higher-developed nervous systems.

In the development of the human brain, a considerable space problem was encountered. How should all the neural hardware that is required for such differentiated functions be stored in such a limited space? The same location that houses the human language system hosts very different functions among other mammals. Where should the place for the language system come from? The much greater formation of folds and wrinkles in the human cerebral cortex, compared to other mammals, constitutes one solution to this problem. Another solution is a greater specialization of the two brain hemispheres on different functions. I have so far addressed this lateralization only briefly, in as far as it was relevant for the subtopic under discussion. However, much more could be said about the relations between the two hemispheres. In fact, this is one of the most fascinating examples of labor division and coordination. The most important role in the coordination between the two brain halves is played by the corpus callosum, that thick nerve bundle that connects the two hemispheres with each other. If it is severed, the right hand literally no longer knows what the left hand does. LeDoux (2002) reported how he observed a patient, whose corpus callosum had been severed, while the patient was putting on his trousers. One of his hands was pulling the trousers up while the other hand pulled them down. This is certainly an extreme and rare example of incongruence in mental functioning, but it can be used to comprehend more concretely how much inconsistency can interfere with the ability to engage effectively with the environment. The corpus callosum is one of the basic consistency-assurance mechanisms of the brain. As long as it functions normally, this function will not be evident. Its function is one of the givens of the brain, and its function is only apparent when its integrity is compromised.

Another consistency-assuring mechanism is usually taken for granted in a similar fashion: that of activity-dependent inhibition. If one bends an arm, the arm-stretch muscles are inhibited. In spastic disorders, this basic mechanism of activity-dependent inhibition does not function properly. Bend and stretch muscles are simultaneously activated and inhibit each other. The resulting spastic disorder is a very graphic example of the inconsistency tension at the lowest level of the nervous system. However, the principle of activity-dependent inhibition holds true on all the levels of the nervous system. It also holds true for entire cell hierarchies. When one cell hierarchy is activated, others that are incompatible with it must be inhibited. It is impossible to simultaneously eat the cake and store it for later.

If the activity-dependent inhibition does not function on the level of the cell hierarchies—in other words, when mutually incompatible cell hierarchies are simultaneously activated—a conflict is present. The simultaneously activated cell hierarchies mutually inhibit each other, leading to *conflict tension*, another form of inconsistency tension. Conflicts have two direct consequences: First, they are responsible for the situation that none of the cell hierarchies can function without interference. Because of this, goal-attainment is undermined. The result is incongruence with regard to the goals involved in both cell hierarchies. Incongruence is linked with the secretion of stress hormones (see section 4.5.2) and with negative emotions. The second immediate consequence is the conflict tension. In the case of someone with a spastic disorder, the inconsistency tension resulting from the simultaneous activation of bend and stretch muscles can be observed with the naked eye. The inconsistency tension resulting from the simultaneous activation of incompatible cell hierarchies can be experienced, on the one hand, as a feeling of tension; on the other hand, this state can be objectified via indicators such as muscle tension and other psychophysiological, hormonal, and neural parameters. The research on these objectifiable indicators of conflict tensions that result from the simultaneous activation of important goals is still in its infancy, even though all of the methods for such research have been developed. Neither experimental nor correlational conflict research has so far seriously addressed the phenomenon of conflict tension as an empirical phenomenon. This area promises to be a domain in which many aspiring researchers can yet earn their scientific fame and glory. I will discuss later the initial attempts to study what happens in the brain during conflict situations.

Which neural mechanisms have developed, then, to avoid the simultaneous activation of incompatible cell hierarchies that is so detrimental for goal attainment? This question brings us to one of the most important functions of the explicit, conscious mode of functioning. We have seen earlier, when we discussed cognitive dissonances, that conscious awareness is particularly intolerant with regard to inconsistency. Only content that is compatible with whatever happens to be currently activated in the conscious mode of functioning has any chance of reaching the conscious working memory. This is particularly true for the cell hierarchies. If conscious behavior is currently oriented toward the attainment of a particular goal, attention will be directed toward whatever might be relevant for the pursuit of that goal. The degree to which goal-relevant perceptions, memory content, and cognitions can be easily activated, is heightened; other perceptions, memory content, cognitions, emotions, and especially also other goals are actively inhibited. The terms *concentration, will, selective attention, perceptual defense,* and *suppression* all refer to processes that shelter the currently activated cell hierarchy from disturbances. On the neural level, this means that perceptions are actively inhibited somewhere on their way from sensory input to working memory. This inhibition can occur only after the meaning of the sensory input for the motivational goals of the person has been established. The sensory input passes through several steps of neural transformation, then, before it is actively inhibited and prevented from entering into consciousness. The perceptions, whose significance has already been evaluated, might well be able to exert influence on the lower levels of other neural processes, without this whole sequence ever becoming conscious. For example, threatening cues can trigger intense firing of the amygdala. If attention at this moment is fully absorbed by an important goal that is consciously pursued with strong conviction and willpower, the signals that the amygdala continuously sends to the PFC are blocked on their way to the ACC and to other brain regions that are involved in attentional direction. The neurotransmitter gamma amino butric acid (GABA), which inhibits the activation transmission at the synapses, plays an important role in this context.

The inhibition of neural signals that are not conducive to the currently pursued agenda enables humans to act in a goal-oriented way even in life-threatening situations.

This ability is illustrated, for example, in the combat soldier who sees and hears shell explosions all around and nevertheless emerges from the trenches to conquer a strategically important position, thereby acting against blocked, consistency-conflicting perceptions and impulses (to hide or flee). Another example is the surgeon who operates all morning with focused concentration, even though his wife informed him the same morning that she wants to seek a divorce. Not just technical but also creative activities profit greatly from this ability to shelter conscious awareness from interfering influences. The artist who is deeply emerged in his work on a sculpture while bills are piling up on his desk is just one example for this principle.

The identification with a goal as well as the resulting intrinsic motivation, ability to concentrate, determination, and willpower are all means of ensuring the consistency of consciously pursued goal hierarchies via activity-dependent inhibition. These have their own specific neural underpinnings, located primarily in the PFC. When the PFC is damaged, these functions are impaired to a greater or lesser degree. This ability— just like language—constitutes a unique human capacity. It is the basis of the great cultural achievements of which humans are capable. It is not just the extent of their ability for conscious thought that distinguishes humans from animals but their ability to ensure the consistency of intentional action. Without this capacity, we would pursue one activity now and another a moment later—impulse-governed—depending on what happens to be activated in each respective situation. Therefore, the capacity for conscious goal-directed activity had to be accompanied by the development of neural mechanisms that would ensure the continuity of consistency on this highest level of action control.

These highly developed neural mechanisms that ensure the continued consistency of processes transpiring in consciousness are a valuable human capacity, then, that can contribute considerably to the fulfillment of important needs. However, the mechanisms also harbor risks. They can remove content from conscious awareness that actually is very important for the individual's well-being. Also, the processes that are being guarded or sheltered by this consistency–continuity protective mechanism may not always be particularly valuable, as would be the case with prosocial or creative activity. If the amygdala continues to fire incessantly because of incoming—albeit unconscious—threatening perceptions, then this activity can have long-term harmful effects on the mental system even though the person would not consciously experience any anxiety. The signals flowing from the amygdala may not reach working memory, but the amygdala also has various efferent connections that cannot be inhibited by working memory. The autonomic nervous system would react, among other things, with elevated blood pressure and increased heart rate. Adrenaline would be released and, if the anxiety circuit remains activated for longer periods, corticosteroids would be secreted, whose harmful consequences were already discussed earlier (see section 4.5.2). Via the close connections of the amygdala with the nucleus basalis in the brainstem, a massive release of acetylcholine would follow, which also would lead to a nonspecific arousal increase. We have previously encountered acetylcholine as the opponent of dopamine (section 4.7.6), the "fuel" of motivated action.

In section 2.12, we also saw that conditioning processes do not require conscious awareness. Therefore, it is likely that more and more cues and situations will acquire a threatening quality, even though the person would not be consciously aware of these processes.

If the removal of threatening or incongruent perceptions from consciousness is not just temporary, to protect a particularly important conscious plan, but when this becomes more of a chronic pattern, in which threatening signals, perceptions, memories, and thoughts are habitually repressed or inhibited from entering into working memory, the consequence will be a dissociation of the processes transpiring in the explicit versus implicit modes of functioning. At this point, the short-term advantage of consistency–continuity becomes a long-term disadvantage. Because the person does not engage with, or confront, the situations that are unconsciously perceived as threatening, the situations maintain their threatening character and continue to exert their effects on the nervous system in the implicit mode of functioning. The processes in the explicit and implicit modes are no longer in tune with one another—they dissociate. As a consequence, there is no cortical inhibition of the anxiety and the continuous hyperactivation of the anxiety circuit can lead to regulatory failures, such as the ones we reviewed in chapter 3 in the sections on neural correlates of generalized anxiety disorder, panic disorder, and obsessive–compulsive disorder. The neural models of all three of these disorders regard a dysregulation of negative emotions as a core component of the disorder and construe the lack of inhibition originating from the orbitofrontal cortex as a main causal factor. Because of the close communication in the PFC, such an inhibition would necessarily involve working memory. However, if working memory is not informed about the fact that something has registered that ought to be dealt with, then it also will be unable to initiate the needed inhibitory signals. In the case of posttraumatic stress disorder, the removal of traumatic experiences from consciousness, and the resulting memory dissociation between the explicit and implicit systems has even been conceptualized as a main aspect of the disorder.

The removal of threatening or incongruent perceptions from consciousness, which serves the purpose of consistency–continuity in the short term, can have precisely the opposite effect over the long term, because the resulting dissociation and dysregulation can lead eventually to a chronic increase in overall inconsistency. Consistency-protective mechanisms can therefore be adaptive as well as maladaptive.

Temporary denial and suppression of very threatening perceptions can initially ensure the person's continued ability to function, which can be highly useful during reactions to objectively threatening events. However, when perceptions that are simply habitually threatening or incongruent are kept from entering consciousness, the danger exists that such perceptions can lead via the anxiety circuit to dysregulations in the implicit system because of insufficient inhibition from the PFC. Such dysregulations can then firm up in the process of repeated facilitation, which can result in the formation of psychopathological syndromes.

Individuals who habitually remove threatening or goal-incongruent perceptions from consciousness are termed *repressors* in the literature. The phenomenon of repression has been studied in quite a bit of detail, and there can be no serious doubt that the phenomenon does in fact exist (Holmes, 1990; Kihlstrom & Hoyt, 1990; Shevrin, 1990).

According to G. Schwartz (1990), the redirection of attention away from the incongruence between goals and perceptions undermines the normal feedback mechanisms in the mental system that are essential for the regulation of mental processes.

> Both negative and positive feedback serve to interconnect components so that the system can function (behave) in a self-regulating (self-moderating or self-amplifying), and therefore predictable, manner. If the information is distorted or misperceived, or, in extreme cases, disconnected altogether, the self-regulation engendered by the negative and/or positive feedback will be impaired or cease altogether. ...
>
> Integrating all of the above, it follows that repression (and self-deception) includes disattention to negative feedback that are essential for self-regulation and, therefore, healing. Disattention promotes a state of relative disconnection (e.g., a functional disconnection of the left and right hemispheres as proposed by Galin, 1974). This state of neuropsychological disconnection induces a state of psychophysiological dysregulation, which is expressed as disorder in biological, psychological, and social functioning. This disordered biopsychosocial functioning is hypothesized to contribute to physical, mental, and social disease. (G. Schwartz, 1990, pp. 408–409)

A relationship between habitual repression and elevated degrees of physical and psychopathological disorders can be regarded as empirically substantiated (Blatt, 1990; Bonanno & Singer, 1990; G. Schwartz, 1990; G. E. Schwartz, 1983; Weinberger, 1990). Beyond that, the type of disorder a person develops appears to depend, in part, on the extent to which he or she relies on habitual repression or, instead, tends to turn toward the inconsistencies that elicit negative feelings (Blatt, 1990; Bonanno & Singer, 1990).

Based on what is discussed here, the avoidance of inconsistency, dissociation, and dysregulation has structural as well as functional neural bases. The corpus callosum is a particularly obvious structural foundation for the avoidance of dysregulation between that which the right versus the left hemisphere does. Antagonistic inhibition is a common functional principle of the nervous system with pervasive structural underpinnings. On the higher levels of complex processing, attention and consciousness fulfill the functions of consistency maintenance. They, too, have structural neural foundations that have formed—and had to be formed—over the course of evolution in order to prevent in this complex system the kind of interference and competition of simultaneously activated processes that would render the system inefficient as it engages with the challenges posed by the environment.

A particularly important role within consistency regulation appears to be played by the ACC. The ACC responds to various forms of inconsistency. For example, it is activated when something itches somewhere or when we feel a sudden pain (Hsieh et al.,

1994; Jones, Brown, Friston, & Frackowiak, 1991). It also responds to feedback about mistakes that have been made (W. Miltner, Braun, & Coles, 1997). Both are forms of incongruence—a specific form of inconsistency. The ACC is also activated, however, by situations that simultaneously trigger competing, conflicting reaction tendencies. This has been studied in great detail with regard to the Stroop principle (see previous). The involvement of the ACC in such conflict-related situations is so clearly established that Botvinick et al. (2001) explicitly called this area the *conflict monitor*. Using the terminology of consistency theory and including as well the types of incongruence reviewed previously, it would perhaps be even more appropriate in this context to use the term *inconsistency monitor*.

The fact that in the course of evolution a certain part of the brain was entrusted with the special task of monitoring whether inconsistencies appear somewhere in mental functioning shows once again how important it is for mental activity to prevent the occurrence of enduring inconsistencies. So far, there have been no studies that explicitly investigated whether the ACC is also activated during motivational conflicts, but it would be very surprising if this were not the case, because motivational conflicts also contain incompatible cognitions and behavioral tendencies, and the ACC is known to respond reliably to such situations. In addition to this monitoring task, the ACC functions to mobilize control reactions that lead to reductions of inconsistency. This would include primarily the redirection of attention toward the source of the inconsistency, which would then mobilize the options associated with conscious control. It is an important task of the ACC to direct attention toward the areas in which the inconsistency that now requires conscious intervention originated.

We have seen in section 3.2.2 that the ACC is deactivated in depressed individuals. This means that a very important function is disabled among depressed people. They can no longer rely on a functional inconsistency monitor that mobilizes resources for conscious engagement with the difficulties the person encounters. The fact that the ACC no longer responds in depressed individuals could be related to the fact that neural signals indicating inconsistency are being blocked on their way to the ACC, or that it is the ACC itself that normally generates such signals but is being inhibited in this function. Because the ACC also plays an important role for the conscious experiencing of emotion, and because this function is also impaired among depressives, there is good evidence that the ACC is being actively inhibited in depressed individuals. Conceivably, this inhibition originates from a hyperactivated avoidance system in the right PFC.

The functions of the ACC must be regarded as very important for the purpose of active engagement with the environment. If inconsistency-generating signals do not reach or do not activate the inconsistency monitor of the ACC, then the person is unable to respond with the options provided by the conscious mode of functioning. This is the great risk associated with the use of repression as a consistency-ensuring mechanism. If repression via inconsistency avoidance has been sufficiently frequently facilitated and, thereby, has become a habitual reaction tendency, the person cannot employ his or her most effective tools for the pursuit of important goals—the tools of consciously directed perceptions, of reflecting and planning, and of deliberate action. This will certainly have the consequence that the person's motivational goals will be poorly realized in his or her concrete life circumstances. Poor well-being results, and, on the basis of unresolved inconsistency tensions, mental disorders are more likely to emerge, as discussed in more detail in the next section. This is how the empirically confirmed relationship between the habitual use of repression and poor mental and physical health can be explained.

When motivational goals and associated reactions, such as anxiety, that are already inconsistent with the current content of consciousness that is activated, they will—as detailed previously—be prevented from becoming conscious and exert their influence on mental activity in the implicit mode of functioning. This can, of course, also be the case when several motivational goals are being simultaneously activated, such as, for instance, an approach and an avoidance goal. This would be the kind of situation in which the ACC, as the inconsistency monitor, should become active. However, if the neural signals representing the motivational goals do not even reach the ACC because their path toward it is being blocked, then the ACC cannot direct attention toward the conflicted situation and cannot mobilize resources that would enable the successful resolution of the conflict. What I said earlier about goals that have been removed from consciousness is all the more true for such "unconscious conflicts"; that is, the goals have little chance of being attained, which then leads to a chronic incongruence with respect to both conflict components. Moreover, the mutually inhibiting conflict components engender an enduring state of conflict tension, which cannot be reduced via conscious action because consciousness is preoccupied with entirely different content.

As we saw earlier, the mental system tends to seek reductions of elevated inconsistency tension. In a situation of high inconsistency tension, therefore, a high "reinforcement potential" is created: Anything that achieves a reduction in inconsistency tension will be facilitated particularly strongly. Thus, if the mental system during a time of such inconsistency tension somehow attains a state that momentarily reduces the tension, these new states will be facilitated and will be likely to recur in the future. This facilitation does not depend on whether these states of mental activity are useful or harmful with regard to the attainment of motivational goals. The facilitation is only contingent upon the extent to which the new states effectively reduce the existing inconsistency tension. Inconsistency tension can therefore lead to the formation of adaptive as well as unfavorable new patterns of mental activity. This question—of how the formation of new neural activation patterns should be understood—is of central importance for learning, change, and development. It is addressed in the next section.

4.8.3 Inconsistency Reduction as the Engine of Mental Development

Imagine being back in high school. You are in the process of doing your mathematics homework. You are faced with a problem type that you have not encountered previously. The teacher said that the students should try on their own to find a solution to the problem, that the better ones would manage to accomplish this somehow. You think of yourself as one of the better students. You just spoke on the phone with a fellow student who told you that it wasn't easy, that one would have to be really good, but that he managed to solve the problem. Your motivational eagerness to solve this problem is strongly activated. However, you have been sitting there with the problem in front of you for the past 10 minutes, with no solution anywhere in sight. You have tried all kinds of things, but nothing worked. You start to feel frustrated. Why did the other student manage to solve it? What was the teacher thinking, to demand something impossible? Suddenly you are jolted by an idea: What if I tried it the other way around? Ah yes, it might work like this, I'll give it a go. The paralysis and frustration have vanished; you are following a hot trail. Yes, of course, this is how it works! I have solved it! I'll show the teacher! No way he can trick me with this kind of problem! You manage to solve the problem. You attempt a second problem of the same type. It takes some time until you figure out the trick this time. You have to retrace, at first, how exactly you managed to solve the first problem. Now the principle emerges more clearly. This is how the second problem can also be solved. The third one is solved even more easily, and after that you can solve one problem after another without much difficulty. You got it! Problems of this type will no longer pose difficulties.

Now let us reflect on what might have happened within your brain during this process. When you were faced with the new problem, motivational goals and self-efficacy expectancies were both activated, and solving the problem would have meant satisfaction and confirmation in this regard. Your mental system was oriented toward the achievement of the corresponding perceptions. For that purpose, brain regions that are responsible for the processing of logical relationships and numbers were activated. Relevant memory content—that is, previously facilitated neural connections—which had a particularly high excitability in this context (including the problem type, topics previously discussed in this math class, etc.) was activated. After the activation of these neural activation patterns did not lead to perceptions that were congruent with the activated goals—and, therefore, no reduction in incongruence was achieved—new neural activation patterns were activated, including eventually even those with relatively little excitability. This quick, sequential activation of different, overlapping neural activation patterns constitutes an unstable state as activation fluctuates among all kinds of possible patterns. If one of these intermittently briefly activated states had led to a perception that was concordant with the motivational goals, then this state would have been shifted into the focus of conscious attention and would have been maintained in working memory in the same way that it is generally possible to keep memory content that is relevant for attaining a currently pursued goal in working memory (see also the discussion in chapter 2 on goal-oriented action). The longer this neural activation pattern is maintained, the better it is facilitated. In this case, none of the previously facilitated activation patterns was effective in reducing the existing motivational incongruence. To the contrary: The continued incongruence conflicted strongly with your self-efficacy expectancies and, thereby, led to the activation of thoughts (what was the teacher thinking, etc.) that did not contribute to solving the problem. None of the neural activation patterns was able to stabilize. Your neural activity oscillated among the various possibilities.

In the course of such fluctuations, it is possible that patterns are created for fleeting moments that have never previously existed and certainly have not been facilitated as such. The fact that these new, fleeting, never before seen patterns can come into existence at all is an effect related to the previously facilitated patterns. Without them, the neural activity would change back and forth in the same situation and fluctuate among patterns that have no chance of contributing to a solution. Given this previously facilitated preknowledge, however, it is possible that in the process of these unstable fluctuations, new, never previously attained brain patterns are formed, which can suddenly reduce to some degree the existing inconsistency within the current context (Stop! It might work like this!).

I had reported in chapter 2 in the section on goal-oriented action that goal-oriented activity is accompanied by differential activity of dopaminergic neurons. If an expected positive perception is not encountered, dopamine neurons are actively inhibited, according to Hollerman and Schultz (1998). This is quite useful because the activation of dopamine neurons leads via the release of dopamine to second messenger processes at the currently activated postsynaptic neurons and thereby facilitates the strengthening of synaptic transmission. It would be counterproductive, however, if even the currently transpiring processes that do not lead to goal-attainment were to be facilitated in this way. Dopamine neurons tend to fire predominantly when pleasant consequences occur surprisingly (Mirenowicz & Schultz, 1994, 1996). This is the case in our current example in just the moment in which you suddenly experience a mental jolt that allowed you to see the solution. The neural constellation existing in just that moment is now shifted into working memory via attentional control. This means that the neurons involved in this pattern fire continuously. The NMDA receptors are unblocked, which opens the door for the second messenger cascade described in chapter 2, which then ultimately leads to long-term potentiation. Because of the simultaneous firing of dopamine neurons, dopamine is released into the synaptic gap and binds to the dopamine receptions of neurons participating in this specifically new activation pattern. This triggers another second messenger cascade, which—along with the cascades initiated by the NMDA-receptors—leads via the transcription of genes to a permanent strengthening of the transmission at the synapses involved in this activation pattern. As a result, the activation pattern can eventually be reactivated with greater ease. The newly acquired solution path can ultimately be easily integrated with other thought patterns, such as frequently used terms.

With this example, I aimed to show how entirely new neural activation patterns—which are associated with an entirely new quality of experience or which open up opportunities for entirely novel behavior—can come into existence. The processes transpiring at the synapses that facilitate these new neural connections could be termed a differential strengthening of newly formed cell assemblies via consistency gain, which in this case is reflected in the reduction of incongruence. In his theory of the selection of neural groups, Gerald Edelman (1987) has conceptualized the formation of emergent experiential qualities in a very similar way. Among the many possible neural groups that can be formed in the brain at the respective moment, only those that are followed by inconsistency reduction are selected via differential strengthening or reinforcement.

This learning mechanism—learning via inconsistency reduction—is very common in daily life. When a small child sees something that he or she wants to have, he or she initially simply reaches to obtain it. This could be, for instance, a tasty pudding while the family is gathered around the dinner table. The mother or father mandates restraint: "Stop! You first have to ask" or "Stop, you have to wait until everybody has been served," and then they remove from the child the pudding that he or she had already obtained. They thereby create inconsistency tension. The impulse to obtain the desired object and the fear of losing the parents' affection or of being punished are simultaneously activated. The activation of important goals triggers the release of a great deal of dopamine, such that a high reinforcement potential is created. If the child says, at this point: "May I please have some pudding?" the inconsistency tension is effectively reduced. The child receives the pudding and the displeasure of the parents changes into approval. After repeated facilitation of similar sequences, the child will not simply reach for an object when this impulse arises but, instead, will ask courteously if he or she may have some pudding. Even later, this impulse might even be further transformed into, "Would anyone like to have some pudding?" In his book *On the Process of Civilization*, Norbert Elias (1982) has provided a wealth of fascinating examples that show the gradual transformation of impulsive, "uncivilized" behavior, via the help of initially explicit threats but soon via shame and guilt feelings, into "civilized" behavior that satisfies the original wish but simultaneously complies with social demands. The mechanism by which such complex forms of behavior emerge always involves the creation of inconsistency tension, which then is reduced by emitting new, "more civilized" behavior. The reduction of inconsistency tension reinforces the synapses that participate in this new behavior.

In these examples, the newly learned behavior also serves the satisfaction of basic needs. It widens the repertoire the individual has at his or her disposal for the satisfaction of basic needs. What is happening can be described as a differentiation of the person's motivational schemas. Reduction of inconsistency can lead to the emergence of new and adaptive forms of behavior, then. The same mechanism can enable the emergence of new avoidant forms of behavior, then, which limit rather than improve the person's options for basic need satisfaction. The short-term gain in consistency that is created by the avoidance behavior is produced at the expense of further opportunities for need satisfaction.

Reduction of inconsistency can also result in the formation of new patterns of mental activity that initially do not seem to have any recognizable function in terms of need satisfaction. An example would be the formation of attitudes and persuasions in the process of dissonance reduction, as examined a thousand times in research on cognitive dissonance. These are often not instrumental behaviors that directly affect the environment but rather cognitions that might never be expressed in behavior if they had not been explicitly studied. Attitudes, beliefs, and persuasions that are created in the process of dissonance reduction are the best examples of the idea that inconsistency reduction must be regarded as an independent motivational force in mental functioning. If one were to regard only the striving for basic need satisfaction as a source of motivation, their function in many activities would remain somewhat unclear. What has been learned via the reduction of conflict tension and dissonance reduction is from then on an important factor in mental activity, and one could wonder endlessly about the origin and function of the pattern if one didn't already know that its genesis must one day have served the purpose of inconsistency reduction.

Earlier I discussed neural mechanisms of consistency maintenance that have a structural basis in the brain. In addition, functional consistency-ensuring mechanisms can be formed based on the same learning mechanism that was just described here. Everything that serves to prevent the appearance of high inconsistency tension or that down-regulates such tension once it has appeared has the potential to be facilitated via inconsistency reduction. In this way, mechanisms can evolve that will be habitually employed as regulators of inconsistency. The literature uses various terms about such mechanisms for consistency maintenance, such as *coping, defense mechanisms, emotion regulation*, and *stress resolution* (Znoj & Grawe, 2000). All of them serve the purpose of keeping the inconsistency tension on a tolerable level. In this context, too, one can differentiate among more adaptive versus less adaptive forms. Consistency-ensuring mechanisms that actually elevate the inconsistency level over the long term can be regarded as unfavorable, as would be the case, for example, in repression.

One of the core features of borderline personality disorder is the patients' tendency to repeatedly enter states of intolerable inconsistency tension, in which self-mutilation, aggressive outbursts, or dissociations can occur. These "symptoms" of borderline personality disorder can be regarded as attempts by the person to somehow reduce the intolerable inconsistency tension. It is a logical consequence, therefore, that Linehan's dialectical behavior therapy devotes considerable attention to the construction of adaptive consistency-maintenance mechanisms. The patients learn to use strategies such as "mindfulness," stress tolerance, and emotion regulation (Bohus, 2002). All of these have the goal of avoiding the occurrence of extreme inconsistency tension, so that the extremely destructive forms of inconsistency reduction that are a core part of the borderline syndrome will not be used. One attempts to reduce these symptoms by removing from them their inconsistency-reducing function.

This example of borderline personality disorder shows a principle that can be applied to other disorders as well. Certain components of mental disorders can be construed as direct attempts to down-regulate an acutely intolerable inconsistency tension. These components are originally learned because the reduction of inconsistency tension that was caused by them led to negative reinforcement. They are later reactivated whenever a particularly high inconsistency tension is present, and they are maintained by repeatedly leading to further short-term reductions in inconsistency tension. They often serve the purpose of regaining some degree of control in a situation that is marked by an enormous amount of tension, as is typical for the state of inconsistency tension. Inconsistency tension cannot be clearly linked by the person to specific objects because this tension results from processes that the person is generally not fully aware of. In anxiety disorders as well, important components of the disorders serve the purpose of consistency maintenance. The worrying in generalized anxiety disorder; the counting, controlling, or cleaning in obsessive–compulsive disorder; and the avoidance in agoraphobia can all be regarded as desperate attempts to gain control over a situation that is experienced as intolerable but that is hard to define. Regardless of how the patients describes this state subjectively, its original cause is related to the fact that mutually inconsistent processes are activated simultaneously—processes that are highly relevant for the motivational goals of the person but that at this moment are not consciously represented. Over time, the control attempts that have been facilitated via negative reinforcement will occur earlier or sooner, even prior to the emergence of a fully fledged inconsistency tension.

Learning via inconsistency reduction is a particularly important form of learning, but it is not the only way to learn. Whenever neural activation patterns are formed repeatedly, the synapses among the simultaneously activated neurons are facilitated. It is sufficient to see a new object repeatedly in order to be able to recognize the object more easily later on. The longer the object is present in our working memory, the better the neural activation pattern corresponding to its perception will be facilitated. This is explained by the second messenger processes that originate at the NMDA receptors.

The facilitation will be even more effective if other second messenger processes join in, for example, those originating from dopaminergic or noradrenergic receptors. Neurotransmitters; hormones; and peptides such as dopamine, norepinephrine, acetylcholine, adrenalin, and oxytocine can massively increase the transmission at the synapses and, thereby, can improve the capacity to learn. In such cases, we speak of "motivated learning." Neural activation patterns that are activated conjointly with important motivational goals are particularly well facilitated because the activation of motivational goals leads to increases in dopamine release and, often, also to increased adrenaline release (see sections 4.5.2.1 and 4.7.6). This is true not only for instrumental behavior. A conditioned reaction such as saliva flow in response to a previously neutral tone that has repeatedly accompanied the appearance of food will form more quickly and enduringly in a hungry dog, compared to a satiated one.

Strong inconsistency tensions always include important motivational goals. The activation of such goals—even or perhaps especially when they are being blocked—is always associated with an increased release of neurotransmitters and hormones, which shift the mental system into a particularly high state of heightened learning readiness. We have seen in section 4.7.6 that the reduction of an aversive state is accompanied by a massive release of dopamine. The result—easier and more thorough facilitation—is the basis of the mechanism of negative reinforcement. A strong inconsistency tension is an aversive state that people typically aim to avoid. In section 4.5.2.1 we have seen that adrenaline is released in various brain areas during stress situations, which also strengthens the signal transmission at the synapses. A strong inconsistency tension is such a stress situation. Neural activation patterns that are followed or accompanied by a reduction of strong inconsistency tension are, therefore, particularly thoroughly facilitated. This also holds true for simultaneously activated activation patterns that did not directly cause a reduction in inconsistency. Such patterns can also be activated more easily in the future or will more likely occur spontaneously.

The noninstrumental components of psychopathological disorders—such as uninhibited autonomic arousal and its consequences, in the case of anxiety disorders—will be triggered automatically when important motivational goals are activated but cannot be attained. According to Devinsky, Morrell, and Vogt (1995), the anterior executive region permanently monitors the motivational significance of internal and external stimuli (see section on generalized anxiety disorder in chapter 3). It generates context-dependent emotions via the activation of some and the inhibition of other neural connections. There are many feedback loops within the anterior executive region that normally keep emotional reactions within a particular range of variability. If the inconsistency tension lasts for a long time, however, processes are initiated that were described in section 4.5.2.2 as consequences of uncontrollable incongruence. The normal feedback processes are overwhelmed and an escalating cycle rather than the inhibition of the stress and anxiety reaction that was elicited by the uncontrollable incongruence is observed. This leads to a permanent hyperarousal with all its consequences, which were described in the sections on the neural correlates of anxiety disorders.

These dysregulated autonomic reactions occur simultaneously with the inconsistency tension that ultimately caused them. At the same time, the mechanisms discussed previously for the reduction of inconsistency tension are activated. The various components of the disorder, the motivational inconsistency, the autonomic hyperarousal, and the mechanisms for the down-regulation of inconsistency tension (worrying, controlling, etc.) are simultaneously activated and soon grow together into an interconnected disorder pattern. The "glue" responsible for this joining of simultaneously activated neural groups is the reduction of current inconsistency tension, which is caused by some of the disorder components. All of the synapses that are activated at this moment—and not just the ones that actually contributed to inconsistency reduction—will be facilitated more thoroughly under the influence of the neurotransmitters and hormones that are currently released. Over time, an increasingly firm syndromal pattern is created, whose different components are increasingly strongly connected to each other, such that ultimately the activation of only one component can trigger the entire disorder pattern. Thus, it is possible that eventually even the fear of fear can elicit a full-blown panic attack.

Once this state of a good facilitation of the disorder pattern has been reached, it can be activated via its single components even without the presence of current motivational incongruence. The disorder pattern becomes disconnected from its engendering conditions. It gains partial functional autonomy. It will continue to be activated in situations of strong inconsistency tension, but it can also be activated even when no inconsistency tension is in place, for example, by situations or cognitions to which it is linked via conditioning.

Inconsistency tensions play a central role in the pathogenesis of mental disorders. They can also play an important role in their maintenance, whenever a situation is present in which a high level of motivation exists. Disorder patterns can also persist, however, without the requirement of being continuously reinforced via inconsistency reduction. In those cases, they have become a well-established memory component that can be activated via all kinds of trigger points. When I speak of memory in this context, I refer primarily to implicit memory. The disorder components have their neural foundations in implicit memory systems. They can be triggered bottom up via perceptions or memories that do not necessarily have to be represented in working memory. This is the main reason why it is not, or is only partially, possible to control them volitionally.

It is critically important in the context of treatment to know whether a disorder is still being reinforced by inconsistency reduction. If that is the case, the inconsistency tension must be reduced in order to remove the disorder pattern from its function in the context of inconsistency regulation. However, this will not eliminate the neural foundations of the disorder pattern in implicit memory. These foundations are well-facilitated neural activation patterns that must be actively inhibited so that they are less likely to recur in certain situations.

The generation of active inhibition is also necessary when the disorder pattern has become autonomous and no longer serves the reduction of inconsistency tension. In those cases, the creation of such inhibition becomes a central therapeutic task. A con-siderable repertoire of disorder-specific interventions has been developed based on decades of experience and research. These interventions can be employed here for the purpose of inhibiting the specific disorder patterns. Some of the interventions could perhaps be further improved and be selected and administered even more efficiently, once more knowledge about the neural foundations of each of the disorders has been uncovered, and once the mechanisms of action of these interventions have been clarified from a neural perspective. For the time being, the disorder-specific treatment research has progressed further than the neuroscientific research on the various spe-cific mental disorders and, therefore, this research should for now continue to guide therapeutic decision making if one aims to effectively inhibit the neural bases of mental disorders.

4.8.4   Inconsistency and Mental Health

Based on the discussions in the previous three sections, inconsistency has an acceler-ative function in mental activity. It motivates the formation of new neural activation patterns and stabilizes such patterns via reinforcement. This can result in the creation of new options to enable need satisfaction. This learning mechanism is indispensable for the differentiation of motivational schemas and their adaptation to ever-changing life circumstances and situational demands.

If it is not possible to reduce states of inconsistency and the inconsistency remains in place over longer periods, it tends to have negative effects. It compromises well-being, reduces the effectiveness of mental processes, and can lead over the long term to the creation of mental and physical disorders. From this perspective as well, inconsistency can be said to have an accelerative function, by engendering the formation of new patterns of mental activity that can become a source of suffering. If the formation of such harmful new patterns is maintained over longer periods, this inconsistency can even cause ill health. Because the formation of disorders also in itself constitutes a violation of the needs for pleasure maximization and distress avoidance, as well as a violation of the needs for control and self-esteem, the disorders, once formed, contribute further to the overall level of inconsistency in mental functioning. This encourages the formation of additional unfavorable mechanisms for the down-regulation of inconsistency and increases the probability that additional disorders will form. An escalating cycle that leads eventually to high levels of comorbidity is set into motion. The individual is caught in this vicious cycle of short-term inconsistency reduction by formation of new symptoms and long-term inconsistency increases due to just these same symptoms. It is therefore plausible to assume that patients with many comorbid disorders will exhibit high levels of inconsistency and a poor satisfaction of basic needs.

These are assumptions that follow logically from a consistency-theoretical perspective on mental functioning. But what about the empirical evidence for these assumptions? In the following section, I will review relevant findings with regard to three types of inconsistency: dissonance, motivational conflicts, and incongruence.

4.8.4.1 Continuous dissonance undermines physical health

The first form of inconsistency I will review in the context of physical health is somewhat outside the frame I have articulated up to now. That is, this section will first address the topic of musical dissonances. Thirty years ago, Fuhrmeister and Wiesenhütter (1973) showed in a study of orchestra musicians that too much dissonance is experienced as aversive by our mental system and can cause physical illness over the long term. The researchers conducted a semistructured clinical interview, which took up to three hours, with 208 professional musicians from three different orchestras. The orchestras differed in terms of the music that they played regularly. Orchestra A played exclusively contemporary music. The proportion of contemporary music in orchestra B was about 25%. Orchestra C did not play any contemporary music at all but focused exclusively on classical music, opera, and operetta. The musicians were asked in detail about all kinds of physical and psychological complaints and about how they experienced the music that they played. They did not differ in their mean age (46), and the analyses took into consideration whether the respective complaints had occurred prior to their joining the orchestra or only later on.

The study yielded an unambiguous picture: The playing of contemporary music was universally hated by the musicians because of the dissonances it contained, which they thought created an intolerable mental and physical state. This opinion was shared by the musicians in all three orchestras, even those in Orchestra C, which did not ever play contemporary music. A comparison of the health status of the musicians showed that, "the physical as well as the mental health status of the predominantly 'contemporary orchestra' (A) was significantly the worst (Fuhrmeister & Wiesenhütter, 1973, p. 75, translated from German)." It was clear that the medication consumption of musicians in Orchestra A was significantly higher than that in the two other orchestras; it was regarded as natural that almost every musician in this orchestra took relaxation, pain, or sleeping pills, and such pills were sometimes even 'officially' distributed prior to practice sessions. The musicians, in turn, justified all of this by referring to the adverse consequences of contemporary music" (p. 78).

Figure 4–18 shows the percentage of musicians that developed certain illness symptoms after joining each respective orchestra

Following their joining the orchestra, the musicians in Orchestra A tended to suffer primarily from gastrointestinal disorders, severe sleep disorders, headaches, and severe forms of nervousness, at greater frequencies than the musicians in Orchestra C, with those in Orchestra B taking a middle position. These results were concordant with the statements made by the musicians. The musicians were more physically ill to the degree that they were exposed to contemporary music with its frequent dissonances. Constant listening to, or creation of, dissonant music facilitates illness. The validity of this result was further confirmed by the finding that the musicians in Orchestra B tended to complain more about symptoms that had to do with the frequent playing of their instruments—the typical symptoms found among professional musicians. The results can therefore not be explained by a simple exaggerated tendency to complain or to assume the sick role among musicians in Orchestra A. It really appears that it has to do with the type of music they play.

This interpretation of the findings is also supported by the finding that the musicians tended to develop particular symptoms more frequently after practice sessions in which contemporary music was played, compared to sessions in which classical music was played. Figure 4–19 shows the results in this context for musicians of Orchestras A and B. Orchestra C is not included here because this orchestra never played any contemporary music at all.

This comparison, which addresses the immediate consequences of playing contemporary music, shows that musicians in Orchestra B were even more negatively affected than those in Orchestra A. This also supports the idea that the symptoms are, in fact, a direct consequence of the dissonance. More than 80% of the musicians felt nervous and irritable if they had just previously played contemporary music, and about one third complained directly afterward about headaches, sleep disturbances, or depression. It is therefore not surprising that the musicians who had been exposed the most frequently to these short-term effects (those in Orchestra A) also developed the most severe and enduring health impairments.

The results of this study strongly support the idea that our mental system has a natural preference for harmony, consonance, and consistency. If this preference or tendency cannot be realized continuously, or when frequent opposing effects are encountered, the system responds with dysregulations of the otherwise smoothly coordinated systemic processes, which ultimately will be reflected in the form of mental or physical complaints.

According to a PET study by Blood, Zatorre, Bermudes, and Evans (1999), who studied participants' brain activity while they listened to six different versions of the same music piece, musical dissonances activate primarily the parahippocampal gyrus, which is closely connected to the amygdala. The unpleasant perceptions that are triggered by dissonant music apparently activate brain structures that have been implicated in negative emotions. If such dissonances recur frequently, they can repeatedly elicit negative emotions via this amygdala-mediated pathway. Hyperactivity of the amygdala has been documented in anxiety disorders and in depression. The pathway

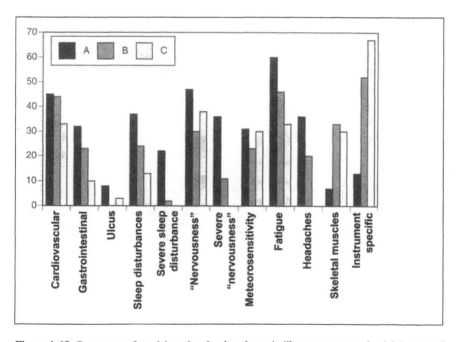

Figure 4–18. Percentage of musicians that developed certain illness symptoms after joining one of three orchestras (A, B, or C). Orchestra A = exclusively contemporary music; Orchestra B = 25% contemporary music; Orchestra C = classical music, opera, and operetta only. (Data from Fuhrmeister & Wiesenhütter, 1973).

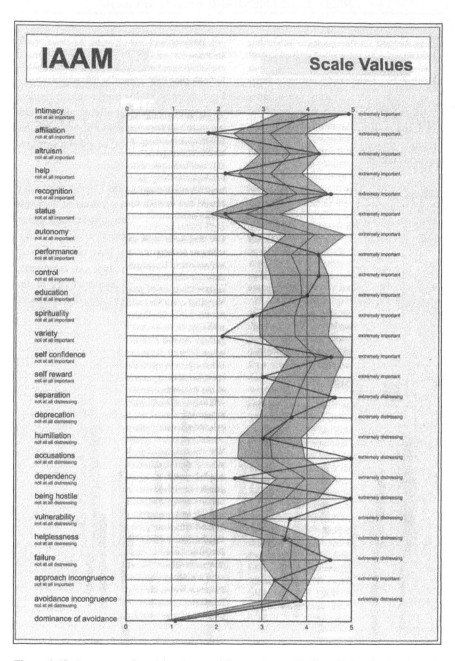

Figure 4–19. Percentage of musicians in one of three orchestras (A, B, or C) that developed certain somatic and psychological symptoms immediately after practice sessions. Orchestra A = exclusively contemporary music; Orchestra B = 25% contemporary music; Orchestra C = classical music, opera, and operetta only. (Data from Fuhrmeister & Wiesenhütter, 1973.)

via the amygdala could be the route by which inconsistencies exert their harmful effects on mental functioning.

Listening to dissonant music constitutes yet another form of inconsistency, then, in addition to the forms that are already listed previously. This form of inconsistency is not of primary importance in the context of psychotherapy, but it shows powerfully that inconsistency is not merely unpleasant but can also have harmful consequences for well-being and physical health.

4.8.4.2 Negative consequences of motivational inconsistency

Many studies in the experimental conflict research in the 1940s used animals to experimentally create certain kinds of conflict (see section 4.8.1). A regular finding was that the animals developed "experimental neuroses" when they were confronted with unsolvable conflict situations. Their behavior resembled that which Seligman observed and described in his studies on the paradigm of learned helplessness. An unsolvable conflict inherently contains the total loss of control and, therefore, can indeed be viewed as a state of helplessness.

I will not discuss these studies in detail here because other, newer studies have examined the effects of conflicts in human. The results of these newer studies are undoubtedly easier to apply directly to questions relevant for psychotherapy.

Emmons and King (1988) asked 40 college students to list 15 of their most important personal strivings. Examples of personal strivings are, for example, "to be attractive for people of the other sex," "to be a good listener," or "to be better than others." A Striving Instrumentality Matrix was then used in which each participant estimated how compatible or incompatible each possible pair of strivings was. In addition, the degree to which the strivings were ambivalent or clear-cut was rated. When a goal is being pursued but negative consequences are also anticipated when the goal is attained, this ambivalence can also be regarded as one form of inconsistency. Significant correlations were found for both indexes—the total degree of conflict among goals, as well as the total ambivalence across goals. Both indexes correlated significantly with the severity of negative emotions as well as the scales assessing somatization, anxiety, and depression of the SCL-90-R. The correlations were of a magnitude between .30 and .40. In a second study with 48 students the correlations were somewhat lower but nevertheless significant. In the second study, the amount of goal conflict also correlated with the "hard" data—that is, not self-report based—of the number of visits to the student health center and the number of diagnoses given by the center physicians. Both indexes were based on the files of the health center and not just self-reports of the participants. Students who experience their most important personal strivings as conflicted, then, have an objectively worse physical health.

Building on the personal strivings approach by Emmons as well as Deci and Ryan's (1985, 2000) self-determination theory, Sheldon and colleagues (Sheldon & Elliot, 1999; Sheldon & Houser-Marko, 2001; Sheldon & Kasser, 1995) developed a model surrounding the concept of self-concordance. Self-concordance is present when the goals that a person pursues are in line with his or her highest values and interests; that is, when the person can identify with the goals. This is the condition corresponding to the pursuit of intrinsic goals. Not all "personal strivings," as described by Emmons are necessarily self-concordant in this sense. They can also be pursued because of external pressure or in order to placate a guilty conscience. Sheldon and Kasser (1995) differentiate between vertical and horizontal concordance of goals. Horizontal concordance is present when goals on roughly equivalent levels support each other. Vertical coherence is present when the goals on lower hierarchy levels are consistent with higher-level goals. On the highest level, vertical coherence is identical with self-concordance. Two individuals might pursue the same goals on lower levels but might differ in terms of their overarching goals, as shown in Figure 4–20.

The goals of example Person A in the figure have a higher degree of vertical coherence than those of Person B. Person A will, therefore, pursue his or her concrete personal strivings with a higher degree of intrinsic motivation, will devote greater effort, and will consequently be more likely to attain the goals. Person A will thus enjoy a higher level of need satisfaction than person B and, in turn, will benefit from greater well-being. Figure 4–21 presents a graphical depiction of the complete self-concordance model with its assumptions and predictions.

The self-concordance model also makes another prediction, beyond what has already been conveyed: that the attainment of self-concordant goals results in a greater need satisfaction than the attainment of goals that are less self-concordant.

The specific assumptions of this model were tested in several studies, most of which were correlational rather than experimental in their design. The various constructs of the model were operationalized with different assessment devices and questionnaires. In most studies, samples of college student participants were followed over the course of a semester, and the data were analyzed using correlations or structural equation modeling. To provide an example, Figure 4–22 shows the results of a study by Sheldon and Elliot (1999) among 73 college students.

This structural equation model supported all of the assumptions of the model. The specific assumption of the self-concordance approach—first, that concordant goals would be pursued with particularly strong intrinsic motivation and, second, that the attainment of self-concordant goals would contribute more to well-being than the attainment of less self-concordant goals—was supported not only in this one study but in all the studies that were conducted, at an alpha significance level set at the conventional .05. These results documented that the vertical coherence of concrete goals (goals that are closer to specific behavioral acts) represents a specific form of consistency, whose relevance for the attainment of well-being can be regarded as soundly empirically substantiated. These results can be integrated very easily into consistency theory, even though the specific assumptions are not made explicitly in that model.

There is another point in which the self-concordance model is remarkably compatible with consistency theory. Both approaches assume that goal attainment affects well-being via the pathway of need satisfaction. Sheldon and Elliot derive their understanding of need satisfaction from self-determination theory, as articulated by Deci and Ryan (2000), in which three basic psychological needs are postulated: The need for autonomy, competence, and (interpersonal) relatedness. In one of their studies, Sheldon and Elliot asked their students at several time points over the semester to estimate, on a scale from 1 to 7, how much they had encountered experiences that would satisfy these needs. The respective items were, "feeling generally competent and able in what I attempt," "feeling generally autonomous and self-determined in what I do," and "feeling generally related and connected to the people I spend time with." These ratings correlated for each of the three needs at around .50 with well-being measured at the same time. The assumption that this type of goal attainment is linked to need satisfaction, as operationalized in this study, was clearly supported in the structural equation model.

Most of these and similar findings from Sheldon's work group have been replicated several times and complement important assumptions made by consistency theory. Although I have some reservations as to whether the three items could appropriately measure what I mean by basic need satisfaction and I also regard the basic needs postulated in consistency theory as better substantiated than those articulated by Deci and Ryan, it is notable that Sheldon and Elliot at least attempted to operationalize the

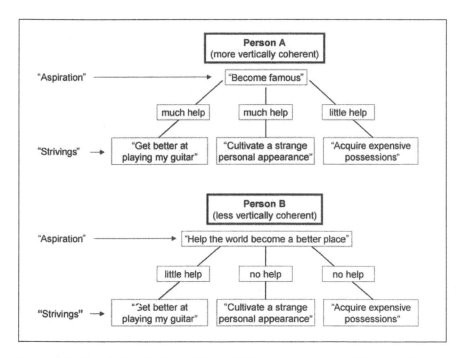

Figure 4–20. Example from Sheldon and Kasser (1995) for goal structures with higher and low vertical coherence (Adapted from Sheldon & Kasser, 1995; p.532)

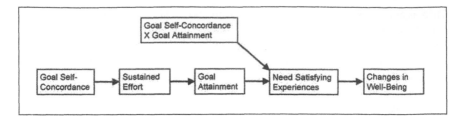

Figure 4–21. The self-concordance model according to Sheldon and Elliot (1999) (Adapted from Sheldon & Elliot, 1999; p.483).

construct of need satisfaction and to show the correlation indicating that need satisfaction is not simply identical with goal attainment.

Brunstein and colleagues (Brunstein, 1993; Brunstein, Schultheiss, & Grässmann, 1998; Brunstein, Schultheiss, & Maier, 1999) have conducted several studies to examine the influence of the consistency of explicit and implicit motive dispositions. In line with McClelland (1985), these authors regard motive disposition as enduring motives that are similar to basic needs, which determine pervasively what a person regards as important. These dispositions generally influence behavior outside of conscious awareness. McClelland differentiated four such motives: The striving to be influential (power motive), to be able to do something better (achievement motive), to be together with others (affiliation motive), and to have romantic relationships (intimacy motive). Influenced by Bakan's (1966) distinction between two "fundamental modalities in the existence of living forms, agency for the existence of an organism" (pp. 14–15), which he termed *agency* and *communion,* Brunstein classified the striving for power and achievement as agency motive dispositions, and the striving for affiliation and intimacy as communion dispositions. Because it is assumed that these basic motives influence mental activity in the implicit mode of functioning—also following McClelland—they were assessed not with questionnaires but via stories that the participants were asked to tell in response to Thematic Apperception Test-like pictures. The scoring was performed in a way similar to modern TAT analyses with a coding manual, which enabled the researchers to achieve high reliability levels. In this way, Brunstein and colleagues were able to categorize their participants into predominantly communion-oriented versus predominantly agency-oriented groups. The participants did not know, of course, that these characteristics were ascribed to them. They were asked to name four personal goals that they intended to pursue in the near future. Four categories were presented:

- Striving for intimacy and interpersonal closeness.
- Striving for affiliation and friendly social contacts.
- Striving for performance and competence.
- Striving for independence and influence.

Participants were asked to name a personal goal that they explicitly intended to pursue in the near future. In the following 2 weeks, participants rated their well-being

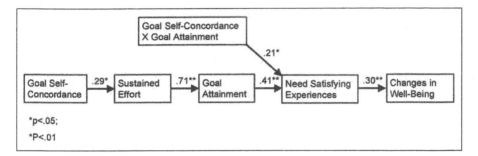

Figure 4–22. Structural equation model of study 3 in Sheldon and Elliot (1999) to test assumption underlying their self-concordance model (Adapted from Sheldon & Elliot, 1999; p. 492).

every other day as well as, at the end of the period, how much progress they had made with respect to each of their goals. In a second study, their goal commitment and estimated goal attainability were also measured.

The data showed that the attainment of goals led to improvements in well-being only if the goals were consistent with the individual's basic motivational orientation. Whenever an agency-motivated participant attained a communion goal, for example, this had no effect on his or her well-being. The same held true when a communion-oriented participant attained an agency goal. The positive effect on well-being was particularly strong when a motive-congruent goal with high commitment was pursued under conditions of high attainability. Motive-incongruent goals that were pursued with high commitment were less likely to be attained than motive-congruent goals. High commitment to a goal is not sufficient, then, if the goal is not being pursued with intrinsic motivation; that is, when it is not concordant with the person's basic motivational orientation. In such conditions, high goal commitment was even associated with decrements in well-being.

In a study with 72 psychotherapy patients, Michalak, Klappheck, and Kosfelder (2004) initially measured the most important personal strivings of the patients. They then asked the patients to estimate the importance of each goal and the expectancy as to whether or not they would attain these goals with or without the aid of therapy. They formed a product of expectancy × value and thereby obtained an index for the *volitional strength* (Heckhausen, Gollwitzer, & Weinert, 1987) of goal pursuit. Beyond that, they asked each patient to indicate whether each goal was pursued primarily for external, imposed reasons or because they personally identified with the goal and were intrinsically motivated. In addition, various measures of psychopathological symptom severity were obtained, as well as the average session outcome for the five therapy sessions that followed the therapy goal assessment. Volitional strength as well as intrinsic goal motivation correlated significantly and strongly inversely (around −.50) with the various measures of symptom severity. The patients who had not set clear goals that they pursued with intrinsic motivation, then, were feeling much worse. In addition, both goal aspects correlated with session outcome. The patients with intrinsically motivated goals and those who had a high volitional strength for their goals achieved more desirable changes in the session than patients who were more ambivalent about their motivation and who pursued goals more for external reasons. Both goal aspects, volitional strength as well as intrinsic motivation, refer to aspects of motivational consistency in mental functioning. A high level of motivational consistency is accompanied by better psychological health and better prospects for a good therapy outcome. In their previous study, among psychotherapy patients with anxiety disorders, Michalak and Schulte (2002) had already reported that a greater extent of conflict among personal strivings at the outset of therapy was linked with a worse therapy outcome. Their relatively worse therapy result was mediated by the fact that patients with high a high goal-conflict level work with less motivation in therapy, were less likely to open up to the therapist, were more reluctant to exert effort, and generally showed more resistance in the therapy process.

If we consider all these results in combination, a clear conclusion emerges for psychotherapy: Therapists should not support patients in the pursuit of therapy goals that are not consistent with their most important, overarching motivational goals and their self-chosen, intrinsically pursued personal goals, even if the patient declares the goal as very important. One should try whatever possible to encourage the patient to set goals that are truly consistent with his or her own needs. Goals that have the character of "I should," "I must," or "one can expect from me that I ..." will have little chance of being realized, and even if they are attained, they are unlikely to lead to real improvements. It is by no means trivial to inquire about the vertical coherence of goals the patient intends to pursue. It was a common finding in the studies by Sheldon and Elliot that participants set goals that were not self-concordant. It is also common that psychotherapy patients set goals whose attainment is actually inconsequential for their most important implicit motives. Even if such goals are pursued with high dedication, the chances of attaining the goal are slim, and even if the goal is achieved, little satisfaction results. The targets that therapist and patients must collaboratively strive for in treatment must be in the service of already existent important motivational goals of the patient. If that is the case, the prospects for goal attainment are good and goal-attainment will be accompanied by marked improvements in well-being.

When we treat a patient with a specific disorder by following a treatment manual that has proven to be effective, on average, then we thereby impose the goal of therapy. This goal, however, is not automatically compatible with the most important basic motives and the person's idiosyncratic goals. The goal suggested by the manual has the advantage of seeming reasonable, and it would be difficult for the patient to reject this goal because that would make it appear as if he or she is not interested in improving his or her state. However, the pressure that is implicit in the therapy situation for the adoption of this goal does not by itself render the goal motive-congruent. The answer to the question of which therapy goals therapist and patient should pursue together is not simply provided by the diagnosis itself. This answer must be reflected and decided against the backdrop of the patient's existing motivational structure, because the actual motor of therapeutic change is always the motives of the patient. He or she cannot simply reject or accept these motives. They are the underlying foundation of therapy, even for the patient, and the therapy must work with this foundation if it is to be successful. It is a considerable therapeutic mistake for a therapist to regard the goals that he or she deems correct as the foundation of therapy, without previously clarifying whether these goals are, in fact, consistent with the most important basic motives of the patient.

There are several other studies on the relationships between conflicts in mental functioning on the one hand and well-being and mental health on the other. These studies address not the inconsistency of goals but inconsistencies in the cognitive representation of important life domains. These studies have all utilized a computerized conflict-assessment developed by Lauterbach (1996). In this test, participants are asked to evaluate a number of items with regard to the positive or negative relationship between item pairs, and between the items and the person him or herself. For that purpose, three terms are always presented in triads, and one of the terms in the triad is always "I." The test analyzes the number of balanced and unbalanced relationships in the triad. Unbalanced triads are regarded as conflicted. The total degree of conflict expresses how much conflict tension is present in the participant's cognitive representation of important life domains.

Hoyer, Frank, and Lauterbach (1994) used this method to study a sample of alcoholic patients and a sample of patients at a psychosomatic clinic. The following terms were presented to them:

1. I.
2. My mother.
3. Steady relationship.
4. Steady work-place.
5. Spare-time.
6. Contact with others.
7. Independence.
8. Moderate alcohol consumption.
9. Relapse.
10. Therapy.

In addition, psychopathological symptom severity was measured with the SCL-90-R. Among alcoholics as well as psychosomatic patients (i.e., psychiatric patients, to use a term more commonly used for such groups in North American studies), the extent of conflict tension correlated highly significantly with symptom severity. Because this is a correlational association, it remains unclear for now how this link ought to be interpreted; whether the conflicts led to the symptoms or the symptoms led to an inconsistent cognitive structure. Consistency theory would regard both causal directions as plausible.

Renner and Leibetseder (2000) applied the conflict test developed by Lauterbach to a sample of psychotherapy patients and a sample of adults who participated in a psychological training course at their workplace. Psychopathological symptom severity was measured in both samples with the SCL-90-R. As expected, higher conflict scores were observed in the sample of therapy patients. Both samples were then pooled for the calculation of correlation coefficients. For the pooled sample ($N = 139$), a significant correlation of .32 was observed between the total conflict score and the average symptom severity. In another analysis, the authors formed one group with a high and one with a low degree of conflict, and the correlations between conflict and symptoms were then computed separately within these groups. A clear difference emerged between the groups. In the group with the lower conflict scores, the still existing variability in symptom severity correlated highly—at .62—with conflict. In the group with high conflict values, by contrast, the corresponding correlation was only .02. This finding suggests that conflicts facilitate the formation of psychopathological symptoms; however, once a certain level of conflict has been reached, further increases in conflict are not necessarily accompanied by additional increases in symptom severity. This result is quite plausible. The simple finding of "the more conflict, the more severe the symptoms" would hardly be compatible with the multitude of causal factors that influence symptom genesis, such as predispositions, consistency-maintenance mechanisms, and so forth. The results of this study are also in line with the assumption that inconsistency tensions play an important role in the formation of mental disorders.

The studies reviewed here, as mentioned previously, did not examine motivational conflicts but conflicted cognitive representations. Grosse Holtforth and Grawe (2003b) attempted to adapt Lauterbach's conflict test to measure the degree of conflict among patients' most important motivational goals. Instead of predetermined words, the patients' own motivational goals were presented, as previously assessed with the IAAM. The patients were asked to evaluate each motivational goal dyad in terms of their compatibility or incompatibility. The therapists were also asked to use this method to estimate the degree of compatibility among their patients' goals. The implementation of this method, however, was fraught with problems. The patients often required more than an hour for these evaluations and experienced them as quite exhausting, difficult, or even overwhelming. The therapists also experienced the evaluation of patient goal conflicts as more difficult than any other measures they were asked to complete over the course of therapy. In short, the evaluation of conflict among motivational goals with this method was distinctly unpopular among patients as well as therapists. Moreover, it became evident that there were not only practical problems with the measurement of conflicts; the analyses also showed that this method apparently does not produce valid assessments of motivational conflict strength. At the beginning of therapy, the total score for the extent of conflict among motivational goals, as measured by therapist and patient ratings, did not correlate with well-being or with symptom severity, as had been expected based on the pattern of findings reported in the literature (Betschart, 2002; Nydegger, 2003). It was only by the end of therapy that the therapist-rated, still-remaining degree of goal conflict correlated surprisingly highly (.62, $N = .32$) with the severity of the remaining symptoms (Nydegger, 2003). It appeared that the therapists were able to validly estimate the extent to which their patients' goals conflicted only by the end of therapy, after they had formed a much deeper and more detailed impression of the patient. The magnitude of the correlation between the observer-rated conflict extent and self-rated symptom severity suggests that a real association between symptom severity and conflict extensiveness does, in fact, exist, but that it may be possible only after many sessions for therapists to provide a valid judgment about the degree of this conflict. For patients, this was still not possible even at the end of therapy, or, at least, it was clear that by the end of therapy there was still no correlation between their self-rated conflict level and symptom severity.

Our experiences with this method suggest that it is doubtful whether people can provide a valid estimate of the degree to which their motivational goals are conflicting with one another. In cases of unconscious conflicts, it seems immediately obvious that this would pose a problem. For these reasons, we have now abandoned this method for measuring inconsistency and have opted to pursue an approach that proved to be more fruitful. This will be discussed in the next section.

4.8.4.3 Incongruence and mental health

The listing of the various forms of inconsistency in section 4.8.1 demonstrates that there are indeed many forms of inconsistency, all of which can contribute to the total level of inconsistency in mental functioning. The listing certainly cannot be regarded as complete. Each of these forms of inconsistency can exert negative effects on well-being and mental health. If one measures one of these aspects of inconsistency and then estimates its relationship with measures of mental health, the typical finding will be a significant but only moderately strong correlation.

In order to obtain a more thorough estimate of the total extent of inconsistency in mental functioning, one would have to assess all the different forms of inconsistency simultaneously. This is not possible, for feasibility constraints alone. In clinical practice such measurement efforts would also be completely unrealistic and impossible. However, it is particularly in clinical contexts that a reasonably reliable and valid measurement of inconsistency levels is of interest.

These considerations prompted us to limit our assessments of incongruence to one particular form—the form of inconsistency that perhaps best represents the other types of incoherence: The incongruence between motivational goals and actual perceptions. This decision was justified by the following reasoning: the various forms of inconsistency in mental functioning—interferences, motivational conflicts, dissonances, repression, dissociation, self-discrepancy, vertical incoherence, and so forth—all compromise the person's ability to engage effectively with the environment. Incompatible processes that are not coordinated well with one another tend to inhibit or hinder each other. This leads to the consequence that the person's motivational goals will not be satisfied effectively. High inconsistency in mental functioning should therefore be reflected in a high level of incongruence between motivational goals and the person's perceived reality. The incongruence level of a person—which in itself is an important source of inconsistency—can therefore be regarded as a sort of collective pool in which all the consequences of other inconsistency forms are reflected.

Another considerable advantage of incongruence as a total measure of inconsistency has to do with the fact that incongruence can be measured particularly efficiently. It is difficult for a person to report on his or her motivational conflicts, especially when one assumes that motivational conflicts are partially unconscious. This also holds true for other forms of inconsistency. However, it is by contrast relatively easy to report whether one has recently experienced certain events or situations, or not.

I mentioned in section 4.7.5 that we had used the IAAM, a questionnaire that assesses patients' approach and avoidance goals, as the basis for the development of another questionnaire, the incongruence questionnaire. This questionnaire takes each item of the IAAM and asks patients to estimate the extent to which each goal has been attained or, respectively, successfully avoided in recent times. Figure 4–23 contains an example of the items in this incongruence questionnaire.

Similar to the IAAM, the items are collated into scales that correspond to the motivational goals measured by the IAAM. This method yields 14 scores for approach incongruence (the extent to which desired goals have been attained) and 9 scores for avoidance incongruence (the extent to which feared outcomes have been encountered). In addition, like the IAAM, the incongruence questionnaire yields sum scores for approach and avoidance incongruence. Figure 4–24a shows a patient test profile resulting from the analysis of this questionnaire.

Figure 4–24a shows the incongruence pretest profile of a patient who had lost all his money because of his gambling addiction and now sought therapy for this reason. The total degree of incongruence is at a very high level, even when compared to that of other psychotherapy patients. This is true for approach goals as well as avoidance goals. The three least-satisfied approach goals of the patient are control, enjoying life, and self-reward. The most strongly violated avoidance goals are dependency/autonomy-loss, being vulnerable, helplessness and failure, and debasement/embarrassment.

Incongruence in the avoidance scales means that patients did not manage to avoid that the dreaded outcome has come true. The patient feels raw, dependent, helpless, totally devalued, and like a loser. He has lost control over his life, and there is nothing left that he experiences as pleasurable.

Figure 4–24b shows the posttest profile of the same patient and the profile at a 6-month follow-up. After therapy, all of the incongruences are substantially reduced. The only areas in which the patient continues to experience slight incongruences are the domains of being superior/impressing others, and accusations/criticism. The incongruence reduction remained largely stable even after 6 months. New questions were evident, however, due to the increase with regard to being alone/separation at the follow-up assessment. In a routine follow-up call, the patient indicated that just prior to completing the questionnaire, he had lost a close acquaintance due to a car accident.

This example shows that the incongruence questionnaire is well suited for the routine use in clinical practice. It provides suggestions about the patient's area of most acute maladjustment, about which basic needs are the most violated. If the IAAM is also administered to determine the importance of motivational goals, and this is then compared with the incongruence questionnaire, the profile comparison will quickly clarify in which areas changes must occur in order for the patient's needs to be satisfied more satisfactorily. The incongruence questionnaire does not just yield information about the degree of motivational incongruence but also about the sources of the incongruence, which can then be pursued further in clinical interviews. The IAAM and the incongruence questionnaire are generally highly valued by our therapists because they are relatively efficient means to obtain suggestions that are useful for therapy planning.

The example also shows that the incongruence questionnaire is useful for the measurement of therapy effects. When the pre- and posttest profiles of the patient are compared with each other, it is immediately obvious that the patient has improved greatly after therapy. It is not just the psychopathological symptoms that are reduced in therapy; rather, successful therapies also contribute to marked improvements in need satisfaction, as expressed by the incongruence profile. This is, of course, true only for cases in which incongruence was present at the outset of therapy. The pretest profile in Figure 4–24a shows that psychotherapy patients, compared to normal control samples, have markedly higher levels of incongruence.

If the purpose is simply to determine very quickly whether any incongruence is present for a particular patient, it is also possible to administer the short form of the incongruence questionnaire. This form includes only 23 items and requires no more than 5 minutes for completion. (The English versions of the incongruence question-

Part 1	1 = not at all
	2 = slightly
How satisfied have you recently been with regard to the following experiences?	3 = moderately
	4 = quite a bit
	5 = a lot
1. I've been productive	1 ... 2 ... 3 ... 4 ... 5
2. I've been independent	1 ... 2 ... 3 ... 4 ... 5
3. I've had faith in myself	1 ... 2 ... 3 ... 4 ... 5
... ...	1 ... 2 ... 3 ... 4 ... 5
13. I've stuck up for the weak or needy	1 ... 2 ... 3 ... 4 ... 5
14. I've lived a life full of variety	1 ... 2 ... 3 ... 4 ... 5

Part 2	1 = hardly ever
	2 = rarely
	3 = sometimes
Recently, I've experienced what the item says:	4 = often
	5 = very often
15. I've had to show my weaknesses to others	1 ... 2 ... 3 ... 4 ... 5
16. I've been inadequate	1 ... 2 ... 3 ... 4 ... 5
... ...	1 ... 2 ... 3 ... 4 ... 5
22. I've been left by a spouse, partner, or significant other (e.g., husband, girlfriend)	1 ... 2 ... 3 ... 4 ... 5
23. I've not received recognition	1 ... 2 ... 3 ... 4 ... 5

Figure 4–23. Sample items of the Incongruence Questionnaire (short version).

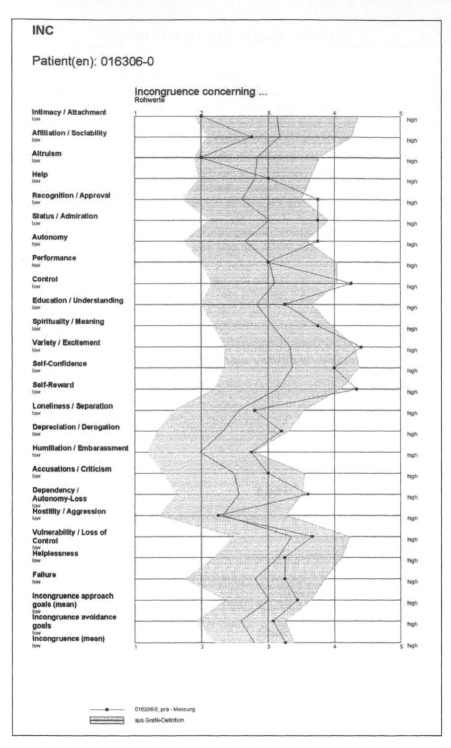

Figure 4–24a. Incongruence profile of the patient described in the text prior to the treatment (pretest). The thin continuous line in the middle of the gray-shaded area is the incongruence mean score of a mixed sample of psychotherapy patients on each respective scale. The gray-shaded area corresponds to the area between one standard deviation above and below that mean. Further explanations in text. (Adapted from Grosse Holtforth, Grawe, & Tamcan, 2003).

naire can be obtained from M. Grosse Holtforth, grosse@ptp.unibe.ch) The total score of the short form correlates with the long version at a magnitude of .97. Both forms yield equivalent results with regard to the overall level of incongruence.

I noted earlier that incongruence can be regarded as the most important form of inconsistency because all other forms of inconsistency are reflected within it. If this is correct, incongruence should be linked even more strongly with problematic well-being and poor mental health. In order to test this idea, we administered the incongruence questionnaire along with well-being and mental health measures to various samples of normal participants and clinically disordered patients. Detailed information on these results is provided in Grosse Holtforth, Grawe and Tamcan (2004). Figure 4–25 summarizes the correlations that we observed in a sample of psychotherapy patients.

The correlation between incongruence and well-being was indeed unusually strong. In a sample of 283 psychotherapy patients, the observed correlation was –.78. In a mixed sample of normal participants and various clinical groups ($N < 1,000$), the correlation was even –.87. This suggests that well-being depends almost entirely on the degree to which individuals manage to attain their motivational goals. The correlation with respect to approach incongruence was found to be even slightly higher than that for avoidance incongruence (–.77 vs. –.63). Similar patterns were observed for other measures of well-being. The well-being of a person is even more linked to the extent to which he or she manages to realize personal approach goals, compared to the extent to which he or she manages to avoid dreaded outcomes.

The correlation of .67 with psychopathological symptom severity is also very high in this large sample. In a sample of 200 normal participants, this correlation was even .75. A person who does not manage to realize his or her motivational goals, then, has a greatly elevated risk for the development of psychopathological symptoms. This correlation should, of course, not be interpreted in only one direction. Psychopathological symptoms bind mental energies, which also means that motivational goals will be more poorly realized. However, on the basis of what was said earlier about the significance of inconsistency for the formation of mental disorders, such a high correlation—as predicted by consistency theory—can also be interpreted as confirming the causal role of inconsistency for the emergence of mental disorders.

We also observed a very substantial correlation of .50 between the avoidant attachment style (as measured with the Measurement of Attachment Qualities (MAQ) by Carver, 1997) and the level of incongruence (this is not included in Figure 4–25). We know from previous research that attachment styles are very stable characteristics. A directional interpretation is therefore justified: An avoidant attachment style leads to elevated incongruence. This can also be regarded as strong confirmation of a relationship predicted by consistency theory.

Incongruence correlates positive (at .65) with neuroticism and with a tendency to experience negative affect (at .63, as measured with the PANAS scale by Watson,

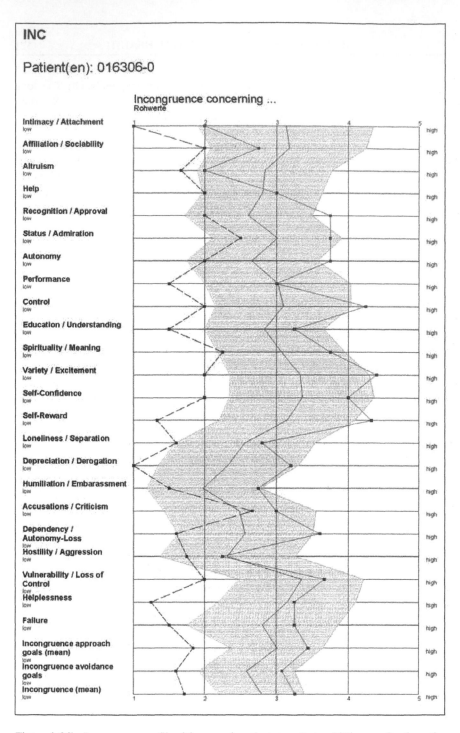

Figure 4–24b. Incongruence profile of the example patient at posttest and follow-up after 6 months. Further explanations under Figure 4.24a and in text. K1 = Follow-up assessment after 6 months. (Adapted from Grosse Holtforth, Grawe, & Tamcan, 2003).

322

Clark, and Tellegen, 1988). It also correlates negatively with extraversion (at −.25) and with a tendency to experience positive affect (at −.43). In the context of that which was discussed in section 4.7.3 about approach and avoidance temperaments, these correlations are also consistent with theoretical expectations.

These very clear findings substantiate that incongruence plays a very negative role in mental activity. It is a form of inconsistency that links the person with his or her environment. It corresponds to the presence or absence of very specific life experiences, thereby linking the state of a person to his or her experiences. If one alters the concrete life experiences of a person in such a way that the person is more congruent with his or her motivational goals, then this will have immediate effects on the person's mental state. Based on the findings reported here, we can regard this association—which is one of the core assumptions of consistency theory—as very strongly substantiated. The implication for psychotherapy is that if therapy can enable the patient to encounter experiences that are congruent with his or her most important motivational goals, then we can expect that the patient's well-being will improve and his or her psychopathological symptoms will decrease. This constitutes a second approach by which we can attain patient improvement, in addition to the disorder-specific approach, which aims to influence the neural foundations of the mental disorder directly; that is, by using disorder-specific interventions. This second approach aims to alter the inconsistency in mental functioning by facilitating the presence of need-

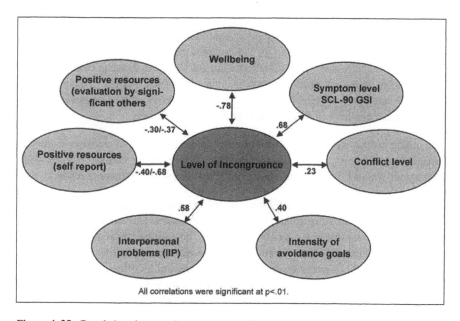

Figure 4–25. Correlations between incongruence total score and clinically meaningful other measures among psychotherapy patients. The sample sizes vary somewhat because not all measures were available for all patients. They range from $N = 185$ to $N = 283$. All measures were administered prior to the beginning of therapy.

satisfying life experiences. To the degree that such efforts are successful, we can expect that the well-being and health status of the patient will improve.

We have been able to test this conclusion, which has emerged from the previously discussed findings, in a sample of completed psychotherapies, in which incongruence was measured both at the beginning and the end of a therapy. The sample is not as large as was the case for the associations involving the incongruence questionnaire because this questionnaire simply has not been available for very long and, thus, it took some time before a sufficient number of therapies with complete measurements had been completed. Figure 4–26 shows a selection of correlations between changes in incongruence and other changes that occurred in therapy.

The correlations involving the incongruence change values are on the same magnitude as those observed for the incongruence state scores. A reduction of incongruence is accompanied by improvements in well-being ($r = .76$), a reduction in psychopathological symptoms ($r = .64$), a reduction in depression ($r = .64$), and improvements in interpersonal problems ($r = .47$). Again, these are bidirectional associations: If one manages to create real perceptions that are congruent with motivational goals, the other problems decrease. At the same time, if one manages to reduce the incongruence surrounding these problem areas, overall incongruence also reduces. These problem areas can all be viewed as possible leverage points for a reduction of incongruence. A better utilization of existing resources, better competency expectancies, improvements in positive coping strategies (constructive thinking in the sense of the constructive thinking inventory by Epstein, 1989), and a reduction of avoidance goals all contribute to reductions of incongruence and this, in turn, leads to improved well-being and less intense symptoms. It was notable that no correlation was observed between the change score in avoidant attachment and incongruence changes. Given the overall stability of attachment styles, one would also expect that attachment should correlate only with incongruence state values but not with change values. This was reflected precisely in the observed empirical correlations.

According to this perspective, incongruence takes on a central position in mental functioning. Changes in incongruence have far-reaching effects. Figure 4–27 attempts to graphically illustrate this central functional role of incongruence.

Figure 4–27 aims to make clear the various factors that can influence incongruence. Some of the arrows are directional, which intends to show that one direction of influence predominates in these cases. For example, objectively unfavorable life conditions hinder a person's efforts to attain positive experiences in line with his or her motivational goals. Conversely, incongruence has less of an effect on harsh life conditions such as poverty, war, poor education, and so forth. Other possible sources of incongruence involve bidirectional causality, as expressed here via bidirectional arrows. Unfavorable attachment behavior, problematic consistency-maintenance mechanisms, dysfunctional cognitions, motivational conflicts, excessive avoidance, high symptom severity, and poor well-being are, on the hand, potential sources of incongruence. On the other hand, they can amplify in strength when the incongruence level in mental functioning increases. Everything that has been learned through the reduction of inconsistency tensions, such as consistency-maintenance mechanisms, psychopathological symptoms, dysfunctional cognitions, and avoidance behavior, should become stronger as the incongruence level in mental functioning increases because this engenders an activation of its function for the down-regulation of inconsistency tensions. However, if it is somehow possible to reduce the incongruence level in mental functioning—for example, because the patient is guided, without having to exert personal effort, to encounter need-satisfying experiences—then these problems should also diminish because their inconsistency-reducing function is activated to a lesser degree. It is less needed because the incongruence level has been reduced by other, more favorable means.

If a therapeutic intervention somehow manages to reduce incongruence in mental functioning, then one should expect, based on these empirically validated associations, that an overall improvement in all functionally related problem areas will be observed, including improvements in psychopathological symptoms. This principle can also be used to explain how therapies can achieve symptom reductions even though they may not specifically target the symptoms at all.

If it is of such importance to reduce incongruence in psychotherapy, the question arises as to what might be the concrete possibilities for doing so. Is it indeed possible to

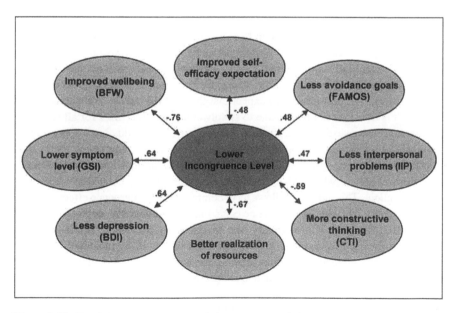

Figure 4–26. Correlations between changes in incongruence and changes in other measures over the course therapy from intake to termination in a sample of psychotherapy patients at the outpatient clinic of the University of Bern ($N = 71$).

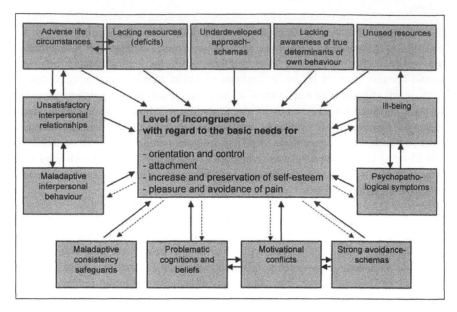

Figure 4–27. Functional role of incongruence in mental functioning.

reduce incongruence, in the same way that symptoms and problems can be reduced? We were able to compute the pre–post effect sizes for the pre–post incongruence assessments at our outpatient clinic in Bern, in order to test the extent to which incongruence changes in comparison to other parameters over the course of therapy. One difficulty was initially encountered. That is, a patient normally will not have high levels of incongruence across all of the incongruence scales. If no incongruence is present, however, it is not sensible to expect that this would reduce over the course of therapy. It is therefore not reasonable to simply compute effect sizes for all scales of the incongruence questionnaire. Instead, we opted for an individualized approach. We computed effect sizes only for individual patients whose incongruence scores on a particular scale were significantly higher, compared to a normal, nonclinical sample. For these scales, for which this requirement was met, we calculated average effect sizes per patient. This approach allows us to compare the incongruence effect size relatively directly with the effect sizes one can compute among, say, depressed patients on the BDI, because it is normally one of the study inclusion criteria that only patients whose scores exceed a certain threshold level will be included. The effect sizes that we computed with this method for approach incongruence, avoidance incongruence, and overall incongruence are presented in Figure 4–28.

For approach incongruence, an effect size of more than 2 was estimated; for avoidance incongruence, this value was still more than 1.5. This shows that it is possible to reduce incongruence by about the same magnitude as what is possible for anxiety symptoms or depressive symptoms. The question is, then, whether such incongruence reductions would also be achieved spontaneously, without therapy, in a similar way as is true for depression (see section 4.4.4). It might be that the contribution of specific therapeutic interventions in the reduction of incongruence is larger than it is in the reduction of symptoms. This question cannot be answered at present because we have not yet collected the relevant data with regard to incongruence changes among control groups. The fact that incongruence can be reduced effectively in psychotherapy is also confirmed in an independent study by Berking, Grosse Holtforth, and Jacobi (2003b) in a sample of inpatients who were treated with a behavioral therapeutic approach. In this study, changes were observed that were similar to the ones we had measured in our own outpatient sample.

These findings demonstrate that it is quite realistic for therapists to aim for a reduction in incongruence, as long as marked incongruence is present at the beginning of therapy. The fact that it is apparently easier to reduce approach incongruence than avoidance incongruence is fortunate from a therapeutic perspective, because reductions in approach incongruence correlate more highly with other favorable therapeutic changes than do reductions in avoidance incongruence.

If one computes separate correlations involving approach- and avoidance-goal incongruence vis-à-vis other change measures, the picture presented in Figure 4–29 results.

The subscales of the four tests are ordered according to the magnitude of the correlation coefficients. Thus, the differences between approach and avoidance incongruence are larger for the scales that are listed first. The same pattern holds true for every test: A reduction of approach incongruence is more tightly linked with other positive changes than is a reduction in avoidance incongruence. This means that it is important and useful to help patients avoid additional psychological injuries or disappointments, but it is even more important to support patients in such a way that they can more effectively realize their approach goals.

This finding may look relatively unremarkable at first glance. However, a message for therapists is contained in it that is everything but self-evident. Psychotherapy necessarily deals with the fears, anxieties, and insults the patient has suffered, as well as with the avoidance strategies he or she has developed for the purpose of self-protection. Most of this corresponds to avoidance incongruence. The correlations in Figure 2–27 suggest, however, that therapists should devote greater attention to approach incongruence and its causes, rather than avoidance incongruence. The therapist should focus more on the facilitation of approach behavior than on the reduction of avoidance behavior. Greater positive changes will be attained when one concentrates on the positive goals of the patient than when one focuses predominantly on fears and anxieties. This does not imply, of course, that therapists should not devote attention to fears and anxieties. Rather, it means that one should ensure that this does not become the predominant emphasis in therapy. In strongly simplified terms, achieving something positive is more important for therapy success than alleviating something nega-

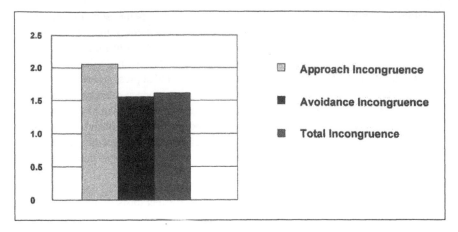

Figure 4–28. Pre–post effect sizes for patients who had elevated scores in the individual incongruence scales.

tive. This simple principle has far-reaching therapeutic implications, which will be elaborated in chapter 5.

The view of the role of incongruence shown in Figure 4–27 creates new, very positive perspectives for psychotherapy. There are almost always multiple leverage points by which one can attain positive therapeutic changes. It is possible, therefore, to choose precisely those points of attack for which the patient is highly motivated and for which he or she has the prerequisite abilities and resources. These access points promise to yield the greatest success because they take advantage of existing patient strengths and because the patient will engage with intrinsic motivation, which has been shown—as reviewed earlier—to be crucially important for successful goal attainment. It is not true that the type of patient problems determine the best avenues for therapeutic intervention. There are multiple possible therapeutic access points for all those patients who present with heightened levels of inconsistency in mental functioning—in other words, for the vast majority of all psychotherapy patients. The patients who can be treated successfully only by disorder- or problem-specific approaches are in the minority. They do exist, however—patients with specific phobias can be mentioned in this context—but the majority of patients therapists encounter in treatment also have, beyond their specific symptoms, marked deficits in need satisfaction of the kind that results from elevated incongruence in mental functioning. For them, it is not immediately obvious how one should proceed in order to best help them. In order to find this out, one must first determine the sources of the incongruence, which will then yield information about potential therapeutic leverage points. Which of these points is pursued initially depends primarily on resource-related considerations and not on the type of patient problems. This view suggests that most therapy cases can be successfully treated via several potential avenues; therefore, this perspective opens up creative space that the therapist can fill with an individualized treatment plan.

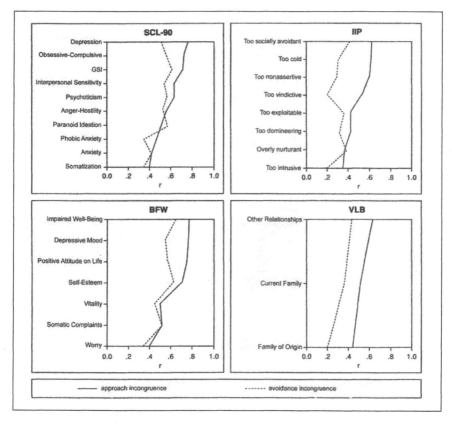

Figure 4–29. Comparison of the correlations for the changes in approach and avoidance incongruence with changes in the subscales of the SCL-90 R (Symptom Checklist; Franke, 1995), the IIP (Inventory of Interpersonal Problems; Horowitz, Strauß, & Kordy, 2000), the Bern Well-Being Inventory (Grob, 1993), and the VLB (Change of Optimism; Grawe, 2004).

4.9 THE DEVELOPMENT OF MENTAL DISORDERS FROM A LIFETIME DEVELOPMENTAL PERSPECTIVE

4.9.1 Development of Disorder Foundations

4.9.1.1 The significance of early childhood life experiences

Even before a person is born, the trajectory that determines whether he or she is likely to develop mental disorders later on in life has already been partially determined. Some individuals are born with a tendency to experience negative emotions and to have a sensitive, highly responsive avoidance system (Tellegen et al., 1988). This is caused by multiple genes. The strength of this innate tendency depends on the number of such genes a person is equipped with (Baker et al., 1992). One of these genes is the serotonin-transporter gene identified by Lesch et al. (1996). Persons who have inherited the less efficient "short" HTT-allele will tend to experience a lesser expression of the gene and, consequently, will be characterized by a lower level of serotonergic functioning (Bennett et al., 1998). Serotonin has an placating, calming effect. An infant with a short allele of the serotonin transporter gene is, therefore, much harder to pacify. Such an infant poses higher demands for his or her caregiver; that is, in most cases, the infant's parents.

The characteristics and abilities of the infants' parents are also already determined prior to the child's birth. They are equipped with characteristics that make them more or less ideal attachment persons. Individuals who had an insecurely attached mother are themselves more likely to develop insecure attachment and to pass this on to their own children (Main, Kaplan, & Cassidy, 1985; Ricks, 1985). Having an insecurely attached mother is just as much of a serious risk factor for a child as the inherited tendency toward an avoidant temperament. A second parent-related serious risk factor is the presence of mental disorders among the parents, especially severe depressions (Gaensbauer et al., 1984; Radke-Yarrow et al., 1985; Zahn-Waxler et al., 1984).

If both risk factors are present—an inherited avoidance temperament and a mother who, for one reason or another, has difficulty providing the child with secure attachment experiences—then the trajectory for subsequent development points in an unfavorable direction. The mother, who is already dealing with her own difficult developmental experiences or her present negative state, is confronted with a child who would pose difficulty even for very competent mothers because such children require extra amounts of empathic care and emotion regulation skills on the part of the mother (van den Boom & Hoeksma, 1994). If the mother—or whoever occupies the role of primary attachment figure—is unable to provide this, the child will not be soothed when he or she exhibits the negative emotions and irritability he or she is naturally predisposed to exhibit. Because the child's arousal is not effectively down-regulated via the interaction with the mother, the child rarely experiences episodes of relaxation, calmness, contentment, and security. Instead, the child is frequently maintained on a high level of negative arousal. With all the means available within his or her repertoire—and often, especially, with crying—the child attempts to attain control and establish secure proximity, but these efforts are rarely crowned by success. The child permanently accumulates incongruence experiences with regard to his or her attachment need, control need, and need for pleasure maximization and distress avoidance. The mother, in turn, is also shifted into a negative emotional state because she also experiences frequent incongruence. She feels overwhelmed, doubts herself as a mother, gets angry at herself and at this frustrating child, which she thought would have been a cause for joy. Negative emotions are triggered on both sides; an escalating cycle is established.

At the same time, the child is continuously in the process of learning. The life experiences the child encounters leave deep traces in his or her implicit memory. The child does not learn, however, that which he or she ought to learn during this period—such as the expectation that someone will reliably be there for him or her when needed;

that intimacy is an attainable possibility when desired; that he or she can effectively influence his or her environment in such a way that will obtain the things that are felt to be needed; and that emotions can to some degree be regulated by one's own action. The interactive regulation of emotions and need states in the context of the attachment relationship is the basis for the establishment and smooth functioning of intrapersonal neurophysiological feedback loops (Hofer, 1984, 1987), which ensure that the autonomic arousal with its different neurophysiological parameters is kept via negative feedback within a range that is beneficial for the organism. A problematic regulation of emotions in the attachment relationship becomes the foundation for a later tendency toward intrapersonal dysregulation. The dysregulation of autonomic arousal—that is, the absence of well-established negative feedback loops, as we saw in chapter 3— is a core characteristic of all anxiety disorders. This foundation is laid within the first months and years of life on the basis of genetic predispositions.

Given the combination of unfavorable dispositions and a poor attachment relationship, the child will not learn many of the things that later constitute good mental health. Instead, different things are being facilitated. All neural activation patterns that are repeatedly activated are facilitated. In this case, then, negative emotions are facilitated. The consequence is that the corresponding neural circuits are ever more thoroughly formed and later can be activated more easily. The child is thereby sensitized to experience negative emotions. Less intense triggers will be needed to elicit these emotions. The frequent facilitations result in the formation of more transmission-conducive synapses. The brain areas governing the control of negative emotions develop particularly strongly. These include, foremost, certain aspects of the amygdala and the ventromedial and dorsolateral part of the right prefrontal cortex. At the age of 10 months, an enduring neural structure has already formed on the basis of this predisposition and these early life experiences, and it is possible to measure this structure in order to predict how the child will respond to demanding situations such as brief separations from the mother (Davidson & Fox, 1989). The structure remains stable over long periods. It is still in place at 36 months and predicts the extent to which the child's behavior is inhibited (Davidson, 1993). The behavioral inhibition system (Gray, 1982) can be activated with particular ease, practically throughout the lifetime, among these children. It tends to be marked by high degrees of stability even in adulthood (Costa & McCrae, 1988).

Beyond the sensitization in the sense of increasingly easily triggered negative emotions, the child also experiences frequent conditioning trials, such that negative emotions will be elicited by an ever greater array of stimuli. This also contributes the pattern that the child will ever more frequently enter into negative emotional states.

During this time, however, it is not just the readiness to experience negative emotions that is being facilitated. In addition, the first foundations for the child's motivational schemas are being formed. What the child experiences during this time is often associated with strong inconsistency tension. This tension is the basis for avoidance learning. The child attempts to avoid or terminate the negative states and, for that purpose, forms mechanisms that are then facilitated via negative reinforcement. The developing avoidance goals have a high "motivational salience" in the sense described by Berridge and Robinson (1998). Their activation is accompanied by dopamine release, which leads to the particularly thorough facilitation of whatever reduces the negative emotions and inconsistency tension (Rada, Mark, & Hoebel, 1998). The avoidance goals acquire situation-specific means for their realization. Thus, avoidance schemas are formed. With this formation of avoidance schemas, the child plays an increasingly active role in the creation of interactions with the environment. To the degree that avoidance schemas develop, the child's mental activity is then no longer oriented toward approach or, in cases in which both avoidance and approach schemas develop for similar situations, the child frequently enters into approach–avoidance conflicts. Both processes are unfavorable for the satisfaction of the child's needs.

Naturally, such developments occur in gradations of severity. Neither the dispositions nor the child's environment operate according to the either–or, all-or-nothing principle. Both can vary from slightly to clearly unfavorable. Correspondingly, whatever the child learns during this period can range from slightly to clearly problematic for subsequent development. For example, most children have only one primary attachment person. The two or three most important caregivers might not all be poor attachment figures, however. This means that the child probably encounters some positive attachment experiences, even when the primary attachment figure is unable to respond well to his or her attachment needs.

There are also children, however, who are severely neglected or even abused and mistreated. Even with favorable genetic dispositions, such children have almost no chance to escape from the unfortunate developmental processes just described here. According to V. Carlson et al. (1989), 80% of them will develop insecure attachment styles, most typically of the disorganized type. This means that they have acquired a severe handicap for their subsequent development and are much more likely to develop serious mental disorders.

As mentioned before, it is not just the neural structures for negative emotions and avoidance schemas that develop during this time. In addition, basic regulatory processes in the neurophysiological system become established at this stage. Severely mistreated children experience continuous uncontrollable incongruence. Such experiences are regularly accompanied by the release of stress hormones. In normal development, negative feedback loop mechanisms are established at an early age that limit the release of glucocorticoids before they reach an excessively high levels. The brain uses this feedback mechanism to protect itself from insults that it would otherwise suffer from a continuously high cortisol level. These insults and injuries include, primarily, the shrinking of the hippocampus, which is associated with an impairment in its functional capacity. In addition to forming explicit memory contents, the most important function of the hippocampus is the contextual anchoring of experience in time and space. Without such a well-functioning anchoring of emotional reactions in time and space, the emotions are likely to recur not just in appropriate situations but, instead, excessive emotional reactions to stress become more likely and stress responses will be observed even in objectively harmless situations, as is the case with anxiety disorders, for instance.

If the continuous uncontrollable incongruence persists, then, an additional negative effect is the sensitization to stress. Abused children who experience such a continues incongruence over which they have no control will eventually respond with excessive stress reactions even to relative minor challenges, and these reactions will last much longer than they normally should. This hyper-responsiveness to stress tends to persist for a lifetime. Among adult women who had been abused as children, Heim et al. (2000) documented stress reactions to a moderate social stressor that were still six times more intense, compared to women without abuse experiences. Not surprisingly, many of the women had developed severe depressions. The reduced hippocampus, which is regularly documented in depression, reflects these histories of depression. However, it is not just abuse histories that can lead to such diminished stress tolerance levels. In a sample of anxious-inhibited 2-year-old children—that is, children with avoidant temperament dispositions—Gunnar et al. (1996) documented excessive cortisol reactions in response to mental stress that were more pronounced among the insecurely than the securely attached children. The attachment experiences encountered by infants or toddlers will be directly reflected, then, in their stronger or weaker stress tolerance levels. Even children with avoidant temperaments will acquire better stress resilience if they have been exposed to positive attachment experiences.

Children's developmental trajectory does not point inevitably in a negative direction, then, even if the children were born with an avoidant temperament disposition. If such a child is lucky, he or she will have a particularly sensitive and competent mother who will be irritated by his frequent crises and negative emotions, but who will nevertheless reliably and affectionately turn toward the child and enable him or her to acquire predominantly positive perceptions with regard to his or her attachment and control needs. Suomi (1987, 1991) has shown in an impressive study that such children will flourish in their development and end up with even more optimal developmental trajectories than children with more favorable inherited tendencies who are raised by average mothers. Even if one has inherited a short HTT-allele, the serotonergic function can still be expressed—it will just be more difficult, not impossible. The genetic handicap can be more than just compensated by a favorable environment. If a particularly caring–attentive attachment figure enables the experience of positive attachment and control situations, positive neural structures and regulatory mechanisms will develop that eventually become self-maintaining. The threat of a genetically determined suboptimal fate can be averted and transformed into a favorable trajectory via positive experiences in early childhood. Such a child will develop good stress tolerance levels and emotion regulation skills and will, thereby, remain more resilient to stress than he or she would have been with the same innate disposition but a less attentive–affectionate mother. Moreover, approach schemas will dominate over avoidance schemas in such a child, which in turn will ensure that motivational inconsistencies are less likely to be encountered later in life. Via this pathway as well, then, the child will be less vulnerable to the emergence of a mental disorder.

Van den Boom (1994) has documented in his experimental intervention study among particularly irritable infants that the harmful interaction between avoidant temperament and negative early childhood experiences can be subverted by involving the children's mothers in a sensitivity training program. A relatively limited intervention administered at the right time point, then, can achieve an effect that for the child makes the difference between a life filled with happiness versus despair. If one considers the suffering and social costs caused by mental disorders, the conclusion seems unavoidable that we should not leave things to their own devices but should intervene early and preventatively in order to help parents to raise their children more optimally.

Children with more favorable genetic dispositions with regard to their propensity for negative emotions are also not permanently protected from negative outcomes. If they have the additional good luck of having a caring–attentive mother, such that they benefit from a positive attachment relationship throughout their childhood, they will likely maintain a lifelong strong resistance to mental stress and will most probably not develop any severe mental disorders. But even children with a favorable genetic disposition can encounter the misfortune of growing up in a harmful environment, of having an unaffectionate, uncaring mother, or of being mistreated or abused. Their positive genetic basis would not be useful under such conditions because the genes would not be optimally expressed. Even though a more favorable trajectory would have been within their reach, they, too, will form neural structures that render them susceptible for the later emergence of mental disorders. They would require more consistent negative life experiences, compared to a child with an avoidant temperament, in order to develop similar levels of poor stress tolerance and a similar predominance of the avoidance system.

4.9.1.2 Conclusions for psychotherapy

The research findings I discussed in this fourth chapter do not leave any doubt that the foundations for mental health are already laid in the very early years of childhood; indeed, in the phase of life that we later cannot recall. These foundations are encoded in the implicit memory system and, by nature, are not recallable, not even after many years on the therapist's couch or under hypnosis. They have not been encoded in such way that they could be processed in the explicit mode of functioning. No human being can recall the process in which the basis of his or her own personality was formed. Our personality—and this includes chiefly our motivational system—is, as LeDoux (2002) puts it—rooted in the *implicit self.* The foundations for this are laid in the first years of life. Their further development constitutes the implicit self of the adult human.

It is the implicit self that ultimately determines our experience and behavior. The way in which it accomplishes this is beyond the reach of our introspective awareness. The subjective opinion of most people is that their conscious "I " determines what they do. This is one of the positive illusions that serve to satisfy our need for control. But what about mental disorders? Who or what determines disordered experience and behavior? Our conscious I as well? This is where the control illusions end. Mental disorders are not experienced as controlled by the conscious I; they are experienced as suffered by this I, as something that is somehow imposed as a burden. Nevertheless, the disorders are a part of ourselves—a part of our implicit self—even when they are rejected vehemently by the conscious I. They are an actual creation of the mind, personality, or psyche, regardless of what we might call this. Their foundations escape our introspective access, but our conscious thoughts about ourselves never encompass more than a small part of our implicit self anyway.

Mental disorders originate from the very fabric of our being. They are an integral part of ourselves, even in cases in which we suffer severely from them. They belong to us like our personal history belongs to us. Their foundations reach all the way back to our individual beginnings. They were a part of us right from the start. When a mental disorder actually surfaces at a given moment, it indicates that this particular part is currently particularly active.

If mental disorders are a part of ourselves, does that mean that one has to feel responsible for them? Of course not! Nobody is responsible for that which he or she encountered in the very first months or years of life. The person was also was not responsible for it back then, in infancy. The foundations for mental disorders are a question of the good luck or bad fortune one has had in life, in the same way that one might have been lucky or unlucky with in terms of one's physical appearance. But even if someone is unlucky in terms of his or her looks, it is nevertheless a good idea to accept one's appearance as a part of oneself and to govern it consciously. By taking good care of our appearance and using appropriate clothing one can still make the best of it. This works similarly for individuals with mental disorders. It is wise to accept them as a part of oneself and to govern them consciously. In this respect, too, one can make the best of it by treating oneself with mindful awareness. *Mindfulness* means in this context that one takes oneself seriously regarding one's needs. At the moment that a mental disorder emerges, it is a warning signal telling us about a high level of inconsistency in mental functioning, and high inconsistency indicates poor need satisfaction. Treating oneself mindfully means taking one's own needs seriously and assuming responsibility to ensure that the needs are met to the best degree. The prerequisite for this is that one accepts how one has come to be that which one now is. Doing so will allow one to "make the best of it."

Ruminating about the past or becoming resentful and angry about the bad things one has encountered has as little utility as resentment about the fact that one was not born with a more beautiful appearance. If one wants to improve one's fate, one has to take fate into one's own hands. If one has a mental disorder, it is not useful to be resentful about it. That won't make it disappear. Instead, one has to take seriously the fact that one is someone whose mental system produces, on a daily basis or in regular intervals, the kind of experience and behavior that is contrary to one's wishes. Something must exist beyond one's personal control that is at the root of this. It is completely understandable, from this perspective, that patients often have a great desire or need to search for the reasons within themselves. This need can be satisfied, if all goes well, in a supportive, clarification-oriented psychotherapy (Sachse, 2003). However, only the kinds of causes that are accessible to introspection or self-reflection can become the focus in such an approach. If the clarification efforts refer to current determinants of inconsistency, they can contribute to the identification of current sources of inconsistency and, thereby, can help create the basis for changing them via conscious action.

Many foundations of mental disorders lie beyond the areas that are accessible to introspection. The amount of glucocorticoids one releases in response to stress, or the state of one's own hippocampus and amygdala are among the things that are beyond self-reflection because one has never experienced anything other than that which the nature of one's own neural structure has enabled one to experience. By comparison, one can determine that one is more fearful than others, but one nevertheless will not know why this might be. The real causes for that escape our introspective access.

It is certainly possible to recall negative or harmful life experiences that were encountered after the maturation of the hippocampus; that is, after the point at which explicit memory contents can be formed. Patients will often trace their misery back to such unfortunate recallable events. The true foundations for their despair, however, reach back much further in most cases. Many psychotherapists devote a great deal of time to searching for traces in the autobiographical memory of the patient that might reveal how it came to be that they are how they are. It makes sense if they do this in order to satisfy the patient's strong need for understanding; however, it would be a fatal mistake to believe that such memory searches would uncover the objective causes of the disorder. In a patient whose need for orientation is strongly activated, it can be very useful for him or her to feel that he or she has understood something more thoroughly. This would in itself constitute a kind of need satisfaction. This insight can also motivate the patient to change something. If he or she develops the efficacy to make such changes and then actually makes them, this can lead to positive control experiences. However, it is by no means certain that the patient would have changed anything having to do with the foundations of his or her experience and behavior in the implicit part of memory. These foundations are deeply facilitated and can be altered only via frequently repeated "corrective experiences," because new neural activation patterns must be facilitated and old ones inhibited. The foundations for well-facilitated mental disorders cannot be altered via introspection and insight but only via new, real experiences that weaken old synaptic connections and facilitate new ones. Which specific neural connections these are must be determined via an analysis of current response patterns, not via a search in the past.

The neuroscientific analysis of mental disorders leads to a somewhat paradoxical conclusion: The most important foundations of mental disorders can be traced to early childhood, but looking back toward the past is not useful in terms of altering their foundations. Change can be achieved only via real experiences made in the present, which can integrate and thereby transform into new patterns the foundations that were laid long ago.

4.9.1.3 From early childhood into adulthood

The experiences of early childhood do not terminate suddenly but typically continue smoothly over time. Anxious and depressed patients tend to recall their parents retrospectively as more controlling and strict but simultaneously less affectionate than healthy control persons (G. Parker, 1979a, 1979b, 1981, 1983). These memories refer to a time period that participants can consciously recall. Therefore, they refer to events that occurred at a later time than the period discussed in the previous section.

Longitudinal studies by Main, Kaplan, and Cassidy (1985); Grossman and Grossman (1991); and Sroufe, Carlson, and Shulman (1993) have shown that insecurely attached children carry their early acquired handicaps forward into all subsequent life stages. They have weakly developed self-confidence and low self-efficacy expectancies, more negative self-esteem, and less resilience (robustness) in response to stress and challenges. They differ markedly from their securely attached peers in their interpersonal behavior and in the quality of their peer relationships. They are evaluated by peers and parents as less social, capable of establishing relationships, empathic, and popular, compared to securely attached children and adolescents. They cannot express their impulses, wishes, and—especially negative—feelings as well as others.

These findings reflect and reveal a great tragedy. Although these children were initially disadvantaged by unfavorable life circumstances and/or by unfavorable genetic dispositions, they now often personally perpetuate these unfavorable experiences from their earlier childhood and carry them forward into other life domains such as school or peer relationships. The reason for this relates to the motivational schemas that have developed at this point. Their avoidance schemas are much more strongly developed than those of children who have previously encountered secure attachment experiences. Children who have acquired an avoidant attachment style in early childhood already do not dare anymore to express their wishes to their teacher in the next phase of life at school. They will do this either indirectly or will wait for an unusually secure situation to present itself. If they are at some point in need or distress—for example, because they have hurt themselves, are disappointed, or feel stressed—they do not muster the courage to turn to the teacher for help and support. Avoidant children often elicit punishing reactions from their teachers. The teachers demand more discipline from them, control them more strictly, expect more willingness from them, and express less warmth and affection but more anger toward them, compared to other children. Thus, in their relationships with these important adults the children reproduce relationship patterns similar to those they knew from home. They provoke with their own behavior the fact that they are now being treated so badly. When they see another child in need, insecure–avoidant children often react with little empathy and support or will behave inappropriately and hurtfully; for example, by teasing a crying child or making fun of a child who has injured himself. The relationships with their peers are often hostile and aggressive or very distanced. In play situations, they tend to exploit others, derogate others, and act hostile toward them, at times even hitting or physically punishing other children. They tend to assume the role of the perpetrator.

In this regard, too, they perpetuate the pattern of rejection and isolation they know so well (all this is based on E. A. Carlson & Sroufe, 1993).

Of course, even these children still have positive wishes that correspond to their basic needs. However, they have developed few situation-appropriate means for the realization of these wishes and, therefore, recurrently encounter incongruence experiences that result in negative emotions. Their already existing tendency toward negative emotions is, therefore, continuously further reinforced. All this is accompanied by strong autonomic arousal and frequent stress reactions involving the release of stress hormones that damage their neural structures and impair their capacities for subsequent positive learning.

The extent to which the structure of the motivational schemas that have developed in infancy and toddlerhood influence subsequent development is evident not only in the comparisons with securely attached children, who have encountered much more positive experiences in their childhood and adolescence, but also in comparisons with children who, although they have also developed avoidance schemas in their attachment relationships, nevertheless have less impoverished approach tendencies. These are children with insecure–ambivalent attachment styles. These children also perpetuate their experiences from infancy and toddlerhood, but the patterns they exhibit look different from those of children in whom avoidance predominates entirely.

Insecure–ambivalent children continuously seek the proximity of the teacher. They do not develop independent initiative but always wait for the teacher to take the initiative. The teachers are more attentive and tolerant toward them, compared to avoidant children, they are more likely to excuse minor transgressions, to help and support them, and so forth. In the context of play with others, these children also leave the initiative to others and gladly allow themselves to be led. When other children tease or exploit them, they allow it to happen. They also perpetuate the pattern that they knew from their home environments.

These differences between securely and insecurely attached children, on the one hand, and between insecurely avoidant and insecurely ambivalently attached children, on the other hand, essentially remain in place across all stages of subsequent development, in kindergarten, preschool, elementary school, middle school, high school, in holiday camps, and even into adolescence and beyond. Early childhood experiences determine to a large extent the kinds of experiences, then, that the children encounter in the following life stages, and this effect appears to be mediated by the motivational schemas the children develop in infancy and toddlerhood. Among somewhat older children, the violations of the self-esteem need will also contribute further to inconsistencies in mental functioning. Given the wealth of negative experiences that children with an unfavorable developmental trajectories encounter in their first 10 to 15 years, it is not surprising that many of them develop mental disorders even in childhood or during their teens. The easy elicitation of negative emotions, autonomic hyperarousal, and dysregulation already form the core of many disorders, and not much more is needed for these patterns to turn into firmly established disorde syndromes.

Indeed, research has confirmed rather high prevalence rates of anxiety disorders and depressions among adolescents. In survey studies, about 20% to 35% of boys and 25% to 40% of girls report having experienced depressive mood episodes within the past 6 months. Clinically significant depressive disorders can be identified among about 7% (Petersen, Leffert, & Hurrelmann, 1993). These numbers are roughly consistent with the values Merikangas and Angst (1995) calculated on the basis of eight epidemiological studies. It is also notable that even at this early age, a high comorbidity rate is already evident with disorders of social relatedness, anxiety disorders, eating disorders, and substance misuse disorders. The comorbidity rates reported in the literature range between 58% (Essau, Karpinsky, Petermann, & Conradt, 1998) and 43% (Lewinsohn, Hops, Robert, & Seeley, 1993). Similar to adults, disorders of social relatedness and anxiety disorders often precede depressive disorders among teenagers (Kovacs, Gatsonis, Paulauskas, & Richards, 1989; Last, Hansen, & Franco, 1997; Rohde, Lewinsoh, & Seeley, 1991).

Based on our knowledge of these developmental trajectories, then, mental disorders do not appear "out of the blue sky." They emerge from backgrounds of highly problem-fraught life histories, which can be traced back step by step to the period before birth. Violations of the attachment and control needs, and later of the self-esteem need, play a decisive role in these unfavorable developmental processes. According to a meta-analysis by Dozier, Stovall, and Albus (1999), almost 90% of all patients with mental disorders have some form of insecure attachment, and the specific form of attachment disturbance can to some degree reflect differences in disorder type. Among 528 inpatients studied by Schauenburg and Strauss (2002), more than 90% had an insecure attachment style. We know a great deal by now about the harmful experiences that engender insecure attachment styles and about the developmental consequences of such styles. There are certainly also opportunities for individuals who have developed insecure attachment styles in early childhood to later encounter positive life experiences that can shift their subsequent developmental trajectory into a more positive direction. Even a single loving relationship can suffice for the purpose, even as late as in adulthood. If a person with negative early attachment experiences somehow manages to engage in an intensive loving relationship with a partner who has the qualities of a good attachment figure, this can lead to far-reaching corrective experiences that can alter his or her motivational schema structure permanently. Not every person with problematic attachment experiences, then, will inevitably develop a mental disorder. But the converse is that almost all patients with serious mental disorders have life histories that involve the kinds of violations of their attachment and control needs in early childhood that are then expressed in insecure attachment styles. Their problems can, therefore, not be reduced to the recently developed disorder. The disorder is only a last link in a long chain of need-violating experiences, which over time have led to such high levels of inconsistency that a qualitatively new mechanism for inconsistency reduction was required.

The patients themselves also know that their psychopathological disorders in a narrow sense constitute but a small part of the problems that ought to be addressed in treatment. Only 9% of all patients list the alleviation of their disorders as their sole therapy goal. Three out of four patients list problems in interpersonal relatedness as one of their three main problems (Grosse Holtforth & Grawe, 2002a). This corresponds to the great importance of interpersonal relationships as the locality around which their problematic life experiences have usually transpired.

My point here was to articulate in a comprehensible way the pathways by which violations of basic needs can lead to the emergence of mental disorders, so that the implications for psychotherapy can be more clearly inferred later on. My examples have usually involved anxiety disorders and depressive disorders, which are also the clinical problems whose neural correlates I reviewed in some detail in chapter 3. It is, of course, true that many important differences among the various disorders could have been considered here, especially if one were to extend this discussion to externalizing disorders as well. More could also be said about the genetic and social contributory factors in development, but such material would be more appropriate for a textbook on mental disorders. My goal here is to explicate the mechanisms of disorder genesis in a coherent perspective. I want to show in particular how in a mental life that was just previously considered nondisordered, something qualitatively new can emerge that does not belong among the normal mental phenomena. The genesis or onset of new symptoms is, therefore, the topic of the next section.

4.9.2   The Onset of Mental Disorders During Times of Heightened Inconsistency

As we have seen in the previous section, individuals who develop a mental disorder differ in many ways from those who do not develop one, even prior to the disorder's onset. On average, they tend to experience more negative emotions. Once negative emotions have been triggered, they are less able to control or down-regulate them. They respond with excessive stress reactions even to moderate demands or challenges, and the stress reactions persist for abnormally long periods. Their autonomic nervous systems are frequently hyperaroused. All this was established in them at an early age, and they have formed corresponding neural circuits. In such an individual, the amygdala is easily and extraordinarily strongly triggered in response to anything that could conceivably pose a threat, and, in turn, triggers these circuits without the involvement of conscious awareness.

The impaired emotion regulation and lowered stress tolerance are accompanied by poor coping and weakly developed adaptive consistency-maintenance mechanisms. There is an elevated risk that strong inconsistency tension will be recurrently encountered, even in situations that would be easily mastered by other people.

Under the influence of the many violations of their needs, a great number of strong avoidance schemas have formed within such individuals, and these schemas also operate automatically because they have been facilitated hundreds and thousands of times. Because of these thoroughly facilitated avoidance schemas, mental activity is not clearly and enduringly oriented toward approach schemas. Avoidance schemas interfere continuously, leading to conflicts or assuming primary command over mental functioning. The conflicted or avoidance-motivated behavior leads only very rarely to truly satisfying experiences. Instead—as we know from the experiences insecurely attached people encounter in their peer relationships—new mental injuries and need violations are encountered repeatedly. Interpersonal relationships are, therefore, more often experienced as the localities of problems rather than the source of satisfactions.

The brain continuously and automatically forms expectancies about how one moment will flow into the next. If a person has encountered many negative experiences, negative outcome, reaction, and self-efficacy expectancies will predominate. The person will, therefore, tend to expect negative events and will broadly lack confidence in his or her abilities. Individuals who have repeatedly experienced violations of their control need do not develop positive control expectancies. They do not approach their environment in a confident and optimistic manner or think that things will turn out all right. Instead, they expect negative outcomes, and their cognitions tend to be pessimistic and avoidant.

A person with such broadly negative expectancies will experience an extremely high degree of inconsistencies across all areas of mental functioning, which also will be reflected in extremely high incongruence levels. The person will be highly vulnerable to the development of a disorder, given that many of the characteristics of an anxiety disorder are already in place.

When reading such descriptions, one gains the impression of a person whose misery would be unmistakably evident even in the briefest of encounters. However, that is not the actual impression one gets of most people with mental disorders, not even after they have suffered for longer periods from a particular disorder. Fortunately, it is rare that individuals suffer to such degrees. I have listed characteristics here that differentiate such individuals on average from others who do not develop disorders. Hardly anyone exists to whom all of these characteristics would apply simultaneously. Under normal circumstances, a person with an unfavorable developmental trajectory would form some of these characteristics, but not all. Moreover, each characteristic can of course be formed to a greater or lesser degree.

What is missing from the previous listing of negative characteristics is the fact that most people who are at high risk for the formation of mental disorders simultaneously also have very positive features and characteristics. Those positive characteristics are completely independent of the negative ones. They might have stunning good looks; be very intelligent; or have outstanding athletic, creative, or academic abilities. They might be witty, lovable, or charming; or they might make a good pal.

It is probable that they have formed personal niches or islands on which they feel somewhat secure and protected from psychological insults and injuries and on which they can engage and cultivate their positive features. They can often retreat to such niches even for longer periods, often enabling them to function moderately well and to feel generally comfortable despite their easily triggered avoidance system, because this system is often less frequently activated in their self-protective niche. Critical situations surface whenever they have to leave these protected areas in order to face life's unavoidable demands. This would be the case, for instance, when new developmental tasks must be mastered, such as when a child enters school; or later, during the transition to adolescence and puberty; when severing the bond from the parental home; in adulthood, at the birth of one's own children; even later, at the departure of the children away from the home; after divorce; in the context of a partner's death; during the transition from work to retirement, and so forth.

Whereas each previous phase was marked by a certain degree of stability in schema structure, which permitted sufficient consistency in mental functioning, the schema structure now encounters new demands for which new schemas must initially be formed. The old schemas no longer function under the new conditions. This is accompanied by incongruence experiences and heightened inconsistency. Before new schemas have become established, a general destabilization of mental order is observed. Old order patterns fail to function but new ones are not yet sufficiently formed. Situations are repeatedly encountered for which adaptive, need-satisfying mechanisms have not yet been formed. All this does not transpire in an emotionally neutral fashion, to be sure, because incongruences with regard to the old motivational schemas are repeatedly encountered. Such a developmental phase of new adjustment and insecurity is characterized by elevated destabilization even for individuals who have so far benefited from favorable development and who have primarily formed approach schemas. The destabilized state is indeed necessary because the stable continued existence of old motivational schemas would interfere with the person's successful adjustment to the new situational demands.

As I discussed in section 4.8.3, novel neural activation patterns will primarily be formed in such situations of heightened inconsistency. The destabilization maintains the mental activity in a hovering or fluctuating state, as it were, in which it is not yet clear how the next steps will unfold. The heightened inconsistency tension simultaneously harbors a high reinforcement potential, which will differentially reinforce those neural activation patterns that engender a reduction in inconsistency tension. The neural activity can fluctuate between only the kinds of states, however, that are possible to be formed on the basis of that which has previously been in place. This basis always includes the already well-facilitated excitability of neural circuits. Whatever is activated in such a situation has good chances of being integrated into the newly developing neural activation pattern.

Individuals with a prior history of many negative attachment, control, and self-esteem experiences enter such critical inconsistency situations with a poor set of tools for adaptive mastery. For them, too, such situations of destabilization and fluctuation, and simultaneous heightened inconsistency tension, must somehow give way to new states. The unstable state cannot persist indefinitely because the external world does not stand still and continuously produces new demands. In this situation of acutely heightened inconsistency tension, the excitability of neural circuits governing excessive negative emotions and stress reactions will be activated because the situation contains strong incongruence with regard to important motivational goals, which so far could not be controlled. In this situation, anything that somehow enables the person to experience control will be negatively reinforced because such control will somewhat reduce the inconsistency tension.

In such a situation of unbearable inconsistency tension, one might, for instance, experience the phenomenon that "crazy thoughts" suddenly flash up, such as the vision that one might stab the eyeballs of one's own child, or the fear that one might be infected with a fatal disease by touching a door handle, or that one might have left the gas stove turned on. Horowitz (1975) showed some time ago in empirical studies that such intrusive thoughts surface primarily under stress and in situations involving negative emotions. Normal individuals will also experience such "crazy" thoughts from time to time, but they will tend to be fleeting and the person will not pursue any action to counter the thoughts. In such a situation of strong inconsistency tension, in which negative emotions with no discernable cause are activated, however, such a thought fits precisely with the negative emotions. The person experiences these emotions as deep-seated uneasiness. He or she feels the urge to do something to avert the looming danger or misery, and indeed, this is within his or her reach. The person could, for instance, check the gas stove repeatedly in order to make sure that danger is averted. The person could clean the door handle or avoid touching it, and he or she could wash thoroughly to make sure that he or she won't be infected. The person could lock away scissors and knives to make sure that his or her aggressive impulses won't be transformed into action. The person could firmly commit to never allow such awful thoughts to surface again (which will most certainly ensure their repeated occurrence).

All this would enable the person to experience a certain degree of control at that moment. At least something can be done. The control experience reduces the inconsistency to some degree, and the control behavior is therefore negatively reinforced; that is, particularly thoroughly facilitated. The behavior will have an elevated likelihood of occurring again whenever the next inconsistency tension appears. It will then again be negatively reinforced, facilitated, and so forth. It is not just the control behavior that is being facilitated, however, but all the synaptic connections that are activated in this particular situation. This includes the situational perceptions, the strong negative emotions, the cognitions and the control behavior. Over time, all these are joined together to form a neural activation pattern that previously did not exist in just this form. It has an emergent quality. Something qualitatively new has appeared—the beginnings of an obsessive–compulsive disorder. If the inconsistency tension remains in place because nothing changes in the constellation of the motivational schemas and in the life situation whose interaction created the inconsistency, then this new method of achieving short-term inconsistency reduction will repeatedly recur and will become increasingly thoroughly facilitated. Eventually, the pattern can be activated not just by current inconsistency tensions but also by situations, thoughts, memories, and emotions that are associated with it. The disorder pattern has become well-facilitated content of memory, which now starts to "live an autonomous life," apart from its original function for the reduction of inconsistency tensions. As long as these tensions remain in place, the disorder pattern will also keep its function for inconsistency reduction and will recur with particular intensity whenever the inconsistency tensions gain in force.

The disorder always leads to no more than short-term reductions in inconsistency tension. The ultimate causes underlying this tension are not changed at all. The opposite is true: The disorder itself is, once again, experienced as a loss of control; it is unpleasant and detrimental to the person's self-esteem. Beyond the short term reduction, then, the disorder actually elevates the inconsistency in mental function and thereby contributes to subsequent symptom recurrences of greater intensity, to their repeated reinforcement, to long-term increases in inconsistency levels, and so forth. The vicious cycle that characterizes many mental disorders is in full swing.

In a large longitudinal epidemiological study in a representative community sample of young women (N = 2,064), Margraf (2001) found that 18.3% of the women were suffering from some form of anxiety disorder as defined by DSM-IV. At the second assessment point, 1-1/2 years later (N = 1,569), only 57% of the women who had an anxiety disorder at T1 were still diagnosable with an anxiety disorder. This shows that anxiety disorders are initially not rigidly fixed or unchangeable. They can sometimes fade away spontaneously (the women in this sample had not yet been treated). What this probably means is that the inconsistency situation among 43% of the women was only temporary or weak in nature, such that the disorder pattern was not thoroughly facilitated. The idea that anxiety disorder patterns are initially not rigidly facilitated is also reflected by the finding that many of the women changed in disorder category between the first and second assessment, over the course of 1-1/2 years. Only 20% of the women who were diagnosed with panic disorder at T1 still had the same diagnosis at T2. The same percentage change was found for generalized anxiety disorder. The other anxiety disorders had slightly higher stability rates, and the highest one was found for social phobia (47%). Some women with an anxiety disorder at T1 did not have any anxiety disorder at all at T2, but a more frequent finding was that they had changed the type of anxiety disorder diagnosis between the two assessment points. This suggests that some disorder components had been replaced by others. The great commonality that characterizes all anxiety disorders—the pronounced tendency to experience negative emotions and the striving for control—remained in place.

All anxiety disorders include an important component related to the striving for control. In generalized anxiety disorder, this is the component of "worrying," the ineffective effort to gain control over diffuse fears via intensive rumination. In panic disorder, the attention directed toward suspicious bodily symptoms has this control function; in social phobia, this is essentially quite similar. In social phobia, as well as in posttraumatic stress disorder, avoidance and escape are the primary means for exerting control. Schulte (2000) differentiates three components that cut across the anxiety disorder diagnostic categories: The avoidance of anxiety, the excessive fighting of anxiety, and the excessive attention toward and recognition of the anxiety. The type of control means a person employs constitutes an important differential criterion of the anxiety disorders.

These control components of the various anxiety disorders can all be acquired via the learning mechanism that I briefly described with the example of obsessive–compulsive disorder. They are initially reinforced by the short-term reduction of inconsistency tension and maintain this function as long as a high degree of inconsistency remains in place. If this function is eliminated because the inconsistency level has reduced for other reasons—before the disorder pattern has been thoroughly facilitated—the anxiety disorder can spontaneously fade away. However, once the anxiety disorder becomes firmly established via frequent and strong facilitation, once it has acquired autonomy, the control components serve the function of anxiety reduction and thereby contribute to the maintenance of the disorder. This explains why anxiety disorders often remain in place even when their function for the reduction of inconsistency no longer exists because the inconsistency has reduced over the long term.

In the anxiety disorders, one observes not just symptom changes across the diagnostic categories but also high levels of general comorbidity. Once an anxiety disorder has formed, 40% of anxiety disorder patients will develop a second anxiety disorder on top of this (Schulte, 2000). That is, they simultaneously exhibit the symptoms of two or sometimes even more anxiety disorders. In such cases, there must be a fertile breeding ground for the development of additional disorders. This breeding ground consists of a continuously high level of inconsistency tension, with its high reinforcement potential for anything that enables slight increases in control. The additional anxiety disorders, with their specific control mechanisms, can emerge based on the same mechanism as was described for the example of obsessive–compulsive disorder. Comorbidity can, therefore, be viewed as a clue that a continuously high level of inconsistency tension exists within the person. Indeed, we confirmed empirically that patients with high levels of comorbidity report particularly elevated incongruence levels.

It is also common that comorbidities develop with disorders other than anxiety disorders; most frequently, these are depressive disorders. The depression develops after the anxiety more often than the other way around. Sixty-six percent to 78% of all obsessive–compulsive disorder patients develop an additional depression. In 85% of cases, the obsessive–compulsive disorder precedes the depression, and in 15% of the cases, both appear at roughly the same time (Schulte, 2000). This path from anxiety to depression (Alloy et al., 1990) can be explained quite well from the perspective of consistency theory.

Anxiety is an alarm signal. It appears when the organism feels threatened by something. In humans, it is particularly likely to appear when a person feels threatened in terms of his or her important motivational goals. The anxiety mobilizes the person's resources to defend against the looming danger. The marked control components characterizing the anxiety disorders show that the patient has not yet surrendered. He or she continues to struggle with the anxiety. He or she has not yet given up on the threatened motivational goals. The ACC, which is important for the active engagement with the environment, continues to function. None of the studies on the anxiety disorders have suggested that this area would be deactivated in the same way as is found in depression.

Depression constitutes a next logical step, when all the efforts one has exerted to defend against the threats are regarded as having failed. When one's important needs, values, and goals remain violated or unsatisfied despite the fact that even desperate control efforts have been attempted, which other response options remain? How should one protect oneself from such uncontrollable threats?

One pathway, however, remains possible: If one surrenders or abandons all wishes and all wanting; if feelings are no longer generated; if one gives up all expectations; derogates oneself; and becomes tiny in one's social role and context; if one reduces all life functions and ceases to eat; if one abandons the pursuit of all that used to be enjoyable; if one retreats from all people; if one gives up all that used to be important; if one refrains from any engagement with the environment—then one will be rather well protected from further insults and injuries. From this perspective, depression can be regarded as a generalized self-protective stance; an exaggerated avoidance of potential harm. The only function that remains active in a depressed person is his or her avoidance system. The amygdala and the ventromedial and dorsolateral PFC—brain areas linked to avoidance and negative emotions—are permanently highly activated. The ACC, which is important for the transformation of emotions into clearly experienced feelings and for the monitoring of inconsistency sources that require attention and action, is completely shut down. The depressed person has given up. He or she has retreated into the shelter of the sick role, turned off his or her approach system, and turned on his or her avoidance system to the highest level. The person achieves, thereby, that he or she will not have to suffer any further mental beatings. This is not a consciously pursued strategy; instead, these changes in experience and behavior are occasioned by changes in neural structures that cannot be attained via conscious, deliberate decisions. The generalized avoidance behavior is negative reinforced by the avoidance of even more incongruence. These are all processes transpiring in the implicit mode of functioning. Total avoidance is the only form of control remaining for the depressed person, and no one can take away from the person this last morsel of control.

A close functional relationship exists between the severity of the depression and current incongruence. The more severe the need violation, the more intense the depression. Therefore, the depression will also respond directly to improvements in incongruence. We observed very high correlations between depression and the state of incongruence as well as for the degree of change in incongruence (see previous). This relationship was not found for anxiety disorders. Their onset is also influenced in important ways by current inconsistency, but once the underlying neural structures are well facilitated, the disorder will remain in place independently of the current extent of incongruence. This is related to the fact that the anxiety itself takes on the functional role of inconsistency. The negative reinforcement of its control-related components does not transpire anymore via inconsistency reduction but via the avoidance and reduction of the aversive anxiety state. In depression, this aversive negative reinforcer that acquires its own motivational function does not exist as such. The depression remains functionally embedded in the person's general motivational mental activity, which includes the person's degree of need satisfaction and the realization of motivational goals.

This explains why anxiety disorders are relatively rarely characterized by spontaneous remissions, whereas this is the rule rather than the exception in depression. Having retreated into the shelter of the sick role, the depressed person is removed from the continuous battleground of repeated need violations. The depressive state has rendered the person invulnerable to such injuries. He or she is no longer engaged with the normal demands and interactions posed by the environment, which previously caused continuous incongruent perceptions with regard to his or her motivational goals. The state of depression is also marked by certain fluctuations, however. Over time, it is inevitable that even the depressed person encounters some positive experiences. For example, this might occur when someone shows caring affection, does something nice for the person, when the person hears a piece of music that he or she likes, and so forth. At that point, a positive feedback mechanism that points in a positive direction is set into motion. The more such positive experiences are encountered, the more the person will once again turn toward the environment, encounter more positive experiences, and so on. This positive feedback process can, of course, be stimulated and facilitated via therapeutic intervention, but in all probability the process would get started over time even without specific expert interventions. That is why we observe such enormously high pre–post effect sizes even among depressed patients who have not received specific therapeutic interventions (see section 4.4.4).

Therapists of all kinds profit from this pronounced tendency toward spontaneous remission and gladly take credit, interpreting the positive changes as specific therapy effects. We have seen in section 4.4.4 that the real therapy effects in depression—those that go beyond spontaneous remission—are rather modest. They could not possibly be much larger, however, because the depressive symptoms cannot more than disappear. The domain in which depression therapies can really prove their value concerns the enduring improvement of consistency in mental functioning. For each depressed person, one must individually determine and treat the sources of incongruence. It is these sources that must be altered therapeutically in order to prevent the patient from becoming depressed again in the future. The real value of depression interventions consists in their ability to attain not just short-term state improvements but longer-lasting protection from relapse and recurrence.

Depression is the disorder for which it is the most obvious that, beyond treating symptoms, inconsistency needs to be addressed in treatment in order to achieve long-lasting effects. Inconsistency cannot be separated from the concrete life endeavors of the person. In order to achieve enduring effects, one has to engage with the content of the life concerns of the depressed patient. Therefore, medication treatment of depression will normally not by itself produce fully satisfactory long-term improvements. The degree of happiness or misery in one's life is not just a function of the correct mixture of neurotransmitters but a function of the life experiences that a person encounters.

Many mental disorders accompany depression. In most cases, this points to a high level of inconsistency among these individuals. This is true for comorbidity more generally. Where there is comorbidity, the conditions favoring the emergence of mental disorders must have been in place for longer periods of time.

Inconsistency creates the space, as it were, within which mental disorders can develop. Individuals who have predominantly encountered need-fulfilling experiences and, therefore, have formed a motivational schema structure that repeatedly leads to similar experiences, by definition, have a low level of incongruence. Their mental processes are characterized by consistency. Such a mental apparatus does not leave room for mental disorders; the foundations for their development are not in place. It is only in situations in which inconsistency predominates in mental processing, in which the mental processes are not determined by powerful order patterns oriented toward positive experiences, that a space is created that can then be occupied by mental disorders. Inconsistency, therefore, is the most important concurrent precondition for the emergence of mental disorders. If this space is made smaller by improving the consistency in mental functioning, less room is left for the mental disorders. Congruence and consistency are the natural opponents of mental disorders. However, *congruence* and *consistency* refer to the motivated aspects of mental functioning. Psychotherapy must concern itself with this motivated mental functioning if it aims to reduce the space for mental disorders. If the efforts to increase consistency in mental functioning are successful, mental disorders are forced into retreat. We have observed high negative correlations between incongruence and many aspects of mental health.

Mental disorders themselves also diminish the degree of consistency in mental functioning. Once the neural foundations for a mental disorder are facilitated so thoroughly that the disorder acquires autonomy—that it becomes detached from motivational functioning—the neural disorder foundations must be altered in order to attain consistency in mental functioning. The alteration of these disorder foundations requires disorder-specific interventions. Efforts to treat disorders and those to treat inconsistency are not at odds with each other; they are therapeutic strategies that can enhance each other. The optimal relationship between these different strategies must be individually determined for each case. The treatment of patients with anxiety disorders will typically require more of an emphasis on disorder-specific interventions, whereas the treatment of depressed patients requires a greater focus on inconsistency reduction, but the precise determination of treatment emphases can only occur when the specific circumstances of the individual case are known. In order to be able to make such treatment decisions, it is necessary in each case to analyze not just the disorder but also the patient's inconsistency profile. The diagnosis of disorders is currently more advanced than the diagnosis of inconsistency. There is great room for further improvement in this domain, and in the domain concerning the therapeutic consequences that arise from the diagnosis of inconsistency. I expect that by pursuing these avenues we will witness a considerable improvement in the effectiveness of psychotherapy, beyond the levels attained today.

We have now arrived at the concrete therapeutic implications flowing from a consistency-theoretical perspective on the origin and treatment of mental disorders. This will be the focus of the next chapter.

CHAPTER FIVE

IMPLICATIONS FOR PSYCHOTHERAPY

My goal in this chapter is to articulate the therapeutic conclusions arising from the content developed in the preceding chapters. These conclusions refer, on the one hand, to the goals that psychotherapy ought to pursue. They flow from the previously developed perspective on how mental disorders originate and which neural structures and processes underlie them. On the other hand, this chapter will elaborate conclusions with regard to our understanding of the mechanisms of change of psychotherapy. This, in turn, will have implications for the further optimization of psychotherapy, both in terms of the choice of therapeutic access points and the conduct of therapy itself.

In the first part of the conclusions, the construct of consistency takes on a central role because it connects our understanding of mental disorders with an understanding of therapeutic change mechanisms. In combination with the sections in chapter 4 on the basic needs, motivational schemas, inconsistency, and incongruence, these conclusions can be regarded as a more thoroughly elaborated and, in some parts, revised version of my consistency theory of psychotherapy, which I first presented in my book *Psychological Therapy* (Grawe, 2004).

A second part of the conclusions drawn in this chapter from the previous discussions has to do less directly with the theory, although it is compatible with it. This second part addresses the neural mechanisms underlying therapeutic changes. Such mechanisms have already been mentioned several times in the preceding sections. Their role in the context of achieving therapeutic change will once more be reflected in detail in this chapter.

The third and last part of the conclusions takes the form of concrete guidelines for therapy practice. I discussed in many different sections various findings that have clear implications for therapeutic practice. On occasion, I have also briefly alluded to their relevance for therapy. However, more detailed discussions in those earlier sections would have interrupted the flow of the prose. Sometimes, therapeutic implications have also surfaced in various different contexts. Therefore, I will take the opportunity to transform the findings here in a concentrated form as therapeutic guidelines—separate sets for therapy planning and for the therapy process.

These guidelines operate outside of the usual thought framework of psychotherapy. They refer neither to therapy methods nor to specific mental disorders. They are, as it were, guidelines that cut across established boundaries, and by following these principles, therapists might be able to raise the effectiveness of their therapies in various contexts.

All three groups of conclusions together form what I called, in the book's title, *Neuropsychotherapy: A Neuroscientifically Informed Psychotherapy*.

5.1 MENTAL DISORDERS RESULT FROM UNSUCCESSFUL INCONSISTENCY REGULATION

As shown in the last section of chapter 4, the new neural activation patterns that underlie mental disorders originate in situations of acutely elevated inconsistency. They are facilitated by short-term reductions in inconsistency tension. Mental disorders form only when heightened levels of inconsistency persist over longer periods because this allows the facilitation processes to be repeated sufficiently so that a rigid disorder pattern can be formed.

Some components of the disorder pattern have already existed previously. They are only now joined together, however, into a single coherent disorder pattern. For example, individuals who develop anxiety have previously already tended to respond to challenges, stresses, or incongruence with excessively strong and long-lasting autonomic activation. Such potentialities can be genetically determined. They will only be expressed, however, if the person has encountered uncontrollable incongruence experiences in early childhood. With their massive glucocorticoid releases, such incongruence experiences can disable the negative feedback loops the brain normally employs to protect itself from elevated stress hormone levels. Other structures involved in this process are the amygdala, which has been hypersensitized by too many and too intense incongruence experiences, and the hippocampus, which either because of its innate disposition or as a consequence of damage inflicted by stress hormones is now compromised in its functional capacity. A well-functioning hippocampus is important for the learning of contextual relationships. It normally ensures that stress and anxiety reactions are experienced only in situations that pose real threats. It also helps ensure that anxiety reactions are effectively inhibited from the orbitofrontal cortex, which evaluates the contextual meaning and significance of situations. Therefore, an elevated tendency for emotional dysregulation has existed even before the emergence of the disorder pattern as such.

The tendency to experience emotional dysregulation is typically accompanied by the tendency to easily feel threatened and to develop defensive and avoidant strategies to control these threats. Avoidance goals bind attention and mental energy and cannot be permanently reached. The more mental activity is influenced by avoidance goals, the more the attainment of approach goals is compromised. Strongly formed avoidance goals lead to approach incongruence and to motivational discordance because approach and avoidance goals mutually hinder each other. As a consequence, situations involving elevated incongruence and inconsistency tension are repeatedly experienced. The negative emotions that go along with this cannot be effectively downregulated because of the tendency toward emotional dysregulation. The individual components—excessive autonomic arousal, poor emotion regulation, predominant avoidance tendencies, incongruence experiences, and inconsistency tensions—interact with one another, and positive feedback loops among the components result in escalating cycles toward ever higher inconsistency levels.

All of these processes transpire primarily in the implicit mode of functioning. Only a part of it will be reflected in conscious awareness—and even if it is, this will only be in dissonance-avoiding form. If a particularly high inconsistency tension then occurs in a particular situation, against a background of an already elevated inconsistency level—and note that this situation would involve the strong activation of control needs—then it can happen that particular behaviors that are new in this context, such as a thought, perception, or action, can momentarily convey a sense of control over the inconsistency by causing a short-term tension decrease. The control-related behavior is negatively reinforced in this process. However, it is not just this act but all of the neural activation patterns that are activated at this moment that are being facilitated particularly strongly and that, based on Hebb's (1949) principle that neurons that fire together wire together, are joined to form a single, coherent activation pattern. As a consequence, they will be more likely to be jointly activated again in the future. If this specific constellation of neural activation patterns, which now also incorporates the addition of the control component, is again accompanied during its next occurrence by a reduction of inconsistency tension, the newly developed neural activation pattern will be facilitated ever more thoroughly. Such activation patterns include multiple components, which are now for the first time joined together into one common pattern. These are not just simple patterns, such as those corresponding to the perception of objects, for instance, but rather, they are complex neural circuits. Autonomic arousal, avoidance reactions, cognitions, and the respective control components, such as "worrying," obsessive thoughts, compulsive behaviors, avoidance, and so forth grow together into an ever more firmly established disorder pattern.

The disorder pattern is ultimately the result of an escalating feedback process that has stretched out over a considerable duration and that has eventually led to such strong inconsistency tensions that it had to be down-regulated in a sort of emergency operation. At the moment of their formation, then, psychopathological disorder patterns serve the purpose of inconsistency regulation. This inconsistency regulation must be regarded as failed inasmuch as it manages to reduce the inconsistency only for short periods; over the medium and long terms, it even contributes to further elevations in inconsistency.

Mental disorders, in turn, constitute violations of basic needs in themselves. Moreover, they bind energy that is then missing elsewhere for the pursuit of approach goals. As a consequence of mental disorders, then, both approach incongruence and avoidance incongruence intensify further. Incongruence can therefore simultaneously be regarded as a cause and a consequence of mental disorders. This also explains the very high empirical correlation between incongruence and symptom severity, which I reviewed in section 4.8.4.3.

The consistency level in mental functioning has continued to deteriorate, then, after the formation of a mental disorder. The even further diminished consistency can become a breeding ground for the development of additional disorders. The mechanism essentially remains the same. The high inconsistency tension is the "glue," the negative reinforcer, for the joining together of largely preexisting—but also in part, entirely new—components, each of which has its own neural foundation, into a new disorder pattern with a new neural circuit that becomes ever more thoroughly established via repeated facilitation.

The escalating cycle toward a negative direction can be broken if something happens that somehow increases the consistency in mental functioning. This could be, in particular, positive, need-satisfying life experiences that lead to better approach congruence levels. In some disorders, such as for instance depression, the effect can be that the disorder's components—the symptoms—diminish in intensity because they have maintained a current function for the regulation of consistency. Indeed, as reviewed in section 4.8.4.3, increases and reductions in incongruence correlate very highly with increases and reductions in the severity of depressive symptoms.

In anxiety disorders—and potentially also in other disorders, which will still have to be examined—the functional relationship between symptom changes and incongruence changes appears to be less direct. The reason might be that, once the disorder pattern has become firmly established, anxiety disorders are not maintained by the reduction of inconsistency tension (negatively reinforced) but—as conceptualized by Schulte (2000)—because of their role in the struggle against anxiety itself. The control components of the disorder continue to play a critical role here as well, but what is being controlled is the anxiety and not the inconsistency tension. Autonomic arousal and anxious feelings are, to be sure, important and integral components of the disorder pattern. When the disorder pattern is activated, these components are activated as well. The reduction or the avoidance of anxiety leads to negative reinforcement of whatever means were employed for that purpose, which then further facilitates the totality of the disorder pattern. The control components that serve the purpose of anxiety reduction or avoidance, therefore, lead to the exact opposite of that which they aimed to achieve. They maintain the anxiety instead of diminishing it.

Anxiety disorders can be maintained, then, even if the inconsistency level in mental functioning is reduced. If the inconsistency level remains high, however, the symptoms—especially the control components of the symptoms—should appear with greater intensity, particularly when the inconsistency tensions flare up, in which case the symptoms retain their inconsistency-reducing function. Furthermore, the danger of the formation of additional disorders exists in such situations, whether they be anxiety disorders, depressive disorders, or entirely different syndromes.

High inconsistency is probably always a contributory cause for the formation of new disorders, but it is not always the cause underlying the maintenance of already existing disorders. Mental disorders—and especially anxiety disorders—can develop functional autonomy and become decoupled from general motivated mental functioning. This will primarily be the case when an anxiety disorder is present despite otherwise low levels of incongruence. In those cases, therapy can retain a focus primarily on the treatment of the disorder.

In cases in which mental disorders and, simultaneously, elevated incongruence levels are present, however, it would be short-sighted to limit oneself to the treatment of only the narrowly defined disorder. Most likely, the disorder continues to function for the purpose of inconsistency regulation and is increasingly thoroughly facilitated because of the short-term inconsistency reductions it enables. In this case, the inconsistency in mental functioning would also have to become a focus of therapy in order to achieve a good outcome.

5.2 PSYCHOTHERAPY WORKS VIA CONSISTENCY IMPROVEMENT

According to the view that was developed in detail in chapter 4 and shortly reiterated in the preceding section, the outcome of psychotherapy depends primarily on the extent to which the therapy manages to achieve consistency improvements in the patient's mental functioning. This does not mean that consistency should be regarded as the ultimate criterion for therapeutic success. The criteria for the evaluation of therapeutic success are already predetermined. The ultimate goal is always to free patients from their suffering. This means, primarily, to improve their mental disorder(s) and to elevate their subjective well-being. Psychotherapy must confront the challenge of being measured with reference to such societally relevant criteria (Schulte, 1993). This holds true for therapies of all kinds. No therapy can escape this necessity by choosing its own success criteria.

The different therapeutic approaches can be differentiated, however, in terms of their assumptions about which kinds of changes in mental functioning ought to be attained in order to facilitate improvements in the disorders and in well-being. According to Beck, dysfunctional cognitive schemas ought to be altered; according to psychoanalytic theory, conflicted motivational constellations should be targeted; and so forth. According to the perspective developed here, it is of primary importance to improve the consistency in mental functioning in order to achieve improvements in the disorders and in well-being. However, a thorough understanding of inconsistency includes also the domain of mental disorders as a possible source of inconsistency. If one improves the disorders, one also improves mental consistency. Disorder improvements and consistency improvements are not alternative options; they serve and support each other. Nevertheless, the construct of consistency goes beyond disorders. It embeds the disorders in the broader context of motivational mental functioning. High consistency—which is equivalent with adequate need satisfaction—protects from the development of mental disorders. At least at the moment of the actual onset of the disorders, individuals who develop mental disorders have a heightened level of inconsistency. Often, however, this inconsistency persists long beyond the onset of the disorder. Not surprisingly, enduring inconsistency is associated with high levels of comorbidity.

What is the therapeutic utility of the construct of consistency? Without such utility, the construct would be superfluous. Ultimately, the utility must consist of the fact that the criteria for psychotherapeutic effectiveness are better fulfilled when therapists orient their work on the assumptions of consistency theory rather than on other assumptions. One conclusion suggested by consistency theory is the assumption that better therapeutic outcomes will be achieved to the extent that consistency in mental functioning can be improved. Consistency improvement, then, is the most important theory-specific therapy goal.

We know from chapter 4 (see section 4.8.1) that many different forms of inconsistency in mental functioning exist. All of them have a more or less strong influence on the person's ability to perform his or her life tasks and on his or her subjective well-being, and some of them also influence mental health. It is impossible in therapy practice to assess all of these forms simultaneously in a single patient. Therefore, when aiming to determine the overall inconsistency level in mental functioning, it seems sensible to concentrate on those forms of inconsistency that most widely reflect all of the other types of inconsistency and that simultaneously inform our knowledge with regard to the patient's need satisfaction. This holds true for motivational congruence; that is, the extent to which the patient manages to attain his or her most important motivational goals—both the approach- and the avoidance-related motivational goals.

In section 4.8.4.3 I introduced the incongruence questionnaire that we developed for this purpose. Its total score correlates with subjective well-being in a mixed sample of more than 1,000 normal individuals and clinically disordered patients at $r = -.87$. In such a large sample, this is an unusually strong correlation coefficient. It indicates the following: If we know the extent to which a person attains his or her motivational goals, then we also know how good or bad he or she feels. Or, conversely, if a person has high levels of overall well-being, then this person is able to attain his or her motivational goals effectively. Because it is the most important goal of psychotherapy to achieve positive changes in the patient's well-being, we must help patients with elevated incongruence to realize their motivational goals more effectively. Such efforts are opposed, however, by various sources of incongruence. If these did not exist, the patient would have been able to realize these goals on his or her own—otherwise, the patient would not be in treatment. It is never easy to remove these sources of incongruence. Otherwise, the patient would have already done it.

From a neuroscientific perspective, it is quite evident why this is so difficult for patients. Ultimately, the patient's own neural structures create the inconsistency and, thereby, are responsible for the incongruence in his or her mental functioning. To change one's own neural structures is not at all easy because most of these structures—and especially those that create difficulties—are very well facilitated and transpire outside of consciousness. A person who wants to change his or her neural structures without external assistance runs the risk of spinning in circles within the labyrinth of his or her well-facilitated neural circuits. The person will only ever become aware of those things that, because of his or her well-facilitated neural activation patterns, can be easily activated, and this might very well be a part of the person's problem. Therefore, the patient is rendered dependent on external input. Getting outside help—or, on a neuroscientific level, influencing one's neural structures via someone with other neural structures (the therapist)—makes good sense.

From this perspective, it is also sensible for the therapist to assume a different view of the patient than the patient himself or herself. Patients' understanding of their problems is important, in part, because it is the origin of their change motivation. Simply because of this fact, the patient's perspective must be properly considered and utilized in the service of the change process. However, the way patients construe their problems can very well be one of the problems that undermine their ability to better satisfy their needs. The patient is, by nature, unable to employ introspection to understand sources of incongruence such as transactions in the implicit mode of functioning or malfunctioning neural structures (e.g., ACC or hippocampus). Thus, the patient needs an expert outside perspective in order to determine the sources of his or her misery. At the same time, it is clear that the patient can be changed only toward such directions for which he or she has already developed motivational tendencies. This does not refer just to the patient's explicit therapy goals but, primarily, to his or her implicit motivational goals.

If the therapist approaches a patient with the understanding that he or she ought to improve the consistency in the patient's mental functioning—that is, if the therapist wants to help the patient—he or she will present the patient with an independent expert perspective in order to uncover access points that enable improved consistency in mental functioning. In a second step, the therapist will test the extent to which these objective access points for the attainment of therapeutic change are congruent with the motivational propensities of the patient. The therapist will then apply appropriate interventions at such access points that, on the one hand, promise to engender substantial consistency improvements and, on the other hand, are supported by already existing motivational leanings of the patient.

Based on the ideas developed in the previous three chapters, a therapist with this orientation will assume three perspectives in order to determine potential access points for effective intervention:

1. The *disorder perspective*, as specified in the DSM and ICD, complements the knowledge of neural underpinnings of individual disorders.
2. The *process perspective:* Which possibilities exist to enable the patient to encounter consistency-elevating experiences in the therapy process?

Independent of other access points for consistency improvement, such experiences should have a direct positive effect on the patient because of their ability to improve consistency.

3. The *inconsistency perspective:* What is the nature and extent of the patient's inconsistency level? Apart from the psychopathological disorder(s), which other sources of inconsistency can be determined, so that a change or improvement in these sources can then lead to overall improvements?

These three perspectives are assumed by the therapist because they refer to the three most promising avenues for attaining consistency improvement. Each of these perspectives reveals concrete therapeutic conclusions that will now be elaborated.

5.3 THE MOST IMPORTANT OPTIONS FOR ENHANCING CONSISTENCY VIA PSYCHOTHERAPY

5.3.1 Improving Consistency via Disorder-Oriented Treatment

Once a mental disorder has formed and has become established, the disorder constitutes an additional source of inconsistency in mental functioning. Such disorders obstruct the attainment of motivational goals and are themselves a violation of important basic needs, because they indicate a loss of control, are experienced as unpleasant, and wound the person's self-esteem. If it is possible to alleviate the disorder by using syndrome-specific interventions, it leads to overall improvements in consistency and clears the way for a better realization of motivational goals. Successful disorder-specific interventions should lead to improvements in incongruence levels, then, and beyond that, to improvements in well-being. The empirical correlation we observed between reductions in psychopathological symptom severity and incongruence reductions was .64. The hypothesized functional relationship really in fact exists, then.

If we once again consider the correlations presented in Figure 4–24 (pp. 420, 422), which show the associations between incongruence changes and changes in other variables, it becomes clear that symptom reductions could lead to even broader positive changes via their effects on incongruence reduction. Such improvements in symptom severity can reverse the escalating cycle that led to the formation of the disorder in the first place. The improvements in the disorder bring about reductions in incongruence levels, which then can positively impact all of those variables that correlate substantially with incongruence in Figure 4–24 (pp. 420, 422). It then becomes likely that that intensity of avoidance goals will diminish, that competency expectancies will strengthen, that positive coping behavior will improve, that resources will better be utilized, and that interpersonal problems will lessen. Because these are all correlational associations, it is impossible to say with precision just how strong the influence in this direction—from incongruence reductions toward improvements in other variables—will be. This correlation also reflects, of course, the opposite causal direction—changes in the other variables also lead to reductions in incongruence. Substantively, however, it seems perfectly plausible to infer from these correlations that reciprocal, bidirectional causality is present among these variables. That means that if incongruence reductions result as a consequence of improvements in disorder severity, then the consequence, in turn, will probably be that some of these other variables will improve. These other changes would then, once again, reduce the inconsistency in mental functioning. Such a positive feedback loop with escalation in a negative direction had originally contributed to the formation of the disorder, as detailed previously. Now improvements in the disorder kick off a positive feedback loop pointing in a positive direction. This explains why disorder-specific therapies can trigger positive changes in very different life domains, even if the interventions focus exclusively on the disorder—a phenomenon that is well known from psychotherapy research. The phenomenon is explained when we regard consistency improvements as the functional links among the various changes.

We saw in chapter 3 that mental disorders are accompanied by structural changes in the brain. Such structural changes are the result of intensive facilitation processes over extended periods of time. It is therefore simply not possible, then, to commit oneself to not feeling anxious anymore, to stop worrying, to stop having obsessive thoughts, to think less negatively, and so forth. Altering neural structures does not work this way. Changes in neural structures can be attained only via very intensive, frequently repeated facilitations. In order to effectively inhibit the well-facilitated processes that underlie specific components of the disorder, more is needed than just willpower or volitional commitment. Instead, this also requires the intensive facilitation of processes that exert an inhibiting influence on the disorder components.

Research on disorder-specific interventions is currently more advanced than the neuroscientific research on the individual disorders. Disorder-specific manuals that extract specific therapeutic guidelines from the intervention research are available and can help therapists identify the access points that can be targeted in order to effectively inhibit and destabilize a particular disorder pattern.

The better our knowledge about the specific neural foundations of the individual mental disorders becomes, the more such knowledge should be incorporated into the disorder-specific treatment manuals. In chapter 3, I discussed neuroscientific research suggesting that different subtypes of major depression probably exist, each of which probably requires different therapeutic approaches. Depressives who have an ACC that can still be relatively easily activated will have a better prognosis with the currently available interventions. By integrating a neuroscientific perspective, novel intervention research could investigate, for example, which interventions could best be employed to reinvigorate the functioning of a deactivated ACC. Neuroscientific research also suggests the existence of subtypes of obsessive–compulsive disorder that might require different therapeutic interventions. I would expect that the effectiveness of disorder-specific interventions could, once again, be substantially improved when interventions are devised that target the specific neural foundations of each respective disorder.

Post-traumatic stress disorder (PTSD) can be regarded as one positive example that shows how neuroscientific research and psychological intervention research can arrive at a mutually consistent conceptualization of the disorder and, in turn, at common therapeutic conclusions that flow from this conceptualization. With this disorder as well, however, the neuroscientific research suggests additional questions: Which therapeutic conclusions should we draw from the fact that the hippocampus appears to be functioning less effectively among persons with PTSD? If it were to be confirmed that this is a consequence of harm inflicted by excessive stress hormones, then would it make sense to conceptualize improvements in stress tolerance as a specific therapy goal, in addition to the integration of implicit and explicit memory? Such questions are only able to be asked when incorporating neuroscientific research—if this research did not exist, nothing would be known about the role of the hippocampus in the development of this disorder.

It can be expected that neuroscientific research will invigorate and inspire disorder-specific interventions in coming years. Even within the foreseeable future, a neuroscientific diagnosis could become a part of routine pretherapy assessment, which would enable therapists to specify the access points for promising interventions with greater precision, and to examine later in therapy the extent to which the targeted changes are being achieved on the neural level. The study by Furmark et al. (2002)—reviewed in section 2.19—on neural changes during the treatment of patients with social phobia was a first step in this direction that already has inspired further research.

The neuroscientific approach to mental disorders, then, creates new perspectives for the further improvement of disorder-specific interventions. Whether the high expectations will be fulfilled remains to be seen. Even today, however, the neuroscientific view of mental disorders suggests certain conclusions with regard to how disorder-specific interventions should be conducted. These conclusions do not just refer to the disorder-specific access points for therapeutic interventions; they also inform how the influence of disorder-specific interventions on the various disorder components can be understood. They clarify the neural mechanisms of such therapeutic changes. The question about the neural mechanisms underlying therapeutic changes, however, extends beyond the domain of disorder-specific interventions. It is of broader significance for psychotherapy and not linked to the construct of consistency. Therefore, I will discuss these neural mechanisms later, in the second section of the conclusions.

5.3.2   Consistency Improvements via Experiences in the Therapy Process

Let us imagine the situation a psychotherapy patient faces at the beginning of therapy. Let's assume the patient has suffered for years from an anxiety disorder and, in recent months, has also developed a moderately severe depression. Such patients practically always have a high inconsistency level as measured by the incongruence questionnaire. In most cases, this means that many of their approach goals are not adequately met. Most commonly, they have several strongly established avoidance goals, as measured by the IAAM. These avoidance goals point to sore spots they have attempted to protect, albeit unsuccessfully. Their elevated avoidance congruence indicates that they are currently unable to avoid that which they fear the most. Depressive people typically have strongly elevated scores on the IAAM fear of vulnerability scale, especially on the item, "showing my weaknesses to others." At the same time, they report that they are unable to avoid weaknesses, vulnerabilities, and psychological injuries (see Table 4–6).

In such a patient, all four basic needs are acutely violated at the beginning of therapy. We do not even need the IAAM and incongruence questionnaire in order to see this in most cases.

Suffering from an anxiety disorder constitutes a loss of control in itself. Anxiety disorders are accompanied by strong efforts to gain control over the anxiety (Schulte, 2000). The need for orientation and control is acutely activated, then, among such patients. This renders the patients very receptive to positive control experiences. If the therapist explains the process and makes therapy transparent; if he or she conveys a plausible understanding of the disorder to the patient and articulates in detail what the patient himself or herself can do in order for things to improve, the patient's need for orientation will be satisfied. If the therapist explicitly includes the patient in all steps and decisions; if he or she sensitively acknowledges and takes seriously all of the patients' ideas, suggestions, and initiates, even when they might objectively have little chances for success; if the therapist consistently presents a menu of several options from which the patient can choose; if he or she structures the session in such a way that the patient gains the feeling that he or she can participate effectively, that he or she can repeatedly experience the small successes of being able to do what he or she is asked to do; if the therapist shows the patient specific ways in which he or she can gain more control over his or her state, then all these can be viewed as positive perceptions with respect to the acutely activated need for orientation and control.

Such experiences are incredibly valuable for the patient, not just because his or her need for control is acutely violated by the anxiety disorder. It is likely that a patient who has developed an anxiety disorder and a depression also has a prior history of violations of his or her control need (see section 4.9.1). In depressions, this loss of control can indeed be regarded as a core of the disorder, if we consider Maier and Seligman's (1976) concept of learned helplessness or Alloy et al.'s (1990) conception of the path from anxiety into depression, which is characterized by increasing loss of control. The control experiences a therapist can enable a patient to have, by engaging him or her in carefully designed interventions, effectively soothes the wounds inflicted by the patient's life history.

The patient might initially have some degree of mistrust toward the therapist because these positive perceptions are met by avoidance schemas that were formed to protect the patient from violations of the attachment need. However, the patient does not have to bare his or her soul and show his or her most vulnerable sides in order to be able to benefit from the control experiences. The experiences are being created by the therapist without having been explicitly requested, and without having to be openly discussed. The patient experiences these situations primarily in the implicit mode of functioning. He or she does not even have to openly declare his or her vulnerabilities and wishes in order to encounter the control-enhancing experiences.

If the patient continuously encounters many such control need satisfying experiences while he or she spends time with the therapist, the incongruence resulting from his or her violated control need will diminish. This will be accompanied by positive emotions. The patient will experience the time with the therapist as positive, and these early experiences will color all subsequent sessions they spend together.

An additional effect is that motivational priming occurs in the implicit mode of functioning. The phenomenon of motivational priming was discussed in section 2.15. It refers to the following: If the approach system is activated and prefacilitated via positive emotional experiences, a tendency to experience further positive emotions and to engage in approach behavior arises. Negative emotions and avoidance reactions, by contrast, grow weaker. The converse is true as well: If the avoidance system is already activated, anxiety reactions to specific events will transpire more strongly, compared to situations in which the person previously is in a positive emotional state. Negative emotional cues facilitate associations, representations, and behavior programs in the avoidance system; positive emotional cues facilitate those in the approach system.

The positive control experiences that the patient continuously encounters prefacilitate his or her approach system. The patient will engage with greater receptiveness and an orientation toward personal approach goals—that is, he or she will operate in the approach mode—as the therapist presents him or her with the next necessary experiences. The patient will be more open to therapeutic interventions, compared to what would have been possible in the avoidance mode. According to the large meta-analysis of process–outcome associations by Orlinsky, Grawe, and Park (1994), patient openness is one of the most important predictors of good therapy outcome.

The positive control experiences that a therapist enables the patient to have in maximally possible dosages—even though these were never explicitly requested by the patient—engender not just an incongruence reduction with all its positive consequences, as discussed in the previous section, but they also improve subsequent within-in-therapy conditions for the processing of problems that need to be addressed. Thereby, the positive experiences contribute indirectly to subsequent incongruence reductions as a consequence of the successful resolution of these problems. There are two pathways, then—a direct path and an indirect path—by which control experiences in therapy can contribute to improvements in consistency.

What I just said here about the incongruence-reducing effects of positive control experiences in the therapy process holds true in an analogue fashion for the experiences that the patient encounters with respect to his or her basic needs in therapy.

Many patients have low levels of self-esteem. The feel ashamed because they feel that they are different from others, that they have problems that they cannot resolve independently. Being mentally ill is experienced by most people as a form of personal failure. Because of this, they feel inferior. By necessity, they also have to show their vulnerable side in therapy. All of these are experiences that are detrimental to their self-esteem.

In such a situation, it feels good to realize that someone else acknowledges one's positive and advantageous aspects, to be recognized for one's strengths and resources. A therapist has many opportunities to guide the patient in such a way that he or she experiences self-esteem enhancing events, for which he or she will very receptive because of the processes explained previously. The therapist can repeatedly provide opportunities for the patient to present himself or herself in a way that emphasizes his or her strengths, by showing genuine interest not only in the patient's problems but also in other aspects of his life—by asking about what else he or she does in life, which kinds of things interest him or her, what he or she can do well or which areas of expertise he or she might have, and so forth. The therapist can also converse with the patient about interests they might share. In this way, the patient will come to feel more equal in status to the therapist.

The therapist can also explicitly ask the patient about life domains that he or she does not want to change, under any circumstances. The therapist can inquire about times in which things seem to go better, and what might make those times different. This directs attention toward resources and positive opportunities that are practically present in every person's life. The therapist raises the patient's interest in what else the patient might have already tried to solve his or her problems. The therapist can acknowledge and appreciate the positive fact that the patient has not given up but has decided to confront his or her problem, to actively do something about it.

If the therapist conducts the process in such a manner that the patient can effectively engage (see previous), this will create additional opportunities to feed back to the patient how well he or she participates, how constructively he or she engages and actively contributes, and how it should not be taken for granted that the patient can do this. The specific way in which the therapist conveys praise, appreciation, respect, and compliments depends on the nature of the receptiveness of the patient. What is important is that the patient experiences the therapist's expressions as genuine; that he or she experiences them as real and positive. This is more a question of therapeutic approach or attitude than one of method or technique. If one's approach is to perceive and recognize the positive in others, and to convey this recognition to others and let them feel it, then one will actually be able to see things that really are positive and worthy of recognition, and one will be able to convey this in a genuine manner. Many opportunities to provide such positive feedback either explicitly or nonverbally arise in almost every session, without the requirement of having to focus the discussion overtly on resource-oriented themes. Self-esteem-enhancing experiences can go along continuously with other work in therapy, if the therapist deliberately aims to conduct the sessions in this manner and to convey such feedback frequently and powerfully.

Enabling patients to encounter positive control experiences and self-esteem enhancing experiences go very well together. Both require a similar attitude from the therapist, which might be best described as *resource-oriented* and can be differentiated from a *problem-* or *deficit-oriented* approach. Many therapists will feel certain that they have such positive attitudes toward their patients. However, in order for the patient's incongruence to be reduced, it is important that he really encounters these concrete perceptions as intensively and frequently as possible in sessions with the therapist. The more often such experiences can be created in each session, the better. In therapeutic reality, however, the patient will not encounter many such situations unless the therapist adopts this as a specific and deliberate session agenda.

This makes sense because the therapy situation naturally seems to call for something very different. The patient often has serious, real problems, shows his or her suffering and conveys his or her misery. This creates enormous pressure on the therapist to attend to these problems and awful experiences that the patient has recently encountered. The more extensive the problems and the greater the patient's suffering, the more the therapist will feel obliged to actively help the patient with his or her problems. This almost forces the therapist to adopt a problem-oriented perspective and, thereby, to become problem-oriented in his or her perception, thinking, and behavior. If one is confronted so intensively with acute problems that urgently require help, it becomes difficult to focus on the patient's positive, healthy aspects or, indeed, to recognize that they exist. With such pressure, it is entirely possible that therapists do not express their appreciation of positive patient aspects in a way that patients notice or feel, even though the therapist might, in principle, agree that it would be a good idea to do so. This is certainly our experience from a large number of process analyses.

In our research group in Bern, we have collected approximately four thousand 10-minute-long videotaped therapy passages with various different kinds of therapists, patients, therapy durations, disorders, and therapeutic methods. We have analyzed these excerpts to examine the extent to which therapists actually convey the kinds of need-satisfying experiences I described with regard to self-esteem and control needs. Depending on the type of research question being asked, the therapy sessions from which these 10-minute segments were selected, were, in turn, chosen from among more than eight thousand videotaped sessions. My colleagues, Daniel Regli, Emma Smith, Andreas Dick, Daniel Gassmann, and I have gone to great effort to analyze these therapy segments with regard to the realization of the mechanisms of action that we assume to be of greatest importance (Grawe, 1995, 1997). The process analysis method we developed specifically for this purpose is known as therapy spectrum analysis (TSA; Grawe, Dick, Regli, & Smith, 1999).

In order to examine the interplay between therapist and patient even more thoroughly, we also developed another method of process analysis, the consistency-theoretical microprocess analysis (CMP; Gassmann & Grawe, 2006). This method allows researchers to analyze therapeutic processes on a minute-to-minute basis. We used this method to analyze 6,800 1-minute-long therapy segments with regard to the mechanisms of action considered to be most important in consistency analysis.

If one reviews such a great number of therapy segments as a representative sample of real therapy practice, the first reaction is typically quite sobering. On average, far fewer need-satisfying experiences are conveyed in the therapies than what actually might be possible. It becomes obvious that the therapists normally have a great deal of other things on their minds besides conveying positive control and self-esteem experiences to the patient at a particular moment. Most of the time this is at least not a currently activated goal that influences their behavior in a noticeable way.

At the same time, these analyses have shown that the extent to which patients experience need-satisfying situations in each session determines the outcome of the session. Across the various specific questions we have examined, and the various research designs we have pursued, a regular finding has been that the extent of need-satisfying experiences in a session contributes more to a positive session outcome than the style with which the respective problems were treated. An extended discussion of these findings and methods is beyond the scope of my aims here, but the relevant details can be found in Dick, Grawe, Regli, & Heim (1999), Gassmann and Grawe (2006), Grawe (1999), Grawe et al. (1999), Regli et al. (2000), E. Smith and Grawe (2001, 2003, 2005, in press), as well as E. Smith, Regli, & Grawe (1999). However, I do want to present some of the findings here, because they show that what happens in therapy with respect to patient needs is of crucial importance for the immediate result of the session and, ultimately, for the success of therapy.

Figure 5–1 shows the basic approach of our process research. It assumes that the patient encounters experiences in every session that have some degree of significance for his or her motivational schemas. They could satisfy his or her needs to a greater or lesser degree, but they could also frustrate or violate these needs. Beyond that, the patient's experiences in therapy have specific functional significance with regard to his or her problems. The experiences can either more or less effectively impact upon these problems. These process experiences determine the effect therapy has on the subjective experience of the patient. The patient might have experienced the session as very positive or, perhaps, as mostly negative. This is termed session outcome. Many session outcomes ultimately lead to a better or worse therapy result at the end of the entire therapy.

The effect of the single therapy session in the experience of the patient is assessed with a questionnaire that the patient completes immediately after each session. The

Figure 5–1. Basic philosophy of the Bern process research approach.

Table 5–1. Operationalization of Session Outcome With Items From the Patient Session Questionnaire That Refer to Positive Coping, Clarification, and Self-Efficacy Experiences

Sum Score Across the Following Items of the Patient Session Questionnaire:

- Today I have the sense that we have really made progress in therapy.
- Today I have gained clarity about the various options available for the solution of my problems.
- I now feel better able to cope with situations that previously felt overwhelming.
- I believe that I am now better able to act in the way I actually want to act.
- I have the feeling that I have a better unterstanding of myself and my problems.
- I have the sense that it is becoming increasingly easy for me to master my problems independently.
- Today I have gained an understanding of things that previously seemed unclear.
- After today's session, I am firmly committed to tackle the problems that we discussed.

session questionnaire contains, among other things, several items that refer to positive coping, clarification, and self-efficacy related experiences the patient might have encountered in that session. The respective items are listed in Table 5–1. The items are rated on 7-point bipolar scales that range from "strongly disagree" to "strongly agree." The summed score of these items constitutes our operationalization of session outcome. If we consider the item content in more detail, it becomes obvious that all of these experiences refer to the needs for orientation and control.

We measured therapy outcome as an integrated effect size, which was calculated across seven therapy measures:

- Goal attainment with regard to the patient's three most important problems.
- The patient's satisfaction with the therapy success.
- The evaluation of therapy success by the patient.
- Increases in optimism and contentment, as measured by Zielke's (1978) Questionnaire to Assess Changes in Experiencing and Behavior (QCEB; in German,VEV).
- Improvements in subjective well-being, as measured by Grob's (1995) Bern Well-Being Inventory.
- Improvements in psychopathological symptom severity, as measured by the SCL-90-R (Derogatis, 1992, 1993).
- Improvements in interpersonal problems in the inventory of interpersonal problems (IIP) by Horowitz, Strauss, and Kordy (1994).

In a sample of 238 completed therapies for which complete process and outcome measures were available, the sum scale of the session outcome, measured across all sessions, correlated at $r = .69$ with overall therapy outcome, as assessed by the seven-measure index described previously. That means that the result of a therapy depends very much on the extent to which a patient encounters positive experiences within the sessions with regard to his or her need for orientation and control, which are acutely activated by the problems the patient is currently experiencing. The items in the session questionnaire do not refer to just any control experiences but, specifically, to control experiences the patient encounters with regard to his or her problems.

The session questionnaire targets the subjective experience of the patient. In our next set of analyses, we were interested in studying what actually happens in the therapy sessions that gives rise to this subjective patient experience. What are the factors that determine the extent to which the patient experiences positive mastery, clarification, or self-efficacy experiences? In order to examine this issue, we asked trained raters to evaluate session activities with regard to which specific actions therapists exhibited to enable patients to experience perceptions in line with their needs for control and self-esteem. All raters viewed 10-minute therapy-session segments, without prior knowledge about which types of therapy were shown, and then rated each segment on various rating scales. Specific details of these ratings, such as reliability, are not reiterated here; they can be found in the original publications.

The left side of Table 5–2 shows the characteristics measured by the rating scales, and the right side shows the question that the raters were asked to assess each characteristic.

Table 5–2. Scales for the Evaluation of Positive Therapy Experiences With Regard to the Need for Control and Self-Esteem

Scale	Evaluation
Explicit reinforcement of patient's goals and values	To what extent can the patient feel himself/herself supported and reinforced in his/her own goals and values?
Explicit reinforcement of patient's competences and abilities	To what extent are patient's competences and abilities explicitly addressed and reinforced?
Process activation of patient's own positive goals in the therapy relationship	To what extent is the patient's therapy-behaviour determined by his/her own motivated goals?
Process activation of patient's abilities and competences in the therapy relationship	To what extent is the patient capable to do what the therapist wants him/her to do in therapy?
Patient's experience of his/her positive sides positive	To what extent can the patient experience himself/herself from his/her side?
Explicit addressing of patient's positive characteristics	To what extent are positive characteristics of the patient explicitly addressed?

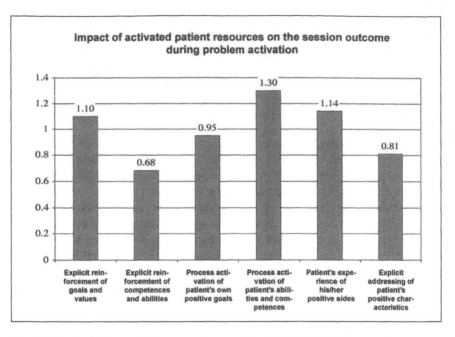

Figure 5–2. The influence of the process characteristics listed in Table 5–2 on session outcome, expressed in terms of effect sizes.

Figure 5–2 shows the influence of these patient experiences on the process of therapy vis-à-vis the outcome of the session. The perspective of an outside observer (objective process experience) is thereby related to the subjective perspective of the patient (session outcome). The magnitude of the association is expressed in effect sizes (Cohen's d).

If one considers that these associations reflect data from different evaluator sources, which are being related "blindly" to one another, then these effect sizes must be regarded as extraordinarily strong. The effect sizes shown in Figure 5–2 are taken from the quasi-experimental comparison in the study by E. Smith, Regli, and Grawe (1999). The design of this study is shown in Figure 5–3.

From among more than 8,000 therapy sessions, we initially selected those in which patients had reported on the session questionnaire that they had experienced considerable psychological pain or anguish; that is, sessions in which negative emotions had been strongly activated. From these sessions, we selected 30 that had a particularly good and 30 that had a particularly poor session outcome. The three 10-minute segments from the middle of each session were analyzed. Thus, 90 therapy segments from one condition were compared with 90 from the other. The comparisons were analyzed with t tests, and the results were converted into effect sizes.

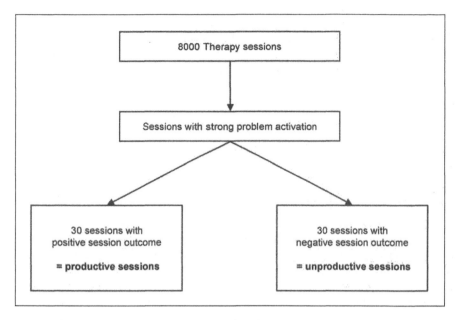

Figure 5–3. Design of the study by E. Smith, Regli, and Grawe (1999).

The magnitude of effect sizes measured in this way should not be compared to the magnitudes of pre–post effect sizes that are observed in outcome studies—those effect sizes average 1.21 (Grawe, Donati, & Bernauer, 1994). Instead, these effect sizes should be evaluated on the basis of J. Cohen's (1969, 1970) guidelines for interpreting the magnitude of effects of independent variables on dependent variables. In this context, .80 can be regarded as a decidedly strong effect. The effect sizes we observed were all clearly above this value. Thus, we found that need-satisfying experiences within the session exert a very strong effect on the outcome of that particular session. The experiences of the patients in the session process were coded in the midparts of the sessions—prior to the time point at which the patients evaluated the session outcome at the end of each session. Therefore, the effect sizes can be interpreted as being consistent with the idea that these process experiences exerted a causal effect on session outcome.

The concrete reality of therapy sessions matches the theoretical assumptions, then: the actual experiences that a patient encounters in therapy with respect to his or her need for control and self-esteem strongly influence the extent to which the patient will experience the session as helpful and, ultimately, the extent to which the therapy as a whole will be successful. It appears that it is particularly important that the patient gains the feeling that he or she now has greater control over his or her problems (clarification, coping–mastery, and self-efficacy expectancies). A therapist can strongly influence with his or her actions the degree to which patients do or don't encounter such need-satisfying experiences within sessions.

However, it is not just important that the patient encounters such experiences at all, but that they occur while the patient and the therapist are working together on his or her problems. In this study, we ensured by the special selection of the sessions that only sessions in which painful experiences had actually been targeted were analyzed. In other studies, however, in which sessions had not been selected based on this criterion, we found effect sizes that were similar in magnitude. One can safely assume, however, that the patients' problems were also the dominant focus in such "average" therapy sessions, even though this might not have been associated with the strong negative emotions as reported in E. Smith, Regli, and Grawe (1999).

In order to compare the influence of need-satisfying experiences, Figure 5–4 shows the effect sizes of the influence of various other aspects of problem processing, which were also assessed with the TSA on session outcome.

We can see that these aspects of problem processing have a much weaker effect on session outcome, compared with the powerful effect of need-satisfying experiences. The only aspects that had a notable effect on session outcome were the intensity with which the respective problems were targeted and the focus on change rather than mere analysis of the problem. Certainly, one should not conclude from these results,

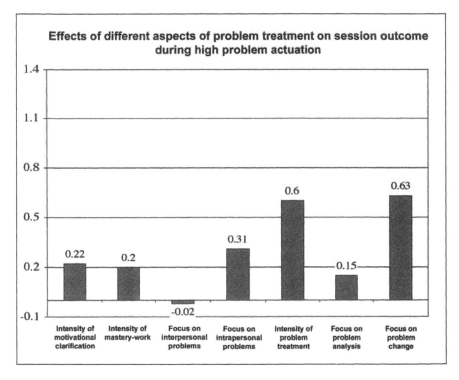

Figure 5–4. The influence of various aspects of problem processing on session outcome in the study by E. Smith, Regli, and Grawe (1999), expressed as effect sizes (Cohen's *d*).

however, that it makes no difference how a problem is approached. This probably is quite important, but the best approach will probably differ for the different types of problems, which explains why one would not find such high effects when averaging across all the different problems being treated. Nevertheless, the comparison with the high effect sizes we observed for need-satisfying experiences is still instructive. After all, these analyses were also not tailored according to the individual need profile of each patient; all of the patients with their different needs were put into one big sample here. This shows that positive experiences with respect to the control need and the self-esteem need are of general importance for good therapy outcomes, regardless of the individual characteristics of individual patients.

The description of how one can enable patients to have positive control and self-esteem experiences, as discussed previously, shows that one ought to utilize and positively emphasize already existing resources, characteristics, and abilities of the patient. Therefore, we have used the term *resource activation* as a summary label for any therapeutic activities oriented toward this idea. If these positive potentialities are currently activated within a patient, this can be viewed as a case of activated resources. Resource activation on the part of the patient means that the patient currently encounters self-esteem enhancing or control enhancing experiences. Resource activation on the part of the therapist, by contrast, means that the therapist strives to convey such need-satisfying experiences to the patient, regardless of whether these efforts are successful.

In a study by Gassmann and Grawe (2006), resource activation on the part of the therapist and activated resources on the part of the patients were evaluated independently by several outside observers. A different set of raters also evaluated the intensity of the patient's emotional engagement in the session while his or her problems were the target of the therapeutic work. This problem targeting will be referred to as *problem activation* in the following. In this study, each therapy was analyzed minute by minute, from beginning to end, with respect to these characteristics. The design of the study by Gassmann and Grawe is shown in Table 5–3.

Table 5–3. Design of the Study By Gassmann and Grawe With Consistency Theory Microprocess Analysis

Therapy Outcome	Session Outcome	Session Outcome High[a]	Session Outcome Low[a]
10 therapies	High outcome	20 sessions	20 sessions
10 therapies	Moderate outcome	20 sessions	20 sessions
10 therapies	Low outcome	20 sessions	20 sessions

Note. From "General Change Mechanisms: The Relation of Problem Activation to Resource Activation in Successful and Unsuccessful Therapeutic Interactions," by D. Gassmann and K. Grawe, 2006, Clinical Psychology & Psychotheraphy, 13. Adapted with permission.
[a]Two sessions per therapy.

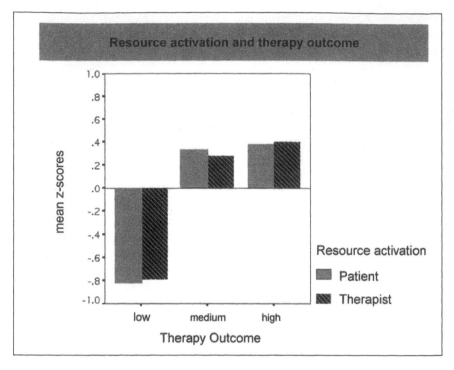

Figure 5–5. Comparison of resource activation in therapy sessions with a very good, moderate, and poor outcome, separately for patient and therapist. Z-transformed scores are shown for all cells of the design. Average therapies, therefore, have a score of zero.

Figure 5–5 shows the comparison among therapies with a very good, moderate, and poor therapy outcome for the resource activation on the part of the patient and the therapist. It is clearly apparent that therapies with a negative outcome are less characterized by resource activation on the part of the patient and, simultaneously, that therapists do less in such therapies in order to activate these resources.

Of the many additional results of this study, I will here present only those on the relations between resource activation and problem activation. CMP also measures other aspects of therapist and patient behavior, which, however, are less relevant in this context. I will also limit my discussion here to the comparison of extreme groups; that is, between therapy sessions with very good versus very poor session and overall outcome. Session and overall therapy outcome were operationalized in a way similar to that used in the study by E. Smith, Regli, and Grawe (1999).

Figure 5–6 shows on the left side the extent to which resources were activated in the patient. On the right side, it shows the extent to which the therapist attempted to activate patient resources. Both sides compare positive and negative therapy sessions with one another. In contrast to the previous figures, this figure represents the activity during an entire therapy session, because every session was analyzed minute by minute.

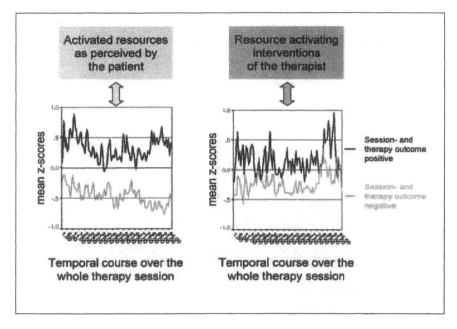

Figure 5–6. Comparison of the resource activation on part of the patient (left) and on part of the therapist (right) between positive and negative therapy sessions. The data are z-transformed. The average therapy has a value of zero at each minute. Because the lengths of therapies differ somewhat, the therapy duration was converted to percentages to facilitate comparability.

This comparison shows that patients in successful therapy sessions clearly encountered more need-satisfying experiences across the entire duration of the session, compared to patients in less successful therapy sessions. After all, successful resource activation on the part of the patient means that need-satisfying situations are experienced. What is also evident here is that the therapist actively works to enable such situations.

Figure 5–7 shows the same data from a somewhat different perspective—with the patient and therapist scores combined within the same figure. The left graphic shows the course of the resource activation for the unsuccessful sessions; the right graphic shows the course of the resource activation for successful sessions. This way of displaying the data shows that poor therapy sessions are characterized by the absence of resource activation on both parts. In the patient, no resources are activated from the very beginning, and this remains the same throughout the session. If anything, things get even worse over time. The therapist does not even attempt at the beginning of therapy to change this picture. It takes until about 10 minutes before the end of therapy that the therapist makes the first efforts to activate resources—probably because the therapist is starting to worry about having to end the session with the patient in such poor state. However, the therapist is now unable to reach the patient. The session ends in disaster. What is notable here is that, even under such poor circumstances, the therapist can still do something to attain the activation of patient resources. This becomes

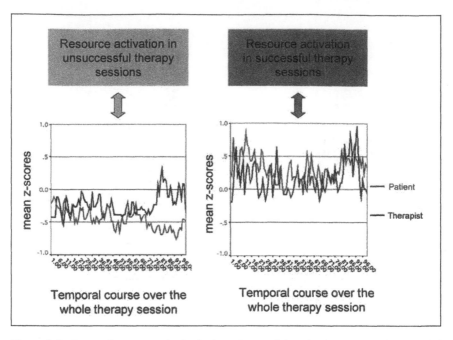

Figure 5–7. Course of resource activation in the patients and therapists between negative (left) and positive (right) therapy sessions.

evident in the last 10 minutes—when it is unfortunately too late. One has to wonder, then, how the session might have unfolded if the therapist had activated resources as actively from the beginning.

Patients can make it very difficult for therapists to activate their resources. This is often the case, for instance, among depressed patients. From chapter 3, we know that the reason for this is related to the state of their brains. They cannot help themselves. Therefore, it is absolutely essential in therapy with such patients that the therapist is persistent with his or her efforts to enable the patient to achieve need-satisfying experiences. This persistence is particularly important, even if it does not initially seem to achieve any effects and is not reinforced by the patient. I regard the poor preparation for this difficult task as one of the major shortcomings of most current therapy training programs. Resource activation is not specifically taught in most such training courses. However, it seems necessary not just to make this an important theoretical focus in therapists' training, but also to intensively train therapists in order to allow them to realize this difficult task even in adverse circumstances. In the supervision of ongoing therapies, an emphasis should also be on directing and supporting therapists so they can structure even difficult therapies in such a way that patients experience need-satisfying perceptions. This is not just a question of what should be an explicit focus in therapy. Resource activation must also be realized, in large part, in the implicit mode of functioning. The therapist must be able to take responsibility for this because many patients are unable to do so by themselves.

AIDS FOR RESOURCE ACTIVATION

My doctoral students Christoph Flückiger, Günther Wüsten, and I took the next logical step, based on these results, and developed a manual for resource activation. We conceptualized resource activation with regard to the patient's experiences in terms of his or her basic needs. The Psychotherapy Outpatient Clinic at the University of Bern is currently testing the practical utility of this manual by training therapists in specific courses to learn the process of resource activation. The manual and the training are designed to focus attention throughout therapy training on the important question of need-satisfying patient experiences. I am convinced that this will be one of the best opportunities to enhance the effectiveness of psychotherapies of all types, because this kind of resource activation can be pursued in many different contexts. It is neither limited to certain disorders nor to particular therapy modalities. Before the manual is available in published form, the version we are currently using can be obtained from our clinic at the University of Bern.

To aid therapists in their efforts to activate patient resources, my doctoral student Anne Trösken and I have developed two questionnaires for the evaluation of patient resources. In one version, the patient self-reports on the means by which he or she has recently been able to achieve positive experiences in different life domains. These means can be regarded as resources for the attainment of need-satisfying experiences. Therefore, the questionnaire assesses current resource realization from the self-evaluation perspective.

In the other questionnaire, patient resources are evaluated from the observer perspective. This questionnaire can be given to significant others (e.g., relatives), in order to obtain information about the patient's resource potential. This also provides an opportunity to find out about aspects the patient is personally unable to report on. The therapist thereby obtains clues about the positive potentialities of the patient—about patient aspects that might never become directly evident in therapy or on which the patient might never report directly. The therapist can also complete this questionnaire in order to clarify for himself or herself which resources can be utilized with a particular patient. Both questionnaires simultaneously also provide information about patient deficits, of course, which can be regarded as sources of incongruence and should, therefore, be considered in the analysis of the patient's incongruence profile (see section 5.3.3).

Figure 5–8 shows the correlations of the self-report questionnaire scales for the assessment of current resource realization vis-à-vis approach and avoidance incongruence as measured by the incongruence questionnaire.

Some of the correlations are rather high and indicate that persons who do not currently have many resources to attain positive experiences with regard to their basic needs—the eight primary scales correspond fairly directly to the four basic needs—also tend to have high levels of incongruence. The observation

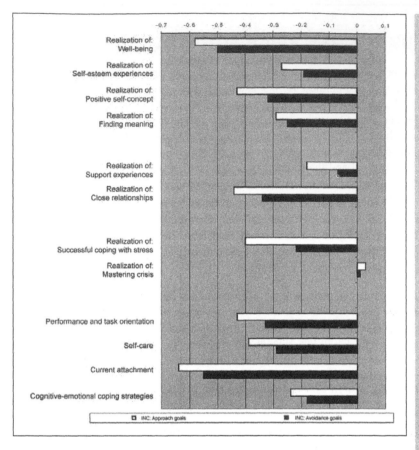

Figure 5–8. Correlations of the current resource realization questionnaire scales with the approach and avoidance incongruence scales. The upper part of the figure shows the correlations with eight primary factors; the lower part those with four secondary factors that emerged from factor analyses. The correlations are based on a sample of 66 psychotherapy patients who completed the resource realization and incongruence questionnaires at the outpatient clinic of the University of Bern.

might seem almost tautological, given that both resource realization and incongruence ratings are based on patient self-reports, and incongruence refers to a poor realization of motivational goals, whereas the resource questionnaire assesses the means a patient has available in the pursuit of his or her motivational goals.

The correlations in Figure 5–9 between the incongruence scales vis-à-vis observer ratings of patient resource potentials—in this case, the observers were close relatives—are less tautological. The upper part of the figure shows the correlations with 16 primary factors, whereas the lower part shows the correlations with three secondary factors and with a total resource score.

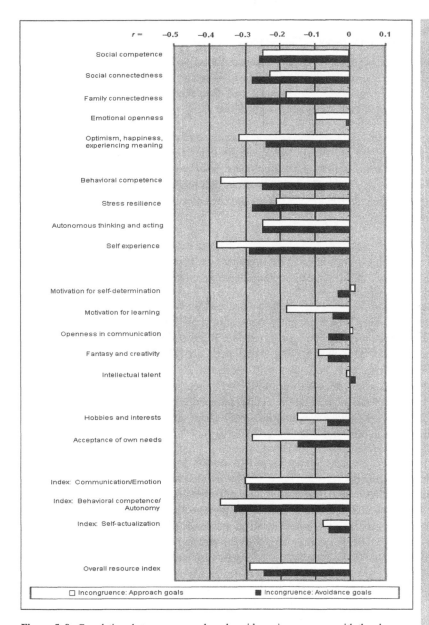

Figure 5–9. Correlations between approach and avoidance incongruence with the observer-based scales of the resource potential questionnaire. The correlations are based on a sample of 115 psychotherapy patients, who were rated prior to therapy by a close relative, and for whom self-report questionnaires were also available. Correlations of $r = .24$ or higher are statistically significant.

These correlations are not as high as those involving the self-ratings, but many of the correlations are nevertheless significant. This means that patients who objectively have poorer-quality resources tend to have higher incongruence levels. A resource deficit, then, must be regarded as a potential source of incongruence during treatment planning, and this should therefore be explicitly assessed prior to the beginning of treatment.

Therapists can utilize these two questionnaires in treatment even without quantitative analysis. By simply looking at the items within the questionnaire and noting what the patient and a significant other say about the resources, they obtain concrete clues about the resources that might be usefully employed in therapy. The resource questionnaires are currently being prepared for publication in German.

Resource activation can convey need-satisfying experiences, based on what was said here, which can then have an incongruence-reducing effect. Via this mediating link, resource activation can lead to other favorable changes.

Beyond that, resource activation also has another very important function in the process of therapy—the function of approach priming. I had discussed the findings on motivational priming in an earlier section. If the patient is continuously led to experience positive, need-satisfying experiences, the approach system is continuously primed. Therefore, the patient will approach his or her problems in more of an approach mode, which will positively influence his or her ability to process these problems, because goals that are being pursued with intrinsic motivation tend to be pursued with greater willingness to exert effort and, in turn, are more likely to be attained (see section 4.7.5).

In addition, the state of the brain that results from these need-satisfying experiences—including the presence of positive emotions and approach tendencies—is an almost indispensable requirement for the formation of processes that inhibit negative emotions such as fear. I will address the therapeutic significance of the formation of inhibitory processes in greater detail in a subsequent section (5.4.3). The formation of effective anxiety inhibitors requires the shifting of the brain into a state of maximally low anxiety susceptibility. This state can best be achieved via approach priming. The most important means the therapist has available for this is the creation of need-satisfying experiences within and outside of therapy.

This perspective is strongly supported by additional findings from the study by Gassmann and Grawe (2006). The left and right sides of Figure 5–10 show the course of the intensity of problem activation, as measured by the patient's emotional involvement in treatment, vis-à-vis the level of resource activation. The left graphic corresponds to negative sessions; the right one to positive sessions. A notable difference was observed between the session types: Although the intensity of problem activation does not differ between positive and negative sessions, the sessions differed massively in terms of the degree to which positive resources are concurrently activated. In

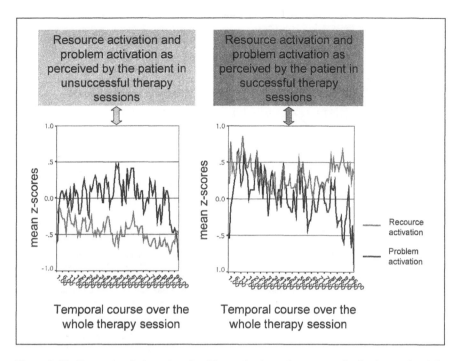

Figure 5–10. Contrasting the intensity of problem activation and resource activation in negative (left) and positive (right) therapy sessions.

negative sessions, the intensity of problem activation, which is often accompanied by negative emotions, is continuously higher than the level of resource activation. In positive sessions, this relationship is the other way around.

Figure 5–11 further clarifies this difference in the relative weighting of resource activation and problem activation. In this case, the values for resource activation and problem activation per patient were subtracted from one another. If resource activation is higher than problem activation, the scores lie above the separation line of $z = 0$, and vice versa. To improve legibility, only the z-transformed values of the first, middle, and last 5 minutes of a session are shown here.

This comparison clarifies dramatically the difference between positive and negative sessions. Positive sessions are always characterized by the relative predominance of resource activation, compared to problem activation. This difference is particularly pronounced toward the beginning and end of a therapy session. In the midsection, problems are being processed intensively, and the degree of resource activation in the patient recedes slightly (left graphic) or is reduced (by the therapist, right graphic). In negative therapy sessions, by contrast, the degree of problem activation always exceeds the degree of simultaneous resource activation. Apparently, it is very important in psychotherapy that the burden associated with the activation of problems is consistently balanced by sufficiently intensive positive, need-satisfying experiences

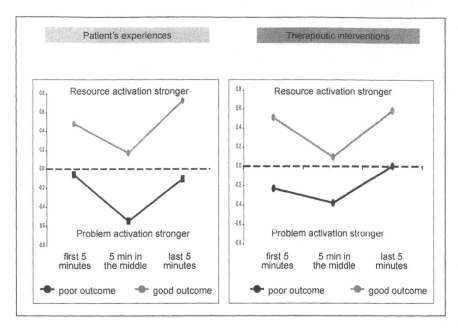

Figure 5–11. Comparison of the relative degree of resource activation and problem activation in positive and negative therapy sessions; left on the part of the patient; right, on the part of the therapist. Explanations in text.

that are encountered simultaneously. These experiences do not just arise naturally; they have to be created, supported, and facilitated by the therapist.

These therapist efforts to create resource activation are clearly evident in successful sessions and successful therapies. Figure 5–12 shows the correlation between resource-activation efforts of the therapist and activated resources within the patients. In negative sessions, from unsuccessful therapies, these correlations are consistently negative! That means that the therapist does not make any efforts to further support patient resources whenever they happen to be activated. The therapist is apparently preoccupied with only the patient's problems; otherwise, he or she would not miss such valuable opportunities. Conversely, it is also true that the therapist, if he or she does attempt to activate resources, does not achieve the intended effect or even achieves the opposite effect in the patient. This coordination between therapist and patient is apparently severely disturbed in negative therapies. They do not function together like a good team; they lack smooth synchronization.

In positive therapies and sessions, by contrast, the findings match our expectations. Resource activation and activated resources consistently go along together and mutually support each other.

These findings suggest that one important reason for disappointing therapy outcomes might relate to the inability of therapists to achieve an effective coordination with the patient. This might be caused by the therapist's excessive focus on patient problems

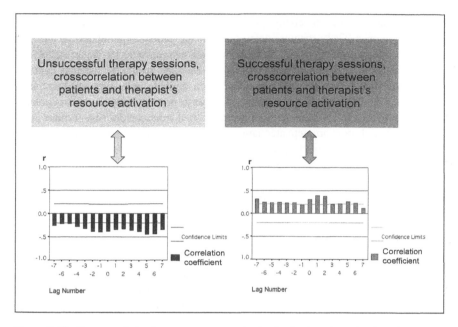

Figure 5–12. Cross-correlation between resource activation by the therapist and activated resources in the patient. At time zero, the correlation of these values during the same minute is shown. Left from the zero-point, the bars refer to the correlations between activated patient resources and therapist-facilitated resource activation one, two, three, and up to seven minutes previously. To the right of the 0 point, the activated resources of the patient precede the resource-activation efforts of the therapist. The left part shows the cross-correlations for negative sessions; to the right, for positive sessions.

and insufficiently active efforts to convey positive, need-satisfying experiences to the patient. Thus, the therapist not only misses the opportunity to improve the state of the patient directly, because these positive experiences would directly reduce incongruence, but he or she also undermines the effectiveness with which problems are being processed in the session because the required approach context is not being established. The consequence is that neither incongruence reduction nor successful problem processing occur, and the therapy is overall unsuccessful.

I have so far omitted in this discussion the need for attachment and the need for pleasure maximization and distress avoidance because the analyses reported here targeted primarily the control need and the need for self-esteem. It is clear, however, that the experiences a patient encounters in therapy with regard to his or her need for attachment are of great importance for the success of therapy.

The studies reviewed in chapter 4 suggest that 80% to 90% of therapy patients with severe disorders have insecure forms of attachment. A therapist has to expect, then, that most of his or her patients will have encountered negative attachment experiences in their early childhoods. The structure of the motivational schemas that have developed based on these experiences differ widely among individuals because everyone has gone through different additional important relationships by the time they reach adulthood, and these later relationships have also affected the development of motivational schemas. These schemas determine how the patient experiences the therapy and the therapist in the here and now. Individuals who have so far experienced few positive events in their relationships will have developed avoidance schemas that are activated in attachment-relevant situations. Such patients will approach therapists with hesitation or perhaps even mistrust, in order to protect themselves from potential harm and injury. The person will have developed strategies for this purpose over the years, which now are activated automatically and cannot be controlled by conscious volition. In therapy, the patient may well be able to fulfill on a conscious level the role expectations of the therapist–patient relationship. In the patient's implicit system, however, all the alarm signals might be activated in terms of his or her ability to trust the therapist, to allow himself or herself to become somewhat dependent on the therapist, or to become vulnerable by opening up to the therapist.

On the other hand, the wish for a sensitive, competent person who understands deeply and supports actively will remain in place. This wish is particularly strongly activated whenever a person is in difficulties or need, and individuals who enter psychotherapy are almost invariably in such difficulties. One can assume, therefore, that in many psychotherapy patients at the beginning of a therapy, both the wishes for a strong attachment relationship and the avoidance schemas, which are intended to protect the patient from harm, are activated simultaneously.

Starting from the very first encounter between patient and therapist, then, everything that transpires will be processed on the implicit level with regard to its meaning for the activated attachment wishes and fears. It is, of course, desirable that a relationship develops between patient and therapist in which both can work together in a climate of trust. We have seen previously how important it is that the patient experiences positive emotions in therapy, and that therapists work actively to prevent the patient from slipping into or remaining in the predominant avoidance modes. For these purposes as well, the experiences a patient encounters with regard to his or her activated attachment need are of paramount importance.

Given the great importance that attachment experiences generally have for mental health (see section 4.4.3), and based on what was just said here, one can really only expect that the experiences a patient encounters in therapy with regard to his or her attachment need must be of great importance in terms of facilitating overall improvements in therapy. The associations between various aspects of the therapeutic relationship and therapy outcome have been studied intensively in psychotherapy research. A recurrent finding has been that the best therapy results are attained by therapists who are experienced as sensitive and empathic, understanding and accepting, as actively promoting patient well-being, trustworthy and reliable, warm and supportive, and as competent (Bohart, Elliot, Greenberg, & Watson, 2003; Horvath & Bedi, 2003; Lambert & Barley, 2003; Orlinsky, Grawe, & Parks, 1994). The same features that characterize a good attachment figure, then, also characterize good therapists. Given the consistently strong relationships that were observed between these therapist qualities and therapy outcomes, for all the various therapy forms and conditions, it is clear that successful therapies are typically accompanied by positive experiences the patient encounters with regard to his or her attachment need. This holds true independently of the specific therapeutic techniques employed in each particular therapy. The experiences a patient encounters with regard to his or her attachment need have a direct and causal influence on the process and outcome of therapies. This can be said because many empirical studies have shown that the way in which therapists are experienced by patients is determined very early on in a course of therapy, and these early perceptions then powerfully predict the eventual outcome of therapy.

In our own studies using the TSA method, we have also been able to confirm the great influence of positive attachment experiences on the outcome of therapy. In the previously discussed study by E. Smith, Regli, and Grawe (1999), the independent raters also evaluated attachment-related aspects of the therapy relationship. Table 5–4 shows the three evaluated characteristics of therapist engagement, therapist competence, and therapy relationship as a resource (exact definition included in table). Figure 5–13 shows the effect sizes that we observed for the influence of these attachment-related characteristics of the therapeutic relationship on session outcome.

The magnitude of these effect sizes supports the predictions articulated previously. Therapy sessions are productive only if the patient experiences the therapist as a positive attachment person; that is, if the patient experiences positive perceptions with regard to his or her activated attachment need while his or her problems are simultaneously being processed. As was true for the control and self-esteem needs, we can also assume that such attachment-related experiences lead directly to incongruence reductions and well-being improvements and, indirectly, that they contribute to approach priming and, thereby, help establish the conditions necessary for successful problem processing.

Table 5–4. Attachment-Relevant Characteristics of the Therapeutic Relationship, Including Scale Descriptions

Experiences Satisfying the Need for Attachment	
Therapist engagement:	To what degree can the patient perceive the therapist as being engaged and really caring for the patient's well being?
Therapist competence:	To what degree can the patient perceive the therapist as a secure base, as trustworthy, supportive and capable of offering effective help?
Therapy relationship as a resource:	To what degree can the patient experience the relationship with the therapist as a positive resource?

Note. From "Wenn Therapie wehtut—Wie Können Therapeuten zu fruchtbaren Problemaktualisierungen beitragen? [When Therapy Hurts—How Can Therapists Contribute to Productive Problem Actuation?]," by E. Smith, D. Regli, and K. Grawe, 1999, *Verhaltenstherapie und Psychosoziale Praxis, 31,* p. 236. Copyright 1999. Adapted with permission.

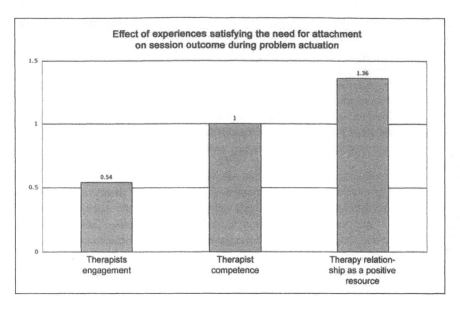

Figure 5–13. Effect sizes for the influence of the attachment-relevant characteristics listed in Table 5–4 on session outcome. (E. Smith, Regli, & Grawe, 1999).

The experiences a patient encounters in therapy with regard to his or her need for pleasure maximization and distress avoidance are also relevant for the process and outcome of therapy. I have already mentioned that conditioning processes are continuously taking place in psychotherapy. If primarily negative emotions are being activated in a patient during therapy, then after some time the therapy situation will by itself elicit negative emotions, which will then tend to shift the patient toward an avoidance mode. Therapists should aim, therefore, to enable the patient to also experience positive emotions in therapy. One method for achieving this is, for example, humor. Laughing together about something can sometimes be a very effective form of reframing. With a good laugh, things suddenly seem much less threatening.

A therapist also has other important technical tools and methods to enable the patient to experience positive states. This will be even more important for patients who hardly remember such positive states. An excellent method for this purpose is hypnosis. In a state of hypnotic trance, the therapist can enable the patient to experience feelings of relaxation, well-being, security, or personal power that he or she normally never experiences in real life. For many patients, this experience alone, that they can indeed feel this way, constitutes a very positive experience whose effect extends beyond the state of hypnosis. The fact that the therapist can shift the patient into such a pleasant state means that the therapist will be viewed as a very competent attachment person, in the same way that this is true for good mothers: They also reinstate pleasant conditions and make the child feel good after he or she has experienced something distressing. Jacobson's progressive relaxation or autogeneous training can also be used for the same purpose. The therapist thereby provides a tool for the patient, which also can be used outside of therapy, whenever the patient wishes to reenter the positive state. This method is much less problematic and conveys more self-efficacy than the alternatives of alcohol, drugs, or medication. Methods such as relaxation and hypnosis, then, can be used in psychotherapy for the purposes of incongruence reduction with regard to the need for pleasure maximization and distress avoidance. They can have the same direct and indirect effects, in turn, that were discussed earlier with respect to positive experiences pertaining to the other basic needs.

If we combine all these different ways of enabling patients to encounter positive, need-satisfying experiences in therapy, we have to conclude that therapists have a rich repertoire of tools by which they can contribute to consistency improvements in the mental functioning of a patient. If these opportunities are not pursued, or if therapists attempt unsuccessfully to realize these opportunities, then it will also be unlikely that consistency can be improved via disorder-specific interventions or via the treatment of other individual sources of incongruence. Consistency improvement via need-satisfying experiences is a necessary condition, then, for successful psychotherapy.

The early experiences that an individual has encountered with respect to the satisfaction or nonsatisfaction of his or her basic needs and the wishes and fears that are currently being activated in his or her relationships are linked together via the motivational schemas that have formed over the course of development. They bestow meaning upon the relationship situations that the patient encounters in his or her present life. It is not just the motivational schemas related to the attachment need that are being activated in current interpersonal relationships. In addition, motivational schemas that have formed for the protection or attainment of other basic needs are being activated. Need satisfaction or violation almost always occurs in the context of interpersonal relationships.

Adult individuals differ, of course, with respect to what constitutes ideal positive perceptions with regard to their basic needs. This depends on the importance of the patient's individual approach and avoidance goals. Therapists aiming to enable their patient to encounter a maximal number of need-satisfying experiences must create a therapy process that complements the patient's most important motivational goals. Complementary processes are those that are compatible with the patient's goals; that is, experiences that do not infringe upon avoidance goals but that do fulfill approach goals. In order to be able to act in a complementary fashion with respect to the patients' most important motivational goals, it is imperative that the therapist know these goals as thoroughly as possible. A method that could help therapists identify these goals, for example, is the IAAM (Inventory of Approach and Avoidance Motivation; see section 4.7.5).

Because motivational goals are primarily pursued in the implicit mode of functioning, it is quite possible that the IAAM does not capture all important motivational goals of a patient. The therapy relationship is in itself a good diagnostic source for implicitly pursued motivational goals. The therapist can turn his or her attention to the nonverbal interaction behavior of the patient, in order to test the hypothesis that this behavior serves the attainment of activated motivational goals in the implicit mode of functioning. This raises questions such as:

- What is the patient trying to achieve with this behavior (approach goals) or, respectively, what is he or she trying to avoid, to prevent, or to make difficult (avoidance goals)?
- How does he or she do this?
- How do I react to this? What do I feel like doing in response? Which kinds of feelings arise within me in response, and what kinds of things could I not at all imagine for this patient?
- Which conclusions can I draw from this with respect to the goals the patient pursues or with respect to which kinds of situations and experiences he or she actively avoids?
- Are these approach and avoidance goals possibly linked to other, overarching goals?

This method of extracting important motivational goals from the processes unfolding within the therapeutic relationship was originally developed by me and Franz Caspar, my colleague for many years (Grawe, 1980; Grawe & Caspar, 1984). We originally termed this method *plan analysis* but later switched to the term *schema analysis* (Grawe, Grawe-Gerber, Heiniger, Ambühl, & Caspar, 1996). The most thorough discussion of this method and its application in clinical practice can be found in Caspar (1989, 1994).

In order to help therapists assess their patients' important motivational goals based on the processes unfolding in the therapeutic process, and to help them formulate individualized treatment plans that will enable them to create complementary therapeutic interactions, we are currently developing and testing a set of diagnostic and intervention-related tools (Stucki, 2004). Along with the manual for resource activation, mentioned previously, these tools can be incorporated into therapy training and routine practice, so that the task of achieving need-satisfying experiences—which was shown to be of such central importance—will be realized on a level comparable to today's manual-based disorder-specific therapy approaches (Grosse Holtforth & Castonguay, 2005). Figure 5–14 once again presents an overview of the central functional role need-satisfying experiences have in terms of leading to better therapy processes and outcomes.

Both of these therapeutic strategies—resource activation and motivationally attuned (complementary) session-interaction patterns—should enable the patient to encounter a large number of positive perceptions with regard to his or her attachment need, control need, need for self-esteem enhancement, and need to experience pleasure and avoid distress. Because these perceptions are in line with the goals of his or her most important motivational schemas, they should, by definition, contribute to the reduction of incongruence with regard to these motivational goals. Given the very

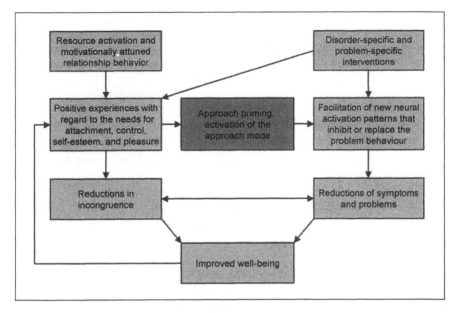

Figure 5–14. Functional role of need-satisfying therapy experiences for the optimization of therapy process and outcome. Explanations are provided in text.

high empirical correlation between incongruence and well-being, which has been found repeatedly across studies, we can expect that incongruence reductions will be directly accompanied by improvements in subjective well-being. Such improvements, in turn, are positive experiences with regard to the need for pleasure maximization and distress avoidance—further instances of need satisfaction. Thus, the previously described positive escalating process is triggered.

The positive need-satisfying experiences that a patient encounters in this escalating process will, in turn, activate his or her approach system and shift him or her, via approach priming, into an approach mode of functioning, in which the entire mental activity is predominately positive and oriented toward approach rather than defense and avoidance. This makes the patient more receptive for disorder- and problem-specific interventions. This is the second great contribution a therapist should make to ensure the success of therapy. The disorder- and problem-specific interventions target the disorder-maintaining components of each respective syndrome directly. Their beneficial effect can only be realized, however, if the patient is currently in a receptive state, oriented toward approach. This is particularly evident in the treatment of anxiety. In such cases, a brain state that is oriented toward approach and positive mastery is particularly urgently required, so that effective mechanisms can be erected to inhibit the arousal originating from the amygdala or other areas such as the insula.

Disorder-specific and problem-specific interventions will lead to reductions in symptoms and problems only if they are delivered in an approach context that has been established and is maintained via the therapeutic facilitation of need-satisfying experiences. If such reductions occur, declines in incongruence and, in turn, improvements in well-being will ensue. This is yet another need-satisfying experience that further fuels the positive escalation, ultimately resulting in good therapy outcome.

From this perspective, a good therapist must preside over a rich repertoire of disorder- and problem-specific interventions, resource-activating interventions, and strategies to individually tailor the therapist–client relationship. These tools allow therapists to fully realize the potential inherent in psychotherapy. The positive escalating cycle that characterizes good therapies willy be established only if this potential is realized simultaneously.

5.3.3 Consistency Improvements via The Treatment of Individual Sources of Incongruence

Mental disorders are the result of excessive inconsistency in mental functioning, and they, in turn, contribute to further increases in inconsistency. They would not have emerged if other sources of incongruence had not already existed prior to their formation. Among individuals with mental disorders, these other sources can vary widely. The disorder itself tells us little about the other processes that might cause incongruence in the person's life. Therefore, the other sources of incongruence cannot be considered in disorder-specific treatment manuals. Examining these sources in details requires what we have termed an *incongruence analysis*.

Incongruence analysis refers to two things: First, determining whether the person's incongruence level is elevated at all and, second, if this is the case, analyzing the sources of the incongruence.

Several indicators point to the likely presence of heightened incongruence levels:

1. A high level of comorbidity. If more than one disorder has developed, it suggests that high levels of inconsistency have probably been present for an extended duration and could not be down-regulated by the first disorder. Based on the mechanism described earlier, this then led to the development of additional disorders. Because each disorder contributes to further increases in incongruence, there is a high probability that a negative escalating cycle is triggered: Excessive current levels of inconsistency led to the development of the first disorder. This increases inconsistency because it binds mental energy that is needed for the realization of motivational goals and because the disorder itself constitutes a need violation. The elevated incongruence level does not only lead to the continuous reinforcement of this initially developed disorder, thereby maintaining it, but it also promotes the development of additional disorders. These, in turn, contribute to further incongruence elevations, and so forth.

This explains the empirical finding that mental disorders are often accompanied by other mental disorders. Approximately 77% of patients with major depression have at least one other mental disorder (Hautzinger, 1998). In anxiety disorders, comorbidity with other anxiety disorders or with a depressive disorder tends to be the rule rather than the exception. According to a longitudinal study by Wittchen (1991), only 14.2% of patients with panic disorder have no additional disorder. Eighty-eight percent of men and 78% of women with PTSD have at least one additional mental disorder (Kessler, Sonnega, Bromet, Hughes, & Nelson, 1995). Patients with somatoform almost always have an additional major depression or dysthymic disorder and about 50% have an anxiety disorder. In addition, about 60% of them fulfill the criteria for a personality disorder. All together, these patients are diagnosable with an average of more than three diagnoses (Rief & Hiller, 1998). Ninety-eight percent of patients with borderline personality disorder will develop an additional depressive disorder, 90% of them an anxiety disorder, and 50% sleep disorders; 60% of women with this disorder also have an eating disorder and 40% suffer from alcohol dependence (60% among men; Bohus, 2002).

These numbers paint a very clear picture. The development of an isolated mental disorder is an exception. Among almost all patients who develop mental disorders, a fertile breeding ground that promotes the development of these disorders appears to exist over long durations. Given the high associations among mental disorders in adulthood vis-à-vis violations of the control and attachment needs in early childhood—which were discussed in chapter 4—all of this should not be particularly surprising. It is very common for individuals with mental disorders to also have a *personality disorder*. This is simply a term to express that they have formed a structure of motivational schemas that leads to difficulties in interpersonal relationships (Benjamin, 1993). This structure is the link between early life experiences and the later emerging mental disorders. *Personality disorders* are defined as enduring patterns of experience and behavior. They do not develop suddenly, as is the case, for instance, with PTSD. Instead, they describe what is typical over the long term for a particular person—they characterize his or her personality. The assignment of patients to distinct disorder categories is much more questionable, compared to syndromal mental disorders. The reliability of these diagnoses can be viewed as much lower than what would normally be considered reasonable for a categorical diagnostic system. There is much to say in favor of a dimensional rather than a categorical diagnostic system for the personality disorders, then. In the same way that Benjamin (1993) has characterized the personality disorders in terms of typical interaction patterns, it would also be possible to characterize them as reflecting specific constellations of motivational schemas. This would clarify that these are not "diseases" but, instead, variants of motivated behavior that lead to typical, recurrent problems in the context of interpersonal relationships.

From this point of view, it is clear that patients with several comorbid disorders—and especially patients who also meet criteria for a personality disorder—will often have elevated incongruence levels. This incongruence should be explicitly considered, in addition to the specific disorders, in the process of treatment planning.

2. A second indicator for a high incongruence level is the number of the treatment issues that a patient brings to therapy. In section 4.4.4, I already reviewed studies conducted by our working group with the Bern Inventory of Therapy Goals (Grosse

Holtforth & Grawe, 2002a). This inventory is structured hierarchically. On the high-est level, five goal categories can be differentiated (see Table 4–4); on the next level, 23 specific subgoals are present, and on the lowest level, the inventory includes 47 concrete treatment issues of the kind that psychotherapy patients typically present at the outset of treatment. It is routine practice in our clinic to give this checklist to patients before they begin therapy so they can indicate which of these issues pertains to them. The therapist can then review this checklist prior to the first session and will thus be able to see quickly whether the patient only seeks to be freed from a specif-ic disorder or, instead, is presenting with a large number of additional concerns. As shown in Table 4–4, this is commonly the case. Even without quantitative analysis, the checklist provides a good basis for the further exploration and discussion with the patient. The number of treatment-related issues a patient marks on this sheet—at a point when he or she is still entirely uninfluenced by any contact with the therapist—correlates at $r = .50$ with the total score of the incongruence questionnaire. Having many treatment concerns or issues can be regarded, then, as an indicator of high incongruence levels. The correlations is not so high, however, that no additional sources of incongruence would have to be considered.

3 *Incongruence assessment with the incongruence questionnaire.* The best esti-mate of incongruence levels can be obtained with the incongruence questionnaire. This can be done very efficiently. The short form of the incongruence questionnaire has only 23 items, and completion requires only about 5 minutes. Because the short form correlates at $r = .97$ with the long form, this is a quick and effective way of determining a patient's overall incongruence level. Given the very high correlations of incongruence with subjective well-being, symptom severity, and many other clin-ical characteristics (see section 4.8.4.3, Figures 4–23 [p. 319] and 4–24 [pp. 320, 322]), this efficient questionnaire enables therapists to know a great deal of informa-tion about the overall state of their patients—even if they have little direct informa-tion from other sources. If this incongruence score is high, the therapy should not be limited to the treatment of only a single disorder pattern; it should aim to lower this incongruence. The question arises, then, of how this high incongruence emerges in the first place. In order to answer it, a thorough incongruence analysis must be per-formed.

Figure 4–25 (p. 323) can be used as a guide in the analysis of sources of elevated incongruence. This figure summarizes the most important possible sources of incon-gruence elevation, including

1. Unfavorable current life circumstances, such as poverty, unemployment, a poor social environment, lack of support, and illness or other physical impairments. These are sources of incongruence that psychotherapy usually cannot change or, if so, only to a very limited extent. If the unfavorable circumstances are extreme, enduring pos-itive change in the patient's state usually cannot be achieved via psychotherapy alone. Nevertheless, it would not make sense to keep such individuals from entering psy-chotherapy because, typically, other sources of incongruence are present as well, and

it is often possible to achieve positive outcomes even under objectively harsh circumstances. It would not be reasonable to ignore or deny the negative influence such objectively adverse circumstances can exert, or to conclude that the person's misery is due to his inadequate utilization of his life circumstances.

2. Unfavorable relationships in which the patient is currently involved. This could be, for example, a conflict-ridden or otherwise disastrous partner relationship, an unsuccessful attempt of a young adult to leave his or her parental home, or someone's relationship with a person with drug and alcohol addictions. Such relationships have their own dynamic. Patients are often unable to change the relationships by themselves; they require interventions that go beyond the individual and target the relationship system directly. Living in such relationships creates perceptions, on a day-to-day basis, that are very incongruent with the person's basic needs. If such relationship circumstances exist, it will also be difficult or impossible to achieve enduring positive change with psychotherapy, unless the circumstances themselves are changed. Appropriate therapeutic modalities for that purpose would be, for example, couples or systemic therapy.

3. Unfavorable relationship behavior on the part of the patient. Individuals who were raised in relationships that repeatedly violated their basic needs will attempt to protect themselves to avoid further violations. Avoidance behavior, however, is not an effective means of establishing need-satisfying relationships. Because need satisfaction tends to occur primarily in the context of interpersonal relationships, individuals with strong avoidance goals will often be oriented more toward avoidance than approach, and toward the exchange of negative rather than positive emotions. Thus, such individuals do not elicit need-satisfying perceptions with their relationship behavior. They themselves contribute to the fact that their relationships continue to be sources of chronic incongruence.

Problematic interpersonal relationship behavior can be assessed, for example, with the Inventory of Interpersonal Problems (Horowitz, Strauss, & Kordy, 1994) or with the Insecurity Questionnaire (Ullrich de Muynck & Ullrich, 1977). Depending on the specific nature of the relationship problems, therapists can attempt to change the relationship behavior with mastery-oriented training methods such as assertiveness training, or with a more general interactional problem-solving therapy (Grawe, 1978, 1980; Grawe, Dziewas, Brutscher, Schaper, & Steffani, 1978; Grawe, Dziewas, & Wedel, 1980), or with clarification-oriented group therapies (Yalom, 1974). From a neural perspective, interpersonal problems can be best addressed in interpersonal settings in which the underlying neural activation patterns are activated "from the bottom up." Group-, couples-, and family-therapy modalities are particularly well suited for this purpose. However, interpersonal problems can, of course, also be addressed in individual therapy if this enables the patient to expose himself or herself to the necessary corrective experiences outside of therapy.

Relationship problems are inevitably linked with the patient's motivational schemas. Therefore, changing relationship behavior always involves attempts to change the patient's approach and avoidance goals and the means that he or she has developed to attain these goals, which might, for instance, involve the balancing of existing behavioral deficits. Goals and deficits are listed as separate sources of incongruence in Figure 4–25. This is indeed correct because they are not limited to the domain of relationships as sources of incongruence. This example shows, however, that the separable sources of incongruence are not independent from each other. By targeting one incongruence source, others are often also influenced. This means that the influence of therapeutic work on the overall incongruence level is potentiated or magnified.

4. Unfavorable consistency-maintenance mechanisms. This includes lacking stress tolerance, poor emotion regulation, deficits in problem-oriented coping, and excessive repression. I discussed the negative consequences of unfavorable forms of consistency maintenance in various sections of chapter 4; most explicitly in section 4.8.2. The sections on the attachment need and the control need clarified that basic forms of physiological and emotion regulation form at very early stages of development. Individuals who, at these stages, are repeatedly exposed to longer episodes of uncontrollable incongruence will respond with autonomic hyperarousal and excessive hormonal stress reactions. The negative feedback processes that normally protect the neural system from this excessive arousal and its negative consequences are not established. Instead, disproportionate stress reactions are facilitated, and the consequence will often be a life-long marked reduction in stress tolerance, which then contributes considerably to the formation of mental disorders.

Patients with markedly elevated incongruence levels tend to be less resilient to stress, tend to have greater difficulties with the regulation of negative emotions, and tend to have more maladaptive coping strategies. Questionnaires have been developed to assess problem- and emotion-oriented coping (Lazarus & Folkman, 1984) as well as emotion regulation (Znoj, 2000; Znoj & Grawe, 2000), but we have to assume that these early acquired forms of stress response and emotion regulation are typically highly automatized, subverbal processes that are beyond conscious awareness and therefore cannot always be assessed via self-report. At best, the individual can gain conscious awareness of the consequences of these processes. In order to ascertain whether excessive stress reactions are an important source of incongruence for an individual patient, it would be much better to observe his or her reactions in a standardized stress situation and to measure the stress hormones that are released in such a situation. This is not commonly done in psychotherapy today. However, because excessive stress reactions can be regarded as important sources of incongruence and, therefore, can stimulate the development of mental disorders, it is important to leave the verbal level in this case and to include in our pretherapy assessment batteries things like endocrinological tests that measure cortisol reactions in response to stress. This does not mean that verbal instruments such as coping questionnaires should no longer be used, but that neurophysiological measures and the observation of automatized, nonverbal reaction patterns would seem more appropriate for the target phenomena.

Therapeutic elements that could be used when such sources of incongruence are present might be coaching interventions to improve stress tolerance and emotion regulation. This might include, for example, Meichenbaum's (1973) stress inoculation, relaxation training, autogeneous training, meditative techniques, self-hypnosis, and others. Conceivably, biofeedback methods could also be developed for this purpose.

Linehan's dialectical behavior therapy for borderline personality disorder already includes such elements. For example, abilities such as "mindfulness," stress tolerance, and emotion regulation are being trained in that therapy (see Bohus, 2002, pp. 77–93). Based on the discussions in chapters 3 and 4, it would make sense to also include such elements in the treatment of disorders other than borderline personality disorder. In anxiety disorders and mood disorders, one could expect that poor stress and emotion regulation also frequently play an important role. In order to reduce the incongruence level permanently, it would seem important to make the "defusing" of this important source of incongruence a major focus of therapy. The concept of mindfulness, which has received increasing attention in recent cognitive –behavioral therapies, can be viewed as a step in this direction. Consistent with our ideas developed here, mindfulness is now also more generally recognized as relevant for many disorders, and its great potential for various areas of clinical psychology has been articulated (Baer, 2003; S. Hayes & Wilson, 2003; Kabat-Zinn, 2003). Teasdale, Segal, and Williams (2003) emphasize the importance of tailoring "mindfulness coaching" to the particular characteristics of different disorders. It is generally recognized that the mindfulness concept opens up a truly new perspective in clinical psychology and psychotherapy, but most questions with regard to its mechanisms of action, areas of application, methods of administration, necessary training, and so forth must still be clarified (Dimidjian & Linehan, 2003; Roemer & Orsillo, 2003)—with which one can only concur. The attractiveness this concept has gained within a very short time shows that the principles I described as consistency maintenance are increasingly recognized as important even by approaches built on other theoretical foundations. From the perspective I developed here, however, it would seem particularly promising to explicitly also include the neurophysiological level of analysis when addressing the topic of consistency maintenance. This has not yet occurred with regard to the mindfulness approach. Similarly, it will be quite some time until the effectiveness of different specific modalities for consistency maintenance will have been thoroughly empirically evaluated.

Another area that will require a great deal of future research concerns the question of which forms of consistency maintenance might be particularly ineffective. For example, the extent to which unfavorable forms of consistency maintenance, such as excessive repression, can be changed by specific interventions is not yet clear. In general, it seems to me that there is still much room for the development of innovative therapeutic interventions that could target this source of inconsistency. From the neural point of view, this is a particularly important source of incongruence, which, because it is almost entirely beyond self-awareness, has been relatively neglected in traditionally verbally oriented psychotherapy.

5. Unfavorable cognitions. This includes all that a patient thinks in relation to his or her problems. This might include, for example, a problematic understanding of his or her condition, which obstructs his or her path toward successful change. This might also include problematic goals that the patient has set for himself or herself even though they probably cannot be attained. Because cognitions are also determined by motivational goals, problematic cognitions are often influenced by avoidance goals or by conflicted motive constellations. Cognitions can also serve to avoid dissonance,

which then maintains the cognitions. These inconsistency-avoiding cognitions can, in turn, contribute in important ways to the maintenance of the problem because they shift the patient's attention in the wrong direction. Such problematic views can be identified, for example, in clarification-oriented therapeutic discourse; they can be questioned and altered step by step.

Dysfunctional cognitions also include irrational assumptions as delineated by Ellis (1977) or cognitive distortions as proposed by Beck (1979). These include, for example, all-or-nothing thinking, overgeneralizations, jumping to conclusions, or unfavorable attributional styles. This is the domain of cognitive therapy, which has developed a rich arsenal of therapeutic techniques.

Expectancies can also be regarded as cognitions. Low self-efficacy expectancies, negative outcome expectancies, and response expectancies, in the sense that one expects to feel uncontrollable anxiety or other negative emotions in a specific situations, are, on the one hand, a central component of the specific psychopathologic disorder patterns. On the other hand, they also play an important role in determining what a patient does and avoids. Expectancy changes initially require that one clarifies the exact nature of the expectancies. This can be achieved, for example, with cognitive techniques. The actual change, however, is then attained via reality testing, in which the patient is confronted with the respective situation and encounters the experience that the negative expectancies do not turn into reality. This is the domain of behavior therapy, with its large repertoire of problem-specific interventions that have been shown to be effective in creating such corrective experiences.

6. Excessively strong avoidance schemas. A dispositional avoidance temperament and violations of basic needs are the bases on which avoidance goals and avoidance strategies can form easily. Because avoidance goals are oriented to keep certain perceptions, thoughts, and emotions away from experience, these goals are usually more difficult to identify than approach goals. They often play the role of the actual villain in mental functioning, however, because they block activity oriented toward approach and need fulfillment, and their pursuit binds excessive amounts of mental energy and attention. Avoidance goals often cannot be named clearly by the patient, and their function with regard to his or her problems is often not recognized. It is the therapist's task, therefore, to examine the extent and nature of the patient's avoidance. Tools for this task include, for instance, the IAAM with its nine avoidance scales or, as described earlier, the methods of plan analysis or schema analysis that can be used to clarify which negative experiences the patient tries to avoid with his or her behavior.

In a patient who tends to avoid a great deal—beyond the avoidance that is already part of the disorder pattern itself—a better degree of need satisfaction can be attained only if the role of avoidance in experience and behavior is weakened. This will then create new space within which more approach-oriented activities can unfold. In order to achieve this, the therapist will need to know in detail what the patient avoids and how he or she avoids it. The profile of avoidance scales in the IAAM provides initial clues to clarify which types of experiences the patient avoids with particular intensity. The means the patient uses for that purpose have to be uncovered gradually via exploratory discussion, observation, or clarification-oriented therapeutic discourse.

In order to be able to expose himself or herself to the hitherto avoided experiences, and thereby be enabled to encounter corrective experiences, the patient first needs to develop a conscious awareness for that which he or she has previously avoided. Once this awareness is in place, he or she can volitionally seek out new experiences and then repeat these experiences until his or her feelings and behavioral tendencies in the respective situations have really changed. The change of avoidance behavior is not a question of insight but, instead, of repeated experiences that do not match previous experiences. Having conscious awareness of one's previous avoidance, however, is a prerequisite for the conscious, deliberate confrontation with the previously avoided experiences. This conscious awareness normally has to be built up in collaboration with the therapist. Clarification-oriented therapeutic discussions are an indispensable tool for this purpose.

The activation of neural patterns that are incompatible with the previous avoidance will trigger acute inconsistency. The natural tendency would be to avoid this inconsistency. In order for that not to happen, it is crucially important that the therapist conducts an approach priming before and during such inconsistency-creating work on avoidance-related problems. The patient must encounter the new experiences while he or she is in an approach mode of functioning, so that the previously dominant avoidance behavior and the associated negative emotions can be effectively inhibited.

A different constellation is present if the patient has strongly formed avoidance goals but has been unable to avoid that which he or she fears; for instance, if the patient actually has failed, does feel publicly humiliated, and so forth. This would be reflected in a high level of avoidance incongruence on the incongruence questionnaire. This would be akin to the treatment of open wounds, as it were. The therapeutic strategies can differ considerably in this context. They can aim to establish a different meaning context with regard to the control and self-esteem needs, or they can aim to equip the patient with better tools with which he or she can protect himself or herself from similar harm in the future. One can also try to reduce the importance of the violated avoidance goals, in which case the negative meaning of the experienced injuries will diminish.

The weakening of avoidance goals and reduction of avoidance incongruence are typically demanding, difficult therapeutic tasks that require a combination of clarification-oriented, confrontational, context-altering, and repeated practice related interventions. The constellations in individual cases differ to such an extent that it would not be sensible to streamline the interventions with the use of a manual. The therapist must have a clear understanding of the patient's problem and of potential solutions, however, because only this will enable the therapist to show a coherent set of ideas to the patient, which is of such great importance for the patient's need for orientation and control. Such clear understanding will also allow the therapist to exude a sense of security, conveying that the therapist knows what he or she does, which is very important for the patient's attachment need. On this basis, the therapist can ensure that new, desired neural activation patterns are repeatedly activated and undesired ones more and more effectively inhibited. Because this therapeutic work is always accompanied by an activation of the avoidance system, it will also be particularly important to utilize resource activation and to create optimally complementary therapeutic relationship constellations. It is fair to say that the changing of motivational avoidance schemas is a skill of highest difficulty within the broad domain of psychotherapy. I do not think that any one therapeutic approach could rightfully claim having developed the best repertoire for this purpose.

7. Weakly formed of underutilized approach schemas. Compared to normal individuals, psychotherapy patients do not have less pronounced approach goals. On the IAAM, they even report that approach goals are slightly more important for them. The reason is, presumably, that the approach goals are even more strongly activated among them because they are not being attained satisfactorily. The difference between patients and normal persons is that the patients are much less able to attain their approach goals. On average, they have a high level of approach incongruence. Their problem is that their approach goals largely remain wishes because they lack the means to turn them into reality. This might be so because their current life circumstances make it difficult to achieve the goals. If someone's partner has recently died, for example, he or she might experience strong incongruence with regard to his or her approach goals of intimacy and closeness, even though this might generally be an individual with a good repertoire for establishing close relationships. In most cases, however, approach incongruence has to do with the fact that the person does not engage in the actions required to realize the goals. The patient may lack the behavioral repertoire because he or she might never have learned it. In such cases, a true resource deficit would be present. The necessary behavioral repertoire might also be present, in principle, however, but for one reason or another, the person might not utilize it. The person might feel as if motivation and decision-making ability are lacking. Another reason might be that the person is at present mostly in an avoidance mode of functioning. Also, it could be that the attainment of the approach goals is being blocked by specific fears or worries.

Depending on the nature of these difficulties, the therapeutic approaches would differ. If a true resource deficit is present, the therapy should aim to build up the required behavioral repertoire; for example, via role playing or repeated practice in realistic situations. If approach behavior is being blocked by specific fears, therapy should primarily target these fears. If an existing repertoire is not being utilized, therapy ought to aim to activate these resources. Jacobson, Martell, and Dimidian's (2001) behavioral activation therapy would be one example of such an intervention, for example. It is usually not difficult for a therapist to differentiate among these options because he or she can find out via questioning or observation whether a true deficit is present or whether resources that are, in principle, available, are not being utilized because they are deactivated or actively inhibited.

8. Motivational conflicts. These are typically a combination of excessively developed avoidance schemas and insufficiently developed approach schemas. If relatively strong approach and avoidance schemas are simultaneously activated, the behavior programs triggered by them tend to interfere with one another, such that neither the approach goal nor the avoidance goal can be reached effectively. Approach–avoidance conflicts, therefore, will lead to strong approach incongruence as well as strong avoidance incongruence. The resolution of such conflicts always requires a strengthening of the approach component and a weakening of the avoid-

ance component. In neural terms, the approach component must be facilitated more strongly and the avoidance component must be actively inhibited. What I said earlier about the avoidance schemas must once again be repeated here. Resolving conflicts is not a question of insight but a question of repeated experiences that can achieve that the avoidance component can be inhibited and the approach component can be more strongly facilitated. In order for that to happen, however, the patient must encounter such experiences repeatedly, consciously, and volitionally. This, in turn, requires the presence of conscious awareness for that which the patient has previously avoided, for the fact that he or she has, in fact, avoided at all, and for that which he or she genuinely desires. Gaining this awareness is a prerequisite for achieving the needed change experiences because these experiences require the patient's active collaboration. Achieving the awareness is not the ultimate goal, however. The ultimate goal is the resolution of the conflict, which requires the facilitation of new neural activation patterns via maximally frequent experiences. Once the nature of a conflict has been identified via motivational clarification work, the therapeutic work is not yet done. This constitutes only an initial step; the next steps are the weakening of the avoidance component via repeated confrontation with appropriate experiences, and the better facilitation of the approach component, so that better levels of need attainment can be achieved.

5.4   CONSISTENCY IMPROVEMENTS VIA CHANGES IN NEURAL STRUCTURES?

Neither Figure 4–25 nor the preceding section mentioned neural structures as sources of inconsistency. Is this not an omission, given that we know, for example, that the hippocampus and PFC are shrunken in depression and that the ACC is severely disabled? Aren't these changes among the most important causes of the patient's inability to encounter need-satisfying experiences? Such questions are entirely legitimate. Once the hippocampus has been damaged by stress hormones, this damage becomes one of the causes of the fact that the patient no longer behaves in the way he or she should in order to satisfy basic needs. Therefore, the damaged hippocampus, for instance, or a variety of other neural structures, could indeed be listed as independent sources of incongruence.

The damaged hippocampus is a part of the depressive disorder, however, and disorders, with their own dynamic, are already listed in Figure 4–25 as sources of incongruence. The unique dynamic of each individual disorders has its own neural foundations, as we saw in chapter 3. A therapist should be aware of what these neural foundations are, and should consider them when attempting to change the disorder dynamic via disorder-specific interventions. The therapist's understanding of depression should include its neural bases. Therefore, the neural bases do not have to be listed as a separate source of incongruence here.

This would hold true not just for most other mental disorders but also for most of the sources of incongruence listed in Figure 4–25. Disproportionate stress reactions, poor emotion regulation, excessive avoidance, and so forth all have their own specific neural foundations. The more we known about them, the more these neural foundations can be considered in the context of diagnosis and treatment of the various sources of incongruence. Presenting the individual sources of incongruence as separated from their neural foundations, as if they led a separate existence, would not be consistent with the general view of mental activity that was developed in this book. The neural foundations of the sources of incongruence should be regarded as an integral part of our knowledge about these sources of incongruence.

I think it is highly probable that, over time, our improved understanding of neural activity will influence the interventions we use to alter the individual sources of incongruence. For the treatment of depression and anxiety disorders, I have discussed in some detail how this might look. The integration of specific neuroscientific findings into the process of psychotherapy will occur one step at a time. But even today, the general recognition that all mental processes correspond to specific neural activation patterns can already influence the conduct of psychotherapy. According to the perspective developed here, changes can be achieved via the facilitation of new neural activation patterns. Today, such a view of psychotherapy as the targeted and systematic facilitation of new neural activation patterns is not yet obvious or self-evident to most. However, this is one inevitable conclusion we can draw based on all that was discussed in the previous chapters. The second part of the conclusions will revisit this point once again. Before getting to that, however, the first part will conclude with a brief summary of how psychotherapy can be optimized from the perspective of consistency.

5.5 IMPLICATIONS FOR AN EFFECTIVENESS-OPTIMIZED PSYCHOTHERAPY

The effects of psychotherapy critically depend on the extent to which a patient encounters experiences relevant to his or her basic needs. Everything that he or she experiences or does in therapy will have a positive or negative meaning with regard to his or her motivational schemas. This process begins even prior to the therapy as such. Everything the patient finds out about the therapy setting or about the therapist, how others in the clinic treat him or her prior to the first session, and especially the very first impressions of the therapist and of the therapy setting, can all acquire more or less positive meanings for the patient's activated need for orientation and control, his or her attachment need, self-esteem need, and need for pleasure maximization and distress avoidance. All of this can be arranged in such a way that the patient will tend to encounter positive rather than negative experiences with regard to these needs. For example, if the patient is being bossed around or if he or she is not told how long he or she will have to wait, these are negative experiences in terms of the patient's basic needs. They trigger negative emotions and prime the avoidance system. Conversely, structuring these conditions carefully can contribute to effective approach priming.

From the very first minute of therapy, the therapist can use the principles of motivational attunement and resource activation to enable the patient to experience need-satisfying events and, thereby, to prime the approach system. This is one of the most important tasks of the therapist at the beginning of therapy and it remains important throughout the duration of therapy, even if the processing of problems soon becomes the more obvious focus of the sessions.

The differences between process and content, between explicit and implicit mode of functioning, and between approach and avoidance modes should be internalized by psychotherapists in the early stages of their training. Although the patient's problems are the overt focus of the therapeutic discussions—that is, attention is directed toward the explicit mode of functioning—it is necessary that the patient continuously encounters need-satisfying experiences in the implicit mode of functioning. These experiences in the implicit mode lead to a reduction of the incongruence level, shift the patient toward an overall better state, and thereby have positive effects in terms of contributing to a better therapeutic outcome. In addition, these need-satisfying events activate the approach system and, thereby, increase the patient's receptiveness and willingness to engage with problem-specific interventions, which counterbalances the negative emphasis associated with problem processing. Moreover, the activation of the approach system is a prerequisite for the effective inhibition of negative emotions such as anxiety and should therefore be regarded as an inherent component of many problem- and disorder-specific interventions.

The success of therapy is, to a large extent, dependent on the degree to which therapists succeed in activating patient resources and in structuring and tailoring the therapeutic relationship in a complementary manner, such that patients encounter need-satisfying experiences. Beyond that, the success of therapy depends primarily on the kinds of goals that are pursued in therapy and on the degree to which these goals are attained.

The question of which goals should be pursued in therapy is by no means trivial. The patient's treatment goals and the presenting psychopathologic symptoms may partially determine these goals, because both aspects are "predetermined" criteria for the evaluation of therapy outcome. The patient's subjective well-being at the end of therapy is another predetermined criterion. The question of which access points the therapy should target in order to fulfill these criteria, however, is by no means clear at the outset of treatment. Subjective well-being is not the kind of goal that can be pursued directly; it is the result of the way in which the individual maneuvers life more generally. According to the perspective developed here, subjective well-being depends primarily on the satisfaction of basic needs. Basic needs will be better satisfied, in turn, to the extent that mental processes are consistent with one another. From this point of view, therapy should primarily strive to improve the level of consistency in mental functioning, which is equivalent to saying that therapy should aim to achieve optimal congruence between real experiences and motivational goals.

In order to be able to do this, it is first necessary to determine which factors hinder or impair the congruence between current perceptions and motivational goals. If this is only an isolated disorder or a circumscribed problem, the goal of the therapy is clear: Alleviation of the disorder or the problem. If several sources of incongruence are present, however—which would be reflected in generally higher incongruence levels—the question must be posed differently. It now becomes: How can the level of incongruence be lowered most effectively? Thus, an initial question concerns the possible access points for incongruence reduction. If incongruence levels are high, several sources of incongruence tend to be present. In such cases, for example, several disorders tend to exist simultaneously. Each of these would be an access point via which incongruence could be reduced, but it is not yet clear which of the potential access points would be the best. It is not even clear that the disorders would at all be a good access point in an individual case. Every one of the sources of incongruence listed in section 5.3.3 constitutes a possible access point for a reduction of incongruence and could, therefore, become a primary therapy goal. Given the high correlations we observed between reductions in incongruence and improvements in sources of incongruence such as symptom severity, well-being, interpersonal problems, self-efficacy expectancies, coping, avoidance goals, and so forth (see section 4.8.4.3), each of these sources of incongruence can be regarded as a primary access point for the reduction of this incongruence. The likely consequence would be that improvements of any single sources of incongruence would trigger broader improvement in all other sources of incongruence.

It is possible, therefore, to select from among all the potential therapeutic access points those that seem the most promising in terms of treatment goal attainment. Those goals would be the ones for which patients are highly motivated and for which strategies exist that the patient would be able and willing to engage with. That is, strategies that take advantage of existing patient resources would be utilized. The competencies and motivational propensities of the therapist can also influence such therapy goal related considerations. Interventions that the therapist can administer with a greater sense of security, for which he or she has more positive competency expectancies or that he or she simply personally prefers are more likely to lead to therapy goal attainment, compared to those that the therapist administers reluctantly or without high levels of competence.

It is possible, then, to use these criteria to select successive therapy goals from among all the possible therapeutic access points (i.e., all the sources of incongruence). The immediate goal is the removal or alleviation of the source of incongruence, but the overarching goal is the reduction of overall incongruence. Because of the tight interconnectedness of mental processes, it is probably not really necessary to treat all of these access points in order to achieve satisfactory incongruence reductions. The length with which one continues with the treatment of sources of incongruence depends on the extent to which reductions in incongruence have already been achieved.

The criteria for the evaluation of therapy outcome remain untouched by all these considerations. Therapy success is only attained when the disorders of the patient are markedly improved, when his or her treatment goals are largely achieved, and when his or her well-being is clearly better. The evaluation of therapy success, then, is largely independent of the construct of incongruence. It is a specific assumption of consistency theory, however, that these criteria will be achieved to the degree to which the sources of inconsistency or incongruence can be removed or reduced.

If one considers all of the potential sources of incongruence and the therapeutic methods by which they can be altered, one notices that they represent the entire spectrum of psychotherapy: coping- and clarification-oriented forms, individual and systemic forms, cognitive, behavioral, and conflict-oriented (i.e., psychodynamic) modalities. One could link the various sources of incongruence with different therapeutic approaches that tend to target the respective access points more explicitly and that have developed special repertoires for their treatment. In this view, the different therapeutic approaches are not mutually exclusive; they complement and mutually support each other. They are different options to reduce the inconsistency in mental functioning, and there are no reasons not to combine them as needed. On the contrary: It should be possible to achieve better therapy outcomes if one systematically utilizes the various options for attaining consistency improvement that were discussed in the first part of the conclusions. This would require, of course, that therapists are trained appropriately. Specific elements that have previously been neglected could be added to augment existing therapy training courses. Different approaches could unite to pursue the common goal of utilizing all the various psychotherapeutic methods in order to optimize the treatment effectiveness.

5.6 NEURAL MECHANISMS OF THERAPEUTIC CHANGES

5.6.1 Changes via Inhibition of Neural Excitability

All therapeutic changes rely ultimately on changes in synaptic transmission. The signal transmission can be reinforced via facilitation or it can be weakened via nonutilization or via active inhibition. I will discuss the neural mechanisms associated with facilitation and reinforcement in the next section. Changes due to nonutilization do not play a significant role in psychotherapy. An example for this would be the process of ordinary forgetting. Weakening of synaptic connections due to forgetting is a question of the duration for which the synapses have not been activated. In psychotherapy, it is hardly possible to wait until the patient forgets his problems. The neural activation patterns underlying these problems are repeatedly activated and, thereby, more thoroughly facilitated. A different neural mechanism is required to prevent the activation of undesired neural activation—that of active inhibition.

Because most psychotherapies commonly aim to reduce something unwanted or to alleviate something undesired so that it occurs less intensively or less frequently, the neural mechanism of inhibition is very important for achieving change in therapy. The most important inhibiting neurotransmitter is gamma-amino-butric acid (GABA). The firing of a cell can be actively inhibited via the GABA receptors, such that the neuron no longer transmits its activation (see section 2.3). A neural circuit can be disrupted at various locations via this process. In the context of discussing the motivating function of dopamine (section 4.7.6), we also encountered the modulatory function of acetylcholine as an opponent of dopamine. Serotonin also exerts a calming, arousal-inhibiting role in neural functioning.

Inhibition exists not just on the level of the single synapses and neurons, but on all levels of the nervous system. The process of inhibition is absolutely indispensable for consistency maintenance in the nervous system (see section 4.8.2). For example, it is impossible to pursue multiple goal hierarchies simultaneously in the conscious mode of functioning. The hierarchy with the highest motivational salience is being shielded from the competing influence of other goal hierarchies via active inhibition. Attention is also accompanied by the active disruption of other activated circuits before their signals reach working memory. Many stimuli are being perceived, processed, and even triggering reactions, even though they do not become conscious because working memory is occupied by other contents. In the nervous system, inhibition typically occurs because the transmission of signals is being actively inhibited at specific locations.

In everyday language, *inhibition* tends to have somewhat negative connotations. Being inhibited tends to be viewed as a problem and, indeed, excessively inhibited behavior can be a therapeutic problem. This is primarily true in cases in which too little approach behavior is present. In psychotherapy, however, a deficit in inhibition is often the most serious problem, for instance, if negative emotions such as anxiety, aggression, slight, disappointment, grief, or depression are not being effectively inhibited.

Emotional dysregulation plays an important role in many mental disorders. Strong negative emotions occur in situations in which they are not appropriate. In the discussion of the neural correlates of anxiety disorders in chapter 3, it became clear that practically all neuroscientific disorder models posit that the impaired inhibition of situation-inappropriate emotions plays a major functional role in the etiology and maintenance of these disorders. Accordingly, much of the therapy of anxiety disorders involves the construction of situation-appropriate inhibitory mechanisms. Improved control of emotions is also an important issue in the treatment of other disorders such as borderline personality disorder, eating disorders, depressive disorders, aggressive disorders, and so on. I will show the importance of inhibition here by discussing the example of the anxiety disorders in a bit more detail. These considerations can also be applied—with appropriate modifications—to many other therapeutic problems and tasks.

An example of the role that inhibition can play in the therapy of anxiety disorders was already discussed in section 2.10.4, in the context of the question of whether anxiety can be extinguished. Gutberlet and Miltner (1999) found among successfully treated spider phobics that, after treatment, they no longer experienced subjective anxiety feelings and avoidance behavior, and the autonomic arousal (heart rate and skin conductance) that had been triggered earlier upon confrontation with spiders was also diminished, but the amygdala still showed the same degree of activation as it had before treatment. The anxiety reaction had not been "extinguished," then, but the transmission of the arousal to other circuits had been inhibited or disrupted.

Le Doux (2001, 2002) also argued that anxiety reactions cannot be extinguished as such, but that in each successfully treated case, the transmission of the arousal from the amygdala to other brain regions is disrupted by the facilitation of inhibitory circuits. When conditioned anxiety reactions are being "extinguished," then, the reality is that an inhibitory activation pattern is being built. The process of extinction is, in fact, a process of facilitation. The fact that inhibition must also be facilitated is reflected in the fact that when conditioned anxiety reactions are being extinguished, several learning trials without the presence of the unconditioned stimulus are always required in order for the anxiety reactions to diminish.

There are many clues suggesting that the inhibition of anxiety reactions originates from the medial PFC, which, in turn, is closely interconnected with the hippocampus. Among rats whose medial PFC had been lesioned—especially in the orbitofrontal area—conditioned anxiety reactions were observed to be excessive, and it was almost impossible to achieve any degree of reduction. The medial PFC and amygdala appear to be in a reciprocal relationship with one another. Strong activation of the amygdala switches off the medial PFC—and strongly irrational behavior follows. Conversely, the arousal of other brain regions originating from the amygdala can be inhibited via the medial PFC.

In order to effectively inhibit anxiety, then, neural activation patterns that can inhibit the anxiety must be established via facilitation in the medial PFC. This brain area is not identical with working memory. Inhibiting anxiety does not work by creating reasonable thoughts in working memory, such as, "You don't need to feel anxious; the situation is not really dangerous." Instead, the implicit situation evaluations that are continuously automatically created in the orbital PFC must be changed in order to inhibit the transmission from the amygdala to other brain regions. Implicit learning is not a question of insight but of repeated experiences that show the person that nothing truly dangerous is transpiring. These experiences are not typically encountered by a person with an anxiety disorder. On the contrary, his or her experience tends to be that something very unpleasant is transpiring when he or she is confronted with a phobic stimulus or is having an anxiety attack. The person experiences real anxiety, which is in itself so aversive that he or she seeks to avoid experiencing it again. The avoidance prevents the facilitation of new implicit evaluations. The experiences that facilitate such new evaluations must be encountered repeatedly and forcefully. It is only after sufficiently strong activation by real experiences that the new activation patterns representing the new implicit evaluations will be so thoroughly established that they can be activated more or less automatically. It is only then that the inhibitory impact is strong enough to block the further transmission of arousal originating from the amygdala.

In order to build up this amygdala inhibition, the person must expose himself or herself to the situations that trigger the anxiety. The neural activation patterns representing the perception of the situation and the neural activation patterns underlying the anxiety are now activated simultaneously. They have been coactivated like this many times, which is why they are now joined together into a common neural activation pattern. The point now is to once again unravel this neural activation pattern, which has been established by many concurrent activation episodes. This requires the additional of a third neural activation pattern that inhibits the anxiety while the neural activation patterns representing the situation–perception are activated. This third, new neural activation pattern must be maximally incompatible with the anxiety. If this new neural pattern is repeatedly activated conjointly with the patterns representing the perception of the situation, the situation–perception and the anxiety patterns will be decoupled. This decoupling is based on the inhibition of the neural activation pattern representing the anxiety reaction, via the new, anxiety-incompatible activation pattern.

All three neural activation patterns must at first be activated simultaneously. Because the new activation pattern is being more and more thoroughly facilitated and coupled with the situation perception by repeated conjoint activation, it will be more and more likely to automatically occur along with the perception in the future. The anxiety-inhibition will only occur to the extent to which this new pattern is truly incompatible with anxiety. The thought, "I don't have to be anxious—nothing terrible is happening" is not sufficient for this purpose because the simultaneously activated implicit evaluations tell a different story: Something terrible is, in fact, happening at just that moment—anxiety is present. It is only at the point that the patient repeatedly encounters the "bottom-up" experience that the feared event does not come true, that a neural activation pattern will be formed in his or her implicit memory and can then inhibit the anxiety on this implicit level. The explicit thought, "I don't have to feel anxious; nothing terrible is happening" must be supported by implicit convictions before the thought can acquire anxiety-inhibiting properties. Implicit learning, however, requires bottom-up learning via real experiences. It is only the repeated real experience of having been in the actual situation, but without any anxiety, that a corresponding new neural activation pattern will be facilitated in the implicit mode. Once this neural activation pattern is firmly established, it will effectively inhibit the occurrence of anxiety. The critical processes in the formation of anxiety inhibition transpire in the implicit mode of functioning. Whatever information the therapist and patient exchange consciously will lead to anxiety inhibition only to the extent that it leads the patient toward a state in which he or she is ready and willing to expose himself or herself to the necessary implicit experiences.

The formation of anxiety inhibition requires, then, that the patient exposes himself or herself to the situations that have hitherto triggered anxiety, until the experience is encountered that he or she is in the situation without having the anxiety. The fact that the patient exposes himself or herself at all to these situations becomes possible due to the previous and continuous presence of approach priming. The activation of the approach system also contributes to the ability to keep the intensity of the anxiety within tolerable limits (e.g., via graduated exposure in the sense of anxiety hierarchies) and/or that the patient develops the required willpower and willingness to exert effort to stay in the situation sufficiently long, until the experience occurs that the anxiety subsides without having to utilize any means to control the anxiety. If they are encountered often enough, both types of experiences are capable of changing the implicit evaluations in the orbital PFC, such that the anxiety signals, as they are passing through the "slow loop" on their way to the amygdala (see section 2.10.1) are evaluated as nonthreatening. The corresponding signals are being sent to the basal and lateral nuclei of the amygdala and, at that location, inhibit the arousal that has already been triggered via the "fast, direct route."

Appropriate therapeutic methods can be used actively to ensure that the implicit evaluation of "nonthreatening" actually is made in the medial PFC. The most important of these methods, as mentioned previously, is motivational priming. Anxiety reactions and other negative emotions can be regarded as belonging to the avoidance system (see section 4.7.5). If the avoidance system is already activated, anxiety reactions to a specific event will be more intense than they would have been if the person had been in an emotionally neutral state when the event occurred. Negative emotional stimuli facilitate associations, representations, and behavioral programs in the avoidance system; positive emotional stimuli do this in the approach system. If the approach system is activated, the tendency to experience negative emotional reactions and defensive reactions is much weaker, compared to situations in which the individual is operating in the avoidance mode (see section 2.15).

Motivational priming must be regarded as a critical component of effective anxiety treatment. If the approach system of a patient is strongly activated because positive emotions, expectancies, and goals have been activated within the patient, then his or her neural system is shifted into a state of low anxiety-readiness. If the patient is in this approach state and then, in the service of the pursuit of his or her own goals, deliberately exposes himself or herself to situations that have previously elicited anxiety, the anxiety reactions will be markedly weaker, compared to how they would have been if the patient had experienced the situations passively in the avoidance mode. Under these approach conditions, much stronger inhibitory signals will pass from the PFC to the amygdala. The formation of anxiety inhibition transpires more easily and more quickly, and the process is less aversive for the patient.

Because therapy generally deals with the treatment of aversive problems that are linked with negative emotions, and because psychotherapy patients tend to be more avoidance- than approach-motivated, a natural tendency exists for them to be in a negatively primed motivational state; that is, they are more likely to operate in an avoidance mode rather than an approach mode at that point. If aversive problems are being processed while the patient is in such a state, it is highly likely that strong negative emotions along with no mastery experiences will be elicited. The probability that such therapy sessions achieve positive outcomes is rather low (see previous). Instead, the therapy context—the therapist, the room, the clinic, and the entire setting—is being continuously conditioned to negative emotions, such that the patient will soon anticipate therapy with a sense of dread and associate the entire situation with negative emotions. These would be extremely unfavorable conditions to achieve anything positive in therapy. The danger exists that the patient will drop out of therapy or that the therapy will be unsuccessful.

I have emphasized repeatedly that the processes discussed here largely transpire in the nonconscious mode of functioning. At a conscious level, the patient might experience the therapist as friendly and as someone who tries hard to help, which makes it difficult to accuse him or her of anything negative. Nevertheless, the patient might continuously experience negative emotions and lose all hope of achieving improvement. Such constellations and therapy processes are, unfortunately, not rare exceptions, as we have found in the process analyses of thousands of therapy segments that are discussed previously.

It would not be possible to establish an effective inhibition of negative emotions, such as anxiety, in a motivationally unfavorable therapy situation of the kind described previously. If therapeutic interventions such as exposure exercises are conducted in such a motivational constellation, the anxiety will not be inhibited. The neural context from which the inhibition would originate is insufficiently established.

One of the most important tasks of the therapist, prior to the administration of any specific interventions such as anxiety-related exposure, concerns the creation of an appropriate neural context for such interventions. If one recognizes that anxiety cannot be extinguished but must be inhibited, then it becomes clear that the creation of an anxiety-inhibiting brain state will be a core aim of all therapeutic activity. This must be an important part of the therapy manual that instructs therapists on how to conduct anxiety treatments. Some such manuals already include such instructions, but others do not pay much explicit attention to the creation of a maximally anxiety-incompatible motivational state. I am not referring here to a one-shot motivational therapy preparation or to explicit instruction giving. What I mean is a continuous motivational priming that must be accomplished primarily in the implicit mode of functioning. In section 5.3.2, under the keywords of complementary therapist behavior and resource activation, I already discussed how such a therapeutic context can be established.

The inhibition of negative emotions and the facilitation of neural activation patterns from which the inhibition originates refer to neural mechanisms that play an important role in almost any psychotherapy, because most therapies deal, at least in part, with the issue of seeking improved control or regulation of negative emotions. To the extent that negative emotions are an important incongruence source for the patient, and to the extent that their inhibition succeeds, improvements in the consistency of mental functioning will follow, and these improvements will in turn have other positive consequences, as described earlier.

5.6.2 Changes via Activation of Existing and Facilitation of New Neural Activation Patterns

We have seen in the previous section that the inhibition of existing neural circuits occurs via the facilitation of other circuits from which the inhibition originates. Facilitation can, therefore, be regarded as a core principle of all changes that occur via learning. The facilitation of a new neural activation pattern requires its activation, and each activation simultaneously constitutes facilitation. Activation and facilitation, then, are inseparably linked.

Facilitation occurs according to Hebb's (1991) principle that neurons that fire together wire together. This principle holds true not just on the level of single neurons but also on the level of neural groupings and neural circuits. A prerequisite for and beginning of each facilitation is the activation of that which is to be facilitated. In order to facilitate something thoroughly, it must be activated repeatedly and enduringly. Anything that the patient should think, feel, or do with greater intensity or frequency, therefore, must be elicited maximally frequently and for a maximum period of time. Only then will the NMDA receptors be opened and the second messenger cascades triggered that lead to the enduring fortification of the synaptic connections participating in the activated neural activation patterns.

Even these basic neural principles suggest concrete therapeutic conclusions. If a therapist wants to achieve that a patient thinks, behaves, or feels in a new way, the therapist should not wait for these events to occur spontaneously; instead, he or she should do whatever is in his or her power so that these thoughts, feelings, or actions occur at all. As soon as the respective behavior occurs even in small parts, it should be maintained or reactivated as soon and as often as possible. The longer and the more often the new behavior is activated, the more thoroughly it is established, and the more easily it can be activated again in the future.

Given the principles of activation and facilitation, the focus of therapy should be oriented as intensively and frequently as possible toward thoughts, behaviors, and feelings that constitute approach with respect to the therapy goals. At the beginning of treatment, it is of course also necessary that the therapy addresses the problematic status quo. This will often be accompanied by an activation of undesired neural activation patterns. However, from this necessary initial stage, the therapeutic focus should swiftly shift toward that which is to be newly acquired—toward approaching the desired goal. Therapies that remain stuck at the stage of identifying and analyzing problems tend to have worse outcomes than therapies that shift their focus earlier toward the process of change (see Figure 5–4). This result, which we have frequently replicated in our studies, is entirely compatible with the idea that therapies should not aim for the frequent activation of old patterns that are to be changed, but of the new, goal approach related neural activation patterns. The second finding represented in Figure 5–4, that the intensity of problem processing correlates with good session outcome, is also entirely compatible with neural principles. In order for something to be facilitated thoroughly, it must be activated as intensively and lastingly as possible.

These ideas did not explicitly integrate the motivational aspect of mental functioning. Learning will be all the more effective, however, if it is accompanied by high motivation. We have already seen in section 2.3 that enduring improvements in synaptic transmission depend primarily on the extent to which second messenger processes in the activated neurons can be triggered. The second messenger processes originating from the NMDA receptors are just one type of such processes. They depend primarily on the intensity of the stimulation. The intensity is high when AMPA receptors are stimulated with high frequency via glutamate, and when the neuron is stimulated simultaneously via several AMPA receptors. The ion channels of the NMDA receptors, which are blocked by magnesium in their resting state, are opened when the membrane of the postsynaptic neuron is depolarized by the summation of several activations. The therapeutic considerations up to this point have all referred to these types of second messenger processes that transpire via the intensive and frequently repeated stimulation of neurons.

The influence of motivation on learning is mediated by the so-called neuromodulators, which bind to so-called non-NMDA receptors. The most important non-NMDA receptors, in the context of motivational learning, are the noradrenergic and dopamine receptors. The neurotransmitters docking to these receptors lead to a temporary increase of a second messenger transmitter in the postsynaptic neuron, which increases the excitability of the previously activated ionotropic "fast" receptors over a period of seconds to minutes. If the neuron continues to be stimulated intensively via the same synapses during this period of high excitability, the activation (or inhibition, if GABA receptors are being stimulated), will become progressively stronger. At this point, additional second messenger cascades are triggered and, via gene transcription, lead to the formation of additional synapses, to the formation of neurotrophins, and to the production of retrograde messengers. That is, the cascades trigger the processes that underlie enduring learning (Abel & Kandel, 1998; see also the discussions in section 2.3 on these points).

Adrenaline and dopamine are, of course, primarily released when important goals are being activated (see sections 4.5.2.1 and 4.7.6). The secretion of these neurotransmitters causes the neurons that are immediately subsequently activated to be more easily depolarized and, thus, more ready to learn. In principle, this holds true for approach goals as well as for avoidance goals. This means that approach as well as avoidance learning is being potentiated if important goals are activated simultaneously. Because therapeutic activities primarily revolve around approach goals, the facilitation of desired behavior primarily requires that important positive goals of the patient are activated while the thoughts, behaviors, and feelings to be learned are also activated. Effective therapeutic learning requires two things:

- The activation of important positive goals, in order to shift the simultaneously activated (or shortly thereafter activated) neurons into a state of greater learning readiness via the release of adrenaline and dopamine.
- The intensive—frequent and enduring—activation of the neural patterns that are to be fortified in this state of increased learning readiness.

Once again, this points to the importance of the principle of motivational priming, which has been mentioned many times now, for the creation of need-satisfying experiences in the process of therapy. If approach priming has occurred, it will be much easier to activate positive goals. Their activation, in turn, is essential for the effective and enduring facilitation of neural activation patterns that contribute to the person's ability to attain these goals.

The activated goals might also include therapy goals that are to be achieved. However, adrenaline and dopamine releases in the presence of such goal activation will occur only when these goals are compatible with important motivational goals held by the patient. Adrenaline and dopamine are not released in response to conscious, rational decisions; they are secreted when the continuous, implicit evaluations that occur in the orbital prefrontal cortex indicate the presence of goal-significant meaning. Thus, important motivational goals must be activated if effective learning is to occur. It is hard, if not impossible, to use therapeutic interventions to change someone in a direction that is incompatible with his or her current motivational goals. When therapy goals are being determined, then, it is particularly important to ensure that these goals actually have a positive meaning with regard to the already existing motivational goals held by the patient. This is another important reason why the therapist must not just understand the patient's current problems but also his or her most important motivational goals.

These considerations with regard to the neural mechanisms underlying therapeutic changes now lead to a set of very concrete conclusions:

- In every therapy session, the therapist should have a clear idea of the goal that he or she seeks to attain together with the patient. This awareness will allow the therapist to specifically activate behavior that serves the approach toward these goals. This must occur systematically, lastingly, and repeatedly in order for the new neural activation pattern to become easily activated later on.

    The goal pursued in the session must be positively significant for at least one important motivational goal of the patient. Therapy goals that do not fulfill this criterion are not likely to be attained.
- The respective motivational goal must be maximally strongly activated at the time when the to-be-established neural activation patterns (actions, thoughts, feelings, etc.) are activated. Therapists should not assume that this is the case; they must work actively to create such activated states. This can be achieved, for example, by using imagery exercises, therapeutic dialogue, and so forth. Most of all, however, this can be achieved by enabling the patient to encounter powerful implicit perceptions that are consistent with the relevant motivational goal(s).

- The activation of positive goals should be prepared via continuous approach priming.
- When a positive goal is activated, the therapist must try intensively to activate thoughts and behaviors that are compatible with this goal.
- It is not necessary to "reinforce" the behavior after its occurrence by praise or overt recognition, as had been assumed in operant learning theory. The reinforcement consists of the fortification of facilitations via the release of specific neuromodulators. The important process is the activation of the positive goal from which the reinforcement originates, and the recognition by the patient that the goal is being achieved, because this recognition activates dopaminergic neurons.
- All the points mentioned previously also hold true for the formation of inhibitory neural activation patterns. This also requires the activation and facilitation of new, positive activation patterns whose subsequent easy activation by the respective situations will automatically inhibit the activation of undesired neural circuits. The newly facilitated activation patterns must be incompatible with the undesired behavior.

The presentation of these conclusions shows that neuropsychotherapy should be regarded as a planned, structured endeavor. Neuropsychotherapy does not refer merely to sitting down with a patient, allowing him or her to talk about his or her problems, responding empathically, stimulating him or her with occasional interventions to try out new thoughts and behaviors, or encouraging the patient to express his or her true feelings. Neither the choice of the therapy goals nor the conduct of the individual sessions can be left by the neuropsychotherapist to the patient. Although the therapist must ensure that the patient consents to everything he or she does, that all therapeutic events are compatible with the patient's motivational goals, the therapist must still assume responsibility for the most important therapeutic decisions. The therapist cannot simply delegate responsibility to the patient in a misguided effort to recognize the patient as an autonomous and equally valuable person, because the patient would be overwhelmed by this responsibility. The therapist must be like a good advocate acting on behalf of the patient. He or she must assume responsibility in areas that currently overwhelm the patient, but he or she must inform the patient and include him or her in decision making as much as possible, given the current progress in therapy. This holds true for the process of therapy planning in general as well as for the structuring of each individual session. Neuropsychotherapy requires a very disciplined, structured approach. Beginning a session with the position, "Let's see what the patient brings up today," and allowing the session content and structure to be determined by the patient's current concerns and the current state of the patient would be incompatible with conclusions that I discussed in the first and second part of this chapter.

The therapist must consider many aspects simultaneously, then. This makes it difficult to determine in advance, via a general algorithm, how therapy should proceed at various decision points over the course of treatment, in a way similar to what is done in some disorder-specific manuals. Nevertheless, there are several concrete and strongly supported rules that a therapist should follow if he or she wants to optimize the process and outcome in line with the principles discussed previously.

The first part of the implications chapter will, therefore, address the concrete therapeutic rules that flow from the discussions in the previous chapters. The guidelines overlap to some extent with the conclusions elaborated in the first two parts of this chapter. I have accepted this redundancy because otherwise several important points that have already been discussed would not have resurfaced in the concrete therapy guidelines, and this would have resulted in a biased picture of the principles that are particularly important for the attainment of optimal therapy outcomes.

5.7 GUIDELINES FOR THERAPY PRACTICE

The following contains a set of concrete guidelines for the practice of therapy. These guidelines flow from the material discussed in previous chapters, but they are once again accompanied by discussions of specific supporting evidence. Each of the guidelines is strongly empirically substantiated. The relevant evidence is discussed in many places throughout the previous chapters. Those who have read these chapters will immediately comprehend the basis of these guidelines.

I will subdivide the guidelines into those that refer to the content of therapy planning and those that refer to the process or the actual conduct of therapy. I formulate these guidelines based on the conviction that therapists who follow these rules will, on average, attain particularly good therapy successes. This conviction is based on the fact that each rule is substantiated by a solid empirical foundation. In recent years, therapy researchers and practitioners have engaged in a lively and at times controversial debate about the clinical importance of "empirically validated treatments." Such treatments are those for which solid efficacy evidence has been reported. The guidelines that are formulated here are based on a different understanding of empirically validated psychotherapy. It is not the methods that have been validated here, but a set of guidelines for therapeutic practice that cut across specific disorders and methods. In my understanding, therapists who conduct their treatments in line with these rules will engage in empirically validated psychotherapy.

For now, this should be viewed as no more than a well-substantiated claim. In order to properly validate this claim, therapies that follow these guidelines would have to be conducted and then compared with other therapies that are based on different rationales. This is a very strict requirement, however. For the guidelines that the American Psychiatric Association has published for the treatment of specific disorders (e.g., APA, 2000), such validation methods have been neither performed nor even requested. An empirical test of the clinical utility of the subsequent guidelines could clearly be performed. A considerable advantage of the kind of understanding of empirically validated psychotherapy associated with these guidelines is that these rules are not inseparably tied to specific disorders or methods. The guidelines could be applied to an unselected patient sample in routine clinical practice. Thus, such validation studies could be conducted in mixed patient samples of the kind that are normally encountered in actual clinical practice. This would also increase the clinical representativeness and relevance of such studies, compared to the more common efficacy studies with their homogeneous patient groups and specific therapy methods. Conducting such clinical tests of therapeutic guidelines would constitute a new type of therapy research. Given that such a test has never been attempted for any set of guidelines, I can confidently say—with a clear scientific conscience—that the following guidelines are recommended for clinical application, even in the absence of additional explicit validation.

5.7.1   Guidelines for Therapy Planning

1. Three perspectives should be assumed in the determination of therapy goals: The disorder perspective, the inconsistency perspective, and the perspective of the patient's treatment goals.

2. If a specific disorder pattern has been established for a longer duration, use an empirically validated manual for the determination of disorder-specific therapy goals and for the conduct of disorder-specific interventions.

3. If signs suggest the presence of elevated incongruence (comorbidity, many treatment goals, high scores on the incongruence questionnaire), conduct an incongruence analysis and determine possible access points that enable the changing of the most important sources of incongruence.

4. Find out for which of these possible access points the patient shows the most optimal motivation and has the best resources.

5. Use both of these criteria to rank-order the possible access points to achieve change, and then select those goals as the most important ones for which the patient has the most optimal prerequisites.

6. When working together with the patient on the formulation of therapy goals, ensure that only approach goals are being formulated, and only such goals as can be realistically attained by the patient's own efforts. Do not permit the formulation of avoidance goals or negatively expressed goals.

7. Only select goals for which the patient is intrinsically motivated; that is, goals that are highly significant for the patient's motivational goals.

8. Regardless of the patient's problems, determine the nature of the patient's most important approach and avoidance goals.

9. Conduct an explicit resource analysis with each patient. Determine the nature of the patient's strengths and positive aspects that can then be utilized, activated, and incorporated in therapy.

10. In each therapy, determine which motivational goals of the patient are being activated in the therapeutic relationship.

5.7.2   Guidelines for the Therapy Process

The guidelines for therapy planning refer to content-related considerations and therapeutic behavior that transpires in the explicit mode of functioning. They can be become automatized over time, but it will be possible, in principle, to deliberately call upon and use these guidelines once the required skills have been learned. What happens in the explicit mode of functioning of the participants is, however, just one part of the processes determining the course and outcome of therapy. Very important processes also transpire implicitly, without the participants' current attention and without their conscious awareness of the processes. Many of the neural processes that influence these changes will, by their very nature, remain unconscious. Nevertheless, they can and should be deliberately influenced by the therapist.

All of the perceptions a patient encounters in the therapy process—including the implicit perceptions—affect the process of therapy. A part of these perceptions is influenced by behavior the therapist can govern volitionally. Other perceptions of the patient, however, cannot be influenced volitionally by the therapist at each respective moment. This includes, for example, the impression conveyed to the patient by the person of the therapist. It is only partially possible, at best, to convey that one is a warm-hearted, extraverted, optimistic, and confident person, if this is not, in fact, true by nature. All of these are characteristics of successful psychotherapists. Individuals who naturally possess these traits will find it much easier to become good therapists. Those who are not equipped with these features by nature will have to work much harder on themselves in therapy training if they want to achieve equally good therapy outcomes.

The following guidelines for the therapy process refer in large part to processes that transpire in the implicit mode of functioning and that, at each moment, can be consciously controlled by the therapist to only a limited extent. The required behavior can be trained in the context of therapy training programs, however. Examples are the bodily posture and the tone of the therapist's voice. Therapists should slightly lean in toward the patient, with their arms open and their hands relaxed on their lap rather than arms crossed. As the patient speaks, they should nod repeatedly. Their own statements should be emphasized by lively gestures. It is generally better to leave the legs open rather than crossed. All of these behavioral characteristics have been examined repeatedly. They are important nonverbal signals of empathy, and empathy, in turn, is an important characteristic of positive attachment relationships. Therapists who show these characteristics are evaluated much more positively by patients, compared to therapists who lean back and have their arms crossed in front of their chest (Harrigan & Rosenthal, 1986). The voice of the therapist, regardless of what he or she says, should be warm, professional–competent, and free from fear. These are the most important voice characteristics of therapists who were judged by their supervisors to be particularly competent in their interactions. Male therapists must primarily learn in their therapy training to speak with sufficient warmth in their voices; female therapists must primarily learn to convey competence with their voices (Blank, Rosenthal, & Vannicelli, 1986). These voice qualities are also important characteristics of positive attachment relationships. Even though these characteristics are automatized behavior patterns, they can, in principle, be volitionally altered and changed in targeted training programs (Robbins et al., 1979).

Some of the guidelines articulated following—such as the one that therapists should actualize the qualities of a good attachment relationship—are composed of many more fine-grained characteristics. If a therapist does not have these characteristics at his or her disposal, it will be impossible to consciously decide at each specific therapeutic moment that he or she should now enact these features. It is necessary that therapists who do not possess these characteristics by nature acquire these features in the course of their training. The corresponding neural activation patterns must be easily triggered as a whole. If that is the case, the therapist can consciously actualize these characteristics in the course of treatment.

I am assuming in the following guidelines, then, that the required behavioral repertoire has been acquired by the therapist—either by nature or in training. If this prerequisite is not in place, which will often be the case among therapists who have not been specifically trained in this respect, the therapist will not be able to actualize the guideline simply by consciously deciding that he or she wishes to do so. An appropriate course of training would be required as a preliminary preparation in such cases. In a sense, then, several of these guidelines can be regarded more as suggestions aimed at training programs and trainers, and not just as rules for psychotherapists' conduct of the therapeutic process. Given this general background of understanding, I nevertheless formulated the following guidelines as if the therapists were able to generate the respective behavior volitionally.

1. Ask yourself repeatedly in a therapy session which implicit perceptions are currently experienced by the patient. What is the significance of the current events for the patient's motivational goals and for his or her basic needs? Never attend just to the contents of what is being said between you and the patient. Shift your attention repeatedly from the content level to the process level. Events that transpire in the implicit mode of functioning will not become an explicit focus but they will nevertheless have a critical effect on the mental processes of the patient.

2. Think carefully not just about what you say and do, but also think about and carefully structure how you say and do it.

3. Always remember that the patient's attachment need is activated in therapy, that the patient encounters, at each moment in therapy, perceptions with regard to his or her attachment need, and that it is critically important that positive attachment experiences are encountered in the patient's relationship with you as the therapist. Clarify in your own mind that you are probably facing an individual who has experienced negative attachment events in childhood, and that this is someone who will respond very sensitively to behaviors suggesting that you are occupied with things other than him or her, that you feel differently toward him or her than you say or claim, that you might reject him or her when he or she has difficulties, that he or she is just one among many patients, or that you do not genuinely like him or her or think much of him or her. Make sure to avoid all behavior that might be interpreted by the patient in such a way—especially nonverbal behavior. Be aware that the perceptions with regard to the attachment need transpire primarily in the implicit mode of functioning, beyond the focus on currently discussed topics, and that it is therefore particularly important how you behave nonverbally. Fully attend to the patient; let the patient experience that he or she is at the center of your undivided attention; convey to the patient that his or her difficulties and problems are in good hands with you; that

you feel and understand his or her important concerns. Let the patient feel that you are ready and willing to fully engage on his or her behalf. Exude warmth, confidence, competence, and readiness to take action. Do not show signs of helplessness in therapy, even if you might occasionally feel clueless or helpless. Attend to such feelings outside of therapy, not within therapy.

4. Enable the patient to experience a maximum number of positive perceptions with regard to his or her need for orientation and control. Structure each therapy session as transparently as possible. Ensure that the patient has a clear understanding of the goals of each session, and of how he or she can contribute toward the achievement of those goals. Involve the patient in the considerations, planning, or decision making with regard to these issues. Allow the patient to make a maximum number of decisions himself or herself. Explain the important points to the patient and, whenever possible, present him or her with a menu of options. Plan and structure the session in such a way that the patient confidently believes that he or she can engage and participate effectively, and that he or she has the sense of pursuing a personally important goal. If you notice reservations, hesitations, or signs of resistance, take these signs seriously, listen carefully to the patient, and allow him or her once again to choose freely how the session should proceed. Always let the patient feel that you are invested in helping him or her achieve his or her goals. Do not push the patient to do things that he or she genuinely does not want to do. If you have to do something that will elicit negative emotions in the patient, ensure that he or she is ready and prepared for this.

5. Actively create situations, and pursue them whenever possible, that allow the patient to experience self-esteem enhancing perceptions. Ask directly about interesting, positive aspects in the patient's life—inquire about his or her normal life activities. Allow him or her to convey to you what interests him or her and how he or she likes to spend his or her time. Show the patient that you are interested in getting to know these sides as well. Create room for these aspects in the therapy session. This is not a waste of time; it is approach priming. Repeatedly incorporate into your sentences words, expressions, and images that have positive, self-esteem enhancing significance for the patient, and do this without explicitly directing attention toward the process. Repeatedly bring up things that have positive meaning with regard to the patient's important motivational goals. Search for ways in which you can genuinely convey respect to the patient, express this clearly and stand behind the fact that you find these things respectable or admirable, even if the patient defends against this feedback. Do not permit the patient's defensive or minimizing reactions to change your course; continue to actively and repeatedly convey self-esteem enhancing perceptions to the patient. These patient reactions serve a self-esteem protective function, and the therapist should experience that this protection is no longer necessary in this environment.

6. Use all possible opportunities to ensure that the patient will also experience pleasant states in the therapy session. This might include laughing together about something, appreciating a success, and so forth. The patient should experience as many positive emotions in the session as is possible, given the simultaneous necessity of having to intensively process his or her problems. If the patient has difficulty in

experiencing any positive states at all, use a part of the sessions to create such experiences by using methods such as hypnosis, autogeneous training, relaxation exercises, positive imagery, or meditation, even if such interventions do not contribute directly to the processing of problems.

7. Ensure that every therapy session has a clear focus. It is possible to have more than one focus, but stick for some time with the processing of one problem once you have started it. Avoid jumping back and forth from one problem to another. Remember that changes at the synapses will occur only if they are lastingly, intensively, and repeatedly activated. When you are processing a problem, you are activating the neural activation patterns underlying this problem. Once they are activated, the important point is to change them; that is, to thoroughly facilitate new or still weakly formed activation patterns. Facilitation requires lasting activation and repetition. Therefore, recreate each change step repeatedly; allow the process to transpire repeatedly. Do not be swayed by the patient who says that he or she now understands what the important point is. Explain to him or her the difference between explicit and implicit learning. For implicit learning, the presence of insight is not important; what is important is the intensive and repeated facilitation. The newly learned activation patterns must be activated automatically later on, so it is necessary that they are very thoroughly facilitated.

8. Whenever you bring up, activate, or process a problem that is relevant for the therapy goals, push the activity forward toward a change event. Especially at the beginning of therapy, this might be a better understanding but, even more important, it might be an improved ability to cope with or master certain difficulties. Do not be content with the mere activation or discussion of a problem; always strive to attain change events. Every problem activation should lead directly to concrete mastery or clarification experiences.

9. Prepare each activation and processing of a problem by first conducting approach priming. Shift the patient into an approach mode by first creating need-satisfying experiences and activating positive goals and emotions. This also holds true for the microlevel: If you want to tell the patient something that is likely to elicit negative emotions (such as a painful piece of information, a difficult confrontation or interpretation), refer in the first part of the sentence to a relevant positive goal of the patient, or create a clearly need-satisfying perception, which increases the patient's receptiveness for the intervention and which makes the negative emotion easier to tolerate. Administer the intervention only when this opened state is in place.

10. Ensure that the motivational goal whose attainment the intervention serves is activated just prior to and during a specific problem-focused intervention. In the section on the guidelines for planning the contents of therapy, it was already said that only those therapy goals should be formulated for which the patient is highly motivated in terms of their relevance for important motivational goals. This goal should be explicitly activated again prior to the administration of an intervention. For example, the patient can be guided to imagine how it might be if the goal were to be attained, or, conversely, how awful it would be if things remained as they are. The therapist could also use a role-play exercise in which the patient defends the goal against questioning, because this can activate the motivational goal behind it. The

activation of the motivational goal is important because it is accompanied by the release of adrenaline and/or dopamine, which, in turn, increase the effectiveness with which new or weakly formed neural activation patterns are being facilitated. Keep referring to the approach goal; reintroduce this goal into the therapeutic dialogue, in order to maintain the patient's high willingness to exert effort in the pursuit of the goal and, thereby, to fortify the facilitation that occurs in the process of the intervention.

11. Do not stop at the point of facilitations that occur only within the therapy sessions. Ensure as much as possible that the new, to-be-learned neural activation patterns are also activated and facilitated under the conditions of the patient's concrete life reality. Use a sufficient part of every therapy session to plan and discuss in detail how such experiences that would lead to the needed facilitations can be created outside of therapy ("homework"). Begin each therapy session with a detailed discussion of the experiences encountered by the patient in this respect since the last session. Do not simply accept it if the patient reports that such experiences have not occurred in the interim. Discuss in detail the reasons for this; question again the corresponding therapy goal; activate the motivational goals that stand behind the immediate therapy goal; and, if the patient sticks with the goal, create a strong sense of commitment for the pursuit of the necessary experiences. At the end of the session, decide on specific methods, such as the scheduling of telephone calls, that further ensure that the planned exposures, exercises, conversations, and so forth actually will occur.

12. Do not insist on the pursuit of therapy goals with which the patient does not truly engage, and do not administer an intervention for which no true patient motivation is discernable. Instead, strive with the patient toward the attainment of other goals. Avoid, however, continuous or excessively frequent goal changes. If the process repeats itself and no progress is evident, be willing to question your basic case conceptualization and consult experienced colleagues to plan the next steps.

It should be obvious that guidelines cannot be mechanically worked through, from one algorithmic step to the next. They can only be actualized if they are internalized— if they become automatized—such that the mere thought of the rule activates the entire repertoire necessary for its realization. However, even if this requirement is not yet met, one can reflect upon one's own therapies from the perspective of these rules, and one can strive to orient one's own work in line with the rules. Supervisors can test with their supervisees the extent to which the guidelines are being followed in a particular course of therapy, and they can consider and plan together how the rules could be applied and realized in individual cases.

The most realistic way in which such guidelines could be incorporated into therapy practice would be to structure training programs in such a way that the rules are taught and the methods for their realization are trained within the programs, in addition to other important knowledge and competencies that are already being trained, such as, for example, those pertaining to disorder-specific interventions. This would also require the further development and validation of appropriate training elements and procedures. The manuals we developed for resource activation and for complementary therapist behavior can be seen as steps in this direction. Other elements for the realization of these guidelines can already be found in various existing therapy approaches. It is obvious, however, that the guidelines cut across the boundaries among established therapeutic approaches; they cannot be categorized as belonging primarily to one specific school or approach. Even the individual guidelines often cannot be classified as belonging to one therapy approach more than another. For some rules this is perhaps more easily possible than for others.

The reason for this is that the guidelines are based upon a very different set of foundations, compared to the traditional therapy approaches. Most of the studies from which the rules were derived have been conducted within the past 10 years. Many of them are neuroscientific studies. Because the guidelines are intended to be used in therapy practice, I formulated them in a way that their links to brain science are not necessarily immediately apparent. Anyone who has carefully read the preceding chapters, however, will be able to easily reconstruct these links. Most of the rules are not based on the results of single studies but on the conclusions distilled from a larger number of research findings. Such distillation always implies a degree of interpretation, of course. The guidelines are my personal interpretation of the current research in this area. In this context, I have not differentiated too rigidly among findings originating from neuroscientific, psychological, or psychotherapy research. The proportion of neuroscientific research as the basis of these guidelines appears so large, however, that I decided to emphasize this aspect specifically in the title of this book.

The detailed connections with the domain of neuroscientific research most clearly differentiate this book from most other books about psychotherapy. This is what the title *Neuropsychotherapy* intends to convey. However, I have also chosen this title because I am convinced that this book is only the beginning of a development in which the neuroscientific component in psychotherapy will become increasingly influential. The term *neuropsychotherapy* will then be even more fitting. I want to conclude this book now with some brief reflections on how such future forms of neuropsychotherapy might look.

CHAPTER SIX

SUMMARY AND FUTURE PROSPECTS

In the previous five chapters, I have discussed findings from the neurosciences that I found relevant for the domain of psychotherapy, and I have derived from this body of knowledge some conclusions for our understanding of mental disorders and of psychotherapy. The empirical evidence suggests that we ought to conduct a very different form of psychotherapy than what is currently practiced. Indeed, it would be more than surprising if a psychotherapy derived from a body of literature that has been constructed only within the last 1 to 2 decades would resemble the therapy modalities whose decade-old theoretical foundations have practically nothing in common with these new findings.

In my intensive studies of the neuroscientific literature, I did not attempt to find evidence to support, post hoc, existing psychotherapeutic assumptions and methods. Because neuroscience has a good reputation and is regarded as progressive, this post hoc supporting of psychotherapeutic methods with neuroscientific findings is quite popular these days. This allows one to build a more modern and prestigious foundation to support that which one has always regarded as true, without actually having to change anything. For example, Rohde-Dachser (2003) uses the neurosciences to support his conviction that psychoanalysis has always had it right with its conceptualization of transference and countertransference. He argued that the neuroscientific findings simply lead to a better understanding of why this was, in fact, always true. In Rohde-Dachser's (2003) words:

> The further development of the psychoanalytic method in Germany since the 1970s initially occurred independent from the progress made in neuroscientific research. *By contrast, today these findings are employed primarily for the scientific substantiation of the psychoanalytic method* [italics added]. Gaining insight into unconscious feelings, fantasies, and pathologic relationship patterns means, translated into the language of neuroscience, to direct the psychoanalyst's attention from the patient's autobiographical memory to his or her procedural memory in order to achieve a change there. In contrast to what was postulated by Freud, this does not require the uncovering of repressed memories; this can occur only via the autobiographical memory. The relationship patterns stored in procedural memory cannot be remembered. Their alteration, therefore, must occur via the analysis of transference, in which these early relationship patterns are reflected. The primary emphasis in psychoanalytic treatment today is, therefore, on the

processing of the transference–countertransference constellation. (p. 416; translated from German original)

It is indeed correct that a neuroscientific perspective leads to a different understanding of the relationship patterns between therapist and patient than what was postulated by Freud, but the neuroscientific findings have considerably more far-reaching implications for psychotherapy than simply their ability to translate established psychotherapeutic concepts such as transference–countertransference into a new language. It is, of course, true that humans transfer their previous experiences onto later experiences. This is the core principle of neural and mental functioning. It is impossible not to transfer. If not on the basis of previous experiences, which are reflected in different memory systems, then how should an individual establish any kind of relationship with his or her surroundings? The fact that this also holds true for experiences that have been encountered early in life goes without saying because there are no exceptions from this core principle of mental functioning. If one has encountered painful experiences in one's early relationships, then this will be reflected in memory and will affect how one feels and behaves in later relationships. From a neuroscientific perspective, this idea is so self-evident that the concept of transference really does not have any specific information value, except in terms of the therapeutic conclusions one may derive from this. However, it is just these—the psychoanalytic conclusions in this domain—that cannot at all be supported neuroscientifically. The therapeutic conclusions that Rohde-Dachser and other psychoanalysts draw from the disconnection between autobiographical and procedural memory actually conflict directly with the neuroscientific findings on how neural activation patterns can effectively changed.

Violations of the attachment and the control needs in relationships with our early attachment figures leave deep traces in the neural system. The most important ones are excessive stress reactions in response to even minor emotional challenges, dysfunctional regulation of autonomic arousal, and an easily triggered avoidance system. This has subsequent negative consequences in all domains of life and, throughout life, reduces one's room for positive experiences. In psychotherapy, it is important to actively inhibit these early acquired problematic neural circuits with other, more need-satisfying neural patterns that must be facilitated in therapy and that will allow the person to structure his or her interpersonal relationships more effectively. In order to achieve this, situations must be actively created in a clearly need-satisfying context, as described in chapter 5, in which the new neural patterns are intensively, lastingly, and frequently activated and facilitated; not just in the therapy situation but also in real-life situations experienced by the patient. It is especially the neuroscientific research that clarifies that the psychoanalytic "analysis of transference" of the therapist–patient relationship would not be an appropriate method for this purpose. With regard to the need-satisfying context, the active inhibition of problematic neural activation patterns, and the facilitation of new patterns, such transference analyses are incompatible with the basic requirements necessary to achieve effective changes in problematic memory contents and their effects on current experience and behavior.

It is possible, then, to ignore the implications of neuroscientific research for effective therapeutic conduct and, instead, to pick out only those parts that seem to fit with premises that were originally justified very differently. The example mentioned here is simply one of many.

In order to clearly differentiate my project from such attempts to use neuroscientific research for the post hoc legitimization of older interventions, I used the term *neuropsychotherapy* as a fresh, programmatic expression. My point is not to gather neuroscientific evidence in order to support an already existing therapeutic approach; instead, I am interested in the question of which kind of psychotherapy emerges if one views the problems associated with therapy from a neuroscientific perspective. The result—neuropsychotherapy—is of course influenced not only by neuroscientific research but also by more than 30 years of my personal experience as a psychotherapist and psychotherapy researcher. If someone else, without this specific background, were to conceptualize psychotherapy from a neuroscientific perspective, he or she would surely arrive at a different set of conclusions. However, no one else has yet made such an attempt. It would also be very difficult for a neuroscientist to create a feasible system of psychotherapy, unless the person were intimately familiar with the specific therapy-related problems, practical parameters, and therapy research findings. My selection and interpretation of neuroscientific findings has also been influenced, of course, by ideas and concepts that were already familiar to me before I began my intensive study of the neuroscientific literature. If one compares my statements in chapters 4 and 5 with my understanding of psychotherapy more than 10 years ago (Grawe, 1995; Grawe, Donati, & Bernauer, 1994), it will be evident that these neuroscientific discoveries have indeed markedly influenced my ideas on what constitutes maximally effective psychotherapy.

One view that remains consistent and, if anything, has become even stronger, is my conviction that the concepts of the traditional therapy schools can no longer be regarded as adequate foundations of psychotherapy. I do not miss anything since I have stopped thinking in these old-school terms. The psychological sciences as well as the neurosciences have by now established a solid foundation that enables us to relinquish these outdated concepts. The results uncovered in psychotherapy research can also be interpreted more effectively on this new basis, compared to the traditional bases of distinct therapy school concepts. This also casts massive doubt on the utility of these older concepts.

The consistency-theoretical view of mental disorders and their treatments, as delineated in chapters 4 and 5, is open to further developments and revisions that result from the discovery of new research findings. For example, if research were to show that my focus on specifically those four basic needs is not optimal, that other basic need models can be more thoroughly substantiated, then such changes could be integrated with the theory without having to completely revise the conclusions with regard to psychotherapy. Other research findings that suggest the utility of additional therapy guidelines might also be added to the system, or research suggesting slight revisions, more specific elaborations, or reformulations of individual guidelines might become available. All of these would be welcome improvements within this open system because they would potentially further increase the effectiveness of psychotherapy.

The system is particularly open and receptive for additional findings and methods originating from the neurosciences. I expect that many innovative impulses for psychotherapy will come from this direction in the future. My very personal opinion is, indeed, that the most important impulses for psychotherapy over the next years and decades will come from the domain of the neurosciences. At present, psychotherapists are generally poorly prepared for this growing influence of neuroscientific findings and methods. Chapters 2 and 3 in this book aimed to help practicing and new psychotherapists access the world of the neurosciences. The specific findings across the various domains within the neurosciences will grow and multiply in coming years, but it is unlikely that such cumulative progress would mean that all previous findings become invalid. It will be possible to integrate the new findings within the framework that I described in these chapters, and, at times, the frame itself will have to be readjusted. The study of these chapters should enable readers, at least, to evaluate the significance of novel neuroscientific findings.

What will be completely new for psychotherapists in the near future is the use of neuroscientific methods in the planning, conduct, and evaluation of psychotherapy. I expect that such methods will soon be introduced in this context. Several therapy studies are already under way in which structural and functional brain changes are considered as outcome criteria for the pre–post evaluation of effectiveness. Studies like the landmark study by Furmark et al. (2002), described in chapter 3, are examples of such novel forms of psychotherapy outcome research. Neuroscientific research on the characteristic brain activity patterns in individual disorders is in full swing. I anticipate that in about 10 years, the knowledge of these disorder-specific neural characteristics will be expected to be part of psychotherapists' standard knowledge base.

It also seems likely that this neuroscientific knowledge about individual disorders will include, for some disorders, criteria that can be used for differential treatment planning. It will be routine practice in such disorders, then, to examine brain activity patterns in response to certain test situations, in order to decide whether the patient should be treated with Intervention A or Intervention B. It will also be more common to examine at the end of therapy whether the changes targeted in therapy are reflected in the expected changes in brain activity. Given that until recently, many psychotherapists regarded the use of any type of assessment as a personal affront, these new methods will not be greeted with open arms in all quarters. It seems clear to me that these innovations will arrive on the scene, however, and they will do so in the very near future, such that most currently practicing therapists will witness these changes. It is clearly timely, then, to start preparing for these changes. Training institutes that don't want to miss this train will have to start preparing their therapist trainees for these developments.

The natural scientists have a very similar view of these issues. In a discussion forum on the topic of "Biological Psychology 2010—visions of the future of the field of psychology," published in *Psychologische Rundschau*, Güntürkün writes from the perspective of the year 2010:

> Now, in the year 2010, neuroscience is considered a natural part of any clinical–neurological intervention. If human behavior is generated on the basis of brain processes, then behavior and experience can only be optimally modified when the research on these brain processes is integrated into our interventions. This simple fact has taken a long time to be recognized, and many colleagues are still regarding it with skepticism. The synthesis of classical behavior therapy and biological psychology is now more evident than ever before. (Güntürkün, 2003, p. 122)

Sooner or later, neuroscientific methods will also be integrated directly in psychotherapeutic interventions. This will probably initially occur in areas that currently are beyond the reach of psychotherapeutic intervention. At the World Congress of Neurology in 2003 in Hamburg, so-called brain pacemakers were introduced as a great innovation that can be used primarily in the treatment of movement disorders such as Parkinson's disease. These instruments administer impulses to certain brain regions that no longer generate these impulses naturally. In the case of Parkinson's disease, the loss of dopaminergic neurons plays a causal role. As we saw in chapter 4, these neurons also play an important role in motivated behavior, in addictive disorders, and so on. If it is not possible, then, to activate or inhibit neural patterns implicated in certain problem behaviors via the creation of specific life experiences—that is, via purely psychological means—would it not make sense to stimulate these neural groups directly via electrodes, magnetic fields, or other methods?

We do not hesitate today to use chemical interventions to influence the ease with which certain neurotransmitters are activated or inhibited. The main problem users associate with the somewhat unsatisfactory results of this approach is a technical, not an ethical problem. The influence of the chemical agents is not sufficiently specific. In each case, processes that are not the target are also being altered, because the neurotransmitters simultaneously serve different functions in different brain areas. Thus, it is not surprising that most psychopharmacological medications still have unwanted side effects, which sometimes can be very severe. One can anticipate already today that, and how, this problem can be fixed in the near future. For example, one could implant very small magnets in specific, precisely identified brain areas. Their magnetic power would suffice to increase the rate by which certain molecules that circulate in the blood and that are equipped with specific markers will dock at exactly the targeted brain region. In this way, neurotransmitter metabolism could be altered at exactly the locations where the change is desired, so that very specific effects can be achieved without unwanted side effects. The more precisely one knows which brain regions are implicated in the specific mental processes that are to be altered, the more precisely one could alter the activation and facilitation of certain neural patterns in these brain regions. For example, a specific brain region could be sensitized for particular learning processes. One could, for instance, create higher dopamine concentrations in target areas in which a new neural activation pattern is to be facilitated. Normally, dopamine—which can reinforce the facilitation of lasting synaptic transmission—is released in response to the activation of important motivational goals. If this is not possible for some reason, the dopamine release could be achieved via this artificial method. If certain activation patterns are repeatedly activated simultaneously, they will be more thoroughly facilitated and will subsequently be more easily activated, compared to the state prior to this artificial reinforcement of the facilitation.

It would also be technically possible to activate a particular brain area via implanted electrodes (pacemakers) or via magnetic stimulation, or to shift the brain area into a state in which it will be more easily activated by other stimuli. Such electric stimulation by itself does not create new neural activation patterns, but if the artificial stimulation of the brain area is combined with the repeated creation of specific experiences, the new neural activation potentials or inhibitions could be effectively facilitated via this process. It also does not seem unrealistically utopian to imagine that therapists could use improved brain-imaging methods to receive continuous feedback while conducting their interventions, in order to examine online changes in the activity of specific brain areas in the patient. They could check, thereby, which immediate effects their interventions have on the brain activity patterns, or they could assess online whether the brain is currently in a state in which it would be particularly receptive for specific interventions.

Such ideas about possible future forms of neuropsychotherapy might at first seem shocking. However, if we think of the human suffering and societal costs that are caused by some psychopaths who commit repeated "cold-blooded" crimes as soon as they are released again, then the question about novel intervention approaches seems obvious. We know today that the brains of such individuals do not function as they normally should. In situations in which others experience anxiety or empathy, which is reflected in the activation of specific brain areas, the same brain areas among the psychopaths remain completely silent. They do not feel what other people experience in such situations. They do not even know such feelings.

Societies have evolved different solutions for such problems. In some countries, such individuals are being killed sooner or later; that is, they receive the death sentence. Some of them are being castrated. Some will voluntarily undergo brain surgeries in which important parts of their brains are destroyed. Most of them end up being incarcerated with life-long sentences in prisons or other institutions, even though this often does not result in permanent protection of society from such people. If it were possible to use artificial stimulation of specific brain regions, accompanied by the creation of certain experiences, in order to stimulate experiences such as empathy or fear in appropriate situations, then such interventions could contribute to a relatively humane solution to this grave societal and human problem. Because it is possible to objectively evaluate the state of the brain with imaging methods, one could also be certain whether lasting changes have indeed been attained. This would be important for the question of whether such people should be permanently incarcerated or not.

The neuroscientific findings that show that these individuals have a different brain, which does not enable them to experience certain human sentiments, also raise questions with regard to the responsibility such people have for their own actions. This also changes one's perspective on how such interventions should be evaluated, once they have been developed. One could also argue that it would be ethically questionable to deprive such persons of the opportunity to benefit from interventions that would change their entire lives for the better. The question of whether one should temporarily implant small electrodes in specific brain areas in order to support specific psychological interventions will then seem different from the initial thought of manipulating human brains via electric or magnetic stimulation.

What I have described here remains, for now, utopian. Given our common experience that humans do not shy away from using whatever is technically possible—the hotly debated topic of genetic manipulation of humans comes to mind in this context—we should prepare ourselves to expect that such questions will one day arise for us. It seems likely to me that the technical capabilities will be developed soon to allow for the combination of psychological, mechanical, and/or pharmacological intervention approaches. This will raise questions for psychotherapists that cannot be answered easily. What if it could be shown that severely depressed or schizophrenic patients, who have been treatment-resistant so far, could be effectively helped by such combinations of psychotherapy and artificial stimulation of specific brain areas? Could we deprive them of such help? I feel certain that, in the long term, we could not, and should not, keep such treatments away from patients. Such situations are already similar to what we encounter in normal psychotherapy clinics today. Psychotherapists and psychiatrists are very commonly confronted with such cases in their practice.

Once we have grown accustomed to the thought that we, as psychotherapists, alter the brain to the extent that we conduct effective treatments, then the question of how the brain could be altered even more effectively by combining psychological and neuroscientific methods does not seem unreasonable. At that point, one could use the term *neuropsychotherapy* in an even more concrete sense. I feel certain that neuropsychotherapy will continue to develop in this direction.

I can also see a danger, however. That is, the danger exists that such neuropsychotherapists would direct all their attention toward the problematic brain areas while demoting the human being, with his or her whole life context, developmental history, wishes and fears, to a background role. I have tried to carefully demonstrate in this book that a focus on neural activity does not mean that the person, with all his or her idiosyncrasies and individuality, would have to be neglected. The disorders of mental life cannot be disconnected from the whole human being and his or her life context. If one thinks in the terms of consistency theory, the question always arises as to whether the person's basic needs are sufficiently satisfied. The point is never to simply identify and remove specific disorders. This should also never become the sole point in the future, when new neuroscientific methods enrich today's version of neuropsychotherapy. I regard the concept of neuropsychotherapy that was developed in this book as a good foundation to ensure that such technical–reductionistic developments will be averted.

In my understanding, neuropsychotherapy aims to change the brain, but it does not directly target primarily the brain but focuses on the life experiences encountered by the person. The brain specializes in the processing of life experiences. Life experiences are meaningful with regard to the needs that are embedded within the brain structures of each human being. Neuropsychotherapy strives to shift the brain into a state that enables these basic needs to be fully satisfied. The best method for improving the health of the brain, then, is to ensure basic need satisfaction.

REFERENCES

Abel, T., & Kandel, E. (1998 May 13, 2006). Positive and negative regulatory mechanisms that mediate long-term memory storage. *Brain Research Reviews, 26,* 360–378.

Abercrombie, H. C., Schaefer, S. M., Larson, C. L., Oakes, T. R., Lindgren, K. A., Holden, J. E., et al. (1998). Metabolic rate in the right amygdala predicts negative affect in depressed patients. *Neuroreport, 9,* 3301–3307.

Adam, K. S. (1994). Suicidal Behavior and attachment. In W. H. Sperling & M. B. Berman (Eds.), *Attachment in adults—Clinical and developmental perspectives.* (pp. 275–298). New York: Guilford.

Adler, A. (1920). *Praxis und Theorie der Individualtherapie* [Praxis and theory of individual therapy]. Munich, Germany: Bergmann.

Adler, A. (1927). *Studie über Minderwertigkeit von Organen* [Study of organ inferiority and its psychical compensation]. Munich, Germany: Bergmann.

Adolphs, R., Tranel, D., & Damasio, A. R. (1998). The human amygdala in social judgment. *Nature, 393,* 470–474.

Ainsworth, M. D., Blehar, M. C., Waters, E., & Wall, S. (1978). *Patterns of attachment: A psychological study of the strange situation.* Hillsdale, NJ: Lawrence Erlbaum Associates, Inc.

Alexander, G. E., Mentis, M. J., van Horn, J. D., Grady, C. L., Berman, K. F., Furey, M. L., et al. (1999). Individual differences in PET activation of object perception and attention systems predict face matching accuracy. *Neuroreport, 10,* 1965–1971.

Alloy, L. B., Kelly, K. A., Mineka, S., & Clements, C. M. (1990). Comorbidity of anxiety and depressive disorders: A helplessness-hopelessness perspective. In J. D. Maser & C. R. Cloninger (Eds.), *Comorbidity of mood and anxiety disorders* (pp. 499–543). Washington, DC: American Psychiatric Press.

Alsaker, F. (1997, April). *Isolation as a powerful victimization technique.* Paper presented at the Biennial Meeting of the Society for Research in Child Development, Washington, DC.

Alsaker, F., & Flammer, A. (1996, March). *Social relationships and depression in adolescents: Social causation and social selection.* Paper presented at the Biannual Meeting of the Society for Research in Adolescence, Boston.

Alsaker, F., & Olweus, D. (2003). Stability and change in global self-esteem and self-related affect. In T. M. Brinthaupt & R. P. Lipka (Eds.), *Understanding early adolescent self and identity* (pp. 193–223). New York: State University of New York Press.

Amaral, D. G., Price, J. L., Pitkänen, A., & Carmichael, S. (1992). Anatomical organization of the primate amygdaloid complex. In J. P. Aggleton (Ed.), *The amygdala: Neurobiological aspects of emotion, memory, and mental dysfunction.* New York: Wiley-Liss.

Amini, F., Lewis, T., Lannon, R., Louie, A., Baumacher, G., McGuiness, T., et al. (1996). Affect, attachment, memory: Contributions toward psychobiologic integration. *Psychiatry, 59,* 213–239.

Anderson, P., Beach, S., & Kaslow, N. (1999). Marital discord and depression: The potential of attachment theory to guide integrative clinical intervention. In T. Joiner & J. Coyne (Eds.), *The interactional nature of depression* (pp. 271–297). Washington: American Psychological Association.

Andreasen, N. (2001). *Brave new brain : Conquering mental illness in the era of the genome.* New York: Oxford University Press.

Andrews, M. W., & Rosenblum, L. A. (1991). Security of attachment in infants raised variable- or low-demand environments. *Child Development, 62,* 686–693.

Angst, J., Vollrath, M., Merikangas, K. R., & Ernst, C. (1990). Comorbidity of anxiety and depression in the Zurich Cohort Study of Young Adults. In J. D. Maser & C. R. Cloninger (Eds.), *Comorbidity of mood and anxiety disorders* (pp. 123–137). Washington, DC: American Psychiatric Publishing.

American Psychiatric Association. (2000). *Practice guideline for the treatment of major depressive disorder* (2nd ed.). Washington, DC: Author.

Arrindell, W. A., Pickersgill, M. J., Merckelbach, H., Ardon, A. M., Cornet, F. C. (1991). Phobic dimensions: III. Factor analytic approaches to the study of common fears; an updated review of findings obtained with adult subjects. *Advances in Behaviour Research and Therapy, 13,* 73–130.

Asaad, W. F., Rainer, G., & Miller, E. K. (1998). Neural activity in the primate frontal cortex during associative learning. *Neuron, 21,* 1399–1407.

Asaad, W. F., Rainer, G., & Miller, E. K. (2000). Task-specific neural activity in the primate prefrontal cortex. *Journal of Neurophysiology, 84,* 451–459.

Bachmann, T. (2000). *Microgenetic approach to the conscious mind.* Amsterdam: Benjamins.

Baddeley, A. (1986). *Working memory.* Oxford, England: Clarendon.

Baer, R. (2003). Mindfulness training as a clinical intervention: A conceptual and empirical review. *Clinical Psychology: Science and Practice, 10,* 125–143.

Bakan, D. (1966). *The duality of human existence.* Chicago: Rand McNally.

Baker, L. A., Cesa, I. L., Gatz, M., & Grodsky, A. (1992). Genetic and environmental influences on positive and negative affect: Support for a two-factor theory. *Psychology and Aging, 7,* 158–163.

Baltensperger, C., & Grawe, K. (2000). Psychotherapie unter gesundheitsökonomischem Aspekt [Psychotherapy from a health-economics point of view]. *Zeitschrift für Klinische Psychologie und Psychotherapie, 30,* 10–21.

Bandura, A. (1977). *Social learning theory.* Englewood Cliffs, NJ: Prentice Hall.

Barlow, D. H. (1988). *Anxiety and its disorders: The nature and treatment of anxiety and panic.* New York: Guilford.

Bartels, A., & Zeki, S. (2000). The neural basis of romantic love. *Neuroreport, 11,* 3829–3834.

Barth, J. A. (1987). Max Wertheimer in Frankfurt—über beginn und aufbaukrise der Gestaltpsychologie. II. Strukturgesetze der Bewegungs- und Raumwahrnehmung (1911–1914) [Max Wertheimer in Frankfurt—The beginning and initial crisis of Gestalt psychology II. Structural principles of motion and spatial perception]. *Zeitschrift für Psychologie, 195,* 403–431.

Baumeister, R. F. (1993). *Self-esteem: The puzzle of low self-regard.* New York: Plenum.

Baxter, L. R., Ackermann, R. F., Swerdlow, N. R., Brody, A., Saxena, S., & Schwartz, J. M. (2000). Specific brain system mediation of obsessive-compulsive disorder responsive to either medication or behavior therapy. In W. K. Goodmann & M. V. Rudorfer (Eds.), *Obsessive-compulsive disorder: Contemporary issues in treatment* (pp. 573–609). Mahwah, NJ: Lawrence Erlbaum Associates, Inc.

Baxter, L. R., Phelps, M. E., Mazziotta, J. C., & Guze, B. H. (1987). Local cerebral glucose metabolic rates in obsessive-compulsive disorder: A comparison with rates in unipolar depression and normal controls. *Archives of General Psychiatry, 44,* 211–218.

Baxter, L. R., Saxena, S., Brody, A. L., Ackermann, R. F., Colgan, M., Schwartz, J. M., et al. (1996). Brain mediation of obsessive-compulsive disorder symptoms: Evidence from functional brain imaging studies in the human and non-human primate. *Seminars in Clinical Neuropsychiatry, 1,* 32–47.

Baxter, L. R., Schwartz, J. M., Bergmann, K. S., Szuba, M. P., Guze, B. H., Mazziotta, J. C., et al. (1992). Caudate glucose metabolic rate changes with both drug and behavior therapy for obsessive-compulsive disorder. *Archives of General Psychiatry, 49,* 681–689.

Beauregard, M., Leroux, J. M., Bergman, S., Arzoumanian, Y., Beaudoin, G., Bourgouin, P., et al. (1998). The functional neuroanatomy of major depression: an fMRI study using an emotional activation paradigm. *Neuroreport, 9,* 3253–3258.

Bech, P. (1999). Pharmacological treatment of depressive disorders: A review. In M. Maj & N. Sartorius (Eds.), *Depressive disorders* (pp. 89–127). New York: Wiley.

Bechara, A., Tranel, D., Damasio, H., Adolphs, R., Rockland, C., & Damasio, A. R. (1995). Double dissociation of conditioning and declarative knowledge relative to the amygdala and hippocampus in humans. *Science, 269,* 1115–1118.

Beck, A. T. (1967). *Depression: Clinical, experimental, and theoretical aspects.* New York: Harper & Row.

Beck, A. T., Rush, J. A., Shaw, B. F., & Emery, G. (1979). *Cognitive therapy of depression.* New York: Guilford.

Beckwith, L., Cohen, S. E., & Hamilton, C. E. (1999). Maternal sensitivity during infancy and subsequent life events relate to attachment representations at early adulthood. *Developmental Psychology, 35,* 693–700.

Behbehani, M. M. (1995). Functional characteristics of the midbrain periaquaeductal gray. *Progress in Neurobiology, 46,* 575–605.

Beitman, B. D., Viamontes, G. I., Soth, A. M., & Nittler, J. R. (2006). Toward a neural circuitry of engagement, self-awareness activation and pattern search. *Psychiatric Annals.*

Bench, C. J., Frackowiak, R. S., & Dolan, R. J. (1995). Changes in regional cerebral blood flow on recovery from depression. *Psychological Medicine, 25,* 247–261.

Benjamin, L. S. (1974). Structural analysis of social behavior. *Psychological Review, 81,* 392–425.

Benjamin, L. S. (1993). *Interpersonal diagnosis and treatment of personality disorders.* New York: Guilford.

Bennett, A. J., Lesch, K. P., Heils, A.,Long, J., Lorenz, J., Shoaf, S. E., et al. (1998). Serotonin transporter gene variation, strain, and early rearing environment affect CSF 5-HIAA concentrations in rhesus monkeys (Macaca mulatta). *American Journal of Primatology, 45,* 168–169.

Benoit, D., & Parker, K. C. (1994). Stability and transmission of attachment across three generations. *Child Development, 65,* 1444–1456.

Bentham, J. (1948). *Principles of morals and legislation.* New York: Hafner. (Original work published 1789)

Berking, M., Grosse Holtforth, M., & Jacobi, C. (2003a). Reduction of incongruence in inpatient psychotherapy. *Clinical Psychology & Psychotherapy, 10,* 86–92.

Berking, M., Grosse Holtforth, M., & Jacobi, C. (2003b). Veränderung klinisch relevanter Ziele und Therapieerfolg: Eine Studie an Patienten während einer stationären Verhaltenstherapie [Changes in clinically relevant goals and therapy outcome: A study with inpatients undergoing cognitive behavioral therapy]. *Psychotherapie, Psychosomatik, Medizinische Psychologie, 53,* 171–177.

Berking, M., Grosse Holtforth, M., Jacobi, C., & Kröner-Herwig, B. (in press). Sage mir deine diagnose und ich sage dir, was du willst: Inwieweit sind therapieziele störungstypisch? [Tell me your diagnosis, and I will tell you what you want: To what extent are therapeutic goals disorder specific?] *Zeitschrift für Klinische Psychologie, Psychiatrie und Psychotherapie.*

Berking, M., Jacobi, C., & Masuhr, O. (2001). Therapieziele in der psychosomatischen Rehabilitation [Treatment goals in psychosomatic rehabilitation]. *Verhaltenstherapie und Psychosoziale Praxis, 33,* 259–272.

Berridge, K. C. (1999). Pleasure, pain, desire, and dread: Hidden core processes of emotion. In D. Kahneman, E. Diener, & N. Schwarz (Eds.), *Well-being: The foundation of hedonic psychology* (pp. 525–557). New York: Russell Sage Foundation.

Berridge, K. C., & Robinson, T. E. (1998). What is the role of dopamine in reward: Hedonic impact, reward learning, or incentive salience? *Brain Research Review, 28,* 309–369.

Berridge, K. C., & Valenstein, E. S. (1991). What psychological process mediates feeding evoked by electrical stimulation of the lateral hypothalamus. *Behavioral Neuroscience, 105,* 3–14.

Betschart, T. (2002). *Inkonsistenzanalyse: Eine Methode zur Messung von intrapsychischen Konflikten und deren funktionale Zusammenhänge im psychologischen Therapieprozess* [Analysis of inconsistency—Assessment of intrapsychic conflicts and functional relationships]. Unpublished master's thesis, Institut für Psychologie, University of Bern, Switzerland.

Beutel, M. E. (2002). Neurowissenschaften und Psychotherapie. Neuere Entwicklungen, Methoden und Ergebnisse [Neurosciences and psychotherapy: Recent developments, methods, and results]. *Psychotherapeut, 47,* 1–10.

Beutel, M. E., Stern, E., Silbersweig, D. A. (2003). The emerging dialogue between psychoanalysis and neuroscience: Neuroimaging perspectives. *Journal of the American Psychoanalytic Association, 51,* 773–801.

Beutler, L. E., & Malik, M. L. (Eds.). (2002). *Rethinking the DSM. A psychological perspective.* Washington, DC: American Psychological Association.

Bichot, N. P., Schall, J. D., & Thompson, K. G. (1996). Visual feature selectivity in frontal eye fields induced by experience in mature macaques. *Nature, 381,* 697–699.

Blanck, P. D., Rosenthal, R., & Vannicelli, M. (1986). Talking to and about patients: The therapist's tone of voice. In P. D. Blanck, R. Buck, & R. Rosenthal (Eds.), *Nonverbal communication in the clinical context.* University Park: The Pennsylvania State University Press.

Blatt, S. J. (1990). Interpersonal relatedness and self-definition: Two personality configurations and their implication for psychopathology and psychotherapy. In J. L. Singer (Ed.), *Repression and dissociation* (pp. 299–336). Chicago: The University of Chicago Press.

Blood, A. J., & Zatorre, R. J. (2001). Intensely pleasurable responses to music correlate with activity in brain regions implicated in reward and emotion. *Proceedings of the National Academy of Sciences of the United States of America, 98,* 11818–11823.

Blood, A. J., Zatorre, R. J., Bermudes, P., & Evans, A. C. (1999). Emotional responses to pleasant and unpleasant music correlate with activity in paralimbic brain regions. *Nature Neuroscience, 2,* 382–387.

Bock, J., & Braun, K. (2002). Frühkindliche Emotionen steuern die funktionelle Reifung des Gehirns: Tierexperimentelle Befunde und ihre mögliche Relevanz für die Psychotherapie [Influence of infant emotions on functional maturation of brain structures. Animal models and their relevance for psychotherapy]. *Psychotherapie, 7,* 190–194.

Bohus, M. (2002). *Borderline-Störung* [Dialectic-behavioral therapy for borderline personality disorder]. Göttingen, Germany: Hogrefe.

Bonanno, G. A., & Singer, J. L. (1990). Repressive personality style: Theoretical and methodological implications for health and pathology. In J. L. Singer (Ed.), *Repression and dissociation* (pp. 435–470). Chicago: The University of Chicago Press.

Bontempi, B., Laurent-Demir, C., Destrade, C., & Jaffard, R. (1999). Time-dependent reorganization of brain circuitry underlying long term memory storage. *Nature, 400,* 671–675.

Botvinick, M. M., Braver, T. S., Barch, D. M., Carter, C. S., & Cohen, J. D. (2001). Conflict monitoring and cognitive control. *Psychological Review, 108,* 624–652.

Bowlby, J. (1969). *Attachment and loss. Vol 1: Attachment.* New York: Basic Books.

Bowlby, J. (1973). *Attachment and loss. Vol 2: Separation. Anxiety and anger.* New York: Basic Books.

Bowlby, J. (1980). *Loss, sadness and depression.* London: Hogarth.

Bowlby, J. (1988). *A secure base: Parent-child attachment and healthy human development.* New York: Basic.

Braun, A. K., Bock, J., Gruss, M., Helmeke, C., Ovtscharoff, W., Schnabel, R., et al. (2002). Frühe emotionale Erfahrungen und ihre Relevanz für die Entstehung und Therapie psychischer Erkrankungen [Early emotional experiences and their relevance for the development and therapy of mental disorders]. In B. Strauss, A. Buchheim, & H. Kächele (Eds.), *Klinische Bindungsforschung* (pp. 121–129). Stuttgart, Germany: Schattauer.

Breiter, H. C., Etcoff, N. L., Whalen, P. J., Kennedy, W. A., Rauch, S. L., Buchner, R. L., et al. (1996). Response and habituation of the human amygdala during visual processing of facial expression. *Neuron, 17,* 875–887.

Breiter, H. C., Gollup, R. L., Weisskoff, R. M., Kennedy, D. N., Madris, N., & Berke, J. D. (1997). Acute effects of cocaine on human brain activity and emotion. *Neuron, 19,* 591–611.

Bremner, J. D. (1999a). Alterations in brain structure and function associated with posttraumatic stress disorder. *Seminars in Clinical Neuropsychiatry, 4,* 249–255.

Bremner, J. D. (1999b). Does stress damage the brain? *Biological Psychiatry, 45,* 797–805.

Bremner, J. D. (2001). Hypotheses and controversies related to effects of stress on the hippocampus: An argument for stress-induced damage to the hippocampus in patients with posttraumatic stress disorder. *Hippocampus, 11,* 75–81.

Bremner, J. D., Randall, S., Scott, T. M., Bronen, R. A., Seibyl, J. P., Southwick, S. M., et al. (1995). MRI-based measurement of hippocampal volume in patients with combat-related posttraumatic stress disorder. *American Journal of Psychiatry, 152,* 973–981.

Bremner, J. D., Randall, S., Vermetten, E., Staib, L. H., Bronen, R. A., Mazure, C., et al. (1997). Magnetic resonance imaging-based measurement of hippocampal volume in posttraumatic stress disorder related to childhood physical and sexual abuse—A preliminary report. *Biological Psychiatry, 41,* 23–32.

Bremner, J. D., Southwick, S. M., Johnson, D. R., Yehuda, R., & Charney, D. S. (1993). Childhood physical abuse and combat-related posttraumatic stress disorder in Vietnam veterans. *American Journal of Psychiatry, 150,* 234–239.

Brisch, K.-H. (2002). Psychotherapeutische Intervention für Eltern mit sehr kleinen frühgeborenen: Das Ulmer modell [Psychotherapeutic interventions for parents of prematurely born children: The Ulm model]. In B. Strauss, A. Buchheim, & H. Kächele (Eds.), *Klinische Bindungsforschung* (pp. 191–195). Stuttgart, Germany: Schattauer.

Brodman, K. (1909). *Vergleichende Lokalisationslehre der Grosshirnrinde*. Leipzig, Germany: Barth.

Brody, A. L., Saxena, S., & Stoessel, P. (2001). Regional brain metabolic changes in patients with major depression treated with either proxetine or interpersonal therapy. *Archives of General Psychiatry, 58,* 631–640.

Brown, J. D. (1993). Motivational conflict and the self: The double bind of low self-esteem. In R. F. Baumeister (Ed.), *Self-esteem: The puzzle of low self-regard* (pp. 117–130). New York: Plenum.

Brown, J. D., Collins, R., & Schmidt, G. W. (1988). Self-esteem and direct versus indirect forms of self-enhancement. *Journal of Personality and Social Psychology, 55,* 445–453.

Brown, J. D., Novick, N. J., Lord, K. A., & Richards, J. M. (1992). When Gulliver travels: Social context, psychological closeness, and self-appraisal. *Journal of Personality and Social Psychology, 60,* 717–727.

Brown, T. A., Chorpita, B. F., & Barlow, D. H. (1998). Structural relationships among dimensions of the DSM-IV anxiety and mood disorders and dimensions of negative affect, positive affect, and autonomic arousal. *Journal of Abnormal Psychology, 107,* 179–192.

Bruder, G. E., Stewart, J. W., Mercier, M. A., Agosti, V., Leite, P., Donovan, S., et al. (1997). Outcome of cognitive-behavioral therapy for depression: Relation to hemispheric dominance for verbal processing. *Journal of Abnormal Psychology, 106,* 138–144.

Bruder, G. E., Stewart, J. W., Tenke, C. E., McGrath, P. J., Leite, P., Bhattacharya, N., et al. (2001). Electroencephalographic and perceptual asymmetry differences between responders and nonresponders to an SSRI antidepressant. *Biological Psychiatry, 49,* 416–425.

Brunstein, J. C. (1993). Personal goals and subjective well-being: A longitudinal study. *Journal of Personality and Social Psychology, 65,* 1061–1070.

Brunstein, J. C., Schultheiss, O. C., & Grässmann, R. (1998). Personal goals and emotional well-being: The moderating role of motive dispositions. *Journal of Personality and Social Psychology, 75,* 494–508.

Brunstein, J. C., Schultheiss, O. C., & Maier, G. W. (1999). The pursuit of personal goals: A motivational approach to well-being and life adjustment. In J. Brandtstaedter & R. M. Lerner (Eds.), *Action and self development:Theory and research through the life span.* (pp. 169–196). Thousand Oaks, CA: Sage.

Buchheim, A., & Strauss, B. (2002). Interviewmethoden der klinischen Bindungsforschung [Psychotherapeutic interventions for parents of prematurely born children: The Ulm model]. In B. Strauss, A. Buchheim, & H. Kächele (Eds.), *Klinische Bindungsforschung* (pp. 27–53). Stuttgart, Germany: Schattauer.

Buchsbaum, M. S., Wu, J., Siegel, B. V., Hackett, E., Trenary, M., Abel, L., et al. (1997). Effect of sertraline on regional metabolic rates in patients with affective disorders. *Biological Psychiatry, 41,* 15–22.

Cacioppo, J. T., Crites, S. L., & Gardner, W. L. (1996). Attitudes to the right: Evaluative processing is associated with lateralized late positive event-related brain potentials. *Personality and Social Psychology Bulletin, 22,* 1205–1219.

Cacioppo, J. T., Gardner, W. L., & Berntson, G. G. (1997). Beyond bipolar conceptualizations and measures: The case of attitudes and evaluative space. *Personality and Social Psychology Review, 1,* 3–25.

Cacioppo, J. T., & Petty, R. E. (1979). Attitudes and cognitive response: An electrophysiological approach. *Journal of Personality and Social Psychology, 37,* 2181–2199.

Cacioppo, J. T., Priester, J. R., & Berntson, G. G. (1993). Rudimentary determinants of attitudes: Arm flexion and extension have differential effects on attitudes. *Journal of Personality and Social Psychology, 65,* 5–17.

Cahill, L., Haier, R. J., Fallon, J., Alkire, M. T., Tang, C., Keator, D., et al. (1996). Amygdala activity at encoding correlated with long-term free recall of emotional information. *Proceedings of the National Academy of Sciences of the United States of America, 93,* 8016–8021.

Campbell, K. W., & Sedikides, C. (1999). Self-threat magnifies the self-serving bias: A meta-analytic integration. *Review of General Psychology, 3,* 23–43.

Cappas, N. M.; Andres-Hyman, R., & Davidson, L. (2005). What psychotherapists can begin to learn from neuroscience: Seven principles of a brain based psychotherapy. *Psychotherapy, 42,* 347–383.

Carlson, E. A., & Sroufe, L. A. (1993). Contributions of attachment theory to developmental psychopathology. In D. Cicchetti & D. Cohen (Eds.), *Developmental psychopathology: Vol. 1. Theory and methods* (pp. 581–617). New York: Wiley.

Carlson, V., Cicchetti, D., Barnett, D., & Braunwahl, K. (1989). Disorganized/disoriented attachment relationships in maltreated infants. *Developmental Psychology, 25,* 525–531.

Carter, C. S., Botwinick, M. M., & Cohen, J. D. (1999). The contribution of the anterior cingulate cortex to executive processes in cognition. *Review of Neuroscience, 10,* 49–57.

Carver, C. S. (1996). Some ways in which goals differ and some implications of those differences. In P. M. Gollwitzer & J. A. Bargh (Eds.), *The psychology of action: Linking cognition and motivation to behavior* (pp. 645–672). New York: Guilford.

Carver, C. S. (1997). Adult attachment and personality: Conveying evidence and a new measure. Personality and Social Psychology Bulletin, 23, 865–883.

Carver, C. S., Lawrence, J. W., & Scheier, M. F. (1996). A control-process perspective on the origins of affect. In L. L. Martin & A. Tesser (Eds.), *Striving and feeling: Interactions among goals, affect and self-regulation.* (pp. 11–52). Hillsdale, NJ: Lawrence Erlbaum Associates, Inc.

Carver, C. S., & Scheier, M. F. (1998). *The self-regulation of behavior.* Hillsdale, NJ: Lawrence Erlbaum Associates, Inc.

Carver, C. S., & White, T. L. (1994). Behavioral inhibition, behavioral activation, and affective responses to impending reward and punishment: The BIS/BAS scales. *Journal of Personality and Social Psychology, 67,* 319–333.

Caspar, F. (1989). *Beziehungen und Probleme verstehen: Eine Einführung in die psychotherapeutische Plananalyse* [Understanding relationships and problems. An introduction to the psychotherapeutic analysis of plans]. Bern, Switzerland: Huber.

Caspar, F. (1994). *Plan analysis. Toward optimizing therapy.* Seattle, WA: Hogrefe Huber.

Caspar, F. (2003). Psychotherapy research and neurobiology. Challenge, chance, or enrichment? Psychotherapy Research, 13, 1–23.

Caspar, F., Rothenfluh, T., & Segal, Z. V. (1992). The appeal of connectionism for clinical psychology. *Clinical Psychology Review, 12,* 719–762.

Cassidy, J., & Shaver, P. R. (Eds.). (1999). *Handbook of attachment: Theory, research, and clinical applications.* New York: Guilford.

Chambless, D. L. (1996). In defense of dissemination of empirically supported psychological interventions. *Clinical Psychology: Science and Practice, 3,* 230–235.

Chambless, D. L., & Hollon, S. (1998). Defining empirically supported psychological interventions. *Journal of Consulting and Clinical Psychology, 66,* 7–18.

Champoux, M., Byrne, E., Delizio, R. D., & Suomi, S. J. (1992). Motherless mothers revisited: Rhesus maternal behavior and rearing history. *Primates, 33,* 251–255.

Chen, G., Raikoska, G., Du, F., Sraji-Bozorgzad, N., & Manji, H. K. (2000). Enhancement of hippocampal neurogenesis by lithium. *Journal of Neurochemistry, 75,* 1729–1734.

Chen, M., & Bargh, J. A. (1999). Consequences of automatic evaluation: Immediate behavioral predisposition to approach or avoid the stimulus. *Personality and Social Psychology Bulletin, 25,* 215–224.

Chorpita, B. F., & Barlow, D. H. (1998). The development of anxiety: The role of control in early environment. *Psychological Bulletin, 124,* 3–21.

Churchland, P. S. (1986). Neurophilosophy. London: MIT Press.

Cialdini, R. B., Borden, R. J., Thorne, A., Walker, M. R., Freeman, S., & Sloan, L. R. (1976). Basking in reflected glory: Three (football) field studies. *Journal of Personality and Social Psychology, 34,* 366–375.

Cialdini, R. B., & De Nicolas, M. W. (1989). Self-presentation by association. *Journal of Personality and Social Psychology, 57,* 626–631.

Clark, D. A., & Beck, A. T. (1999). *Scientific foundations of cognitive theory and therapy of depression.* New York: Wiley.

Clark, L. A., & Watson, D. (1991). Tripartite model of anxiety and depression; Psychometric evidence and taxonomic implications. *Journal of Abnormal Psychology, 100,* 316–336.

Coats, E. J., Janoff-Bulman, R., & Alpert, N. (1996). Approach versus avoidance goals: Differences in self-evaluation and well-being. *Personality and Social Psychology Bulletin, 22,* 1057–1067.

Coffey, C. E., Wilkinson, W. E., Weiner, R. D., Parashos, I. A., Djang, W. T., Webb, M. C., et al. (1993). Quantitative cerebral anatomy in depression: A controlled magnetic resonance imaging study. *Archives of General Psychiatry, 50*, 7–16.

Cohen, J. (1969). *Statistical power analysis for the behavioral sciences.* New York: Academic.

Cohen, J. (1970). Approximate power and sample size determination for common one-sample and two-sample hypothesis tests., *Educational and Psychological Measurement 30*, 811–831.

Cohen, J. D., Perlstein, W. M., Braver, T. S., Nystrom, L. E., Noll, D. C., Jonides, J., et al. (1997). Temporal dynamics of brain activation during a working memory task. *Nature, 386*, 604–608.

Cohen, N. J., & Corkin,S. (1981). The amnestic patient H.M.: Learning and retention of cognitive skills. *Abstracts of Social Neuroscience, 7*, 517–518.

Cohen, N. J., & Squire, L. R. (1980). Preserved learning and retention of pattern-analyzing skills in amnesia: Dissociation of knowing how and knowing that. *Science, 210*, 207–209.

Colvin, C. R., & Block, J. (1994). Do positive illusions foster mental health? An examination of the Taylor and Brown formulation. *Psychological Bulletin, 116*, 3–20.

Cooper, J., & Fazio, R. H. (1984). A new look at dissonance theory. In L. Berkowitz (Ed.), *Advances in experimental social psychology* (Vol. 17, pp. 229–266). New York: Academic.

Coplan, J. D., Andrews, M. W., Rosenblum, L. A., Owens, M. J., Friedman, S., Gorman, J. M., et al. (1996). Persistent elevations of cerebrospinal fluid concentrations of corticotrophin releasing factor in adult nonhuman primates exposed to early-life stressors: Implications for the pathophysiology of mood and anxiety disorders. *Proceedings of the National Academy of Sciences of the United States of America, 93*, 1619–1623.

Corkin, S. (1968). Acquisition of motor skill after bilateral medial temporal lobe excision. *Neuropsychologia, 6*, 255–265.

Costa, P. T., & McCrae, R. R. (1980). Influence of extraversion and neuroticism on subjective well-being: Happy and unhappy people. *Journal of Personality and Social Psychology, 38*, 668–678.

Costa, P. T., & McCrae, R. R. (1988). Personality in adulthood: A six year longitudinal study of self-reports and spouse ratings on the NEO Personality Inventory. *Journal of Personality and Social Psychology, 54*, 853–863.

Costa, P. T., & McCrae, R. R. (1992). *Revised NEO Personality Inventory (NEOPI-R) and Five Factor Inventory (NEO-FFI) professional manual.* Odessa, FL.: Psychological Assessment Resources.

Courtney, S. M., Ungerleider, L. G., Keil, K., & Haxby, J. V. (1997). Transient and sustained activity in a distributed neural system for working memory. *Nature, 386*, 608–612.

Crick, F., & Koch, C. (1990). Towards a neurobiological theory of consciousness. Seminars in the *Neurosciences, 2*, 263–275.

Crick, F., & Koch, C. (2003a). A framework for consciousness. *Nature Neuroscience, 6*, 119–126.

Crick, F., & Koch, C. (2003b). What are the neural correlates of consciousness? In L. van Hemmen & T. J. Sejnowski (Eds.), *Problems in systems neuroscience.* New York: Oxford University Press.

Critchley, H. D., Mathias, C. J., & Dolan, R. J. (2001). Neural activity in the human brain relating to uncertainty and arousal during anticipation. *Neuron, 29*, 537–545.

Crites, S. L., & Cacioppo, J. T. (1996). Electrocortical differentiation of evaluative and nonevaluative categorization. **Psychological Science, 7,** 318–321.

Csikszentmihalyi, M. (1990). *Flow: The psychology of optimal experience.* New York: Harper & Row.

Damasio, A. R. (2000). *Descartes' error: Emotion, reason and the human brain.* New York: Harper Collins.

Damasio, A. R. (1999). *The feeling of what happens.* London: Heinemann.

Damsma, G., Wenkstern, D., Pfaus, J. G., Phillips, A. G., & Fibiger, H. C. (1992). Sexual behavior increases dopamine transmission in the nucleus accumbens and striatum of male rats: Comparison with novelty and locomotion. *Behavioral Neuroscience, 1*, 181–191.

Davidson, R. J. (1993). Childhood temperament and cerebral asymmetry: A neurobiological substrate of behavioral inhibition. In K. H. Rubin & J. B. Asendorpf (Eds.), *Social withdrawal, inhibition, and shyness in childhood* (pp. 31–48). Hillsdale, N. J.: Lawrence Erlbaum Associates, Inc.

Davidson, R. J. (2000). Affective style, psychopathology, and resilience: Brain mechanisms and plasticity. *American Psychologist, 55*, 1196–1214.

Davidson, R. J., Coe, C. C., Dolski, I., & Donzella, B. (1999). Individual differences in prefrontal activation asymmetry predict natural killer cell activity at rest and in response to challenge. *Brain, Behavior, and Immunity, 13*, 93–108.

Davidson, R. J., & Fox, N. A. (1982). Asymmetrical brain activity discriminates between positive versus negative affective stimuli in human infants. *Science, 218,* 1235–1237.

Davidson, R. J., & Fox, N. A. (1989). Frontal brain assymmetry predicts infants' response to maternal separation. *Journal of Abnormal Psychology, 98,* 127–131.

Davidson, R. J., Jackson, D. C., & Kalin, N. H. (2000). Emotion, plasticity, context, and regulation: Perspectives from affective neuroscience. *Psychological Bulletin, 126,* 890–906.

Davidson, R. J., Pizzagalli, D., Nitschke, J. B., & Putnam, K. (2002). Depression: Perspectives from affective neuroscience. *Annual Review of Psychology, 53,* 545–574.

Davis, M., & Whalen, P. J. (2001). The amygdala: Vigilance and emotion. *Molecular Psychiatry, 6,* 13–34.

De La Ronde, C., & Swann, W. B. (1993). Caught in the crossfire: Positivity and self-verification strivings among people with low self-esteem. In R. F. Baumeister (Ed.), *Self-esteem: The puzzle of low self-regard* (pp. 147–165). New York: Plenum.

Debener, S., Beauducel, A., Nessler, D., Brocke, B., Heilemann, H., & Kayser, J. (2000). Is resting anterior EEG alpha asymmetry a trait marker for depression? Findings for healthy adults and clinically depressed patients. *Neuropsychobiology, 41,* 31–37.

DeCasper, A. J., & Fifer, W. (1980). Of human bonding: Newborns prefer their mother's voice. *Science, 208,* 1174–1176.

Deci, E. L., & Ryan, R. M. (1985). The general causality orientations scale: Self-determination in personality. *Journal of Research in Personality, 19,* 109–134.

Deci, E. L., & Ryan, R. M. (2000). The "what" and "why" of goal pursuits: Human needs and the self-determination of behavior. *Psychological inquiry, 11,* 227–268.

Dehaene, S., & Naccache, L. (2001). Towards a cognitive neuroscience of consciousness: Basic evidence and a workspace framework. *Cognition, 79,* 1–37.

Deneke, F.-W. (1999). *Psychische struktur und gehirn. Die gestaltung subjektiver wirklichkeiten* [Mental structure and brain. Design of subjective realities] (2nd ed.). Stuttgart, Germany: Schattauer.

Deneke, F.-W. (2001). *Psychische Struktur und Gehirn. Die Gestaltung subjektiver Wirklichkeiten* [Mental structure and brain. Design of subjective realities] (2nd ed.). Stuttgart, Germany: Schattauer.

Derogatis, L. R. (1992). *SCL-90-R, administration, scoring & procedures manual-II for the R(evised) version and other instruments of the Psychopathology Rating Scale Series.* Towson, MD: Clinical Psychometric Research, Inc.

Derogatis, L. R. (1993). *Brief Symptom Inventory (BSI), administration, scoring, and procedures manual, third edition.* Minneapolis, MN: National Computer Services.

Devinsky, O., Morrell, M. J., & Vogt, B. A. (1995). Contributions of anterior cingulate cortex to behaviour. *Brain, 118,* 279–306.

Deutsche Gesellschaft für Psychiatrie, Psychotherapie und Nervenheilkunde. (2000). *Praxisleitlinien in Psychiatrie und Psychotherapie: Band 5: Behandlungsleitlinie Affektive Erkrankungen* [Treatment guidelines for psychiatry and psychotherapy—Affective disorders]. Darmstadt, Germany: Steinkopff.

Di Nardo, P. A., & Barlow, D. H. (1990). Syndrome and symptom co-occurrence in the anxiety disorders. In J. D. Maser & C. R. Cloninger (Eds.), *Comorbidity of mood and anxiety disorders* (pp. 205–230). Washington, DC: American Psychiatric Press.

Dick, A., Grawe, K., Regli, D., & Heim, P. (1999). Was sollte ich tun, wenn...? Empirische Hinweise für die adaptive Feinsteuerung des Therapiegeschehens innerhalb einzelner Sitzungen [What should I do if ...? Empirical indications for the adaptive regulation of therapeutic processes within individual sessions]. *Verhaltenstherapie und psychosoziale Praxis, 31,* 253–279.

Diener, E., & Lucas, R. E. (1999). Personality and subjective well-being. In D. Kahneman, E. Diener, & N. Schwarz (Eds.), *Well-being: The foundations of hedonic psychology* (pp. 213–229). New York: Russell Sage Foundation.

Dill, J., & Anderson, C. (1999). Loneliness, shyness, and depression: The etiology and interrelationships of everyday problems in living. In T. Joiner & J. Coyne (Eds.), *The Interactional Nature of Depression* (pp. 93–125). Washington: American Psychological Association.

Dimidjian, S., & Linehan, M. (2003). Defining an agenda for future research on the clinical application of mindfulness practice. *Clinical Psychology: Science and Practice, 10,* 166–171.

Dollard, J., & Miller, N. E. (1950). *Personality and psychotherapy.* New York: McGraw-Hill.

Dozier, M., Stovall, K. C., & Albus, K. E. (1999). Attachment and psychopathology in adulthood. In J. Cassidy & P. Shaver (Eds.), *Handbook of attachment* (pp. 497–519). New York: Guilford.

Drevets, W. C. (1998). Functional neuroimaging studies of depression: The anatomy of melancholia. *Annual Review of Medicine, 49,* 341–361.

Drevets, W. C. (2001). Neuroimaging and neuropathological studies of depression: Implications for the cognitive-emotional features of mood disorders. *Current Opinion in Neurobiology, 11,* 249–249.

Drevets, W. C., Price, J. L., Simpson, J. R., Todd, R. D., Reich, T., Vannier, M., et al. (1997). Subgenual prefrontal cortex abnormalities in mood disorders. *Nature, 386,* 824–827.

Drevets, W. C., Videen, T. O., Price, J. L., Preskorn, S. H., Carmichael, S. T., & Raichle, M. E. (1992). A functional anatomical study of unipolar depression. *Journal of Neuroscience, 12,* 3628–3641.

Driessen, M., Herrmann, J., Stahl, K., Zwaan, M., Meier, S., Hill, A., et al. (2000). Magnetic resonance imaging volumes of the hippocampus and the amygdala in women with borderline personality disorders and early traumatization. *Archives of General Psychiatry, 57,* 1115–1122.

Ebert, D., & Ebmeier, K. P. (1996). The role of cingulate gyrus in depression: From functional anatomy to neural chemistry. *Biological Psychiatry, 39,* 1044–1050.

Edelman, G. M. (1987). *Neural Darwinism. The theory of neuronal group selection.* New York: Basic Books.

Edelman, G. M. (1989). *The remembered present. A biological theory of consciousness.* New York: Basic Books.

Edelman, G. M. (1992). *Bright air, brilliant fire: On the matter of the mind.* New York: Basic Books.

Edelman, G. M., & Tononi, G. (2000). *A universe of consciousness.* New York: Basic Books.

Ehlers, A. (1999). *Posttraumatische Belastungsstörung* [Posttraumatic stress disorder]. Göttingen, Germany: Hogrefe.

Ehlers, A., & Clark, D. (2000). A cognitive model of posttraumatic stress disorder. *Behaviour Research and Therapy, 38,* 319–345.

Ekman, P. (1993). Facial expression and emotion. *American Psychologist, 48,* 384–392.

Elias, N. (1982). State formation and civilization. Malden, MA: Blackwell.

Elkin, I. (1994). The NIMH treatment of depression collaborative research program: Where we began and where we are. In A. E. Bergin & S. L. Garfield (Eds.), *Handbook of psychotherapy and behavior change* (4th. ed., pp. 114–139). New York: Wiley.

Elliott, A. J., & Devine, P. G. (1994). On the motivational nature of cognitive dissonance: Dissonance as psychological discomfort. *Journal of Personality and Social Psychology, 67,* 382–394.

Elliott, A. J., & Sheldon, K. M. (1998). Avoidance personal goals and the personality-illness relationship. *Journal of Personality and Social Psychology, 75,* 1282–1299.

Elliott, A. J., Sheldon, K. M., & Church, M. A. (1997). Avoidance personal goals and subjective well-being. *Personality and Social Psychology Bulletin, 23,* 915–927.

Elliott, A. J., & Thrash, T. M. (2002). Approach-avoidance motivation in personality: Approach and avoidance temperaments and goals. *Journal of Personality and Social Psychology, 82,* 804–818.

Ellis, A. (1977). *Die rational-emotive Therapie. Das innere Selbstgespräch bei seelischen Problemen und seine Veränderung* [Reason and emotion in psychotheraphy] (B. Stein, Trans.). Munich, Germany: Pfeiffer. (Original work published 1962)

Emmelkamp, P. M., & van Oppen, P. (2000). *Zwangsstörungen* [Obsessive-compulsive disorder]. Göttingen, Germany: Hogrefe.

Emmons, R. A. (1989). The personal striving approach to personality. In L. A. Pervin (Ed.), *Goal concepts in personality and social psychology* (pp. 87–126). Hillsdale, NJ: Lawrence Erlbaum Associates, Inc.

Emmons, R. A. (1997). Motives and life goals. In S. Brygs, R. Hogan, & W. Jones (Eds.), *Handbook of personality psychology* (pp. 485–512). Orlando, FL: Academic Press.

Emmons, R. A., & King, L. A. (1988). Conflict among personal strivings: Immediate and long-term implications for psychological and physical well-being. *Journal of Personality and Social Psychology, 54,* 1040–1048.

Emmons, R. A., & McAdams, D. P. (1991). Personal strivings and motive dispositions: Exploring the links. *Personality and Social Psychology Bulletin, 17,* 648–654.

Engel, A. K. (1996). Prinzipien der Wahrnehmung: Das visuelle System [Principles of perception: The visual system]. In G. Roth & W. Prinz (Eds.), *Kopf-Arbeit*. Heidelberg, Germany: Spektrum Akademischer Verlag.

Epstein, S. (1962). The measurement of drive and conflict in humans: Theory and experiment. In M. R. Jones (Ed.), *Nebraska Symposium on Motivation: Vol. 10*. (pp. 127–206). Lincoln: University of Nebraska Press.

Epstein, S. (1967). Toward a unified theory of anxiety. In B. A. Maher (Ed.), *Progress in experimental personality research* (Vol. 4., pp. 1–89). New York: Academic Press.

Epstein, S. (1978). Avoidance-approach: The fifth basic conflict. *Journal of Consulting and Clinical Psychology, 46*, 1016–1022.

Epstein, S. (1982). Conflict and stress. In S. Goldberg & S. Bresnitz (Eds.), *Handbook of stress* (pp. 49–68). New York: Free Press.

Epstein, S. (1989). *Constructive Thinking Inventory* (CTI) [Unpublished questionnaire]. Amherst University of Massachusettes.

Epstein, S. (1990). Cognitive-experiential self-theory. In L. A. Pervin (Ed.), *Handbook of personality: Theory and research*. (pp. 165–192). New York: Guilford.

Epstein, S. (1993). Implications of cognitive-experiential self-theory for personality and developmental psychology. In D. C. Funder, R. D. Parke, C. Tomlinson-Keasey & K. Widaman (Eds.), *Studying lives through time: Personality and development* (pp. 399–438). Washington, DC: American Psychological Association.

Epstein, S., & Morling, B. (1995). Is the self motivated to do more than to enhance and/or verify itself? In M. H. Kernis (Ed.), *Efficacy, agency, and self-esteem* (pp. 9–29). New York: Plenum.

Erickson, R. F., Sroufe, L. A., & Egelnad, B. (1985). The relationship between quality of attachment and behaviour problems in preschool in a high risk sample. *Monographs of the Society for Research in Child Development, 50*(1&2), 147–166.

Eriksson, P. S., Perfilieva, E., Bjork-Eriksson, T., Alborn, A., Nordborg, C., Peterson, D. A., et al. (1998). Neurogenesis in the adult human hippocampus. *Nat. Medicine, 4*, 1313–1317.

Essau, C. A., Karpinsky, N. A., Petermann, F., & Conradt, J. (1998). Häufigkeit und Komorbidität psychischer Störungen bei Jugendlichen: Ergebnisse der Bremer Jugendstudie [Frequency and comorbidity of psychological disorders in adolescents. Results of the Bremen Adolescent Study]. *Zeitschrift für Klinische Psychologie, Psychiatrie und Psychotherapie, 46*, 105–124.

Etkin, A., Pittenger, C., Polan, H. J., & Kandel, E. R. (2005). Toward a neurobiology of psychotherapy: Basic science and clinical applications. *Journal of Neuropsychiatry and Clinical Neuroscience, 17*, 145–158.

Eysenck, H. J. (1980). *A model for personality*. New York: Springer-Verlag.

Fairbanks, L. A. (1989). Early experience and cross-generational continuity of mother-infant contact in vervet monkeys. *Developmental Psychobiology, 22*, 669–682.

Faller, H., & Gossler, S. (1998). Probleme und Ziele von Psychotherapiepatienten Eine qualitative-inhaltsanalytisch Untersuchung der Patientenangaben beim Erstgespräch [Problems and therapeutic goals of psychotherapeutic patients: A content-analytical study of patients' statements during the initial assessment]. *Psychotherapie, Psychosomatik und Medizinische Psychologie, 48*, 176–186.

Festinger, L. (1957). *A theory of cognitive dissonance*. Evanston, IL: Row, Peterson.

Fiedler, P. (2001). *Dissoziative Störungen und Konversion* [Dissociative disorders and conversion]. Weinheim, Germany: Beltz.

Field, T. (1985). Attachment as psychobiological attunement: Being on the same wavelength. In M. Reite & T. Field (Eds.), *The psychobiology of attachment and separation*. New York: Academic.

Fischman, M. W. (1989). Relationship between self-reported drug effects and their reinforcing effects: Studies with stimulant drugs. *NIDA Research Monographs, 92*, 211–230.

Fischman, M. W., & Foltin, R. W. (1992). Self-administration of cocaine by humans: A laboratory perspective. In G. R. Bock & J. Whelan (Eds.), *Cocaine: Scientific and social dimensions* (pp. 165–180). Chichester, England: Wiley.

Flammer, A. (1990). *Erfahrung der eigenen Wirksamkeit. Einführung in die Psychologie der Kontrollmeinung* [Experience of own efficacy. Introduction to the psychology of control beliefs]. Bern, Switzerland: Huber.

REFERENCES435

Flatten, G. (2003). Posttraumatische Belastungsreaktionen aus neurobiologischer und synergetischer Perspektive [Posttraumatic stress reactions from a neurobiological and synergetic perspective]. In G. Schiepek (Ed.), *Neurobiologie der Psychotherapie*. Stuttgart, Germany: Schattauer.

Flor, H. (2003). Wie verlernt das Gehirn den Schmerz? Verletzungsbezogene und therapeutisch induzierte neuroplastische Veränderungen des Gehirns bei Schmerz und psychosomatsichen Störungen [Injury-related and therapeutically induced neuroplastic changes in the brain in pain and psychosomatic disorders]. In G. Schiepek (Ed.), *Neurobiologie der Psychotherapie*. Stuttgart, Germany: Schattauer.

Fonagy, P., Steele, H., Steele, M., Leigh, K., Kennedy, R., Mattoon, G., et al. (1994). Attachment, the reflective self, and borderline states. The predictive specificity of the adult attachment interview and pathological emotional development. In S. Goldberg, R. Muir, & J. Kerr (Eds.), *Attachment theory: Social developmental and clinical perspectives*. Englewood Cliffs, NJ: Lawrence Erlbaum Associates, Inc.

Ford, M. E. (1992). *Motivating humans*. Newbury Park, CA: Sage.

Förster, J., & Strack, F. (1996). Influence of overt head movements on memory for valenced words: A case of conceptual-motor compatibility. *Journal of Personality and Social Psychology, 71*, 421–430.

Förstl, H. (2002). Biologische Korrelate psychotherapeutischer Interventionen. Psychotherapie, 7, 184–188.

Friedman, B. H., & Thayer, J. F. (1998). Autonomic balance revisited: Panic anxiety and heart rate variability. *Journal of Psychosomatic Research, 44*, 133–151.

Fuhrmeister, M.-L., & Wiesenhütter, F. (1973). *Metamusik. Psychosomatik der Ausübung zeitgenössischer Musik* [Metamusic—psychosomatic aspects of playing contemporary music]. Munich, Germany: J. F. Lehmanns Verlag.

Fujiwara, E., & Markowitsch, H. J. (2003). Das mnestische Blockadesyndrom–hirnphysiologische Korrelate von Angst und Stress [The mnestic blocking syndrome—Brain physiological correlates of anxiety and stress]. In G. Schiepek (Ed.), *Neurobiologie der Psychotherapie* (pp. 186–212). Stuttgart, Germany: Schattauer.

Furmark, T., Tillfors, M., Marteinsdottir, I., Fischer, H., Pissiota, A., Längström, B., et al. (2002). Common changes in cerebral blood flow in patients with social phobia treated with citalopram or cognitive-behavioral therapy. *Archives of General Psychiatry, 59*, 423–433.

Gabbard, G. O. (2000). A neurobiologically informed perspective of psychotherapy. *British Journal of Psychiatry, 177*, 117–122.

Gabbard, G. O. (2002, August). *The revolution in the neurosciences: Implications for psychotherapy research and practice*. Paper presented at the World Congress for Psychotherapy, Trondheim, Norway.

Gaensbauer, T., Harmon, R., Cytryn, L., & McKnew, D. (1984). Social and affective development in infants with a manic-depressive parent. *American Journal of Psychiatry, 141*, 223–229.

Galin, D. (1974). Implications of left-right cerebral lateralization for psychiatry: A neurophysiological context for unconscious processes. *Archives of General Psychiatry, 9*, 412–418.

Gall, S., Kerschreiter, R., & Mojzisch, A. (2002). *Handbuch Biopsychologie und Neurowissenschaften*. Bern, Switzerland: Huber.

Gallati, D. (2003). *Metaanalyse über die Erfolgsmessung in Vergleichsstudien von Depressionsbehandlungen* [Meta analysis of outcome in depression treatment]. Unpublished masters thesis, Institut für Psychologie, Universität Bern, Bern, Switzerland.

Garavan, H., Ross, R. H., & Stein, E. A. (1999). Right hemispheric dominance of inhibitory control: An event-related functional MRI study. *Proceedings of the National Academy of Sciences of the of the United States of America, 96*, 8301–8306.

Garcia, R., Vouimba, R. M., Baudry, M., & Thompson, R. F. (1999). The amygdala modulates prefrontal cortex activity relative to conditioned fear. *Nature, 402*, 294–296.

Gasiet, S. (1980). *Menschliche Bedürfnisse: Eine theoretische Synthese* [Human needs: A theoretical synthesis]. Frankfurt, Germany: Campus.

Gassmann, D., & Grawe, K. (2006). General change mechanisms: The relation of problem activation to resource activation in successful and unsuccessful therapeutic interactions. *Clinical Psychology & Psychotherapy, 13*, 1–11.

George, M. S., Ketter, T. A., Parakh, P. I., Rosinsky, N., Ring, H. A., Pazzaglia, P. J., et al. (1997). Blunted left cingulate activation in mood disorder subjects during a response interference task (the Stroop). *Journal of Neuropsychiatry and Clinical Neuroscience, 9*, 55–63.

Gilbertson, M. W., Shenton, M. E., Ciszewski, A., Kasai, K., Lasko, N. B., Orr, S. P., et al. (2002). Smaller hippocampal volume predicts pathologic vulnerability to psychological trauma. *Nature-Neuroscience, 5,* 1242–1247.

Goldapple, K., Segal, Z., Garson, D., Lau, M., Bieling, P., Kennedy, S., et al. (2004). Modulation of cortical-limbic pathways in major depression: Treatment-specific effects of cognitive behavior therapy. *Archives of General Psychiatry 61,* 34–41.

Gortner, E. T., Gollan, J. K., Dobson, K. S., & Jacobson, N. S. (1998). Cognitive-behavioral treatment for depression: Relapse prevention. *Journal of Consulting and Clinical Psychology, 66,* 377–384.

Goschke, T. (1996a). Lernen und Gedächtnis: Mentale Prozesse und Gehirnstrukturen [Learning and memory: Mental processes and brain structures]. In G. Roth & W. Prinz (Eds.), *Kopf-Arbeit* (pp. 359–410). Heidelberg, Germany: Spektrum Akademischer Verlag.

Goschke, T. (1996b). Wille und Kognition: Zur funktionalen Architektur der intentionalen Handlungssteuerung [Volition and cognition: The functional architecture of intentional action regulation]. In J. Kuhl & H. Heckhausen (Eds.), *Enzyklopädie der Psychologie. Themenbereich C Theorie und Forschung, Serie IV Motivation und Emotion, Band 4 Motivation, Volition und Handlung* (pp. 583–663). Göttingen, Germany: Hogrefe.

Goschke, T. (2001). Voluntary action and cognitive control from a cognitive neuroscience perspective. In S. Maasen, W. Prinz, & G. Roth (Eds.), *Voluntary action* (pp. 49–85). New York: Oxford University Press.

Gotlib, I. H., Ranganath, C., & Rosenfeld, F. (1998). Frontal EEG alpha asymmetry, depression, and cognitive functioning. *Cognition and Emotion, 12,* 449–478.

Gould, E., Tanapat, T., Rydel, T., & Hastings, N. (2000). Regulation of hippocampal neurogenesis in adulthood. *Biological Psychiatry, 48,* 715–720.

Graf, P., & Schacter, D. L. (1985). Implicit and explicit memory for new associations in normal subjects and amnesic patients. *Journal of Experimental Psychology: Learning, Memory, Cognition, 11,* 501–518.

Grant, S., London, E. D., Newlin, D. B., Villemagne, V. L., Xiang, L., Contoreggi, C., et al. (1996). Activation of memory circuits during cue-elicited cocaine craving. *Proceedings of the National Academy of Sciences of the United States of America, 93,* 12040–12045.

Grawe, K. (1976). *Differentielle Psychotherapie. Indikation und spezifische Wirkung von Verhaltenstherapie und Gesprächstherapie* [Differential psychotherapy I. Indication and specific effect of behavior therapy and client-centered psychotherapy. An investigation with phobic patients]. Bern, Switzerland: Huber.

Grawe, K. (1978). Verhaltenstherapeutische Gruppentherapien [Behavior-therapeutic group therapy]. In L. Pongratz (Ed.), *Handbuch der Psychologie* (Band. 8, Klinische Psychologie 2, pp. 2696–2724). Göttingen, Germany: Hogrefe.

Grawe, K. (1980). *Verhaltenstherapie in Gruppen* [Behavior therapy in groups]. Munich, Germany: Urban & Schwarzenberg.

Grawe, K. (1995). Grundriss einer Allgemeinen Psychotherapie [Outline of a general psychotherapy]. *Psychotherapeut, 40,* 130–145.

Grawe, K. (1997). Research-informed psychotherapy. *Psychotherapy Research, 7,* 1–19.

Grawe, K. (1998). *Psychologische Therapie.* Göttingen, Germany: Hogrefe.

Grawe, K. (1999). Wie kann Psychotherapie noch wirksamer werden? [How can psychotherapy become more effective?]. *Verhaltenstherapie und Psychosoziale Praxis, 31,* 185–199.

Grawe, K. (2002, June). *Consistency theory. A neuroscientific view of symptom formation and therapeutic change.* Paper presented at the annual meeting of the Society for Psychotherapy Research, Santa Barbara, CA.

Grawe, K. (2004). *Psychological therapy.* Seattle, WA: Hogrefe & Huber.

Grawe, K., & Caspar, F. (1984). Die Plananalyse als Konzept und Instrument für die Psychotherapieforschung [Plan analysis as a concept and instrument of psychotherapy research]. In U. Baumann (Ed.), *Psychotherapieforschung. Makro- und Mikroperspektiven.* Göttingen, Germany: Hogrefe.

Grawe, K., Dick, A., Regli, D., & Smith, E. (1999). Wirkfaktorenanalyse—ein Spektroskop für die Psychotherapie [Therapy spectrum analysis—A spectroscope for psychotherapy]. *Verhaltenstherapie und Psychosoziale Praxis, 31,* 201–225.

Grawe, K., Donati, R., & Bernauer, F. (1994). *Psychotherapie im Wandel—Von der Konfession zur Profession* [Psychotherapy changing—From confession to profession.] (5th ed.). Göttingen, Germany: Hogrefe.

Grawe, K., Dziewas, H., Brutscher, H., Schaper, P., & Steffani, K. (1978). Assertive Trainingsgruppen vs. interaktionelle Problemlösungsgruppen. Ein empirischer Vergleich [Assertiveness training vs. problem-solving groups]. *Mitteilungen der DGVT, Sonderheft 1,* 63–85.

Grawe, K., Dziewas, H., & Wedel, S. (1980). Interaktionelle Problemlösungsgruppen—Ein verhaltenstherapeutisches Gruppenkonzept [Interactional problem-solving groups]. In K. Grawe (Ed.), *Fortschritte der Klinischen Psychologie. Verhaltenstherapie in Gruppen* (pp. 266–306). Munich, Germany: Urban & Schwarzenberg.

Grawe, K., & Grawe-Gerber, M. (1999). Ressourcenaktivierung. Ein primäres Wirkprinzip der Psychotherapie [Resource activation: A primary effective factor in psychotherapy]. *Psychotherapeut, 44,* 63–73.

Grawe, K., Grawe-Gerber, M., Heiniger, B., Ambühl, H., & Caspar, F. (1996). Schematheoretische Fallkonzeption und Therapieplanung—Eine Anleitung für Therapeuten [Schema theory case concepts and therapy planning: A guideline for therapists]. In F. Caspar (Ed.), *Psychotherapeutische Problemanalyse* (pp. 189–224). Tübingen, Germany: Deutsche Gesellschaft für Verhaltenstherapie.

Gray, J. A. (1981). A critique of Eysenck's theory of personality. In H. J. Eysenck (Ed.), *A model for personality* (pp. 246–276). New York: Springer-Verlag.

Gray, J. A. (1982). *The neuropsychology of anxiety.* New York: Oxford University Press.

Gray, J. A., & McNaughton, N. (1996). The neuropsychology of anxiety. A reprise. In D. A. Hope (Ed.), *Nebraska Symposion on Motivation: Vol. 43. Perspectives on anxiety, panic, and fear* (pp. 61–134). Lincoln: University of Nebraska Press.

Greenberg, L. S., Watson, J. C., Elliot, R., & Bohart, A. C. (2001). Empathy. *Psychotherapy: Theory, Research, Practice, Training, 38,* 380–384.

Grob, A. (1995). Subjective well-being and significant life-events across life-span. *Swiss Journal of Psychology, 54,* 3–18.

Grosjean, B. (2005). From synapse to psychotherapy: The fascinating evolution of neuroscience. *American Journal of Psychotherapy, 59,* 181–97.

Grosse Holtforth, M. (2001). Was möchten Patienten in ihrer Therapie erreichen?—Die Erfassung und Kategorisierung von Therapiezielen mit dem Berner Inventar für Therapieziele (BIT) [What do clients wish to achieve during therapy? The assessment of therapeutic goals using the Bern Inventory for Therapeutic Goals (BIT)]. *Verhaltenstherapie und Psychosoziale Praxis, 33,* 241–258.

Grosse Holtforth, M., & Castonguay, L. G. (2005). Relationship and techniques in cognitive behavioral therapy—A motivational approach. *Psychotherapy: Theory, Research, Practice, Training, 42,* 443–445.

Grosse Holtforth, M., & Grawe, K. (2000). Fragebogen zur Analyse Motivationaler Schemata (FAMOS) [Questionnaire for the analysis of motivational schemas]. *Zeitschrift für Klinische Psychologie und Psychotherapie, 29,* 170–179.

Grosse Holtforth, M., & Grawe, K. (2002a). Bern Inventory of Treatment Goals (BIT), Part 1: Development and First Application of a Taxonomy of Treatment Goal Themes (BIT-T). *Psychotherapy Research, 12,* 79–99.

Grosse Holtforth, M., & Grawe, K. (2002b). *FAMOS Fragebogen zur Analyse Motivationaler Schemata Manual* [Inventory of Approach and Avoidance Motivation (IAAM)/FAMOS Manual]. Göttingen, Germany: Hogrefe.

Grosse Holtforth, M., & Grawe, K. (2003a). Der Inkongruenzfragebogen (INC)—Ein Instrument zur Analyse motivationaler Inkongruenz [The Incongruence Questionnaire (INC). An instrument for the analysis of motivational incongruence]. *Zeitschrift für klinische Psychologie und Psychotherapie, 32,* 315–323.

Grosse Holtforth, M., & Grawe, K. (2003b). Konfliktdiagnostik aus der Perspektive der Konsistenztheorie [Conflict diagnosis from the perspective of consistency theory]. In P. W. Dahlbender, P. Buchheim, & G. Schüssler (Eds.), *Lernen an der Praxis. OPD und Qualitätssicherung in der Psychodynamischen Psychotherapie.* Bern, Switzerland: Huber.

Grosse Holtforth, M., Grawe, K., Egger, O., & Berking, M. (2005). Reducing the dreaded: Change of avoidance motivation in psychotherapy. *Psychotherapy Research, 15,* 261–271.

Grosse Holtforth, M., Grawe, K., & Tamcan, Ö. (2003). *Inkongruenzfragebogen. Manual* [INC—manual]. Göttingen, Germany: Hogrefe.

Grosse Holtforth, M., Reubi, L., Ruckstuhl, L., Berking, M., & Grawe, K. (2004). The value of treatment goals in the outcome evaluation of psychiatric inpatients. *International Journal of Social Psychiatry, 50*(1), 80–91.

Grosse Holtforth, M., Schulte, D., Grawe, K., Wyss, T., & Michalak, J. (2004). Are treatment goals more than reformulated diagnoses? Treatment goal themes in CBT. Unpublished manuscript.

Grossmann, K. (1990). Entfremdung, Abhängigkeit und Anhänglichkeit im Lichte der Bindungstheorie [Alienation, dependency, and attachment in light of attachment theory]. *Praxis der Psychotherapie und Psychosomatik, 35,* 231–238.

Grossmann, K., Grossmann, K. E., Sprangler, G., Suess, G., & Unzne, L. (1985). Maternal sensitivity and newborn orientation responses as related to quality of attachment in Northern Germany. *Monographs of the Society for Research in Child Development, 50,* 233–278.

Grossmann, K. E., August, P., Fremmer-Bombik, E., Friedl, E., Grossmann, A., Scheuerer-Englisch, H., et al. (1989). Die Bindungstheorie. Modell und entwicklungspsychologische Forschung [Attachment theory: Model and developmental psychological research]. In H. Keller (Ed.), *Handbuch der Kleinkindforschung.* Berlin, Germany: Springer.

Grossmann, K. E., & Grossmann, K. (1991). Attachment quality as an organizer of emotional and behavioral responses in a longitudinal perspective. In C. M. Parkes, J. Stevenson-Hinde, & P. Marris (Eds.), *Attachment across the life cycle* (pp. 93–114). London: Tavistock/Routledge.

Grüsser, O. J., Naumann, A., & Seeck, M. (1990). Neurophysiological and neuropsychological studies on the perception and recognition of faces and facial expressions. In N. Elsner & G. Roth (Eds.), *Brain, perception, cognition* (pp. 83–94). Stuttgart, Germany: Thieme.

Guerin, B., Goeders, N. E., Dworkin, S. I., & Smith, J. E. (1984). Intracranial self-administration of dopamine into the nucleus accumbens. *Society of Neuroscience Abstracts, 10,* 1072.

Gunnar, M. R., Brodersen, L., Nachmias, M., Buss, K., & Rigatuso, J. (1996). Stress reactivity and attachment security. *Develomental Psychobiology, 29,* 191–204.

Güntürkün, O. (2003). Biologishe Psychologie 2010—Visionen zur Zukunft des Faches in der Psychologie [Biological psychology 2010—Visions]. *Psychologische Rundschau, 54,* 122–123.

Gurvits, T. V., Shenton, M. E., Hokama, H., Ohta, H., Lasko, M. B., Gilbertson, M. W., et al. (1996). Magnetic resonance imaging study of hippocampal volume in chronic combat-related posttraumatic stress disorder. *Biological Psychiatry, 40,* 1091–1099.

Gutberlet, I., & Miltner, W. H. (1999). Therapeutic effects on differential electrocortical processing of phobic objects in spider and snake phobics. *International Journal of Psychophysiology, 33,* 180.

Haan, N. (1977). *Coping and defending.* New York: Academic.

Haggard, P., & Eimer, M. (1999). On the relation between brain potentials and the awareness of voluntary movements. *Experimental Brain Research, 126,* 128–133.

Hampson, R. E., Simeral, J. D., & Deadwyler, S. A. (1999). Distribution of spatial and nonspatial information in dorsal hippocampus. *Nature, 402,* 610–614.

Harmon-Jones, E., & Mills, J. (1999). An introduction to cognitive dissonance theory and an overview of current perspectives on the theory. In E. Harmon-Jones & J. Mills (Eds.), *Cognitive dissonance: Progress on a pivotal theory in social psychology* (1st ed.; pp. 3–21). Washington: American Psychological Association.

Harrigan, J. A., & Rosenthal, R. (1986). Nonverbal aspects of empathy and rapport in physician-patient interaction. In P. D. Blanck, R. Buck, & R. Rosenthal (Eds.), *Nonverbal communication in the clinical context.* University Park: The Pennsylvanis State University Press.

Hart, A. J., Whalen, P. J., Shin, L. M., McInerney, S. C., Fischer, H., & Rauch, S. L. (2000). Differential response in the human amygdala to racial outgroup vs ingroup face stimuli. *Neuroreport, 11,* 2351–2355.

Hautzinger, M. (1998). *Depression. Psychologie affektiver Störungen* [Depression: Psychology of affective disorders]. Göttingen, Germany: Hogrefe.

Hautzinger, M., & de Jong-Meyer, R. (1996). Depressionen [Depression] (Themenheft). *Zeitschrift für Klinische Psychologie, 25.*

Hayes, S., & Wilson, K. (2003). Mindfulness: Method and process. *Clinical Psychology: Science and Practice, 10,* 161–165.

Hayes, S. C., Wilson, K. G., Gifford, E. V., Follette, V. M., & Strosahl, K. (1996). Experimental avoidance and behavioral disorders: A functional dimensional approach to diagnosis and treatment. *Journal of Consulting and Clinical Psychology, 64,* 1152–1168.

Headey, B., & Wearing, A. (1989). Personality, life events, and subjective well-being: Toward a dynamic equilibrium model. *Journal of Personality and Social Psychology, 57,* 731–739.

Hebb, D. (1949). *The organization of behavior.* New York: Wiley.

Heckhausen, H. (1980). *Motivation und Handeln: Lehrbuch der Motivationspsychologie* [Motivation and action]. Berlin, Germany: Springer.

Heckhausen, H., Gollwitzer, P. M., & Weinert, F. E. (Eds.). (1987). *Jenseits des Rubikon: Der Wille in den Humanwissenschaften* [Beyond the Rubicon: The will in the humanities]. Berlin, Germany: Springer.

Heider, F. (1982). *Psychology of interpersonal relations.* Hillsdale, NJ: Lawrence Erlbaum Associates, Inc.

Heilizer, F. (1977). A review of theory and research on the assumption of Miller's response competition (conflict) models: Response gradients. *Journal of General Psychology, 97,* 17–71.

Heilizer, F. (1978). Approach-withdrawal response competition (AW-RC), displacement, and behavior modification. *Journal of General Psychology, 99,* 181–204.

Heim, C., Newport, D. J., Heit, S., Graham, J., Wilcox, M., Bonsall, R., et al. (2000). Pituitary-adrenal and automatic responses to stress in women after sexual and physical abuse in childhood. *Journal of the American Medical Association, 284,* 592–597.

Heller, W., & Nitschke, J. B. (1998). The puzzle of regional brain activity in depression and anxiety: the importance of subtypes and comorbidity. *Cognition and Emotion, 12,* 421–447.

Henriques, J., & Davidson, R. J. (1990). Regional brain electrical asymmetries discriminate between previously depressed and healthy control subjects. *Journal of Abnormal Psychology, 99,* 22–31.

Henriques, J. B., & Davidson, R. J. (2000). Decreased responsiveness to reward in depression. *Cognition and Emotion, 14,* 711–724.

Henriques, J. B., Glowacki, J. M., & Davidson, R. J. (1994). Reward fails to alter response bias in depression. *Journal of Abnormal Psychology, 103,* 460–466.

Higgins, E. T. (1987). Self-discrepancy: A theory relating self and affect. *Psychological Review, 94,* 319–340.

Higgins, E. T. (1997). Beyond pleasure and pain. *American Psychologist, 52,* 1280–1300.

Higgins, E. T., Shah, J., & Friedman, R. (1997). Emotional responses to goal attainment: Strength of regulatory focus as moderator. *Journal of Personality and Social Psychology, 72,* 515–525.

Higley, J. D., Suomi, S. J., & Linnoila, M. (1996). A nonhuman primate model of Type II alcoholism? (Part 2): Diminished social competence and excessive aggression correlates with low CSF 5-HIAA concentrations. *Alcoholism: Clinical and Experimental Research, 20,* 643–650.

Ho, A. P., Gillin, J. C., Buchsbaum, M. S., Wu, J. C., Abel, L., & Bunney, W. E. (1996). Brain glucose metabolism during non-rapid eye movement sleep in major depression. A positron emission tomography study. *Archives of General Psychiatry, 53,* 645–652.

Hoebel, B. G. (1997). Neuroscience and appetitive behavior research: Twenty-five years. *Appetite, 29,* 119–133.

Hoebel, B. G., Monaco, A. P., Hernandez, L., Aulisi, E. F., Stanley, B. G., & Lenard, L. (1983). Self-injection of amphetamine directly into the brain. *Psychopharmacology, 81,* 158–163.

Hoebel, B. G., Rada, P. V., Mark, G. P., & Pothos, E. N. (1999). Neural systems for reinforcement and inhibition of behavior: Relevance to eating, addiction, and depression. In D. Kahneman, E. Diener, & N. Schwarz (Eds.), *Well-being: The foundation of hedonic psychology* (pp. 558–569). New York: Russell Sage Foundation.

Hofer, M. A. (1984). Relationships as regulators: A psychobiological perspective on bereavement. *Psychosomatic Medicine, 46,* 183–197.

Hofer, M. A. (1987). Early social relationships: A psychophysiologist's view. *Child Development, 58,* 633–647.

Holahan, C., Moos, R., & Bonin, L. (1999). Social context and depression: An integrative stress and coping framework. In T. Joiner & J. Coyne (Eds.), *The Interactional Nature of Depression* (pp. 39–63). Washington: American Psychological Association.

Holland, P. C., & Gallagher, M. (1999). Amygdala circuitry in attentional and representational processes. *Trends in Cognitive Science, 3,* 65–73.

Hollerman, J. R., & Schultz, W. (1998). Dopamine neurons report an error in the temporal prediction of reward during learning. *Nature Neuroscience, 1,* 304–309.

Hollon, S. (2002, June). Paper presented at the annual meeting of the Society for Psychotherapy Research, Santa Barbara, CA.

Holmes, D. S. (1990). The evidence for repression: An examination of sixty years of research. In J. L. Singer (Ed.), *Repression and dissociation* (pp. 85–102). Chicago: The University of Chicago Press.

Horowitz, L., Strauss, B., & Kordy, H. (1994). *Inventar Interpersonaler Probleme: Manual.* Weinheim, Germany: Beltz-Test.

Horowitz, M. J. (1975). A cognitive model of hallucinations. *American Journal of Psychiatry, 132,* 7879–7895.

Horvath, A. O., & Bedi, R. P. (2002). The alliance. In J. C. Norcross (Ed.), *Psychotherapy relationships that work: Therapist contributions and responsiveness to patients* (pp. 37–69). London: Oxford: University Press.

Hoshi, E., Shima, K., & Tanji, J. (1998). Task-dependent selectivity of movement-related neuronal activity in the primate prefrontal cortex. *Journal of Neurphysiology, 80,* 3392–3397.

Hoyer, J., Fecht, J., Lauterbach, W., & Schneider, R. (2001). Changes in conflict, symptoms, and well-being during psychodynamic and cognitive-behavioral alcohol inpatient treatment. *Psychotherapy and Psychosomatics, 70,* 209–215.

Hoyer, J., Frank, D., & Lauterbach, W. (1994). Intrapsychischer Konflikt und Ambiguitätsintoleranz als Prädiktoren klinischer Symptombelastung auf latenter Ebene [Intrapsychic conflict and intolerance of ambiguity as predictors of severity of clinical symptoms: A latent variable approach]. *Zeitschrift für Klinische Psychologie, 23,* 117–126.

Hrdina, P. D., Demeter, E., Vu, T. B., Stonyi, P., & Palkovits, M. (1993). 5-HT uptake sites and 5-HT2 receptors in brain of antidepressant-free suicide victims/depressives: Increase in 5-HT2 sites in cortex and amygdala. *Brain Research, 614,* 37–44.

Hsieh, J.-C., Hagerman, O., Stahle-Backdahl, M., Ericson, K., Eriksson, L., Stone Elander, S., et al. (1994). Urge to scratch represented in the human cerebral cortex during itch. *Journal of Neurophysiology, 72,* 3004–3008.

Hubel, D. H., & Wiesel, T. N. (1959). Receptive fields of single neurons in the cat's striate cortex. *Journal of Physiology, 148,* 574–591.

Hubel, D. H., & Wiesel, T. N. (1962). Receptive fields, binocular interaction and functional architecture in the cat's visual cortex. *Journal of Physiology, 160,* 106–154.

Hubel, D. H., & Wiesel, T. N. (1968). Receptive fields and functional architecture of monkey striate cortex. *Journal of Physiology, 195,* 215–243.

Huether, G. (1998). Stress and the adaptive self-organization of neuronal connectivity during early childhood. *International Journal of Developmental Neuroscience, 16,* 297–306.

Insel, T. R. (1988). Obsessive-compulsive disorder: A neuroethological perspective. *Psychopharmacological Bulletin, 24,* 365–369.

Ito, T. A., & Cacioppo, J. T. (1999). *The psychophysiology of utility appraisals.* New York: Russell Sage Foundation.

Jackson, D. C., Malmstadt, J. R., Larson, C. L., & Davidson, R. J. (2000). Suppression and enhancement of emotional responses to unpleasant pictures, *Psychophysiology, 37,* 515–522.

Jacobson, N. S. (1999). My perspective of psychotherapy integration. An outsider, who's out of it. *Journal of Psychotherapy Integration, 9,* 251–255.

Jacobson, N. S., Martell, C. R., & Dimidjian, S. (2001). Behavioral activation treatment for depression: Returning to contextual roots. *Clinical Psychology: Science and Practice, 8,* 255–270.

Jenkins, W. M., Merzenich, M. M., Ochs, M. T., Allard, T., & Guic-Robles, E. (1990). Functional reorganization of primary somatosensory cortex in adult owl monkeys after behaviorally controlled tactile stimulation. *Journal of Neurophysiology, 63,* 82–104.

Joffe, R., Sokolov, S., & Streiner, D. (1996). Antidepressant treatment of depression: A metaanalysis. *Canadian Journal of Psychiatry, 41,* 613–616.

Jones, A. K., Brown, W. D., Friston, K. J., & Frackowiak, R. S. (1991). Cortical and subcortical localization of response to pain in man using positron emission tomography. *Proceedings of the Royal Society of London: Biology, 244,* 39–44.

Jourard, S. M., & Landsman, T. (1980). *Healthy personality: An approach from the viewpoint of humanistic psychology* (4th ed.). New York: Macmillan.

Kabat-Zinn, J. (2003). Mindfulness-based interventions in context: Past, present, future. *Clinical Psychology: Science and Practice, 10,* 144–156.

Kalin, N. H., Shelton, S. E., & Davidson, R. J. (2000). Cerbrospinal fluid corticotropin-releasing hormone levels are elevated in monkeys with patterns of brain activity associated with fearful temperament. *Biological Psychiatry, 47,* 579–585.

Kampe, K. K., Frith, C. D., Dolan, R. J., & Frith, U. (2001). Reward value of attractiveness and gaze. *Nature, 413,* 598.

Kandel, E. R. (1996). Zelluläre Grundlagen von Lernen und Gedächtnis [Cellular mechanisms of learning and memory]. In E. R. Kandel, J. H. Schwartz, & T. M. Jessell (Eds.), *Neurowissenschaften.* Heidelberg, Germany: Spectrum Akademischer Verlag.

Kandel, E. R. (1998). A new intellectual framework for psychiatry. *American Journal of Psychiatry, 155,* 457–469.

Kandel, E. R. (1999). Biology and the future of psychoanalysis: A new intellectual framework for psychiatry revisited. *American Journal of Psychiatry, 156,* 505–524.

Kandel, E. R., Schwartz, J. H., & Jessell, T. M. (Eds.). (1996). *Neurowissenschaften* [Neuroscience]. Heidelberg, Germany: Spektrum.

Kandel, E. R., Schwartz, J. H., & Jessell, T. M. (Eds.). (2000). *Principles of neural science.* New York: McGraw Hill.

Karni, A., Meyer, G., Jezzard, P., Adams, M. M., Turner, R., & Ungerleider, L. G. (1995). Functional MRI evidence for adult motor plasticity during motor skill learning. *Nature, 377,* 155–158.

Kawasaki, H., Adolphs, R., Kaufman, O., Damasio, H., Damasio, A. R., Granner, M., et al. (2001). Single-neuron responses to emotional visual stimuli recorded in human ventral prefrontal cortex. *National Neuroscience, 4,* 15–16.

Kempermann, G., Kuhn, H. G., & Gage, F. H. (1997). More hippocampal neurons in adult mice living in an enriched environment. *Nature, 386,* 493–495.

Kennedy, S. H., Evans, K. R., Kruger, S., Mayberg, H. S., Meyer, J. H., McCann, S., et al. (2001). Changes in regional brain glucose metabolism measured with positron emission tomography after paroxetine treatment of major depression. *American Journal of Psychiatry, 158,* 899–905.

Kessler, R. C., Sonnega, A., Bromet, E., Hughes, M., & Nelson, C. B. (1995). Posttraumatic stress disorder in the National Comorbidity Survey. *Archives of General Psychiatry, 52,* 1048–1060.

Kiesler, D. J. (1983). The 1982 Interpersonal Circle: A taxonomy for the complementary in human transactions. *Psychological Review, 90,* 185–214.

Kihlstrom, J. F., & Hoyt, I. P. (1990). Repression, dissociation, and hypnosis. In J. L. Singer (Ed.), *Repression and dissociation.* (pp. 181–208). Chicago: The University of Chicago Press.

Kiresuk, T., & Lund, S. (1979). Goal attainment scaling: Research, evaluation and utilization. In H. C. Schulberg & F. Baker (Eds.), *Program evaluation in health fields,* Vol. 2. New York: Human Sciences Press.

Koch, C. (2003). *The quest for consciousness: A neurobiological approach.* Englewood, CO: Roberts and Company Publishers.

Kosslyn, S. M., Cacioppo, J. T., Davidson, R. J., Hugdahl, K., Lovallo, W. R., Spiegel, D., et al. (2002). Bridging psychology and biology. The analysis of individuals in groups. *American Psychologist, 57,* 341–351.

Kosslyn, S. M., Thompson, W. L., Kim, I. J., Rauch, S. L., & Alpert, N. M. (1996). Individual differences in cerebral blood flow in Area 17 predict the time to evaluate visualized letters. *Journal of Cognitive Neuroscience, 8,* 78–82.

Kovacs, M., Gatsonis, C., Paulauskas, S. L., & Richards, C. (1989). Depressive disorders in childhood, IV. A longitudinal study of comorbidity with and risk for anxiety disorders. *Archives of General Psychiatry, 46,* 776–782.

Kraemer, G. W. (1992). A psychobiological theory of attachment. *Behavioral and Brain Sciences, 15,* 493–541.

Krause, R. (1997). *Allgemeine psychoanalytische Krankheitslehre. Band 1: Grundlagen* [General psychoanalytic theory of illness]. Stuttgart, Germany: Kohlhammer.

Kröner-Herwig, B. (2000). *Rückenschmerz* [Back pain]. Göttingen, Germany: Hogrefe.

Kubovy, M. (1999). On the pleasure of the mind. In D. Kahneman, E. Diener, & N. Schwarz (Eds.), *Well-being: The foundations of hedonic psychology* (pp. 134–154). New York: Russell Sage Foundation.

LaBar, K. S., Gatenby, J. C., Gore, J. C., LeDoux, J. E., & Phelps, E. A. (1998). Human amygdala activation during conditioned fear acquisition and extinction: A mixed trial fMRI study. *Neuron, 20*, 937–945.

LaGasse, L., Gruber, C., & Lipsitt, L. P. (1989). The infantile expression of avidity in relation to later assessments. In J. S. Reznick (Ed.), *Perspectives on behavioral inhibition* (pp. 159–176). Chicago: University of Chicago Press.

Lamb, R. J., Preston, K. L., Schindler, C. W., Meisch, R. A., Davis, F., Katz, J. L., et al. (1991). The reinforcing and subjective effects of morphine in post-addicts: A dose-response study. *Journal of Pharmacology and Experimental Therapies, 259*, 1165–1173.

Lambert, M. J., & Barley, D. E. (2002). Research summary on the therapeutic relationship and psychotherapy outcome. In J. C. Norcross (Ed.), *Psychotherapy relationships that work: Therapist contributions and responsiveness to patients* (pp. 17–32). London: Oxford University Press.

Lambert, M. J., Bergin, A. E., & Garfield, S. L. (2003). Introduction and historical overview. In M. J. Lambert (Ed.), *Bergin and Garfield's handbook of psychotherapy and behavior change.* (5th. ed.). New York: Wiley.

Lane, R. D., Reiman, E. M., Axelrod, B., Yun, L.-S., Holmes, A., & Schwartz, G. E. (1998). Neural correlates of levels of emotional awareness: Evidence of an interaction between emotion and attention in the anterior circular cortex. *Journal of Cognitive Neuroscience, 10*, 525–535.

Lang, P. J. (1995). The emotion probe: Studies of motivation and attention. *American Psychologist, 50*, 372–385.

Larsen, R. J., & Ketelaar, T. (1991). Personality and susceptibility to positive and negative emotional states. *Journal of Personality and Social Psychology, 61*, 132–140.

Last, C. G., Hansen, C., & Franco, N. (1997). Anxious children in adulthood: A prospective study. *Journal of the American Academy of Child and Adolescent Psychiatry, 36*, 645–652.

Lauterbach, W. (1996). The measurement of personal conflict. *Psychotherapy Research, 6*, 213–225.

Lazarus, R. S., & Folkman, S. (1984). *Stress, appraisal, and coping.* New York: Springer.

Leckman, J. F., Grice, D. E., Boardman, J., & Zhang, H. (1997). Symptoms of obsessive-compulsive disorder. *American Journal of Psychiatry, 154*, 911–917.

LeDoux, J. E. (2002). *Synaptic self: How our brains become who we are.* New York: Viking Penguin.

LeDoux, J. E. (2004). *The emotional brain.* London: Phoenix.

Leibowitz, S. F., & Hoebel, B. G. (1998). Behavioral neuroscience of obesity. In G. A. Bray, C. Bouchard, & W. P. T. James (Eds.), *Handbook of obesity* (pp. 315–358). New York: Marcel Dekker.

Leon, M. I., & Shadlen, M. N. (1999). Effects of expected reward magnitude on the response of neurons in the dorsolateral prefrontal cortex of the macaque. *Neuron, 2*, 415–425.

Lesch, K. P., Bengel, D., Heils, A., Sabol, S. Z., Greenberg, B. D., Petri, S., et al. (1996). Association of anxiety-related traits with a polymorphism in the serotonin transporter gene regulatory region. *Science, 274*, 1527–1531.

Lewinsohn, P. M., Hops, H., Roberts, R. E., & Seeley, J. R. (1993). Adolescent psychopathology I: Prevalence and incidence of depression and other DSM-III-R disorders in high school students. *Journal of Abnormal Psychology, 102*, 133–144.

Libet, B. (1978). Neuronal vs. subjective timing for a conscious sensory experience. In P. A. Buser & A. Rougeul-Buser (Eds.), *Cerebral correlates of conscious experience.* (pp. 69–82). Amsterdam: Elsevier.

Libet, B., Gleason, C. A., Wright, E. W., & Pearl, D. K. (1983). Time of conscious intention to act in relation to onset of cerebral activity (readiness-potential). *Brain, 106*, 623–642.

Liepert, J., Bauder, H., Miltner, W., Taub, E., & Weiller, C. (2000). Treatment-induced cortical reorganization after stroke in humans. *Stroke, 31*, 1210–1216.

Liggan, D. Y., & Kay, J. (1999). Some neurobiological aspects of psychotherapy: A review. *Journal of Psychotherapy Practice and Research, 8*, 103–114.

Lindzey, G., & Aronson, E. (1968). *The handbook of social psychology* (2nd ed.). Reading, MA: Addison-Wesley.

Loftus, E. F. (1993). The reality of repressed memories. *American Psychologist, 48,* 518–537.

Louie, K., & Wilson, M. A. (2001). Temporally structured replay of awake hippocampal ensemble activity during rapid eye movement sleep. *Neuron, 29,* 145–156.

Lupien, S. J., de Leon, M., de Santi, S., Convil, A., Tarshish, C., Nair, N. P., et al. (1998). Cortisol levels during human aging predict hippocampal atrophy and memory deficits. *Nature Neuroscience, 1,* 69–73.

MacDonald, A. W., Cohen, J. D., Stenger, V. A., & Carter, C. S. (2000). Dissociating the role of the dorsolateral prefrontal and anterior cingulate cortex in cognitive control. *Science, 288,* 1835–1838.

Magnus, K., Diener, E., Fijita, F., & Pavot, W. (1993). Extraversion and neuroticism as predictors of objective life events: A longitudinal analysis. *Journal of Personality and Social Psychology, 65,* 1046–1053.

Maguire, E. A., Burgess, N., Donnett, J. G., Frackowiak, R. S. J., Frith, C. D., & O'Keefe, J. (1998). Knowing where and getting there: A human navigation network. *Science, 80,* 921–924.

Maguire, E. A., Frackowiak, S. J., & Frith, C. D. (1997). Recalling routes around London: Activation of the right hippocampus in taxi drivers. *Journal of Neuroscience, 17,* 7103–7110.

Maguire, E. A., Gadian, D. G., Johnsrude, I. S., Good, C. D., Ashburner, J., Frakowiack, R. S., et al. (2000). Navigation-related structural change in the hippocampi of taxi drivers. Proceedings of the *National Academy of Science of the United States of America, 97,* 4398–4403.

Maier, S. F., & Seligman, M. E. (1976). Learned helplessness. Theory and evidence. *Journal of Experimental Psychology: General, 105,* 3–46.

Main, M., Kaplan, N., & Cassidy, J. (1985). Security in infancy, childhood and adulthood: A move to the level of representation. Monographs of the Society for Research in Child Development 50(1&2, Serial No. 209).

Malberg, J. E., Eisch, A. J., Nestler, E. J., & Duman, R. S. (2000). Chronic antidepressant treatment increases neurogenesis in adult rat hippocampus. *Journal of Neuroscience, 20,* 9104–9110.

Margraf, J. (2001). *Neue Ergebnisse zu Entstehung und Verlauf psychischer Störungen* [New findings on the etiology and course of mental disorders]. Bern, Switzerland: Paper presented at the meeting of the section Clinical Psychology and Psychotheraphy of the German Psychological Association.

Mark, G. P., Rada, P., Pothos, E., & Hoebel, B. G. (1992). Effects of feeding and drinking on acetylcholine release in the nucleus accumbens, striatum, and hippocampus of freely behaving rats. *Journal of Neurochemistry, 58,* 2269–2274.

Markowitsch, H. (2002). Dem Gedächtnis auf der Spur [Chasing memory]. Darmstadt: Primus Verlag.

Martin, S. D., Martin, E., Rai, S. S., Santoch, S., & Richardson, M. A. (2001). Brain blood flow changes in depressed patients treated with interpersonal psychotherapy or venlafaxine hydrochloride. *Archives of General Psychiatry, 58,* 641–648.

Maslow, A. H. (1967). A theory of metamotivation: The biological rooting of the value-life. *Journal of Humanistic Psychology, 7,* 93–127.

Mayberg, H. S. (1997). Limbic-cortical dysregulation: A proposed model of depression. *Journal of Neuropsychiatry and Clinical Neuroscience, 9,* 471–481.

Mayberg, H. S., Liotti, M., Brannan, S. K., McGinnis, S., Mahurin, R. K., Jerabek, P. A., et al. (1999). Reciprocal limbic-cortical function and negative mood: Converging PET-findings in depression and normal sadness. *American Journal of Psychiatry, 156,* 675–682.

McClelland, D. D. (1985). How motives, skills and values determine what people do. *American Psychologist, 40,* 812–825.

McDougall, W. (1932). *The energies of men.* London: Methuen.

McNaughton, B. L. (1998). The neurophysiology of reminiscence. *Neurobiology, Learning, and Memory, 70,* 252–267.

Meichenbaum, D. H. (1993). Stress inoculation training: A 20-year update. In P. M. Lehrer & R. L. Woolfolk (Eds.), *Principles and practice of stress management* (2nd ed). New York: Guilford.

Merikangas, K. R., & Angst, J. (1995). The challenge of depressive disorders in adolescence. In M. Rutter (Ed.), *Psychosocial disturbances in young people: Challenges for prevention* (pp. 131–165). New York: Cambridge University Press.

Merten, J. (1996). *Affekte und die Regulation nonverbalen interaktiven Verhaltens* [Emotions and the regultion of nonverbal interaction behavior]. Bern, Switzerland: Peter Lang.

Merzenich, M. M., Jenkins, W. M., Johnston, P., Schreiner, C., Miller, S. L., & Tallal, P. (1996). Temporal processing deficits of language-learning impaired children ameliorated by training. *Science, 271,* 77–81.

Merzenich, M. M., Nelson, R. J., Stryker, M. P., Cynader, M. S., Schoppman, A., & Zook, M. J. (1984). Somatosensory cortical map changes following digit amputation in adult monkeys. *Journal of Comparative Neurology, 224,* 591–605.

Mesulam, M.-M. (2000). Neural substrates of psychiatric syndromes. In *Principles of cognitive and behavioral neurology* (pp. 406–438). New York: Oxford University Press.

Meyer, B., & Pilkonis, P. (2002). Attachment style. In J. C. Norcross (Ed.), *Psychotherapy relationships that work: Therapist contributions and responsiveness to patients* (pp. 367–382). London: Oxford University Press.

Michalak, J., Heidenreich, T., & Hoyer, J. (2001). Konflikte zwischen Patientenzielen—Konzepte, Ergebnisse und Konsequenzen für die Therapie [Conflicting goals of patients—Concepts, findings, and consequences for therapeutic practice]. *Verhaltenstherapie und psychosoziale Praxis, 33,* 273–280.

Michalak, J., Klappheck, M. A., & Kosfelder, J. (2004). Personal goals of psychotherapy patients: The intensity and the "why" of goal-motivated behavior and their implications for the therapeutic process. *Psychotherapy Research, 14,* 193–209.

Michalak, J., Püschel, O., Joormann, J., & Schulte, D. (2006). Implicit motives and explicit goals: Two distinctive modes of motivational functioning and their relations to psychopathology. *Clinical Psychology and Psychotherapy, 13,*(2), 81–96..

Michalak, J., & Schulte, D. (2002). Zielkonflikte und Therapiemotivation [Goal conflicts and therapy motivation]. *Zeitschrift für Klinische Psychologie und Psychotherapie: Forschung und Praxis, 31,* 213–219.

Miller, E. K., & Cohen, J. D. (2001). An integrative theory of prefrontal cortex function. *Annual Review of Neuroscience, 24,* 167–202.

Miller, G. A., Galanter, E., & Pribram, K. H. (1960). *Plans and the structure of behavior.* Oxford, England: Holt.

Miller, N. E. (1944). Experimental studies of conflict. In J. Hunt (Ed.), *Personality and the behavior disorders*, Vol. 1 (pp. 431–465). New York: Ronald.

Milner, B. (1965). Memory disturbances after bilateral hippocampal lesions in man. In P. M. Milner & S. E. Glickman (Eds.), *Cognitive processes and the brain.* Oxford, England: Van Nostrand.

Milner, B. (1967). Brain mechanisms suggested by studies of temporal lobes. In F. L. Darley (Ed.), *Brain mechanisms underlying speech and language.* New York: Grunne and Stratton.

Milner, B. (1972). Disorders of learning and memory after temporal lobe lesions in man. *Clinical Neurosurgery, 19,* 421–446.

Miltner, W., Braun, C. H., & Coles, M. G. (1997). Event-related brain potentials following incorrect feedback in a time-estimation task: Evidence for a "generic" neural system for error detection. *Journal of Cognitive Neuroscience, 9,* 788–798.

Miltner, W. H., Braun, C., Arnold, M., Witte, H., & Taub, E. (1999). Coherence of gamma-band EEG activity as a basis for associative learning. *Nature, 397,* 424–426.

Mineka, S., Gunnar, M., & Champoux, M. (1986). Control and early socioemotional development: Infant rhesus monkeys reared in controllable versus uncontrollable environments. *Child Development, 57,* 1241–1256.

Mineka, S., & Zinbarg, R. (1996). Conditioning and ethological models of anxiety disorders. In D. A. Hope (Ed.), *Nebraska Symposion on Motivation: Vol. 43. Perspectives on anxiety, panic, and fear* (pp. 135–210). Lincoln: University of Nebraska Press.

Mirenowicz, J., & Schultz, W. (1994). Importance of unpredictability for reward responses in primate dopamine neurons. *Journal of Neurophysiology, 72,* 1024–1027.

Mirenowicz, J., & Schultz, W. (1996). Preferential activation of midbrain dopamine neurons by appetitive rather than aversive stimuli. *Nature, 379,* 449–451.

Modell, J. G., Mountz, J. M., Curtis, G. C., & Greden, J. F. (1989). Neurophysiological dysfunction in basal ganglia/limbic striatial and thalamo-cortical circuits as a pathogenetic mechanism of obsessive-compulsive disorder. *Journal of Neuropsychiatry, 1,* 27–36.

Montague, P. R., Dayan, P., & Sejmowski, T. J. (1996). A framework for mesencephalic dopamine systems based on predictive Hebbian learning. *Journal of Neuroscience, 16*, 1936–1947.

Morgan, M. A., & LeDoux, J. E. (1995). Differential contribution of dorsal and ventral medial prefrontal cortex to the acquisition and extinction of conditioned fear in rats. *Behavioral Neuroscience, 4*, 681–688.

Morris, J. S., Frith, C. D., Perret, I. D., Rowland, D., Young, W., Calder, A. J., et al. (1996). A differential neural response in the human amygdala to fearful and happy facial expressions. *Nature, 383*, 812–815.

Morris, J. S., Öhman, A., & Dolan, R. J. (1998). Conscious and unconscious emotional learning in the human amygdala. *Nature, 393*, 467–470.

Morris, J. S., Öhman, A., & Dolan, R. J. (1999). A subcortical pathway to the right amygdala mediating "unseen" fear. *Proceedings of the National Academy of Science of the United States of America, 96*, 1680–1685.

Muneoka, K., Mikuni, M., Ogawa, T. K. K., & Takahashi, K. (1994). Periodic maternal deprivation-induced potentiation of the negative feedback sensitivity to glucocorticoids to inhibit stress-induced adrenocortical response persists throughout animal's life-span. *Neuroscience Letter, 168*, 89–92.

Murray, H. A. (1938). *Explorations in personality*. New York: Oxford University Press.

Nadasdy, Z. H., Hirase, H., Czurko, A., Csicsvari, J., & Buzsaki, G. (1999). Replay and time compression of recurring spike sequences in the hippocampus. *Journal of Neuroscience, 19*, 9497–9507.

Nadel, L., & Moscovitch, M. (1997). Memory consolidation, retrograde amnesia, and the hippocampal complex. *Current Opinion in Neurobiology, 7*, 217–227.

Nagel, T. (1974). What is it like to be a bat? *Philosophical Review, 83*, 4435–4450.

Nathan, P. E. (1998). Practice guidelines: Not yet ideal. *American Psychologist, 53*, 290–299.

Nathan, P. E., Gorman, J. M., & Salkind, N. J. (1999). *Treating mental disorders: A guide to what works*. New York: Oxford University Press.

Neumann, R., & Strack, F. (2000). Approach and avoidance: The influence of proprioceptive and exterproprioceptive cues on encoding of affective informations. *Journal of Personality and Social Psychology, 79*, 39–48.

Nisbett, R. E., & Ross, L. (1980). *Human inference: Strategies and shortcomings of social judgment*. Engelwood Cliffs, NJ: Prentice Hall.

Nofzinger, E. A., Nichols, T. E., Meltzer , C. C., Price, J., Steppe, D. A., et al. (1999). Changes in forebrain function from waking to REM sleep in depression: Preliminary analyses from (18F) FDG PET studies. *Psychiatry Research, 91*, 59–78.

Noga, J. T., Vladar, K., & Torrey, E. F. (2001). A volumetric magnetic resonance imaging study of monozygotic twins discordant for bipolar disorder. Psychiatric Research: *Neuroimaging, 106*, 25–34.

Norcross, J. C. (Ed.). (2002). *Psychotherapy relationships that work*. New York: Oxford University Press.

Nyberg, L., McIntosh, A. R., Houle, S., Nilsson, L.-G., & Tulving, E. (1996). Activation of medial temporal structures during episodic memory retrieval. *Nature, 380*, 715–717.

Nydegger, S. (2003). *Intrapsychische Konflikte: Vergleich verschiedener Methoden der Konfliktmessung und Untersuchung klinischer Zusammenhänge* [Intrapsychic conflicts—Assessment and clinical correlates]. Unpublished master's thesis. Institut für Psychologie, University of Bern., Bern, Switzerland.

O'Carroll, R. E., Drysdale, E., Cahill, L., Shajahan, P., & Ebmeier, K. P. (1999). Stimulation of the noradrenergic system enhances and blockade reduces memory for emotional material in man. *Psychological Medicine, 29*, 1083–1088.

O'Doherty, J., Kringelbach, M. L., Rolls, E. T., Hornak, J., & Andrews, C. (2001). Abstract reward and punishment representations in the human orbitofrontal cortex. *National Neuroscience, 4*, 95–102.

O'Doherty, J., Rolls, E. T., Francis, F., Bowtell, R., McGlone, F., Kobal, G., et al. (2000). Sensory-specific satiety-related olfactory activation of the human orbitofrontal cortex. *NeuroReport, 11*, 893–897.

Oesterreich, R. (1981). *Handlungsregulation und Kontrolle* [Action regulation and control]. Munich, Germany: Urban & Schwarzenberg.

Ogawa, T., Mikuni, M., Kuroda, Y., Muneoka, K., Miori, K. J., & Takahashi, K. (1994). Periodic maternal deprivation alters stress response in adult offspring, potentiates the negative feedback regulation of restraint stress induced adrenocortical response and reduces the frequencies of open field-induced behaviors. *Pharmacology Biochemistry Behavior, 49,* 961–967.

Olds, J., & Milner, P. (1954). Positive reinforcement produced by electrical stimulation of septal area and other regions of rat brain. *Journal of Comparative and Physiological Psychology, 47,* 419–427.

Öngür, D., Drevets, W. C., & Price, J. L. (1998). Glia reduction in the subgenual prefrontal cortex in mood disorders. *Proceedings of the National Academy of Sciences of the United States of America, 95,* 13290–13295.

Orlinsky, D., Grawe, K., & Parks, B. (1994). Process and outcome in psychotherapy—noch einmal. In A. E. Bergin & S. L. Garfield (Eds.), *Handbook of psychotherapy and behavior change.* (4th ed., pp. 270–376). New York: Wiley.

Osgood, C. E., & Suci, G. J. (1955). Factor analysis of meaning. *Journal of Experimental Psychology, 50,* 325–338.

Overmier, J. B., & Seligman, M. E. (1967). Effects of inescapable shock upon subsequent escape and avoidance behavior. *Journal of Comparative and Physiological Psychology, 63,* 23–33.

Panksepp, J. (1998). *Affective neuroscience. The foundations of human and animal emotions.* New York: Oxford University Press.

Pantev, C., Oostenveld, R., Engelien, A., Ross, B., Roberts, L. E., & Hoke, M. (1998). Increased auditory cortical representation in musicians. *Nature, 392,* 811–814.

Pantev, C., Roberts, L. E., Schulz, M., Engelien, A., & Ross, B. (2001). Timbre-specific enhancement of auditory cortical representation in musicians. *NeuroReport, 12,* 169–174.

Paquette, V., Levesque, J., Mensour, B., Leroux, J.-M., Beaudoin, G., Bourgouin, P., et al. (2003). Change the mind and you change the brain: Effects of cognitive-behavioral therapy on the neural correlates of spider phobia. *Neuroimage 18,* 401–409.

Pardo, J. V., Pardo, P. J., Janer, K. W., & Raichle, M. E. (1990). The anterior cingulate cortex mediates processing selection in the Stroop attentional conflict paradigm. *Proceedings of the National Academy of Sciences of the United States of America, 87,* 256–259.

Pariante, C. M., & Miller, A. H. (2001). Glucocorticoid receptors in major depression: Relevance to pathophysiology and treatment. *Biological Psychiatry, 49,* 391–404.

Parker, G. (1979a). Parental characteristics in relation to depressive disorders. *British Journal of Psychiatry, 134,* 138–147.

Parker, G. (1979b). Reported parental characteristics in relation to trait depression and anxiety levels in a non-clinical group. *New Zealand Journal of Psychiatry, 13,* 260–264.

Parker, G. (1981). Parental representation of patients with anxiety neurosis. *Acta Psychiatrica Scandinavica, 63,* 33–36.

Parker, G. (1983). *Parental overprotection: A risk factor in psychosocial development.* New York: Grune & Stratton.

Parker, R. M. (1992). *Bordeaux.* Bern, Switzerland/Stuttgart, Germany: Hallwag.

Pauli-Pott, U., & Bade, U. (2002). Bindung und Temperament [Attachment and temperament]. In B. Strauss, A. Buchheim, & H. Kächele (Eds.), *Klinische Bindungsforschung* (pp. 129–144). Stuttgart, Germany: Schattauer.

Pawlow, I. P. (1927). *Conditioned reflexes.* New York: Oxford University Press.

Penfield, W., & Perot, P. (1963). The brain's record of an auditory and visual experience. *Brain, 86,* 595–696.

Pennebaker, J. W. (1993). Putting stress into words: Health, linguistic, and therapeutic implications. *Behaviour Research and Therapy, 31,* 539–548.

Perrett, D. I., Mistlin, A. J., & Chitty, A. J. (1987). Visual neurones responsive to faces. *Trends in Neurosciences, 10,* 358–364.

Perrett, D. I., Smith, P. A., Potter, D. D., Mistlin, A. J., Head, A. S., Milner, A. D., et al. (1984). Neurones responsive to faces in the temporal cortex: Studies of functional organization, sensitivity to identity and relation to perception. *Human Neurobiology, 3,* 197–208.

Petersen, A. C., Leffert, N., & Hurrelmann, K. (1993). Adolescence and schooling in Germany and the United States: A comparison of peer socialization to adulthood. *Teachers College Record, 94,* 611–628.

Petrides, M. (1990). Nonspatial conditional learning impaired in patients with unilateral frontal but not unilateral temporal lobe excisions. *Neuropsychologia, 28,* 137–149.

Petrides, M. (1994). Frontal lobes and behaviour. *Current Opinion in Neurobiology, 4,* 207–211.

Pfaus, J. G., Damsma, G., Wenkstern, D., & Fibiger, H. C. (1995). Sexual activity increases dopamine transmission in the nucleus accumbens and striatum of female rats. *Brain Research, 693,* 21–30.

Phelps, E. A., O'Connor, K. J., Cunningham, W. A., Funayama, E. S., Gatenby, J. C., Gore, J. C., et al. (2000). Performance on indirect measures of race evaluation predicts amygdala activation. *Journal of Cognitive Neuroscience, 12,* 729–738.

Piaget, J. (1977). *The development of thought: Equilibration of cognitive structures.* Oxford, England: Viking.

Pilkonis, P. (1988). Personality prototypes among depressives: Themes of dependency and autonomy. *Journal of Personality Disorders, 2,* 144–152.

Pizzagalli, D., Pascual-Marqui, R. D., Nitschke, J. B., Oakes, T. R., Larson, C. L., Abercrombie, H. C., et al. (2001). Anterior cingulate activity as a predictor of degree of treatment response in major depression: Evidence from brain electrical tomography analysis. *American Journal of Psychiatry, 158,* 405–415.

Poe, G. R., Nitz, D. A., McNaughton, B. L., & Barnes, C. A. (2000). Experience-dependent phase-reversal of hippocampal neuron firing during REM sleep. *Brain Research, 855,* 176–180.

Popper, K. R., & Eccles, J. C. (1977). *The self and its brain.* New York: Springer.

Pothos, E., Creese, I., & Hoebel, B. G. (1995). Restructured eating with weight loss selectively decreases extracellular dopamine in the nucleus accumbens and alters dopamine response to amphetamine, morphine, and food intake. *Journal of Neuroscience, 15,* 6640–6650.

Powers, W. T. (1973). *Behavior and the control of perception.* New York: Aldine.

Prabhakaran, V., Narayanan, K., Zhao, Z., & Gabrieli, J. D. (2000). Integration of diverse information in working memory within the frontal lobe. *Nature Neuroscience, 3,* 85–90.

Quintana, J., & Fuster, F. (1992). Mnemonic and predictive functions of cortical neurons in a memory task. *NeuroReport, 3,* 721–724.

Quirk, G. J., Russo, G. K., Barron, J. L., & Lebron, K. (2000). The role of ventromedial prefrontal cortex in the recovery of extinguished fear. *Journal of Neuroscience, 20,* 6225–6231.

Rada, P., Mark, G., & Hoebel, B. G. (1998). Dopamine release in the nucleus accumbens by hypothalamic stimulation-escape behavior. *Brain Research, 782,* 228–234.

Radke-Yarrow, M. (1991). Attachment patterns in children of depressed mothers. In P. Harris, J. Stevenson-Hinde, & C. Parkes (Eds.), *Attachment across the life cycle* (pp. 115–126). London: Routledge.

Radke-Yarrow, M., Cummings, E. M., Kuczynski, L., & Chapman, M. (1985). Pattern of attachment in two- and three-year-olds in normal families and families with patental depression. *Child Development, 56,* 884–893.

Rajkowska, G. (2000). Postmortem studies in mood disorders indicate altered numbers of neurons and glia cells. *Biological Psychiatry, 48,* 766–777.

Rapoport, J. L., & Wise, S. P. (1988). Obsessive-compulsive disorder: Is it a basal ganglia dysfunction? *Psychopharmacology Bulletin, 24,* 380–384.

Rauch, S. L., Dougherty, D. D., & Shin, I. M. (1998). Neural correlates of factor-analyzed OCD symtom dimensions: A PET study. *CNS Spectrums, 3,* 37–43.

Rauch, S. L., Jenike, M. A., Alpert, N. A., Baer, L., Breiter, H. C., Savage, C. R., et al. (1994). Regional cerebral blood flow measured during symptom provocation in obsessive-compulsive disorder using 15O-labeled CO2 and positron emission tomography. *Archives of General Psychiatry, 51,* 62–70.

Regli, D., Bieber, K., Mathier, F., & Grawe, K. (2000). Beziehungsgestaltung und Aktivierung von Ressourcen in der Anfangsphase von Therapien [Management of the therapeutic relationship and activation of resources in initial therapy sessions]. *Verhaltenstherapie und Verhaltensmedizin, 21,* 399–420.

Reid, S. A., Duke, L. M., & Allen, J. J. (1998). Resting frontal electroencephalographic asymmetry in depression: Inconsistencies suggest the need to identify mediating factors. *Psychophysiology, 35,* 389–404.

Reiman, E. M. (1997). The application of positron emission tomography to the study of normal and pathological emotion. Journal of Clinical Psychiatry, 58, 4–12.

Renken, B., Egeland, B., Marvinney, D., Mangelsdorf, S., & Sroeufe, L. A. (1989). Early childhood antecedents of aggression and passive-withdrawal in early elementary school. *Journal of Personality, 57*, 257–281.

Renner, W., & Leibetseder, M. (2000). The relationship of personal conflict and clinical symptoms in a high-conflict and a low-conflict subgroup: A correlational study. *Psychotherapy Research, 10*, 321–336.

Ricks, M. (1985). The social transmission of parental behavior: Attachment across generations. In I. Bretherton & E. Waters (Eds.), In: *Growing points in attachment theory and research. Monographs of the Society for Research in Child Development, 50*, 211–227.

Rief, W., & Hiller, W. (1998). *Somatisierungsstörungen und Hypochondrie* [Somatization and hypochondriasis]. Göttingen, Germany: Hogrefe.

Robbins, A. S., Kauss, D. R., Heinrich, R., Abrass, L., Dreyer, J., & Clyman, B. (1979). Interpersonal skills training: Evaluation in an internal medicine residency. *Journal of Medical Education, 54*, 885–894.

Roberts, J., & Monroe, S. (1999). Vulnerable self-esteem and social processes in depression: Toward an interpersonal model of self-esteem regulation. In T. Joiner & J. Coyne (Eds.), *The interactional nature of depression* (pp. 149–187). Washington, DC: American Psychological Association.

Roemer, L., & Orsillo, S. (2003). Mindfulness: A promising intervention strategy in need for further study. *Clinical Psychology: Science and Practice, 10*, 172–178.

Rogers, R. D., Owen, A. M., Middleton, H. C., Williams, E. J., Pickard, J. D., Sahakian, B. J., et al. (1999). Choosing between small, likely rewards and large, unlikely rewards activates inferior and orbital prefrontal cortex. *Journal of Neuroscience, 20*, 9029–9038.

Rohde, P., Lewinsohn, P. M., & Seeley, J. R. (1991). Comorbidity of unipolar depression: II. Comorbidity with other mental disorders in adolescents and adults. *Journal of Abnormal Psychology, 100*, 214–222.

Rohde-Dachser, C. (2003). Interview zum Thema Psychoanalyse und Persönlichkeitsstörungen. Bestätigung durch Neurowissenschaft [An interview on psychoanalysis and personality disorders]. *Deutsches Ärzteblatt für Psychologische Psychotherapeuten und Kinder- und Jugendlichentherapeuten, 9*, 416–417.

Rolls, E. T. (1984). Neurons in the cortex of the temporal lobe and in the amygdala of the monkey with responses selective for faces. *Human Neurobiology, 3*, 209–222.

Rolls, E. T. (1995). Central taste anatomy and neurophysiology. In R. L. Doty (Ed.), *Handbook of olfaction and gustation* (pp. 549–573). New York: Marcel Dekker.

Rolls, E. T. (2000). Precis of the brain and emotion. *Behavioral and Brain Sciences, 23*, 177–234.

Rosenblum, L. A., & Andrews, M. W. (1994). Influences of environmental demand on maternal behavior and infant development. *Acta Paediatrica Supplement, 397*, 57–63.

Roth, G. (1995). *Das Gehirn und seine Wirklichkeit* [The brain and its reality. Cognitive neurobiology and its philosophical consequences] (3rd ed.). Frankfurt, Germany: Suhrkamp.

Roth, G. (2001). *Fühlen, Denken, Handeln. Wie das Gehirn unser Verhalten steuert* [Cognitions, emotions, and action]. Frankfurt, Germany: Suhrkamp.

Rotter, J. B. (1966). Generalized expectancies for internal versus external control of reinforcement. *Psychological Monographs, 80*.

Rozin, P. (1990). Getting to like the burn of chili pepper: Biological, psychological, and cultural perspectives. In B. G. Green, J. R. Mason, & M. L. Kare (Eds.), *Chemical irritation in the nose and mouth* (pp. 231–269). New York: Marcel Dekker.

Rozin, P. (1999). Preadaptation and the puzzles and properties of pleasure. In D. Kahneman, E. Diener, & N. Schwartz (Eds.), *Well-being: The foundations of hedonic psychology* (pp. 109–133). New York: Russell Sage Foundation.

Rozin, P., & Schiller, D. (1980). The nature and acquisition of a preference for chili pepper by humans. *Motivation and Emotion, 4*, 77–101.

Rumelhart, D. E., & McClelland, J. L. (Eds.). (1986). *Parallel distributed processing: Explorations in the microstructure of cognition.* Cambridge, MA: MIT Press.

Rush, A., & Thase, M. (1999). Psychotherapy for depressive disorders: A review. In M. Maj & N. Sartorius (Eds.), *Depressive disorders* (pp. 161–232). New York: Wiley.

Sachse, R. (2003). *Klärungsorientierte Psychotherapie* [Resolving-oriented psychotherapy]. Göttingen, Germany: Hogrefe.

Sacks, O. (1998). *The man who mistook his wife for a hat.* New York: Touchstone.

Sameroff, A., Seifer, R., & Zax, M. (1982). Early development of children at risk for emotional disorder. *Monographs of the Society for Research in Child Development, 47,* 82–90.

Sapolsky, R. M. (2000). Glucocorticoids and hippocampal atrophy in neuropsychiatric disorders. *Archives of General Psychiatry, 57,* 925–935.

Savin-Wiliams, R. C., & Jacquish, G. A. (1981). The assessment of adolescent self-esteem. A comparison of methods. *Journal of Personality, 49,* 324–335.

Saxena, S., Brody, A. L., Ho, M. L., Alborzian, S., Ho, M. K., Maidment, K. M., et al. (2001). Cerebral metabolism in major depression and obsessive-compulsive disorder occurring separately and concurrently. *Biological Psychiatry, 50,* 159–170.

Schachter, S., & Singer, J. E. (1962). Cognitive, social, and physiological determinants of emotional state. *Psychological Review, 69,* 379–399.

Schacter, D. L. (1987). Implicit memory: History and current status. *Journal of Experimental Psychology: Learning, Memory, Cognition, 13,* 501–518.

Schacter, D. L., & Curran, T. (2000). Memory without remembering and remembering without memory: Implicit and false memories. In M. S. Gazzaniga (Ed.), *The new cognitive neuroscience* (pp. 829–840). Cambridge, MA: The MIT Press.

Schacter, D. L., Norman, K. A., & Koutstaal, W. (1998). The cognitive neuroscience of constructive memory. *Annual Review of Psychology, 49,* 289–318.

Schacter, D. L., Reiman, E., Curran, T., Yun, L. S., Bandy, D., MsDermott, K. B., et al. (1996). Neuroanatomical correlates of veridical and illusory recognition memory: Evidence from positron emission tomography. *Neuron, 17,* 267–274.

Schacter, D. L., Verfaelle, M., & Pradere, D. (1996). The neuropsychology of memory illusions: False recall and recognition in amnesic patients. *Journal of Memory and Language, 35,* 319–334.

Schauenburg, H., & Strauss, B. (2002). Bindung und Psychotherapie [Attachment and psychotherapy]. In B. Strauss, A. Buchheim, & H. Kächele (Eds.), *Klinische Bindungsforschung* (pp. 281–292). Stuttgart, Germany: Schattauer.

Schiepek, G. (Ed.). (2003). *Neurobiologie der Psychotherapie* [Neurobiology of psychotherapy]. Stuttgart, Germany: Schattauer.

Schmidt, S., & Strauss, B. (1996). Die Bindungstheorie und ihre Relevanz für die Psychotherapie. Teil 1: Grundlagen und Methoden der Bindungsforschung [Attachment theory and its relevance for psychotherapy. Part 1: Basic aspects and methods of attachment research]. *Psychotherapeut, 41,* 139–150.

Schramm, E. (1996). *Interpersonelle Psychotherapie* [Interpersonal psychotherapy]. Stuttgart, Germany: Schattauer.

Schulte, D. (1993). Wie soll Therapieerfolg gemessen werden? [How treatment success should be assessed]. *Zeitschrift für Klinische Psychologie, 22,* 374–393.

Schulte, D. (2000). Angststörungen [Anxiety disorders]. In G. Lazarus-Mainka & S. Siebeneick (Eds.), Angst und Ängstlichkeit. Ein Lehrbuch. Göttingen, Germany: Hogrefe.

Schulte, D. (2003). Zur Validität therapeutischer Entscheidungsprozesse [The validity of decision processes in psychotherapy]. In W. Vollmoeller (Ed.), *Integrative Behandlung in Psychoatrie und Psychotherapie* (pp. 89–100). Stuttgart, Germany: Schattauer.

Schulte-Bahrenberg, T., & Schulte, D. (1993). Change of psychotherapy goals as a process of resignation. *Psychotherapy Research, 3,* 153–165.

Schultz, W. (1998). Predictive reward signal of dopamine neurons. *Journal of Neurophysiology, 80,* 1–27.

Schultz, W., Apicella, P., & Ljungberg, T. (1993). Responses of monkey dopamine neurons to reward and conditioned stimuli during successive steps of learning a delayed response task. *Journal of Neuroscience, 13,* 900–913.

Schultz, W., Dayan, P., & Montague, P. R. (1997). A neural substrate of prediction and reward. *Science, 275,* 1593–1599.

Schultz, W., & Dickinson, A. (2000). Neuronal coding of prediction errors. *Annual Review of Neuroscience, 23,* 473–500.

Schwartz, G. (1990). Psychobiology of repression and health. A systems approach. In J. L. Singer (Ed.), *Repression and dissociation.* Chicago: The University of Chicago Press.

Schwartz, G. E. (1983). Disregulation theory and disease: Applications to the repression/cerebral disconnection/cardiovascular disorder hypothesis. *International Review of Applied Psychology, 32,* 95–118.

Schwartz, G. E., Fair, P. L., Salt, P., Mandel, M. R., & Klerman, G. L. (1976). Facial muscle pattern-ing to affective imagery in depressed and nondepressed subjects. *Science, 192,* 489–491.

Schwartz, J. M., Toessel, P. W., Baxter, L. R., Martin, K. M., & Phelps, M. E. (1996). Systematic changes in cerebral glucose metabolic rate after successful behavior modification treatment of obsessive-compulsive disorder. *Archives of General Psychiatry, 53,* 109–113.

Scott, S. K., Young, A. W., Calder, A. J., Hellawell, D. J., Aggleton, J. P., & Johnson, M. (1997). Impaired auditory recognition of fear and anger following bilateral amygdala lesions. *Nature, 385,* 254–257.

Scoville, W. B., & Milner, B. (1957). Loss of recent memory after bilateral hippocampal lesions. *Journal of Neurology and Psychiatry, 20,* 11–21.

Sedikides, C., & Green, J. D. (2000). On the self-protective nature of inconsistency-negativity man-agement: Using the person memory paradigm to examine self-referent memory. *Journal of Personality and Social Psychology, 79,* 906–922.

Seligman, M. E., & Maier, S. F. (1967). Failure to escape traumatic shock. *Journal of Experimental Psychology, 74,* 1–9.

Sheldon, K. M., & Elliott, A. J. (1999). Goal striving, need satisfaction, and longitudinal well-being: The self-concordance model. *Journal of Personality and Social Psychology, 6,* 482–497.

Sheldon, K. M., & Kasser, T. (1995). Coherence and congruence: Two aspects of personality integra-tion. *Journal of Personality and Social Psychology, 68,* 531–543.

Sheline, Y., Sanghavi, M., Mintun, M., & Gado, M. (1999). Depression duration but not age predicts hippocampal volume loss in medically healthy women with recurrent major depression. *Journal of Neuroscience, 19,* 5034–5043.

Sheline, Y. L. (2000). 3 D MRI studies of neuroanatomic changes in unipolar major depression: The role of stress and medical comorbidity. *Biological Psychiatry, 48,* 791–800.

Shevrin, H. (1990). Subliminal perception and repression. In J. L. Singer (Ed.), *Repression and dis-sociation.* (pp. 103–120). Chicago: The University of Chicago Press.

Shin, L. M., Kosslyn, S. M., McNally, R., Alpert, N. M., Thompson, W. L., Rauch, S. L., et al. (1997). Visual imagery and perception in posttraumatic stress disorder. *Archives of General Psychiatry, 54,* 233–241.

Shizgal, P. (1999). On the neural computation of utility: Implications from studies of brain stimula-tion reward. In D. Kahneman, E. Diener, & N. Schwarz (Eds.), *Well-being: The foundation of hedonic psychology* (pp. 500–524). New York: Russell Sage Foundation.

Singer, W. (1999). Time as coding space? *Current Opinion in Neurobiology, 9,* 189–194.

Skinner, B. F. (1969). *Contingencies of reinforcement.* New York: Appleton-Century-Crofts.

Sloboda, J. A. (1991). Music structure and emotional response: Some empirical findings. *Psychology of Music, 19,* 110–120.

Small, D. M., Zatorre, R. J., Dagher, A., Evans, A. C., & Jones-Gotman, M. (2001). Changes in brain activity related to eating chocolate. From pleasure to aversion. *Brain, 124,* 1720–1733.

Smith, E., & Grawe, K. (2001). Die funktionale Rolle von Ressourcenaktivierung für Psychotherapie [The functional role of resource activation for therapeutic changes]. In H. Schemmel & J. Schaller (Eds.), *Ressourcen. Ein Hand- und Lesebuch zur therapeutischen Arbeit* (pp. 111–122). Tübingen: DGVT Verlag.

Smith, E., & Grawe, K. (2003). What makes psychotherapy sessions productive? A new approach to bridging the gap between process research and practice. *Clinical Psychology & Psychotherapy, 10,* 275–285.

Smith, E., & Grawe, K. (2005). Which therapeutic mechanism works when? A step towards the for-mulation of empirically validated guidelines for therapists' session-to-session decisions. *Clinical Psychology & Psychotherapy, 12,* 112–123.

Smith, E., & Grawe, K. (in press). Turning the tables: How to make the best out of difficult condi-tions. *Clinical Psychology & Psychotherapy.*

Smith, E., Regli, D., & Grawe, K. (1999). Wenn Therapie wehtut—Wie können Therapeuten zu fruchtbaren Problemaktualisierungen beitragen? [When therapy hurts. How can therapists con-tribute to productive problem actuation?]. *Verhaltenstherapie und psychosoziale Praxis, 31,* 227–251.

Smith, M. L., Glass, G. V., & Miller, T. I. (1980). *The benefits of psychotherapy.* Baltimore: John Hopkins University Press.

Sollers, J. J., Mueller, C. A., & Thayer, J. F. (1997). Emotional responses, heartrate variability, and physical activity. *Psychosomatic Medicine, 59,* 93.

Spitz, R. (1945). Hospitalism: An inquiry into the genesis of psychiatric conditions in early childhood. *The Psychoanalytic Study of the Child, 1,* 53–74.

Spitzer, M. (2002). *Musik im Kopf* [Music inside your head]. Stuttgart, Germany: Schattauer.

Spitzer, M. (2003). Neuronale Netzwerke und Psychotherapie [Neuronal networks and psychotherapy]. In G. Schiepek (Eds.), *Neurobiologie der Psychotherapie.* Stuttgart, Germany: Schattauer.

Squire, L. R. (1992). Memory and the hippocampus: A synthesis from findings with rats, monkeys, and humans. *Psychological Review, 99,* 195–231.

Squire, L. R., Bloom, F. E., McConnell, S. K., Roberts, J. L., Spitzer, N. C., & Zigmond, M. J. (Eds.). (2003). *Fundamental neuroscience* (2nd ed.). New York: Academic.

Squire, L. R., & Kandel, E. R. (1999). *Memory: From mind to molecules.* New York: Scientific American Library.

Squire, L. R., Knowlton, B., & Musen, G. (1993). The structure and organization of memory. *Annual Review of Psychology, 44,* 453–495.

Squire, L. R., & Zola, S. M. (1996). Structure and function of declarative and nondeclarative memory systems. *Proceedings of the National Academy of Sciences of the United States of America, 93,* 13515–13522.

Squire, L. R., & Zola, S. M. (1998). Episodic memory, semantic memory, and amnesia. *Hippocampus, 8,* 205–211.

Sroufe, L. A., Carlson, E., & Shulman, S. (1993). The development of individuals in relationships: From infancy through adolescence. In D. C. Funder, R. Parke, C. Tomlinson-Keesey, & K. Widaman (Eds.), *Studying lives through time: Approaches to personality and development* (pp. 315–342). Washington: American Psychological Association.

Stahl, S. M. (1996). Essential psychopharmacology: Neuroscientific basis and clinical applications. New York: Cambridge University Press.

Stein, M. B., Koverola, C., Hanna, C., Torchia, M. G., & McClarty, B. (1997). Hippocampal volume in women victimized by childhood sexual abuse. *Psychological Medicine, 27,* 951–959.

Storch, M. (2002). Die Bedeutung neurowissenschaftlicher Forschungsansätze für die psychotherapeutische Praxis. Teil I: Theorie [The significance of neuroscientific research for psychotherapy. Part 1. Theory]. *Psychotherapie, 7,* 281–294.

Straube, T., Glauer, M., Dilger, S., Mentzel, H.-J., & Miltner, W. H. R. (2006). Effects of cognitive-behavioral therapy on brain activation in specific phobia. *Neuroimage, 29,* 125–135.

Strauss, B., Buchheim, A., & Kächele, H. (Eds.). (2002). *Klinische Bindungsforschung* [Clinical attachment research: Theories, methods, results]. Stuttgart, Germany: Schattauer.

Strauss, B., Lobo-Drost, A., & Pilkonis, P. (1999). Einschätzung von Bindungsstilen bei Erwachsenen—erste Erfahrungen mit der deutschen Version einer Prototypenbeurteilung [Assessment of attachment styles in adults—First experiences with a German version of a prototype method]. *Zeitschrift für Klinische Psychologie, Psychiatrie und Psychotherapie, 47,* 347–364.

Strauss, B., & Schmidt, S. (1997). Die Bindungstheorie und ihre Relevanz für die Psychotherapie. Teil 2: Mögliche Implikationen der Bindungstheorie für die Psychotherapie und die Psychosomatik [Attachment theory and its relevance for psychotherapy. Part 2: Possible implications of attachment theory for psychotherapy and psychosomatics]. *Psychotherapeut, 42,* 1–16.

Stucki, C. (2004). Die Therapiebeziehung differentiell gestalten [Custom-tailoring the therapeutic relationship]. University of Bern: Unpublished dissertation.

Sullivan, H. S. (1953). The interpersonal theory of psychiatry. New York: Norton.

Sulz, S. (2002). Neuropsychologie und Hirnforschung als Herausforderung für die Psychotherapie [Neuropsychology and brain research as a challenge to psychotherapy]. *Psychotherapie, 7,* 18–33.

Suomi, S. J. (1987). Genetic and maternal contributions to individual differences in rhesus monkey biobehavioral development. In N. A. Krasnegor, E. M. Blass, M. A. Hofer, & W. P. Smotherman (Eds.), *Perinatal development: A psychobiological perspective* (pp. 397–419). New York: Academic.

Suomi, S. J. (1991). Up-tight and laid-back monkeys: Individual diffrences to social challenges. In S. Brauth, W. Hall, & R. Dooling (Eds.), *Plasticity of development* (pp. 27–56). Cambridge, MA: MIT Press.

Suomi, S. J. (1999). Attachment in rhesus monkeys. In J. Cassidy & P. R. Shaver (Eds.), *Handbook of attachment* (pp. 181–197). New York: Guilford.

Suomi, S. J. (2000). A biobehavioral perspective on developmental psychopathology. Excessive aggression and serotonergic dysfunction in monkeys. In A. J. Sameroff, M. Lewis, & S. M. Miller (Eds.), *Handbook of developmental psychopathology.* (2nd ed., pp. 237–256). New York: Kluwer Academic/Plenum.

Suomi, S. J. (2002). Parent, peers, and the process of socialization in primates. In J. G. Borkowski, S. L. Ramey, & P. M. Bristol (Eds.), *Parenting and the child's world: Influences on academic, intellectual, and social-emotional development* (pp. 265–279). Mahwah, NJ: Lawrence Erlbaum Associates, Inc.

Suomi, S. J., Eisele, C. D., Grady, S., & Harlow, H. F. (1975). Depressive behavior in adult monkeys following separation from family environment. *Journal of Abnormal Psychology, 84,* 576–578.

Suomi, S. J., & Levine, S. (1998). Psychobiology of intergenerational effects of trauma. Evidence from animal studies. In Y. Danieli (Ed.), *International handbook of multigenerational legacies of trauma* (pp. 623–637). New York: Plenum.

Swann, W. B. (1990). To be adored or to be known. The interplay of self-enhancement and self-verification. In R. M. Sorrentino & E. T. Higgins (Eds.), *Motivaton and cognition* (pp. 408–448). New York: Guilford.

Swann, W. B. (1992). Seeking truth, finding despair: Some unhappy consequences of a negative self-concept. *Current Directions in Psychological Science, 1,* 15–18.

Swann, W. B., Hixon, J. G., Stein-Seroussi, A., & Gilbert, D. T. (1990). The fleeting gleam of praise: Cognitive processes underlying behavioral reactions to self-relevant feedback. *Journal of Personality and Social Psychology, 59,* 17–26.

Swerdlow, N. R. (1995). Serotonin, obsessive-compulsive disorder and the basal ganglia. *International Review of Psychiatry, 7,* 115–129.

Taylor, S. E., & Brown, J. D. (1988). Illusion and well-being: A social psychological perspective on mental health. *Psychological Bulletin, 103,* 193–210.

Teasdale, J., Segal, Z., & Williams, J. M. (2003). Mindfulness training and problem formulation. *Clinical Psychology: Science and Practice, 10,* 157–160.

Tellegen, A., Lykken, D. T., Bouchard, T. J., Wilcox, K. J., Segal, N. L., & Rich, S. (1988). Personality similarity in twins reared apart and together. *Journal of Personality and Social Psychology, 54,* 1031–1039.

Thase, M. E. (1997). Integrating psychotherapy and pharmacotherapy for major depressive disorder: Current status and future considerations. *Journal of Psychotherapy Practice and Research, 6,* 300–306.

Thayer, J. F., Friedman, B. H., & Borkovec, T. D. (1996). Autonomic characteristics of generalized anxiety disorder and worry. *Biological Psychiatry, 39,* 255–266.

Thayer, J. F., & Lane, R. D. (2000). A model of neurovisceral integration in emotion regulation and dysregulation. *Journal of Affective Disorders, 61,* 201–216.

Tillfors, M., Furmark, T., Marteinsdottir, I., Fischer, H., Längström, B., & Fredrikson, M. (2001). Cerebral blood flow in subjects with social phobia during stressful speaking tasks: A PET study. *American Journal of Psychiatry, 158,* 1220–1226.

Tomarken, A. J., Davidson, R. J., Wheeler, R. E., & Kinney, L. (1992). Psychometric properties of resting anterior EEG asymmetry: Temporal stability and internal consistency. *Psychophysiology, 29,* 576–592.

Torgersen, S. (1986). Childhood and family characteristics in panic and generalized anxiety disorders. *American Journal of Psychiatry, 143,* 630–632.

Tramo, M. J. (2001). Music of the hemispheres. *Science, 291,* 54–56.

Traue, H. C. (1998). *Emotion und Gesundheit* [Emotion and health. The psychobiological regulation by inhibition]. Heidelberg, Germany: Spektrum Akademischer Verlag.

Tremblay, L., Hollerman, J. R., & Schultz, W. (1998). Modifications of reward expectation-related neuronal activity during learning in primate striatum. *Journal of Neurophysiology, 80,* 964–977.

Tremblay, L., & Schultz, W. (1999). Relative reward preference in primate orbitofrontal cortex. *Nature, 398,* 704–708.

Ullrich de Munck, R., & Ullrich, R. (1977). *Der Unsicherheitsfragebogen* [The insecurity questionnaire]. Munich, Germany: Pfeiffer.

Vaillant, G. (1977). *Adaptation to life*. Boston: Little, Brown.

Valins, S. (1966). Cognitive effects of false heart-rate feedback. *Journal of Personality and Social Psychology, 4,* 400–408.

van den Boom, D. C. (1994). The influence of temperament and mothering on attachment and exploration. An experimental manipulation of sensitive responsiveness among lower-class mothers with irritable infants. *Child Development, 65,* 1457–1477.

van den Boom, D. C., & Hoeksma, J. B. (1994). The effect of infant irritability on mother-infant interaction: A growth-curve analysis. *Developmental Psychology, 30,* 581–590.

van der Kolk, B. A. (1987). The drug treatment of post-traumatic stress disorders. *Journal of Affective Disorders, 13,* 203–213.

van der Kolk, B. A. (1997). *Psychological trauma*. Washington; DC: American Psychiatric Publishers.

van der Kolk, B. A., Burbridge, J. A., & Suzuki, J. (1997). The psychobiology of traumatic memory: clinical implications of neuroimaging studies. *Annuals of the New York Academy of Science, 821,* 99–113.

van der Kolk, B. A., Fisler, R. E., & Bloom, S. L. (1996). Dissociation and the fragmentary nature of traumatic memories. *British Journal of Psychotherapy, 12,* 352–366.

van Ijzendoorn, M. H., & Bakermans-Kranenburg, J. (1996). Attachment representations in mothers, fathers, adolescents, and clinical groups: A metaanalytic search for normative data. *Journal of Consulting and Clinical Psychology, 64,* 8–21.

Viinamäki, H., Kuikka, J., & Tiihonen, J. (1998). Change in monoamine transporter density related to clinical recovery: A case-control study. *Nordic Journal of Psychiatry, 52,* 39–44.

Volkow, N. D., Wang, G. J., & Fowler, J. S. (1997). Imaging studies of cocaine in the human brain and studies of the cocaine addict. *Annuals of the New York Academy of Sciences, 820,* 41–45.

Von Hickeldey, S., & Fischer, G. (2002). *Psychotraumatologie der Gedächtnisleistung: Diagnostik, Begutachtung und Therapie traumatischer Erinnerungen* [Psychotraumatology of the memory performance]. Munich, Germany: Ernst Reinhard.

Wallis, J. D., Anderson, K. C., & Miller, E. K. (2000). Neuronal representation of abstract rules in the orbital and lateral prefrontal cortices (PFC) [Abstract]. *Society for Neuroscience Abstracts.*

Warrington, E., & Weiskrantz, L. (1973). The effects of prior learning on subsequent retention in amnesic patients. *Neuropsychologia, 20,* 233–248.

Watanabe, M. (1990). Prefrontal unit activity during associative learning in the monkey. *Brain Research, 80,* 296–309.

Watanabe, M. (1992). Frontal units of the monkey coding the associative significance of visual and auditory stimuli. *Experimental Brain Research, 89,* 233–247.

Watanabe, M. (1996). Reward expectancy in primate prefrontal neurons. *Nature, 382,* 629–632.

Waters, E., Merrick, S., Treboux, D., Crowell, J., & Albersheim, L. (2000). Attachment security in infancy and early adulthood: A twenty-year longitudinal study. *Child Development, 71,* 684–689.

Watson, D., & Clark, L. A. (1984). Negative affectivity: The disposition to experience negative emotional states. *Psychological Bulletin, 96,* 465–490.

Watson, D., & Clark, L. A. (1993). Behavioral disinhibition versus constraint: A dispositional perspective. In D. M. Wegner & J. W. Pennebaker (Eds.), *Handbook of mental control* (pp. 506–527). Englewood Cliffs, NJ: Prentice Hall.

Watson, D., Clark, L. A., & Tellegen, A. (1988). Development and validation of brief measures of positive and negative affect: The PANAS scales. *Journal of Personality and Social Psychology, 54,* 1063–1070.

Wegner, D. M., & Pennebaker, J. W. (Eds.). (1993). *Handbook of mental control*. Englewood Cliffs, NJ: Prentice Hall.

Weinberger, D. A. (1990). The construct validity of repressive coping style. In J. L. Singer (Ed.), *Repression and dissociation* (pp. 337–386). Chicago: The University of Chicago Press.

Weinberger, J., & McClelland, D. C. (1990). Cognitive versus traditional motivational models: Irreconcilable or complementary? In E. T. Higgins & R. M. Sorrentino (Eds.), *Handbook of motivation and cognition* (pp. 562–597). New York: Guilford.

Weiskrantz, L., & Warrington, E. (1979). Conditioning in amnesic patients. *Neuropsychologia, 17,* 187–194.

Weiss, T., Miltner, W., Huonker, R., Friedel, R., Schmidt, I., & Taub, E. (2000). Rapid functional plasticity of the somatosensory cortex after finger amputation. *Experimental Brain Research, 134,* 199–203.

Weisz, J. R., Hawley, K. M., Pilkonis, P. A., Woody, S. R., & Follette, W. C. (2000). Stressing the (other) three Rs in the search for empirically supported treatments: Review procedures, research quality, relevance to practice and the public interest. *Clinical Psychology: Science and Practice, 7,* 243–258.

Wessa, M., & Flor, H. (2002). Posttraumatische Belastungsstörung und Traumagedächtnis—eine psychobiologische Perspektive [Posttraumatic stress disorder and trauma memory: A psychobiological perspective]. *Zeitschrift für Psychosomatische Medizin und Psychotherapie, 48,* 28–37.

Westen, D., & Gabbard, G. O. (2002a). Developments in cognitive neuroscience: I. Conflict, compromise, and connectionism. Journal of the American Psychoanalytic Association, 50, 53–98.

Westen, D., & Gabbard, G. O. (2002b). Developments in cognitive neuroscience: II. *Psychoanalytic Association, 50,* 99–134.

Whalen, P. J., Rauch, S. L., Ertcoff, N. L., McInerney, M. B., Lee, M. B., & Jenike, M. A. (1998). Masked presentations of emotional facial expressions modulate amygdala activity without explicit knowledge. *Journal of Neuroscience, 18,* 411–418.

Wheeler, R. E., Davidson, R. J., & Tomarken, A. J. (1993). Frontal brain asymmetry and emotional reactivity: A biological substrate of affective style. *Psychophysiology, 30,* 82–89.

White, I. M., & Wise, S. P. (1999). Rule-dependent neural activity in the prefrontal cortex. *Experimental Brain Research, 126,* 315–335.

Wiedemann, G., Pauli, P., Dengler, W., Lutzenberger, W., Birbaumer, N., & Buchkremer, G. (1999). Frontal brain asymmetry as a biological substrate of emotions in patients with panic disorders. *Archives of General Psychiatry, 56,* 78–85.

Willutzki, U. (1999). VEV-VW. *Neue Version des Veränderungsfragebogens des Erlebens und Verhaltens von Zielke* [New version of the questionnaire on change of experience and behavior after Zielke]. Unveröffentlichter Fragebogen (unpublished questionnaire). Institut für Psychologie der Ruhr-Universität Bochum.

Wilson, M. A., & McNaughton, B. L. (1994). Reactivation of hippocampal ensemble memories during sleep. *Science, 265,* 676–679.

Wittchen, H. U. (1991). Der Langzeitverlauf unbehandelter Angststörungen: Wie häufig sind Spontanremissionen? [The long-term course of untreated anxiety disorders: What is the frequency of spontaneous remission?]. *Verhaltenstherapie, 1,* 273–282.

Witte, E. H. (1989). *Sozialpsychologie. Ein Lehrbuch* [Social psychology. A textbook]. Munich, Germany: Psychologie Verlags Union.

Wu, J. C., Buchsbaum, J. S., Gillin, J. C., Tang, C., Cadwell, S., Wiegand, M., et al. (1999). Prediction of antidepressant effects of sleep deprivation by metabolic rates in the ventral anterior cingulate and medial prefrontal cortex. *American Journal of Psychiatry, 156,* 1149–1158.

Yalom, I. (1974). The theory and practice of group psychotherapy. New York: Basic Books.

Yeh, S.-R., Fricke, R. A., & Edwards, D. H. (1996). The effect of social experience on serotonergic modulation of the escape circuit of crayfish. *Science, 271,* 366–369.

Young, J. E. (1994). *Cognitive therapy for the personality disorders.* Sarasota, FL: Professional Ressource Press.

Young, P., & Yamane, S. (1993). Sparse population coding of faces in the inferotemporal cortex. *Science, 256,* 1327–1332.

Yurgelun-Todd, D. A., Gruber, S. A., Kanayama, G., Killgore, D. S., Baird, A. A., & Young, A. D. (2000). fMRI during affect discrimination in bipolar affective disorders. *Bipolar Disorders, 2,* 237–248.

Zahn-Waxler, C., Cummings, E. M., McKnew, D. H., Davenport, Y. B., & Radke-Yarrow, M. (1984). Altruism, aggression, and social interaction in young children with a manic depressive parent. *Child Development, 48,* 555–562.

Zielke, M. (1978). *Veränderungsfragebogen des Erlebens und Verhaltens* (VEV) [Questionnaire to assess changes in experiencing and behavior]. Weinheim, Germany: Beltz Test GmbH.

Znoj, H. J. (2000). *Konsistenzsicherung durch emotionale Regulationsprozesse: Entwicklung und kontextbezogene Validierung eines Beobachtungsinstrumentes und eines Fragebogens zur Theorie der emotionalen Kontrolle* [Emotion regulation as a means for consistency assurance]. Bern Switzerland: Unveröffentlichte Habilitationsschrift, Institut für Psychologie, University of Bern.

Znoj, H. J., & Grawe, K. (2000). The control of unwanted states and psychological health: Consistency safeguards. In A. Grob & W. Perrig (Eds.), *Control of human behaviour, mental processes and awareness* (pp. 263–282). New York: Lawrence Erlbaum Associates, Inc.

AUTHOR INDEX

Z

Zahn-Waxler, C., 188, 330, 454
Zatorre, R. J., 70, 71, 73, 307, 428, 450
Zax, M., 188, 448
Zeki, S., 73, 426
Zhang, H., 158, 442
Zhao, Z., 98, 447

Zielke, M., 364, 454
Zigmond, M. J., 451
Zinbarg, R., 227, 444
Znoj, H. J., 301, 390, 390, 454
Zola, S. M., 58, 451
Zook, M. J., 120, 122, 443
Zwaan, M., 134, 433

SUBJECT INDEX

A

acetylcholine, 32, 62, 83, 272, 293, 302, 400
ACC (*see* anterior cingulate cortex)
acts of will, 104–106
adrenergic receptors, 220
adrenocorticotropic hormone (ACTH), 138
Adult Attachment Interview (AAI), 191, 193, 194, 195, 197
AER (see anterior executive region)
affective neuroscience, 10, 12
affective subregion, 132
agency, 312, 313
alarm reactions, 76
allele, long, 182–183
allele, short, 330, 333
alpha-amino–3-hydroxy–5-methylisoxazole–4-proprionic acid (AMPA), 36, 40, 62, 80, 406
American Journal of Psychotherapy, 9
American Psychiatric Association, guidelines, 409, 426
AMPA (see Alpha-amino–3-hydroxy–5-methylisoxazole–4-proprionic acid)
AMPA receptors, 36, 40, 62, 80, 406
amputations, 120
amygdala, 15, 16/31, 61, 71, 72, 74, 75–90, 95–96, 101–109, 112–118, 128, 136–139, 144–164, 177, 269, 276, 292, 345, 357, 395
anterior cingulate cortex (ACC), 128, 132, 148, 229, 284, 292, 345, 357, 395
anterior cingulate gyrus, 71
anterior executive region (AER), 148–149
antidepressant treatment, pharamacological, 130, 133
antidepressive medication, 6
antigoal, 257
anxiety center, 75, 136–137, 269
anxiety symptoms, 41, 266
appraisal process, 220
approach incongruence, 171, 217, 260, 318, 321, 326–327, 350–351, 394
approach priming, 376, 381, 396, 403

approach-avoidance process, 282
astrocytes, 220
attachment need, 174, 330, 380, 412
attachment patterns, 175, 191, 210, 234, 240
attachment styles, 174, 210, 255–257, 321
attentional process, 71
auditory cortex, 31, 45, 67–68, 71, 119
avoidance incongruence, 171, 217, 260, 318, 321, 326–327, 329, 373–375, 393–394
avoidance process, 162

B

BAS (*see* behavioral activation system)
basal ganglia, 101, 154–157
basic needs, listing of, 168
BDI (*see* Beck Depression Inventory)
Beck Depression Inventory (BDI), 125, 152, 205, 206, 207, 326
behavioral activation system (BAS), 245, 249, 250, 252, 430,
behavioral inhibition system (BIS), 218, 226, 236, 245, 249, 250, 252, 331, 430
belief-disconfirmation paradigm, 286
benzodiazepines, 153, 276
Bern Inventory of Therapy Goals (BIT), 199, 200, 387
BIS (*see* behavioral inhibition system)
BIT (*see* Bern Inventory of Therapy Goals)
blood-brain barrier, 224
Brave New Brain: Conquering Mental Illness in the Era of the Genome, 9
Brief Symptom Inventory (BSI), 267
Broca's area, 146
Brodmann's area, 112, 149
BSI (*see* Brief Symptom Inventory)

C

CA1, 63
CA3, 63
calcium, 37, 38, 80